PLEASE STAMP DATE DUE, BOTH BELOW AND ON CARD

DATE DUE | DATE DUE |

Cenozoic Foreland Basins of Western Europe

Geological Society Special Publications
Series Editors
A. J. FLEET
A. C. MORTON
A. M. ROBERTS

It is recommended that reference to all or part of this book should be made in one of the following ways.

MASCLE, A., PUIGDEFÀBREGAS, C., LUTERBACHER, H.P. & FERNÀNDEZ, M. (eds) 1998. *Cenozoic Foreland Basins of Western Europe*. Geological Society, London, Special Publications, **134**.

PHILLIPPE, Y., DEVILLE, E. & MASCLE, A. 1998. Thin-skin inversion tectonic of oblique basin margins: example of the western Vercors and Chartreuse subalpine massifs (SE France). *In:* MASCLE, A., PUIGDEFÀBREGAS, C., LUTERBACHER, H. P. & FERNÀNDEZ, M. (eds) *Cenozoic Foreland Basins of Western Europe*. Geological Society, London, Special Publications, **134**, 239–262.

GEOLOGICAL SOCIETY SPECIAL PUBLICATION NO. 134

Cenozoic Foreland Basins of Western Europe

EDITED BY

ALAIN MASCLE
(IFP École du Pétrole et des Moteurs, Rueil-Malmaison, France)

CAI PUIGDEFÀBREGAS
(Consejo Superior de Investigaciones Cientificas, Barcelona, Spain)

HANS PETER LUTERBACHER
(Universität Tübingen, Germany)

AND

MANUEL FERNÀNDEZ
(Consejo Superior de Investigaciones Cientificas, Barcelona, Spain)

1998
Published by
The Geological Society
London

THE GEOLOGICAL SOCIETY

The Society was founded in 1807 as The Geological Society of London and is the oldest geological society in the world. It received its Royal Charter in 1825 for the purpose of 'investigating the mineral structure of the Earth'. The Society is Britain's national society for geology with a membership of around 8500. It has countrywide coverage and approximately 1500 members reside overseas. The Society is responsible for all aspects of the geological sciences including professional matters. The Society has its own publishing house, which produces the Society's international journals, books and maps, and which acts as the European distributor for publications of the American Association of Petroleum Geologists, SEPM and the Geological Society of America.

Fellowship is open to those holding a recognized honours degree in geology or cognate subject and who have at least two years' relevant postgraduate experience, or who have not less than six years' relevant experience in geology or a cognate subject. A Fellow who has not less than five years' relevant postgraduate experience in the practice of geology may apply for validation and, subject to approval, may be able to use the designatory letters C Geol (Chartered Geologist).

Further information about the Society is available from the Membership Manager, The Geological Society, Burlington House, Piccadilly, London W1V 0JU, UK. The Society is a Registered Charity, No. 210161.

Published by the Geological Society from:
The Geological Society Publishing House
Unit 7, Brassmill Enterprise Centre
Brassmill Lane
Bath BA1 3JN
UK
(*Orders*: Tel. 01225 445046
Fax 01225 442836)

First published 1998

The publishers make no representation, express or implied, with regard to the accuracy of the information contained in this book and cannot accept any legal responsibility for any errors or omissions that may be made.

© The Geological Society 1998. All rights reserved. No reproduction, copy or transmission of this publication may be made without written permission. No paragraph of this publication may be reproduced, copied or transmitted save with the provisions of the Copyright Licensing Agency, 90 Tottenham Court Road, London W1P 9HE. Users registered with the Copyright Clearance Center, 27 Congress Street, Salem, MA 01970, USA: the item-fee code for this publication is 0305–8719/98/$10.00.

British Library Cataloguing in Publication Data
A catalogue record for this book is available from the British Library.

ISBN 1–86239–015–0

Typeset by Type Study, Scarborough, UK

Printed in Great Britain by
The Alden Press, Osney Mead, Oxford, UK.

Distributors
USA
AAPG Bookstore
PO Box 979
Tulsa
OK 74101–0979
USA
(*Orders*: Tel. (918) 584–2555
Fax (918) 560–2652)

Australia
Australian Mineral Foundation
63 Conyngham Street
Glenside
South Australia 5065
Australia
(*Orders*: Tel. (08) 379–0444
Fax (08) 379–4634)

India
Affiliated East–West Press PVT Ltd
G–1/16 Ansari Road
New Delhi 110 002
India
(*Orders*: Tel. (11) 327–9113
Fax (11) 326–0538)

Japan
Kanda Book Trading Co.
Tanikawa Building
3–2 Kanda Surugadai
Chiyoda-Ku
Tokyo 101
Japan
(*Orders*: Tel. (03) 3255–3497
Fax (03) 3255–3495)

Contents

Foreword   vii

Acknowledgements   ix

MASCLE, A. & PUIGDEFÀBREGAS, C. Tectonics and sedimentation in foreland basins: results from the Integrated Basins Studies project   1

Guadalquivir and Ebro Foreland Basins (Spain)

FERNÀNDEZ, M., BERÁSTEGUI, X., PUIG, C., GARCÍA-CASTELLANOS, D., JURADO, M. J., TORNÉ, M. & BANKS, C. Geophysical and geological constraints on the evolution of the Guadalquivir Foreland Basin, Spain   29

BERÁSTEGUI, X., BANKS, C. J., PUIG, C., TABERNER, C., WALTHAM, D. & FERNÀNDEZ, M. Lateral diapiric emplacement of Triassic evaporites at the southern margin of the Guadalquivir Basin, Spain   49

WILLIAMS, E. A., FORD, M., VERGÉS, J. & ARTONI, A. Alluvial gravel sedimentation in a contractional growth fold setting, Sant Llorenç de Morunys, southeastern Pyrenees   69

VERGÉS, J., MARZO, M., SANTAEULÀRIA, T., SERRA-KIEL, J., BURBANK, D. W., MUÑOZ, J. A. & GIMÉNEZ-MONTSANT, J. Quantified vertical motions and tectonic evolution of the SE Pyrenean foreland basin   107

NIJMAN, W. Cyclicity and basin axis shift in a piggyback basin: towards modelling of the Eocene Tremp-Ager Basin, south Pyrenees, Spain   135

TRAVÉ, A., LABAUME, P., CALVET, F., SOLER, A., TRITLLA, J., BUATIER, M., POTDEVIN, J-L., SÉGURET, M., RAYNAUD, S. & BRIQUEU, L. Fluid migration during Eocene thrust emplacement in the south Pyrenean foreland basin (Spain): an integrated structural, mineralogical and geochemical approach   163

French Western Alps

LICKORISH, W. H. & FORD, M. Sequential restoration of the external Alpine Digne thrust system, SE France, constrained by kinematic data and synorogenic sediments   189

ARTONI, A. & MECKEL, L. D. History and deformation rates of a thrust sheet top basin: the Barrême basin, western Alps, SE France   213

PHILLIPPE, Y., DEVILLE, E. & MASCLE, A. Thin-skinned inversion tectonics at oblique basin margins: example of the western Vercors and Chartreuse Subalpine massifs (SE France)   239

BECK, C., DEVILLE, E., BLANC, E., PHILLIPPE, Y. & TARDY, M. Horizontal shortening control of Middle Miocene marine siliciclastic accumulation (Upper Marine Molasse) in the southern termination of the Savoy Molasse Basin (northwestern Alps/southern Jura)   263

Swiss, German and Austrian Molasse Basin

BURKHARD, M. & SOMMARUGA, A. Evolution of the western Swiss Molasse basin: structural relationships with the Alps and the Jura belt   279

ZWEIGEL, J., AIGNER, T. & LUTERBACHER, H. Eustatic versus tectonic controls on Alpine foreland basin fill: sequence stratigraphy and subsidence analysis in the SE German molasse   299

ZWEIGEL, J. Reservoir analogue modelling of sandy tidal sediments, Upper Marine Molasse, SW Germany, Alpine foreland basin   325

WAGNER, L. R. Tectono-stratigraphy and hydrocarbons in the Molasse Foredeep of Salzburg, Upper and Lower Austria   339

Numerical Modelling

BORNHOLDT, S. & WESTPHAL, H. Automation of stratigraphic simulations: quasi-backward modelling using genetic algorithms   371

DEN BEZEMER, T., KOOI, H., PODLADCHIKOV, Y. & CLOETINGH, S. Numerical modelling of growth strata and grain-size distributions associated with fault-bend folding   381

ANDEWEG, B. & CLOETINGH, S. Flexure and 'unflexure' of the North Alpine German–Austrian Molasse Basin: constraints from forward tectonic modelling   403

Index   423

Foreword

This book results from the Integrated Basin Studies Project (IBS), which ran during 1992–1995 with the support of the European Commission DGXII. Several papers have been added to the ones resulting directly from IBS, in order to offer the reader a more comprehensive overview of Western Europe's Cenozoic foreland basins. I warmly thank the authors for their much appreciated contribution.

Two other books resulting from IBS will be published in the near future by the Geological Society, completing a series of three books devoted to field studies of European Basins: one will be on the Mediterranean extensional basins within the Alpine orogen and the other will be on the Norwegian rifted margin.

A series of papers on compaction of fine-grained sediments, which was also an important theme of the IBS project, has been accepted for publication in *Marine and Petroleum Geology*.

The IBS project was born at a meeting held in Strasbourg in June 1989 on the initiative of Hubert Curien, former French Minister for Research and Technology. The meeting was aimed at defining promising new avenues of research in Geosciences. It was said that, one such avenue would be research on sedimentary basins, not only for its intrinsic scientific interest, but also because it is upstream of strategic economic activities, such as the oil and gas industry, the management of water resources and the storage of wastes. Also, most human activities take place at the surface of sedimentary basins. These activities are developing exponentially, resulting in a rapid growth of risks and environmental problems, which cannot be mastered properly without an improved knowledge of sedimentary basins at all scales, including knowledge of the physical framework of their development.

It was also mentioned at this meeting that deriving concrete applications in this research would be more effective by designing models of basins and sub-basin formation and evolution through to increasing co-operation, not only between 'geological' disciplines such as structural geology, sedimentology, geophysics and geochemistry, but also between 'geological' disciplines and 'non-geological' disciplines such as fluid and rock mechanics and thermodynamics. At this stage, numerical modelling techniques were believed to provide the necessary link to integrate these disciplines toward a quantitative earth model.

In this spirit, a panel of European researchers met several times in Brussels and designed the IBS project, which was proposed to the DGXII of the European Commission. The DGXII decided to support it after some modifications and integrated it into the Geosciences project of the Joule 2 programme. Many thanks are due to DGXII experts and executives and particularly to J. C. Imarsio and J. M. Bemtgen for having accepted this project and having helped to make it realistic and effective. IBS teams are also indebted to M. Rougeaux, Managing Director of Groupement d'Études et de Recherches en Technologie des Hydrocarbures (GERTH) who was of invaluable help in the management of the project.

The main concept of the IBS project was to create methods and techniques of modelling in which the main physical phenomena responsible for formation, development and the infilling of sedimentary basins or sub-basins are linked, starting from deep lithospheric deformation caused by the convective movements of ductile matter in the upper mantle.

Critical components of such modelling were as follows.

- The capacity to make a sound description of the mechanical linkage between deformation and near-surface deformation in order to produce the geometrical evolution of basins and sub-basins. This has to be possible in extensional and compressional tectonic contexts.
- The capacity to couple the above models with simple but realistic models of linked erosion and sedimentation processes.
- The capacity to couple the whole with a model of fine-grained sediment compaction in order to incorporate the effect of sediment loading and water escape on sediment deformation.

The final objective was to obtain practical information such as the architecture of the basin or sub-basin fill (down to the reservoir scale) and its evolution through geological time, evolution of the temperature stresses and fluid-pressure regimes.

Such methods and techniques aim also at a better design of reservoir geological models and therefore contribute to the improvement of field development planning.

In order to reach set goals, IBS teams used the following method:

(1) Starting from prototypes conceived mainly by a team from Vrije University in Amsterdam (S. Cloetingh *et al.*), models have been developed by using the present knowledge of the rheology of crust and sediments and by using the documentation already available on various thoroughly studied basins in different tectonic settings. At the same time a task force, under the leadership of a team from Newcastle University (A. Aplin *et al.*) undertook the revision of the knowledge on the compaction of fine-grained sediments and the modelling of the corresponding phenomena using theoretical, experimental and observational approaches.

(2) A small number of European basins, set in appropriate tectonic contexts, have been taken as natural laboratoires and carefully documented for the interaction of tectonic and sedimentation processes using extensive synthesis of seismic, well and field data.

These basins were as follows.

- Rifted basins within the Alpine orogen: the Gulf of Lions in France and the Pannonian basin in Hungary studied under the leadership of the University of Montpellier (M. Seranne *et al.*) and of the Eötvös–Lorand laboratory in Budapest (F. Horvath *et al.*).
- Foreland basins (this volume: South Pyrenean basins and the Guadalquivir basin in Spain, the molasse basin in Germany and the Barreme syncline in France studied by teams of the University of Barcelona (M. Marso *et al.*) of the Institut de Ciencas de la Terra of Barcelona (M. Fernàndez *et al.*), of the University of Tübingen (H. P. Luterbacher *et al.*), and of the ETH of Zurich (M. Ford *et al.*), associated under the leadership of the Servei Geologic of Catalunya (C. Puigdefabregas *et al.*).
- The Norwegian margins, the northern Viking Graben, the More basin, the Voring basin and the mid-Norwegian margin in Norway, studied by teams of the Norwegian universities and of the Norwegian oil companies under the leadership of Norsk Hydro (A. Nottvedt *et al.*).

IBS teams have already presented their work at four large scientific meetings within the oil industry: EAGE in Vienna (June 1994), Glasgow (June 1995), Amsterdam (June 1996) and AAPG in Nice (September 1995) and also several times to more academic audiences, in particular at the meeting of the International Lithosphere Programme in Sitges, Spain (September 1995), and at the EUG in Strasbourg (April 1995). As a consequence of the IBS programme two workshops were organized, one on the Mediterranean basins (Cergy-Pontoise, 11–13 December 1996; some of the communications have been included in this volume), and the other one on mudrocks (Geological Society, London, 28–29 January 1997).

The co-operation with industry increased during the project and finally, 21 oil companies helped in a significant way. Some, like the Norwegian oil companies were directly involved. In particular, Norsk-Hydro had the leadership of module 3 (Dynamics of the Norwegian Margins). Others contributed less directly by providing documents or helping with the interpretation.

We acknowledge this help and thank these companies which are listed below (in alphabetical order):

Amoco, BEB, BP, Coparex, DEE, EAP, Esso, INA Naftaplin, Norsk-Hydro, MOL, Mobil, ÖMV, Preussag, Repsol, RWE-DEA, Saga, Shell, Statoil, Total, VVNP, Wintershall.

In total more than 200 researchers belonging to 38 institutions and 15 countries (eight EU countries, six non-EU European countries and the USA) have participated in the IBS project. The IBS teams have been strongly connected with those of the task force: 'Origin of Sedimentary Basins' of the International Lithosphere Programme and those of the network EBRO (European Basin Research Organisation), of the 'Human Capital and Mobility' programme of the European Commission.

Through the IBS project, the DGXII of the European Commission has clearly demonstrated its capacity to create a European research environment. This capacity was enhanced by the access given to the IBS Program to a PECO program with Hungary, which resulted in the IBS program on the Pannonian basin, and by the good co-ordination with the following two institutions:

- the Research Council of Norway (NFR), thanks to which it was possible to launch the IBS-DMN program on the Dynamics of the Norwegian margins;
- the Bundesamt für Bildung und Wissenschaft of Switzerland, thanks to which it was possible to launch the IBS-ETH cooperation.

We also wish to acknowledge here the role played by these institutions and we thank them for their financial support.

The European research environment that has been created consists of academic teams who voluntarily worked on geological problems of interest to the oil industry, and who now have a solid capacity in this field. Many of these teams are now associated with the Eurobasins School, where a number of European Universities and Research Institutions cooperate, under the auspices of Academia Europea. It is my hope that IBS has given the impetus for regular co-operation in research on sedimentary basins in Europe. No doubt these teams will now work for the oil industry and other economic sectors in a more direct partnership.

B. DURAND
Project Leader of IBS
January 1998

Bibliography

CLOETINGH, S., DURAND, B. & PUIGDEFÀBREGAS, C. (eds) 1995. Integrated Basin Studies. *Marine and Petroleum Geology*, **12**, part 8.

DURAND, B. & MASCLE, A. 1996. Interest for the European Oil Industry of the Results Obtained by the Integrated Basin Studies JOULE Project n°: CT92-120. *In*: *The Strategic Importance of Oil and Gas Technology*. Proceedings of the 5th European Union Hydrocarbons Symposium, Edinburgh, 26–28 November 1996, **2**, p. 1151–1167.

Acknowledgements

The editors would like to thank the following reviewers for their very significant help:

A. Artoni, D. Burbank, E. Burov, P. Cobolt, B. Colletta, E. Deville, J. L. Faure, M. Ford, D. Grangeon, P. Homewood, M. Jackson, P. Joseph, F. Lafont, I. Moretti, J. L. Mugnier, U. Norlund, Y. Philippe, Ch. Ravenne, F. Roure, M. Séguret, M. Seranne, P. Trémolières, J. Verges, E. Williams.

The following companies are also warmly thanked for supporting this volume: Total, Amoco and Saga.

Tectonics and sedimentation in foreland basins: results from the Integrated Basin Studies project

ALAIN MASCLE[1] & CAI PUIGDEFÀBREGAS[2,3]

[1]IFP School, 228–232 avenue Napoléon Bonaparte, 92852 Rueil-Malmaison Cedex, France (e-mail: alain.mascle@ifp.fr)
[2]Norsk Hydro Research Centre, Bergen, Norway.
[3]Institut de Ciences de la Terra, CSIC, Barcelona, Spain.

Why foreland basins?

Over the last ten years or so, since the Fribourg meeting in 1985 (Homewood et al. 1986), the attention given by sedimentologists and structural geologists to the geology of foreland basins has been growing continuously, parallel to the increase of co-operative links between scientists from the two disciplines. A number of reasons lie behind this development. Attempting to understand the growth of an orogen without paying due attention to the stratigraphic record of the derived sediments would be unrealistic. It would, moreover, be equally unrealistic to construct restored sections across the chain without considering the constraints imposed by the basin-fill architecture, or to describe the basin-fill evolution disregarding the development of the thrust sequence. As in other sedimentary basins, tectonics and sedimentation dynamically interact in foreland basins. Additionally, as foreland basins are incorporated in the growth process of the orogen, they are more likely than extensional basins to be subject to uplift and, therefore, more accessible to direct field observation. Foreland basins observed on the field may help in the understanding of non-observable subsurface analogues. This is essentially why foreland basins have always been traditional field areas for sedimentological research and training, and also why they have recently been considered as ideal field laboratories, best suited to the study and understanding of the interplay between tectonics and sedimentation (Fig. 1).

Further interest in foreland basins has also been triggered off by some spectacular results from investigations of offshore accretionary prisms. As a matter of fact, such prisms are considered to be very similar in many aspects to onshore thrust belts. Extensive seismic and high-resolution bathymetric surveying of a few selected active margins (Barbados, Cascadian, Nankaï, Middle America), calibrated with both deep-sea drilling and deep-sea diving, have lead to a better understanding of some basic interacting tectonic, sedimentary and hydrologic processes (More & Vrolijk 1992; Touret & van Hinte 1992). Additional data have also been obtained through the development of analogue and numerical models (Larroque et al. 1992; Zoetemeijer 1993). The physical parameters controlling the forward propagation of décollements and thrusts (fluid pressure, roughness, sediment thickness, etc.) have been determined and tested. The relationships between rapidly subsiding piggyback basins and growing ramp anticlines have also been imaged, although the lack of deep-sea well control still prevents accurate sedimentological studies. More significant has been the progress in our understanding of the role of fluids and pore pressure in the development of thrust belts. When the fluids escape from the mineral matrix of sediments during sedimentary and tectonic burial, the resulting fluid flow can either be diffused through the sedimentary column if the average permeability is high enough, or be channelled along pathways such as décollements and active faults when the overall permeability is low. They may also remain trapped within the mineral matrix if such pathways are not available. With increasing burial the pore pressure dramatically increases, eventually approaching the vertical minimum stress, and catastrophic events such as mud volcanoes may occur (e.g. Barbados Ridge, Fig. 2). Migrating fluids will not only exert a fundamental control on tectonic processes, but will also contribute to the transfer of mineral solutions, heat and hydrocarbons from the inner part of the thrust belt to the surface along vertical or more tortuous lateral pathways.

Foreland basins are also important from the point of view of hydrocarbon exploration. Source rocks are commonly provided by pre-compressional rift basin sequences, whereas properly structured rock formations within the foreland basin may eventually provide adequate reservoirs. Classical examples from the Urals, Caucasus and Carpathians together with those

MASCLE, A. & PUIGDEFÀBREGAS, C. 1998. Tectonics and sedimentation in foreland basins: results from the Integrated Basin Studies project. In: MASCLE, A., PUIGDEFÀBREGAS, C., LUTERBACHER, H. P. & FERNÀNDEZ, M. (eds) Cenozoic Foreland Basins of Western Europe. Geological Society Special Publications, 134, 1–28.

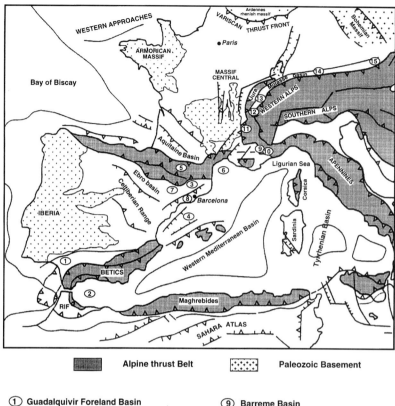

Fig. 1. Location map of the study areas.

① Guadalquivir Foreland Basin
② Alboran Sea
③ San Llorenç de Marunys
④ Valence Basin
⑤ Pyrenees Thrust Belt
⑥ Gulf of lions
⑦ SE Pyrenean Foreland Basin (Ebro Basin)
⑧ Catalan Ranges
⑨ Barreme Basin
⑩ Digne Thrust System
⑪ Chartreux-Vercors subalpine Massifs
⑫ Savoy Molasse Basin
⑬ Swiss Molasse Basin
⑭ German Molasse Basin
⑮ Austrian Molasse Basin

of the Canadian Rockies, have been recently overtaken by the discovery of giant fields in Venezuela and Colombia, with estimated reserves of well over the billion barrels of oil equivalent mark (Duval *et al.* 1995). It is obvious that in such a context, good knowledge of basin evolution and fluid flow in relation to thrust sequence propagation will be required. Every step in the progress of knowledge acquisition might be, directly or indirectly, of prime relevance. Concerning the future of hydrocarbon exploration in Western Europe, while the classical plays are today at a stage of extensive exploration, the present frontier areas are HP-HT prospects in deep stratigraphic intervals (5–8 km) of well-known basins such as the North Sea, and deep-water prospects in areas like the Atlantic continental margins. Apart from the still very speculative potential of non-conventional gas production (coal-bed methane), the next (and possibly the last) frontier will undoubtfully be the deep gas potential of Cenozoic thrust belts: there have already been some recent attempts to define prospective plays in the Pyrenees (Le Vot *et al.* 1996) and Northern Alps (Greber *et al.* 1996). It is there where the Foreland Basin Module of the IBS project hopes to be of most use.

Additional interest in foreland basin geology also stems from the fact that they are often associated with alluvial plains such as the Po plain in Italy, the Parana river in Argentina and the Ganges in India, to mention only three, which support a high human population.

Alluvial plains in foreland basins are very sensitive systems in relation to both predictable and less predictable changes in climate and land use. This is, perhaps, an issue of marginal interest to our project, but of much greater importance to society. We believe that all the newly acquired knowledge, which is of relevance to hydrocarbon exploration, is sooner or later bound to have additional socio-economic applications.

Objectives and research premises

The Foreland Basin Module of the IBS project has been planned to study the interplay, at different scales, between tectonics and sedimentation during the construction of an orogen, and also to study how control is exerted on the architecture of the basin-fill down to the scale of the depositional sequence.

Three main areas of work have been selected: the Guadalquivir basin related to the Betic orogen, the Ebro basin related to the Pyrenees, and the German Molasse related to the Alps. In each of these areas, a number of research activities have been planned to provide an insight to some of the major problems such as the following.

- Is the tectonic load sufficient to account for the observed lithospheric flexure?
- Can we quantify amounts and rates of erosion and topographic growth and incorporate them in our time-step models?
- Can we relate major sedimentary cycles and thrust events?
- Would modelling of growth structures help to understand and predict growth strata geometry?
- Is it possible to discriminate between tectonically controlled supply sequences and eustatic sequences, or is there any purpose even to try?

To address these and other related problems, three scales (crustal, basin-fill and sequence unit) have been adopted in each of the three areas, in the belief that operating processes are linked from one scale to the next. The last two scales (basin-fill and sequence unit) are mostly relevant, and currently considered, in studies of basin evaluation and reservoir characterization, whereas the crustal scale is seldom taken into account. In many cases, however, crustal-scale modelling is the only way to approach some basic parameters such as palaeothermicity or timing and amplitude of erosional events. The three scales should be only considered as mere research strategies. They do not really have a physical entity as rocks themselves do, and it would be misleading to introduce from the start any attempt of order and classification.

It has long been recognized that in most cases, orogens have adjacent elongated foreland basins which are filled with the erosion products of the mountain chain. This observation defines the crustal-scale approach. It is also widely accepted (Riba 1964) that foreland basins are asymmetric in transverse section and that their depocentre axes migrate outwards through time as the thrust sequence propagates – the earlier basin-fill sequences being thus incorporated into the thrust system (Puigdefàbregas et al. 1986). This constitutes the next scale of approach. Surprisingly enough, this evolution of the basin-fill geometry through time, which is the main distinctive feature, is not always taken into account in the definition of the foreland basin (DeCelles & Giles 1996). It is also recognized that, on a third and smaller scale, the propagation of a particular thrust interferes with sedimentation in the adjacent basin at sequence scale. In addition, as topography is created in the orogen, a complex chain of geomorphic and climatic controls is exerted through weathering, erosion and sediment-transport processes, which are not only relevant in predicting the nature of the sediments finally deposited in foreland basins, but also in the further propagation of the orogenic wedge itself.

With these premises in mind, the objectives of the Foreland Basin Module within the Integrated Basin Studies (IBS) project are to contribute to the understanding of how these interactions work, which processes are involved, and at what rates they operate.

Structure and composition of the project group

In order to cope both with the different scales of approach and work areas, a large team was initially set up. Whenever possible and needed, the team spontaneously developed interconnected links. As the project progressed, some of the team members felt the need to enlarge the group and extend collaborative links with new research groups, thus shaping an informal network of research on sedimentary basins within the objectives of the IBS project.

In the final stages of the project, 45 researchers belonging to 11 research institutions actively contributed to the IBS Module on Foreland Basins:

Servei Geològic de Catalunya: X. Berástegui, M. Losantos, J. Cirés, C. Puig, E. Pi, P. Arbués, J. Corregidor, A. Martínez.
Institut de Ciències de la Terra 'Jaume Almera' (CSIC): M. Fernàndez, D. Garcia-Castellanos, C. Taberner, M. Torné.

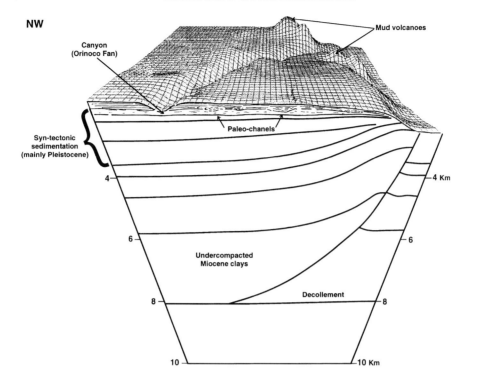

Fig. 2. Ramp anticline and related piggyback basin in the southern Barbados Ridge (Lesser Antilles active margin). The Orinoco deep-sea fan is presently incorporated to the accretionary complex. Consequently, the distributory channels have to find their way through an highly active seabottom morphology. Relative lows, such as the piggyback basins are filled with channel-supplied fine-grained turbidites. Relative highs such as anticlinal crests and mud volcanoes are deeply eroded (locally down to 300 m) by V-shaped segments of the same channels. The turbidite flow ultimately reach the Atlantic Abyssal Plain and form an elongate fan at the foot of the Barbados Ridge deformation front. The subsequent stage will be their incorporation to the accretionary complex when the deformation front will propagate. The numerous mud volcanoes in this area probably originate from undercompacted deep marine clays of early Miocene age. These mud volcanoes and the emerging ramps are the locus of active cold seeps with deep faunal assemblages and ferruginous crust formation (Mascle *et al.* 1990; Jolivet *et al.* 1990; Faugères *et al.* 1993).

Universitat de Barcelona: M. Marzo, J.A. Muñoz, J. Vergés, J. Poblet, J. Giménez, J. Dinarés, M. López Blanco, J. Piña, T. Santaeularia, J. Serra-Kiel, J. Tosquella, A. Travé.
Norsk-Hydro Research Center (Bergen): T. Dreyer, J. Gjelberg.
Rijks Universiteit Utrecht: W. Nijman.
Vrij Universiteit Amsterdam: S. Cloetingh, R. Zoetemeijer, T. Den Bezemer.
ETH-Z, Zürich: M. Ford, A. Artoni, E.A. Williams.
University of Tübingen: H.P. Luterbacher, T. Aigner, J. Jin, J. Zweigel.
Royal Holloway University of London: C. Banks, K. McClay, D. Waltham, S. Hardy, F. Sorti.
University of Bergen: R. Steel, H. Rasmussen.
University of Zaragoza: H. Millán.

To synthesize the results of the research undertaken by such a large group is a difficult task. The present volume 'Cenozoic Foreland Basins of Western Europe' constitutes a compilation of most of the relevant papers (some of them are already published) from the IBS project, enriched with other valuable contributions from recent research in the same, or related areas. The project would be considered even more successful if IBS related papers still continue to be published. Our presentation here builds on the work of all, which means that all members of the group should be considered as contributing authors.

Orogenic growth and basin configuration

Because of the existence of the ECORS-Pyrenees deep-reflection seismic cross-section obtained in 1985 (ECORS-Pyrenees Team

1988), the Pyrenees possibly constitute the area best suited to study the link between crustal and surface geodynamic processes responsible for the orogenic growth which leads to the origin and configuration of foreland basins. The main results of the ECORS-Pyrenees Team have been published by Muñoz (1992), who gives an interpretation of the Pyrenean orogen on the basis of the ECORS data, by Berástegui et al. (1993) with a detailed description of the ECORS-Pyrenees geological section on a full-colour poster format, and by Puigdefàbregas et al. (1992) where a number of stages in the evolution of the foreland basin are distinguished in relation to the propagation of the orogenic wedge. Although modelling and quantification of rates was not attempted, the ECORS-Pyrenees section constitutes a valuable reference for a new balanced and totally restored crustal-scale cross-section constructed about 55 km eastward within the framework of the IBS project (Vergés et al. 1995; Millán et al. 1995).

The deep structure of this Eastern Pyrenees balanced cross-section (Vergés et al. 1995) has been constrained with the available geophysical data, combined with projections from the ECORS section. Since the area has been recently and extensively studied and mapped by various authors (see Vergés et al. 1995 for references), there has been little need for further field observations to produce a good picture of the stratigraphic frame, shallow thrust structure, and timing of successive thrusting. The section (Fig. 3) has been also partially restored at an intermediate (Lutetian) stage.

Both the present-day and the Mid-Lutetian restorations have been flexurally modelled (Millán et al. 1995) in order to evaluate the relative contribution of topographic and subduction loads to the lithospheric deflection, and to obtain additional constraints on the estimation of the topographic load.

Relevant results from this study can be summarized as follows (Vergés et al. 1995; Millán et al. 1995).

(a) Although the topographic load plays the predominant role in the external and intermediate parts of the foreland basin, the subsidence of the internal part is the result of the combined effect of both subduction and topographic loads.

(b) Palaeotopographic elevations between 1 and 2 km appear to be the most plausible values for the Mid-Lutetian restored cross-section.

(c) The total shortening obtained in the Eastern Pyrenees cross-section is 125 km (22 km less than in the ECORS section situated 55 km to the west).

(d) Rates of shortening vary through time from a low rate (less than 0.5 mm a^{-1}) for late Cretaceous and Palaeocene times, to higher rates (up to 4.5 mm a^{-1}) during Early and Mid-Eocene times, and again a lower rate (2 mm a^{-1}) during late Eocene and Oligocene times.

(e) Two stages of orogenic growth are identified: a first stage (up to Mid-Lutetian time) characterized by submarine emplacement of the thrust sheet fronts and widespread marine sedimentation (underfilled stage), and a second stage (from Mid-Lutetian time onwards) where shortening rates decrease as subaerial relief increases and non-marine clastic sedimentation becomes predominant (overfilled stage).

(f) The approximate estimate of the mean erosion rate is 0.5 mm a^{-1}. Moreover, in the Eastern Pyrenees, during the overfilled stage, the rate of erosion was almost three times more than the rates of sedimentation in the foreland basin. This obviously implies that the products of erosion must have been deflected westwards along the basin axis, as it is also indicated by the well documented patterns of sediment dispersal.

(g) According to the calculated deflection in the Lutetian restored cross section, the hydrocarbon source rocks of the Lower Armancies Formation (Lower Eocene) have been subjected to a minimum burial depth of 3 km. Hence, they have been buried deep enough to be associated to the upper part of the oil window (Permanyer et al. 1988; Clavell 1992; Vergés et al. 1995).

The Betics, which are more complex and comparatively less studied than the Pyrenees and without a deep seismic section of reference across the chain, are less suited to attempt successfully a time-step reconstruction of the relationship between the orogenic growth and the basin configuration, except perhaps for the Guadalquivir basin, the latest of its evolutionary stages. **Fernàndez et al.** (this volume) give an extensive account of the present-day knowledge of the area and incorporates the new geological and geophysical data obtained within the framework of the IBS project, with special emphasis on the constraints on the evolution of the Guadaquivir basin. A set of maps including the Bouguer anomalies (10 mGal isoline), the geoid heights (1 m isolines), the depth of the Moho (1 km isolines), and the surface heat flow and basement depth, is presented in the same paper, together with thermometric logs from water wells, and hydrocarbon exploration well logs. The compiled data set introduces important constraints on the possible models of a basin formation. However, the early stages of the development of the Gudalquivir basin and its relation to the history of the orogen still remains

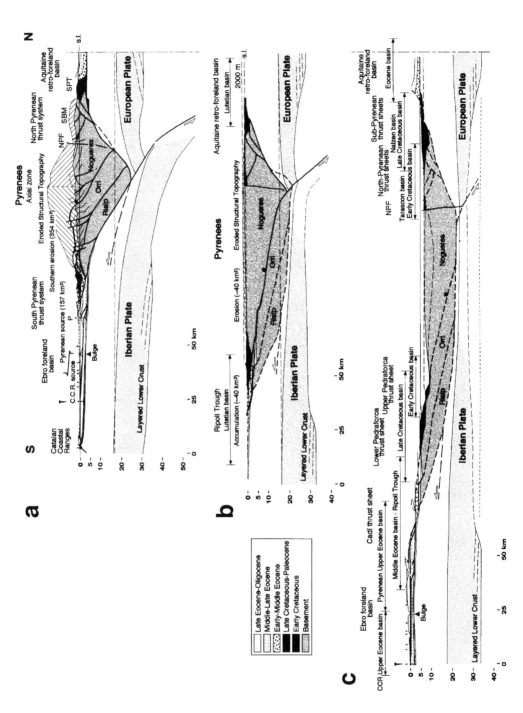

Fig. 3. (a) Post-orogenic balanced cross section through the Eastern Pyrenees. Crustal geometry based on the ECORS-Pyrenees profile. (b) Partially restored cross-section at Middle Lutetian time. (c) Composite restored cross-section showing the unfolding of all the sedimentary basins and basement units. (From Vergés *et al.* 1995, reprinted by kind permission of Elsevier Science-NL.)

obscure. Unresolved issues include: how late Cretaceaous to Langhian foreland basin sequences were incorporated in the northward-propagating thrust wedge, or even if they existed at all.

The stacked tectonic sheets of the inner part of the Betics were later submitted to severe stretching leading to lithospheric thinning and ultimately to the opening of the Alboran sea in Miocene times. This part of the history, related to extensional processes within the Alpine orogen, will be developed in a second volume prepared by the IBS group, to be published by the Geological Society.

Numerical models for the deflection of the lithosphere under time-dependent loads in 2D and 3D are currently being made by D. Garcia-Castellanos (Institute of Earth Sciences, CSIC). The models consider the effects of erosion, sedimentation and lateral variations of the elastic thickness corresponding to a layered lithospheric rheology, and allow a closer interaction between tectonic processes, sedimentary basin-fill and large-scale lithospheric deformation (Garcia-Castellanos et al. in press). The 2D and 3D numerical models have been applied to the Betic foreland basin in order to obtain a detailed view of the regional isostasy in the area (Fig. 4). The basement and sediment geometries determined by **Berástegui et al. (this volume)** and **Fernàndez et al.** (this volume), have been related to the tectonics of the external zone and to the emplacement of a hidden load of undetermined origin but of quantified magnitude.

An additional example of orogenic growth is provided by the 'Digne Nappe', an external segment of the Western Alps well exposed in SE France (**Lickorish & Ford** this volume). Although seismic and well data are not available to constrain the proposed deep section, excellent field exposures and an accurate set of geological maps provided by the French Geological Survey (BRGM) have been used to construct a 100 km long regional section through the thrust belt. Of particular interest, time-step restored sections show the variability of tectonic processes operating through time along this single section, thin-skinned tectonics prevailing during late Eocene and Oligocene times, and linked thick- (at the rear) and thin-skinned (at the front) tectonics in Neogene times. An intervening event of normal faulting in Priabonian times would reflect the flexuration of the eastern segment of the foreland, as a result of tectonic overloading in the inner part of the Alps farther East (Vially 1994). This polyphase history of thrust belts should be properly taken into account for the hydrocabon assessment of such

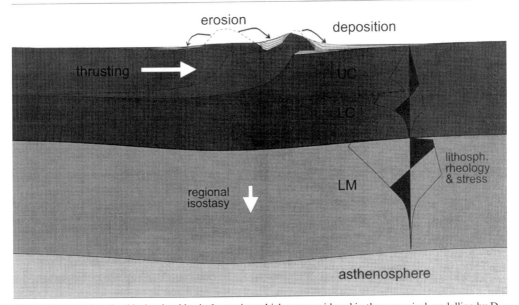

Fig. 4. Processes involved in foreland basin formation which are considered in the numerical modelling by D. Garcia-Castellanos, M. Fernández and M. Torné: Thrust loading, regional isostasy (flexure) with various lithospheric rheologies, and surface processes (denudation and deposition), UC: upper crust; LC: lowe crust; LM: lithospheric Mantle. (From Garcia-Castellanos et al. in press, reprinted by kind permission of Elsevier Science-NL).

areas, especially when migration paths or early generation of hydrocarbons and dismigrations are considered.

Thrust sequences and basin-fill architecture

As we have seen above, in the case illustrated by the Pyrenean orogen, plate convergence, orogenic growth and formation of the adjacent foreland basin are intimately related. We can also infer that a continuum exists between the triggering process of plate convergence down to the final thrust sequence propagation. Because of the presence of syntectonic deposits and good age constraints, it becomes even possible to estimate the rates of thrust displacement between 2 and 4.5 mm/ a^{-1}. Identical values have been obtained for present-day rates of thrust displacement along the front of the offshore Barbados accretionary prism. Compared with the duration of the basin-fill sequences, these figures leave little doubt that thrusting has played a significant role in shaping the sequence architecture of the basin-fill. We will now analyse, in each of the different work areas, the characteristics and details of these relationships.

The Pyrenees

The existence of a close relationship between thrusting and sedimentation is an old observation that goes back to the 1960s. Riba (1964, 1976) recognized the general asymmetry of the Ebro basin, the outward migration of depocentres, and the progressive unconformities with their related syntectonic sedimentary wedges. These observations constituted the starting point of a fruitful line of research. Reille (1971) provided a well-documented field description of the syntectonic wedges along the northern margin of the basin. Soler & Puigdefàbregas (1970) related phases of thrusting and sedimentary packages defined by isochronous and basinwide extended changes in palaeogeography and dispersal patterns. It was, however, Garrido-Megías (1973) who first formulated the concept of a tectonosedimentary unit (TSU): 'A tectonosedimentary unit is a stratigraphic unit made up of strata deposited during a given geological time-span, under sedimentary and tectonic conditions characterized by a specific tendency. The TSU boundaries are basin wide sedimentary discontinuities or their correlative conformities'. It is worth noting that this definition has many common elements with the seismic stratigraphy concept formulated later by Vail & Mitchum (1977); the need was obviously there for new concepts on stratigraphy and, in fact, genetic sedimentary units have long since been in mind of Pyrenean geologists, mainly because of the ubiquitous presence of unconformities associated with observable syntectonic sedimentary wedges.

The generalization of the sequence stratigraphy concept, as essentially controlled by eustasy, resulted in a number of important papers, ranging from those that, although recognizing some tectonic control on the sequence upbuilding, have eustasy as the main influence on the genesis of sequence boundaries (Fonnesu 1984; Crumeyrolle 1987; Mutti et al. 1988; Deramond et al. 1993), to those believing that tectonics predominate in the control of the changes in accommodation needed to generate sequences (Mellere 1993; Mutti et al. 1994; Puigdebàbregas & Souquet 1986).

The acquisition of the ECORS deep seismic profile between Toulouse and Balaguer provided the required ground for the next step in the research. A closer link was established between distinct stages in the basin-fill and stages in the structural evolution (Muñoz 1992; Puigdefàbregas et al. 1992) at the same time that emphasis was given to quantification of rates of processes (Vergés 1993).

Apart from the natural disagreement concerning the number of sequences or cycles, their names and controlling factors, most of the authors will agree that unconformity-bounded sediment packages with a characteristic genetic signature may be distinguished in the southern foreland basin of the Pyrenees. The asymmetric transverse section, the outward migration of depocentres, the presence of stratal growth patterns and thrust-related bounding unconformities, and the axially deflected dispersal patterns are the classical features of these stratigraphic units. They can be distinguished at several scales according to their specific relation to tectonics. A first-order sequence would correspond to the basin itself, considered as a sediment unit related, as we have seen, to the growth of the orogen. Large-scale changes in the style of thrust-sequence propagation determine the piggy-back basin configuration within the foreland basin, and define basin-fill sediment packages that roughly correspond to the classical lithostratigraphic groups, falling within the range of the second-order sequences in conventional sequence stratigraphy.

Third-order sequences are genetic sediment packages essentially controlled by changes in sediment supply (as a response to changes in uplift rates, thrust-related changes of gradient, and efficiency of the fluvial drainage in the hinterland), and changes in the accommodation space (as a response to changes in local tectonic subsidence, thrust-related changes of gradient, and changes of base level or eustatic sea level).

The Tremp–Graus basin, extensively studied by the University of Utrecht since the 1960s, and now incorporated in the IBS project (see **Nijman** this volume), constitutes a classical example that illustrates the basin architecture at that scale of observation. Most of the characteristic features and detailed facies analysis can be found in previously published literature. It can be described as a nappe-top syncline basin (piggy-back basin) with an asymmetrical transverse section and a typical axially deflected dispersal pattern. The antiformal stacking of tectonic units constitutes the northern margin of the basin, while the northern dip slope of the Montsec thrust hanging wall constitutes the southern margin. The basin is about 20 km wide and, over a distance of 50 km grades from alluvial to shallow and deeper marine facies whose turbidite components extend westwards for another 100 km. A number of sequences build up the Tremp–Graus basin-fill (see megasequence and cycle subdivision by **Nijman** this volume).

The sequence elements in the eastern area, dominated by alluvial to delta plain deposits, include (Nijman, this volume, Figs 13 and 16):

(1) a distinct event of degradation and channel incision immediately followed by multiepisodic channel filling, which results in a wider and thicker multi-story fluvial sandstone body;
(2) a marine onlap on top of the former fluvial sandstone, recognized as a thin dark oyster rich mudstone, as soil profiles reflecting the rise of the water table, or even by the presence of tidal structures within the channel sandstone bodies;
(3) a thicker aggradational finer-grained alluvial sequence, which, towards the margin, correlates with major alluvial fan upbuilding.

These three elements are the result of a threefold process which includes:

(1) a rejuvenation phase due to thrust induced gradient steepening,
(2) a subsequent increase of local subsidence;
(3) an increase (and then decrease) of fluvial efficiency of the subaerial drainage network in the hinterland, in response to relief rejuvenation, accounting for the common thickening-up trend of alluvial-fan sequences.

Basin axis shift in the Tremp–Graus basin is described to follow a zigzag pattern (see Nijman this volume, Figs 15 and 16) with a net outward migration. The zigzag pattern would result from the interplay between three main modes of thrust translation:

(1) general uplift due to active floor thrust and major outward displacement of the entire thrust wedge;
(2) one-sided uplift due to the active growth of the antiformal stack in the rear;
(3) active roof thrust displacement.

The distinguished megasequences correlate with the described zigzag pattern of basin axis shift, but show a poor correlation with the eustatic curve (see Nijman this volume, Fig. 13).

In the Ainsa basin, farther to the west, an area of slope environment connects the alluvial and deltaic deposits with basinal turbidites farther west (Mutti *et al.* 1972, 1984; Nijman & Nio 1975). A clear relation can be observed there between thrusting and the generation of large submarine erosion surfaces (Fig. 5), eventually canyons, produced by frequent sediment collapse. Three main collapse surfaces migrate outward with respect to each other in relation to thrust propagation (Fig. 6 and Muñoz *et al.* 1994). Submarine erosion surfaces take their time to form as the collapse front climbs upslope, and often connect to the base of the next transgressive event. It is, therefore, in most of the cases impossible to physically connect surfaces of fluvial incision to surfaces of submarine erosion.

Sequence-bounding surfaces are, therefore, segmented (Dreyer *et al.* 1994) as the result of different processes operating along the depositional profile, from the subaerial hinterland and all the way down to the deeper submarine slope, in response to the same triggering effect of gradient steepening.

It is also important to note that successive piggy-back basins, located on also successive thrust sheets, are not sharply separated in time. One basin may start to be actively filled while the other is still receiving sediments. Consequently, syntectonic sequences may be present in neighbouring basins, as in the case of Ager, Tremp-Graus and Ainsa, and the corresponding dispersal systems may be obliged to follow a rather complicated pattern between rising highs. This is also illustrated in present-day deep marine accretionary prisms where several piggyback basins can be contemporaneously filled, within an overall outward younging of the base of the syntectonic infill.

In the eastern Pyrenees, **Vergés *et al.*** (this

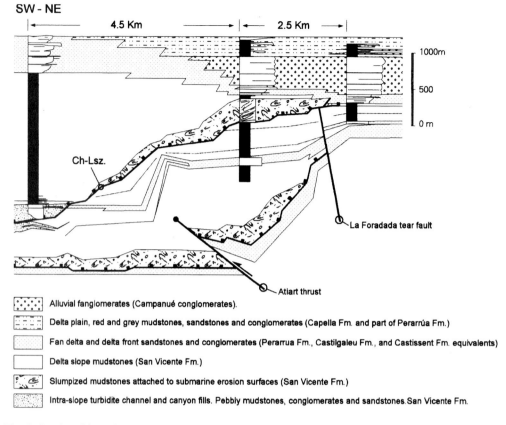

Fig. 5. Stratigraphic section across the transition area between the Tremp–Graus and Ainsa basins. The section illustrates the stratigraphic position of the Charo-Lascorz submarine erosion surface (Ch-Lsz), and the relations to the Atiart thrust. (From P. Arbués, unpublished.)

volume), building on previous work (Puigdefàbregas *et al.* 1986; Giménez 1993), distinguish four third-order transgressive–regressive cycles. The analysis is based on the time-step restored cross-section previously discussed (Vergés *et al.* 1995; Millán *et al.* 1995), on available well and seismic data, field sections and magnetostratigraphic dating (Vergés & Burbank 1996). The main results (**Vergés *et al.*** this volume fig. 8) show that tectonic subsidence started earlier (55.9 Ma) in the Ripoll piggy-back syncline basin, with subsidence rates from 0.13 to 0.38 mm a^{-1}. In the Ebro basin south of the Vallfogona thrust a similar N–S trend is observed, where the maximum of subsidence up to 0.30 mm a^{-1} is shown to start at 41.5 Ma in the Montserrat section, connected with the structure of the coastal ranges, allowing to accommodation of the large amounts of fan delta deposits in the area. These results, further presented and discussed in the paper by **Vergés *et al.*** (this volume), provide a quantification to the generally observed outward migration of the depocentres corresponding to the successive sequences that build up the architecture of the basin-fill.

The German Molasse

The Molasse basin, studied for at least a couple of centuries, is a classic example of a foreland basin. It is there where the main geometric and basin-fill characteristics have been first described and shown to be applicable to most of the basins related to the growth of mountain chains. Within the IBS project, the work group of the University of Tübingen has selected a key area east of Münich between the rivers Inn and Isar. Because of its situation between a western shallower and an eastern deeper sector, the area should be suited to study the sequential upbuilding, and address questions such as the relative role of thrust loading, subsidence, sediment supply and eustasy in the basin-fill architecture. The study, based on about 25 seismic sections

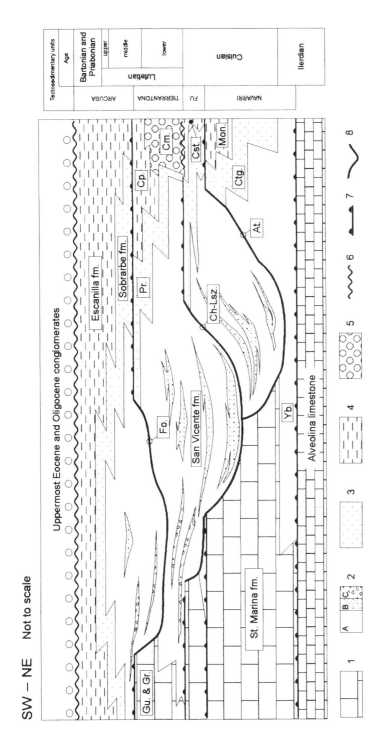

Fig. 6. Synthetic stratigraphy of the Ainsa basin. 1, carbonate shelf; 2, delta and carbonate slope; 2A, mudstones; 2B, siliciclastic turbidites; 2C, resedimented carbonates; 3, delta; 4, alluvial plain; 5, alluvial fan; 6, subaerial unconformity; 7, condensed section; 8, surface of submarine gravitational erosion. Yb, Yeba marls; Gu & Gr, Guara limestone and Grustán mb; Cgl, Castilgaleu fm.; Mon, Montlobat fm.; Cst, Castissent fm.; Cm, Campanué conglomerates; Cp, Capella fm; Pr, Perarrúea fm.; Fu, Fuendecampo tectonosedimentary unit. Submarine erosion surfaces: At, Atiart; Ch-Lsz, Charo-Lascorz; Fo, Formigales. Although the thrust deformation is not represented, the section shows the outward migration of the submarine erosion surfaces with their corresponding depocentres of turbidite deposition. (From P. Arbués, unpublished.)

and over 30 wells, combines three techniques: seismic stratigraphy, subsidence analysis and stratigraphic modelling. Jin *et al.* (1995), and **Zweigel *et al.*** (this volume), give a detailed and precise account of the basin-fill architecture and discuss the results. The sequence stratigraphy analysis has led to an improved reconstruction of the depositional history, and the parameters obtained have been used in stratigraphic modelling. A new back-stripping algorithm has been developed, allowing a two-dimensional subsidence analysis along regional sections. Forward stratigraphic modelling has been calibrated with real data to test and refine the modelling packages, and to quantify and constrain the subsurface interpretations. The PHIL stratigraphic modelling package has been used in longitudinal sections parallel to the basin axis in order to improve the understanding of the interplay between subsidence, sediment supply and eustasy.

The paper by **Zweigel *et al.*** (this volume), distinguishes five depositional sequences on the basis of seismic stratigraphy analysis in longitudinal sections parallel to the basin axis, and to the general shoreline progradation. On the other hand, the subsidence analysis shows two distinct flexural events corresponding to the classic transgressive–regressive cycles. Figure 2 of Zweigel *et al.* (this volume) illustrates the correspondence between cycles and sequences and includes the timing of thrust events.

The authors conclude that shoreline shifts responsible for the definition of the sequence elements may be seen better in longitudinal west–east sections through the basin, whereas transverse sections show better the thrust related changes of subsidence. In other words, there is a longitudinal sequential partitioning controlled by eustasy, and a transverse partitioning related to the northward propagation of the thrust belt. The argument is supported by the good fit between the sea-level curve derived from the seismic stratigraphic analysis and the sea-level curve of Haq *et al.* (1987). On the whole this is an interesting conclusion, which although needs testing on other basins, will certainly stimulate further research.

The Guadalquivir basin

As previously discussed, the Guadalquivir basin constitutes the latest of the evolutionary stages of the Betic foreland basin. The present knowledge of the Betic orogen does not allow the time-step reconstruction of the previous stages. As it is possible to observe the basin-fill architecture and its relation to the thrust sequence propagation, the IBS work group on the Guadalquivir basin has focused the research on the sequence architecture of the Miocene fill of the basin and its relation to the thrust front. **Berástegui *et al.*** (this volume) present the results of this research. Based on the seismic stratigraphic interpretation of more than 1400 km of seismic lines and the study of about 35 exploration wells, the Neogene fill of the Guadalquivir basin is divided into six depositional sequences which correspond to the third-order sequences in the TB2 and TB3 supercycles in Haq *et al.* (1987). (Berástegui *et al.* this volume, fig. 8). In the southern active margin of the basin, an allochthonous body is described that occupies most of the basin and extends laterally all along the Betic front. Because of its chaotic appearance, which mostly consists of Triassic evaporites and mudstones including large extraformational blocks, the allochthonous body has, until now, been interpreted as a large olistostrome fallen into the basin. In order to establish the relationship between the basin-fill sequences and the Betic front, a closer analysis of the body was obviously needed. A number of inconsistencies have been observed, namely its extraordinary size, almost as large as the basin itself, and, most important, the fact that all the rock fragments and blocks included in the Triassic matrix are younger than the matrix itself, immediately ruling out the olistostrome interpretation. Based on three structural transects, the structural analysis of the allochthonous body itself, and supported by analogue and numerical modelling, **Berástegui *et al.*** (this volume, fig. 13), provide us with a new interpretation as a submarine-emplaced lateral diapir of Triassic material displaced from the Intermediate Units, in front of an advancing thrust sheet. As to the timing of the emplacement, sequences 1, 2 and 3 are pre-thrusting, 4 and 5 are deposited during the emplacement of the lateral diapir, and sequence 6 is post-emplacement.

The lateral-diapiric thrust front characterizes the external structure of the Betics. This Betic-type thrust belt front, not often described in literature, might constitute a good model for other foreland basins and should be taken into account in hydrocarbon exploration. Concerning the sequence partitioning it should be noted that the recognized sequences show a good correlation with the eustatic curve of Haq *et al.* (1987), but unlike the German molasse example, they also correspond to well-defined extensional or thrusting events (see **Berástegui *et al.*** this volume, fig. 8).

In comparing the three main basins studied, it

appears that sequence partitioning based on seismic stratigraphy analysis (Guadalquivir and German Molasse basins) shows a good correlation with the global eustatic chart of Haq *et al.* (1987), whereas the same correlation is very poor in basins studied from field observation. There may be a number of explanations for this:

(1) sequence boundaries, because of their segmented character, are more distinct in seismic sections than in detailed field observations;
(2) not enough effort has been made in the Pyrenean foreland basin toward the definition of sequence boundaries;
(3) foreland basins studied at their outer areas, where tectonically induced changes on supply and changes on subsidence are weaker, reveal a stronger eustatic effects in the sequential upbuilding.

This last explanation emphasizes the role of sediment supply and tectonic subsidence in the sequence definition in foreland basins.

Thrust propagation and fluid flow

Intraplate stresses are generally generated along divergent or convergent plate margins. In collision zones, the upper fragile crustal domain is the locus of intense strain which is expressed by shearing, extrusion and the development of localized shear zones and thrusts where important parts of the horizontal shortening take place. Such shear zones develop first at deep crustal levels, following the initial trajectories of the sudducting slabs (oceanic or thinned continental lithospheres). When collision of continents becomes the dominant geodynamic process (in late Cretaceaous times for the Western Alps), shear zones develop between the two colliding continental crusts. These shear zones will rapidely propagate upward and forward, but some backward thrusting may also be initiated, resulting in the classic dual vergence of the Alpine thrust belts (**Vergés *et al.*** and **Burkhard & Sommaruga**; this volume). When propagating to the surface, these major thrusts follow complicated paths, often showing a 3D geometry with 'flats and ramps'. The more-or-less horizontal flat segments (décollements or detachments) can either be localized in ductile layers of the continental basement or be hosted in sedimentary horizons with a strength lower than the surrounding rocks (salt, undercompacted clays, marls, etc.). According to whether the frontal or lateral ramps (segments with higher dips) initiate from great or shallow depth, so-called thick-skinned or thin-skinned tectonic regimes will develop respectively. The term thin-skinned tectonics is usually used to describe structures which develop over décollements within sedimentary layers. But such layers can be at depths in excess of 10 km, as for instance in southeast France (**Philippe *et al.*** this volume). The resulting style of superficial deformation will then be very similar to the deformation related to thick-skinned tectonics, a term better used when the basement is involved. Thick-skinned and thin-skinned tectonics are both present in the western Alps thrust belts and forelands, and both can be linked and synchronous (see **Lickorish & Ford** this volume). From a petroleum exploration point of view, the localization of décollements within the sedimentary package is important as quite distinct structural traps will develop above (synchronous ramp anticlines), or will be preserved below (early tilted blocks for instance). Active flats and ramps will also act as potential pathways for fluids and hydrocarbons (as demonstrated in modern accretionary prisms, Moore & Vrolijk 1992). On the other hand, thick-skinned tectonic will induce the complete uplift of large parts of sedimentary basins, thus immediately freezing oil or gas kitchens, or even leading to the escape of any early accumulation of hydrocarbons (to the surface or to another trap).

The French segment of the Western Alps does not show the development of any significant Neogene foreland basin (or alternatively, the related deposits have been subsequently eroded). As a result, sedimentary layers of the pre-orogenic stage (Mesozoic) are well exposed and the geometry of Neogene structures is relatively well constrained from field data. Several deep wells and a few regional seismic profiles are also available. Well-balanced sections in these areas are proposed in this volume by **Philippe *et al.*; Beck *et al.*; Lickorish & Ford** and **Burkhard & Sommaruga** (including the neighbouring Swiss Molasse Basin). This large amount of data allows the study of lateral variations in the tectonic style of thin-skinned deformation of the western Alps foothills. Over a distance of about 30 km, the broad ramp anticlines of the Vercors Plateau progressively narrow to the North to form an imbricated vertical stack of thin tectonic sheets in the Chartreuse Massif (**Philippe *et al.*** this volume). This tectonic wedge then widens again in the Jura Massif, with the front of deformations being translated to about 100 km towards the west. As a result, a large piggy-back basin, the French and Swiss Molasse Basin, once part of the northwestern Alps foredeep (in Eocene–Oligocene times), has developed in Neogene times (**Beck *et al.*; Burkhard &**

Sommaruga this volume). Can we explain this lateral evolution of the thrust belt? A convincing hypothesis is that these synchronous tectonic styles are primarily controlled by two parameters:

(1) the thickness of the sediment package involved in the deformation,
(2) the efficiency of the major decollement propagating in the foreland basin (here hosted in the upper Triassic interval).

A thick sedimentary package and an efficient (low-friction) décollement will favour the rapid forward propagation of the tectonic wedge. A thin sedimentary column and higher frictions within the décollement will lead to a more vertically imbricated stack and a slower rate of forward propagation. Because of these constrasting tectonic style, the timing of hydrocarbon generation and expulsion will be quite different in these different places (Mascle et al. 1996; Sassi & Deville in press), and consequently the petroleum plays will be different. A late generation of hydrocarbon is possible:

(1) in the Chartreuse and inner Jura Massifs due to the burial of Stephanian, Autunian and Liasic source rocks within or below the verticaly imbricated tectonic stacks;
(2) below the Molasse Basin following the burial of the same source rocks below the thick tertiary sedimentation of the foredeep and piggy-back basin stages. On the other hand only localized tectonic or sedimentary burial of these source rocks have occured in Tertiary times in the Vercors and outer Jura Massifs; as a result, most of the oil generation and expulsion was achieved in early Tertiary times.

During burial and compaction, sediments expel formation water (and source rock hydrocarbons). The behaviour of faults and decollement as seals or pathways during the migration of such fluids is an important and still strongly debated problem. The nature of the rocks and of the damaged zone surrounding the faults must also be taken into account. If most of the sediments are permeable, the nature of the faults has probably little importance as the fluids will be able to circulate to the surface quite easily. With the exception of very sandy accretionary prisms or onshore foredeeps, this is not often the case, and either a fracture network within the sediments, or a few single faults are then preferential pathways. This has been demonstrated in the case of some accretionary prisms such as the Barbados Ridge, where fluids migrations are presently active (Fig. 7). In now inactive thrust belts, fluid circulations along fault zones is shown by the presence of veins, usually calcite, but other mineral assemblages can be found (Roberts 1990). A combined structural and geochemical study of such shear veins has been undertaken in the South Pyrenean Foreland Basin by **Travé et al.** (this volume). Channelized flows along active thrust faults are recorded by calcite shear veins formed by a crack–seal mechanism, suggesting an epidsodic nature of fault slip and related fluid migration. These fluids seem to be a mixture of local formation water mixed with saline fluids from the inner and deeper parts of the belt. Meteoric waters have also locally contributed to the flow of water through the forward-propagating thrust belt. Additional studies of this type will certainly be necessary in the future for a better understanding of interactions of fluids, fluid flows, and fluid pressures with the micro- and macro-tectonic processes occurring in an active tectonic wedge.

Growth folds and related sediment wedges

As previously discussed, numerous examples of synsedimentary growth of folds and thrusts are found in the Pyrenees, especially along the northern margin of the Ebro Basin. Earlier descriptions come from Riba (1964), Ten Haaf (1966), Reille (1971) and Garrido-Megías (1973). Some of them have continuously led to further research and more detailed descriptions have been produced. This is the case of the Sant Corneli anticline and its related Bóixols thrust that affected the sedimentation of the Areny Group (Nagtegaal et al. 1983; Simó 1985; Fondecave-Wallez et al. 1988; Mutti et al. 1994), the Oliana anticline described by Vergés (1993), the Turbon anticline (Pappon 1969; Deramond et al. 1993) and the N–S anticlines at the External Sierras (Millán et al. 1995). Within the IBS project, three examples have been selected: the Vallfogona thrust, discussed by Martínez-Rius et al. (1996) and by Ford et al. (1996), the Bóixols thrust discussed by Arbués et al. (1996) and the Mediano anticline, described and modelled by Poblet et al. (in press).

Syntectonic conglomerates associated to the Vallfogona thrust

According to Martínez-Rius et al. (1996), the emergence of the Vallfogona thrust, together with the emplacement of the lower units in the northern sector, resulted in deformation of the previously emplaced lower Pedraforca thrust sheet (Fig. 8), and produced an imbricated

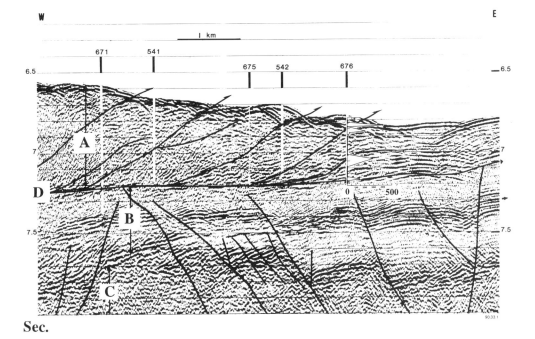

Fig. 7. Barbados Rige, deformation front. 541-542: DSDP leg 75-A; 671 to 676: ODP leg 110; 7 other holes (DSDP leg 75-A and ODP legs 110 and 156) are located a few kilometres to the West and East respectively: A, Miocene to Recent; B, Campanian to Oligocene (the strong reflectors around B represent Eocene sandy turbidites); C, top of the Atlantic oceanic crust. Note that earlier normal faults in the underthrust sequences (B) are preserved below the main décollement (D), while presently active thrust characterized the accreted complex (A). The very low gradient of the sea bottom slope (taking into account a vertical exaggeration of 3.3 for the water and 2.5 for sediments) is a typical feature of décollements with a low rugosity and high pore fluid pressure. Packer tests in nearby holes 948 and 949 (ODP leg 156) suggest that pore pressures in the décollement exceed hydrostatic pressure. The large set of data now available in this area allows us to constrain the present migration paths of fluids expelled from both the accreted and subducted low permeability sediments: pore water geochemistry, distribution of mineral veins and temperature measurements indicate that the main conduits for a lateral migration are:the Eocene sandy turbidites, the décollement and the presently active thrust faults in the accreted complex. For instance at site 676, pore fluids show anomalous contents in methane (here expressed in micromoles per litre) closely related with the location of thrust faults (Behrmann *et al.*, 1988; Gieskes *et al.* 1989; Vrolijk & Sheppard 1991; Labaume *et al.* 1995).

system of hanging-wall thrust slices that propagated out of sequence. Increased relief together with out-of-sequence thrust propagation resulted in the deposition of four conglomerate wedges unconformably related to each of the thrust slices. These relations are comparable to those described by Martínez-Rius *et al.* (1996), Vergés (1993) and Mellere (1993) in other localities of comparable structural setting.

The last of the conglomerate units is time equivalent (Upper Eocene–Oligocene) to the spectacular conglomerate wedge of Sant Llorenç de Morunys (Riba 1976) which developed at the footwall of the Vallfogona thrust. Ford *et al.* (1996) give an outstanding description of that growth structure, based on the construction of three profiles. The geometry is interpreted as a fault propagation fold pair. The growth axial planes form an enechelon pattern on the anticline, and an upward converging pattern on the syncline, together defining the growth triangle (Fig. 9). Sedimentological and structural data are combined with magnetostratigraphy in order to model the structure and to describe the details of erosional and sedimentary responses to the fold growth, where gradient steepening plays a significant role, even at the smallest scale of observation (**Williams *et al.*** this volume).

Relations between the Bòixols thrust and the Areny Group

The Maastrichtian Areny Group is perhaps one of the more studied sedimentary units in the

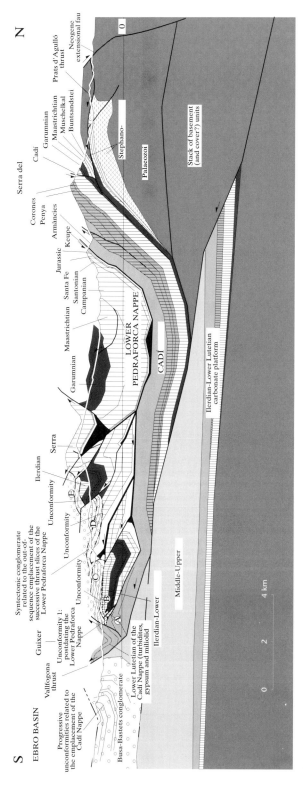

Fig. 8a

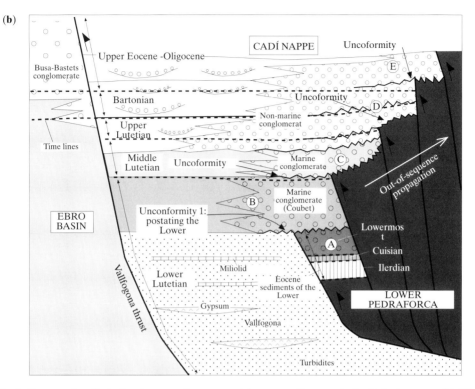

Fig. 8. Cross-section displaying the general structure of the eastern Pyrenees in the Pedraforca area (**a**). The footwall thrust sequence of the lower Pedraforca and Cadí units ends up with a hanging-wall overstep sequence within the Lower Pedraforca thrust sheet. This is shown by the relationships between the syntectonic sediments and related thrust (**b**). The oldest syntectonic sediments are the A conglomerates (Coubet Fm, late Early Lutetian), which overlay the floor thrust of the lower Pedraforca (unconformity 1). The sediments B, C, D and E unconformably overlie the thrust slices (unconformities 2, 3 and 4) which are progressively younger towards the hinterland. (From Martínez-Rius *et al.* 1996, reprinted with permission).

Pyrenees. The IBS project specifically focused on the relations between the growth of the Bòixols anticline and the Maastrichtian sedimentation. Two parallel structural sections have been constructed (Fig. 10), and four sequences have been characterized within the Areny Group (Arbués *et al.* 1996). From the facies and thickness analysis within each of the sequences (Fig. 11), it can be concluded that the generated differential subsidence mostly controlled the distribution of thickness and facies, but did not exert a direct control on sequence partitioning, more likely connected to less local changes in accomodation and gradient steepening. If such sequence bounding surfaces are in relation to local tectonics, and are at the same time of basin wide extent, then a link between basin wide flexure and local thrusting must be assumed. Otherwise, the generation of sequence-bounding surfaces is more likely under eustatic control, with differential subsidence and sediment supply under tectonic control.

Synsedimentary growth of the Mediano anticline

The Mediano anticline is the easternmost anticline within succession of N–S-trending folds that characterize the western segment of the Central Pyrenees. It constitutes the eastern flank of the Ainsa syncline basin. The synsedimentary growth of the Mediano anticline has long been recognized (Garrido-Megías 1973; Nijman & Nio 1975). In a comprehensive paper, Poblet *et al.* (in press) describe and model its geometry, and quantify the rates of growth. Figure 12 shows the geometry of the structure and associated sedimentary wedges, and Figure 13 shows the evolution of the structure on a section perpendicular to the fold axis. Two main unconformities are observed: a lower unconformity between pre-growth and growth strata, and an upper unconformity separating the growth strata onlapping the anticline limbs, from those which overlap the fold crest. It is interesting to

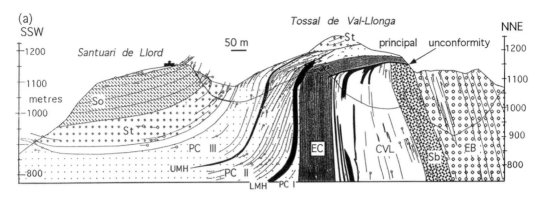

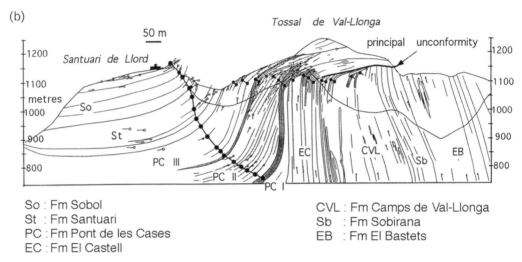

So : Fm Sobol
St : Fm Santuari
PC : Fm Pont de les Cases
EC : Fm El Castell

CVL : Fm Camps de Val-Llonga
Sb : Fm Sobirana
EB : Fm El Bastets

Fig. 9. Section (**a**), and line drawing (**b**) through Tossal de Vall-Llonga showing hinge points in the growth syncline. Note that in the anticline the axial planes at various levels form a discontinuous, en-echelon pattern jumping progressively down to the south. The synclinal and anticlinal growth axial planes converge upward to describe a growth triangle. (From Ford *et al.* 1996, fig. 9, reprinted with kind permission of Elsevier Science-NL.)

note that a typical carbonate platform facies, including *Nummulites* beds and coral and algal reefs, seems to, by preference, develop at the base of the second unconformity around the anticline top.

Reverse modelling (Poblet *et al.* in press) provides estimation of rates of thrust slip (max. 0.58 mm a^{-1}), uplift (up to 0.70 mm a^{-1}), limb rotation and lengthening. The average erosion of the anticline crest is estimated in 0.51 mm a^{-1}. The growth strata were deposited from 47.90 to 43.73 Ma, and from that age onwards, the sediments overlap the anticline crest.

It must also be noted that the growth strata wedges form in response to the relief generation and not directly to changes in compression rates (Hardy & Poblet 1994). An additional result of this survey is that palaeomagnetic data show that the whole Ainsa basin has been rotated by 30° clockwise during sedimentation.

The results of growth-fold modelling are not only relevant to the understanding of the role of tectonics in basin sequence architecture, but also provide a useful tool to predict suitable facies and traps in hydrocarbon exploration.

Sedimentation and tectonics at sequence scale

Two field areas have been selected to illustrate the relations between tectonics and sedimentation at smaller sequence scale. The Sant Llorenç del Munt fan delta, in relation to the

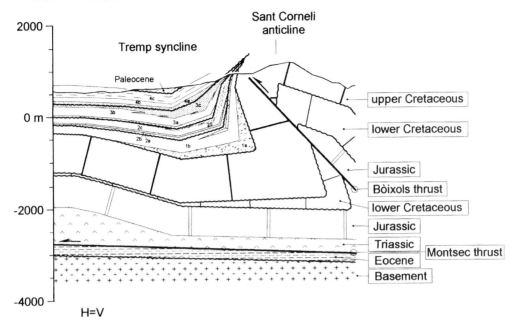

Fig. 10. Geological cross-sections through the Boíxols–Sant Corneli anticline. Numbers 1 to 4 correspond to the successive depositional sequences in the Aren Group. 1a, Puimanyons olistostrome; 1b, slope marls-Salas marls; 2a, bioclastic nearshore sandstones; 2b, offshore sandstones and shales; 2c, offshore shales; 2c′, bioclastic nearshore sandstones; 3a, bioclastic nearshore sandstones; 3b, offshore shales; 3b′: nearshore sandstones; 4a, nearshore sandstones; 4b, lagoonal shales; 4c, alluvial red shales (From Arbués et al. 1996, reprinted with permission.)

structure of the Catalan Coastal Ranges at the southeastern margin of the Ebro basin, and the Ainsa basin-fill as an example of a growth syncline in the central sector of the Southern Pyrenees.

The Sant Llorenç del Munt fan delta offers a unique opportunity to study and discuss the sequence definition in a supply-dominated environment, and in a setting where the marine to non-marine correlation can be guaranteed. **López Blanco et al.** (in press) give a synthesis of several years of study based on detailed mapping of the area. The authors distinguish transgressive–regressive units of three orders of magnitude: fundamental sequences, composite sequences and composite megasequences.

Fundamental sequences are 3–80 m thick and range in duration between 10 000 and 100 000 years. They include a basal transgressive part and an upper regressive part, and they are laterally persistent and mappable. The transgressive surfaces are the fundamental sequence boundaries. The transgressive–regressive couplets correspond to pulses of fan progradation in response to changes in sediment supply.

Composite sequences result from stacking of fundamental sequences. Stacking patterns, defined from shoreline trajectories (**López-Blanco et al.** in press, fig. 7) may be transgressive or regressive. Major surfaces of maximum regression are taken as the composite sequence boundaries. Thicknesses are in the order of one to three hundred meters, they are traceable for a few tens of kilometres, even into the neighbouring fan bodies, and they take between 0.1 and 1.0 Ma to form.

Composite mega sequences result from stacking of composite sequences, and include the same elements. The Sant Llorenç del Munt fan itself constitutes a composite mega sequence, which, with a thickness of more than 1000 m and a duration in the order of 3 Ma, falls in the scale of the previously discussed thrust-related sequences (Puigdefàbregas et al. 1986). In general, the proposed sequence partitioning for the Sant Llorenç del Munt fan delta (**López-Blanco et al.** in press, fig. 17) does not fit to the global eustatic chart of Haq et al. (1987). This misfit, together with the results of subsidence analysis, suggest that composite megasequences are controlled by changes

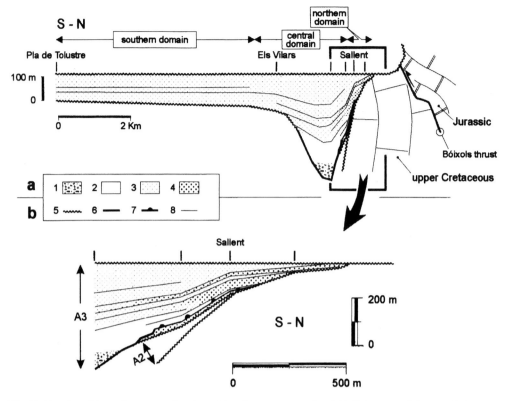

Fig. 11. Stratigraphic section concerning depositional sequence number 3 (A3), between Sallent and Pla de Tolustre localities. 1, Les Vinyes olistostrome, slumps and debris flow deposits; 2, offshore marls; 3, nearshore sandstones; 4, fandeltaic Sallent conglomerates; 5, unconformity; 6, surface of submarine gravity driven erosion; 7, submarine condensed section; 8, time line. (From Arbués et al. 1996, reprinted with permission.)

in sediment supply related to tectonics and, to some extent, to climate.

It must be noted that cycles defined on proximal fan reaches may, or may not coincide with cycles defined on the coastal reaches of the sediment wedge. Proximal fan sequences are more likely to be related to changes in slope gradients in response to tectonic activity. When they do not fit with the sequences defined on the coastal reaches, it may be due to the fact that sediment supply is not equally distributed across the area of deposition. In other words, as previously discussed, different processes acting at different parts of the sedimentation profile result in segmentation of sequence boundaries (Dreyer et al. 1994).

Fundamental sequences, in turn, can be generated by relative sea level changes, by changes in sediment supply or by autocyclic processes or, more likely, by all factors combined. In this case it will not be possible to discriminate between them.

It may be concluded, from this case study, that tectonics and climate are the main controls in supply dominated sequences, climate becoming downscale predominant (Amorosi et al. 1997).

A similar partitioning in three sequential orders was obtained from a detailed study of the Ainsa basin fill (Fig. 14). The higher sequence order corresponds to the concept of unconformity bounded tectono-sedimentary unit (Garrido-Megías 1973). They include a number of transgressive–regressive sequences (second order), each of them resulting from stacking of lower third order sequences. In all cases, sequences start with a transgressive event followed by a regressive trend that includes a swift progradation, which gradually becomes more aggradational and ends up with a prominent progradal maximum.

The distinction between sequence orders is based on the character of the transgressive events. First-order transgressions correspond to

INTRODUCTION

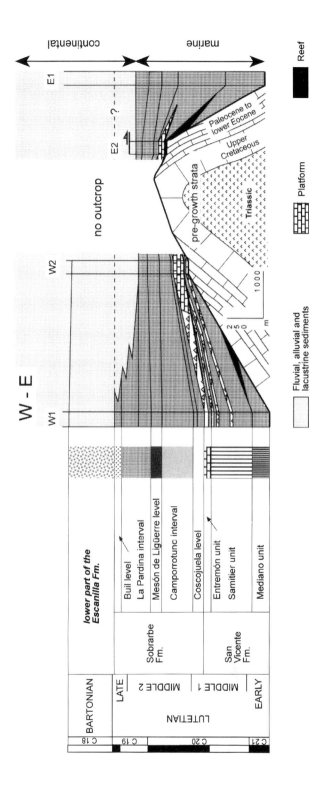

Fig. 12. Correlation of the stratigraphic sections showing growth strata, and location of the sections with respect to the Mediano anticline. The horizontal scale is approximate. (From Poblet *et al.* in press, fig. 6, reprinted with permission.)

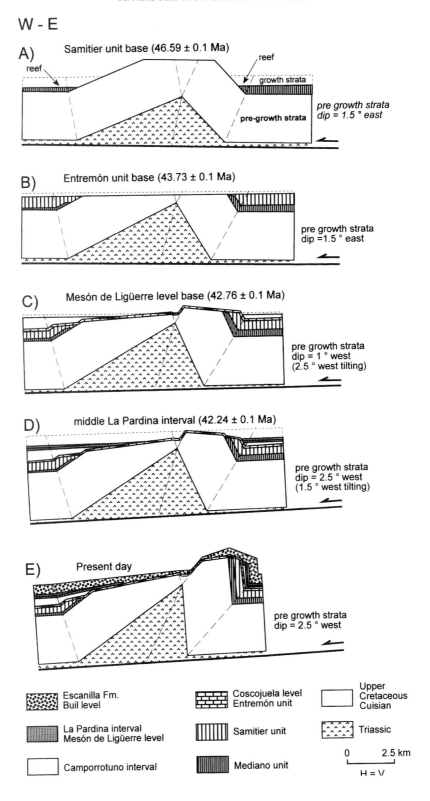

Nummulites carbonate beds whose lateral continuity goes beyond the syncline-basin itself and can be found in neighbouring basins. Second-order transgressions are *Nummulites* and patch-reef carbonate beds whose extent is restricted to the syncline basin, but whose lateral equivalents can be traced to overlay the upper delta plain of the underlying sequence. Third-order transgressions are limited to delta front. They may be allocyclic, but also may be related to autocyclic processes, bounded by abandonment or minor flooding surfaces.

First-order sequences distinguished in the Ainsa basin range in thickness between 300 and 800 m, have an approximate duration of 1.3 Ma (Bentham 1992) and closely compare to the composite depositional sequences of Posamentier *et al.* (1988) and to the composite sequences of **López-Blanco *et al.*** (in press). It should be also noted that delta-plain strata from first-order sequences are often slightly tilted in relation to those of the overlying sequence, and that a distinct component of differential subsidence is appreciated, all suggestive of an important tectonic control on the sequence organisation.

As in the case of Sant Llorenç del Munt, the lower the sequence order, the higher the probability of introduction of new controlling factors other than tectonics, down to the lower order where the autocyclic control predominates.

Discussion of results

The IBS Module on Foreland Basins is characterized by the geographic dispersion of field areas, some of which have been studied from outcrops and some from subsurface data. In addition, a diversity of results has been obtained by the different groups, each of them linked to their own research tradition. Far from an undesired difficulty, this apparent heterogeneity constitutes an important added value to a project that, like the IBS, makes comparison between different basins one of the main issues.

We have been through the main results of the various groups in the different areas, following a down-scale sequence from the orogen to the depositional sequences. Some of the more relevant conclusions should, at this stage, be stressed and eventually discussed.

Time-step restored cross-sections

Construction and modelling of time-step restored cross-sections has been attempted in the Pyrenees. The technique proves to be useful in general to understand the flexural evolution of the orogen and to estimate parameters otherwise difficult to measure, such as palaeotopographic elevations, rates of shortening, topographic growth and erosion, and to constrain better already known concepts such as the definition of the evolutionary stages, and understanding of the linked hydrocarbon evolution.

The Betic thrust front model

The understanding of the role of the salt in thrust fronts, and the description of the Betic thrust front model as an extruded lateral diapir interfering with the foreland basin fill, is an important achievement of the IBS project. The model can be applied in analogous thrust fronts of other mountain chains, and may be of additional use in hydrocarbon exploration in foreland basins.

Growth folds

Growth folds and their related growth strata have been described at various localities (Barrème syncline, Sant Llorenç de Morunys, Bòixols thrust and Mediano anticline). Modelling of these growth folds makes the estimation of rates of growth possible, and enhances the potential of predicting particular types of facies and traps in relation to the growth stratal geometries. The paper by **Williams *et al.*** (this volume), represents a first attempt to analyse the relations between folding and alluvial fan sedimentation at the level of the sedimentary process.

Role of tectonics in sequence stratigraphic partitioning

Four orders of stratigraphic partitioning can be distinguished. The foreland basin in itself constitutes the first order of tectonostratigraphic partitioning in relation to the plate collision, subduction forces and growth of the orogen. The successive piggy-back sub-basin stages constitute the second order, related to the propagation of the thrust system. Tectono-sedimentary units and composite megasequences fall in this category. Composite sequences and fundamental sequences constitute the third and fourth orders, where climate and eustasy combine with all the preceding tectonic controls in defining the

Fig. 13. 2D evolutive diagram for the Mediano anticline in a section perpendicular to the fold axis. The upper continental sediments have not been represented. (From Poblet *et al.* in press, fig. 18, reprinted with permission.)

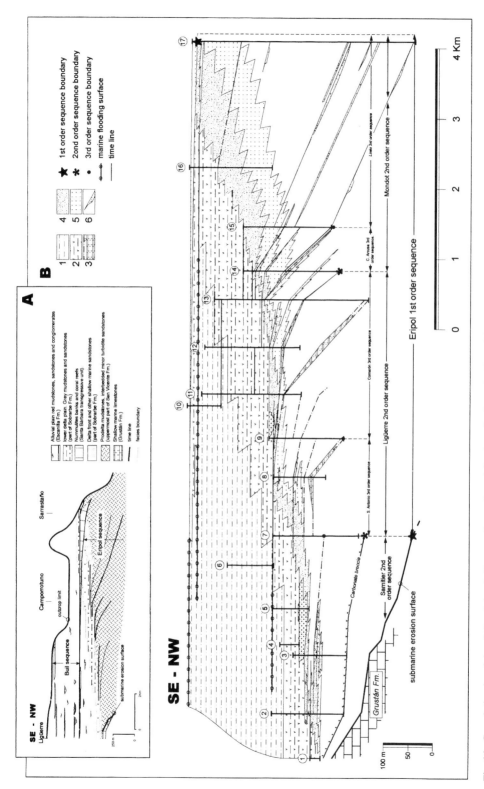

Fig. 14. Correlation involving delta plain, shallow marine and prodelta deposits from the Escanilla, Sobrarbe and San Vicente formations in the Ainsa Basin. (East flank of the Buil syncline). (**b**) is a detailed view of the Ligüerre and Mondot sequences in (**a**). 1, Alluvial and upper delta plain facies; 2, lower delta plain facies; 3, carbonate facies (*Nummulites* beds and coral reefs; 4, coarse-grained (sandy) delta front; 5, finer-grained and burrowed delta front sands and silts; 6, prodelta mudstones with minor turbidites; 7, slope carbonate breccia; 8, marine flooding surface; 9, abandonment surface without obvious flooding; 10, timeline; 11, facies boundary (From P. Arbués & J. Corregidor, unpublished.)

sequence architecture. The first two orders may be referred to as tectono-stratigraphic units, and the last two as sequences. This is an advisable conservative proposal of sequence partitioning. Many other and equally valid orders may be also proposed by adding intermediate subdivisions and further extending the lower-scale end to the autocyclic sequences. It is, nevertheless, our feeling that overclassification should be avoided.

It should be noted that the controls acting on a particular higher-sequence order are also present in all the lower orders. In other words, when downscaling, the controlling factors are added, not substituted. This is contradictory to the concept that high-order sequences are exclusively controlled by tectonics, and low order, exclusively eustatic.

It is also important to bear in mind that classifications of sequence orders are mere research strategies, only meant to help the understanding of the dynamics of the basin-fill. In fact, although sequence boundaries may be sharp, there is a continuum between plate collision, propagation of the orogenic wedge, and motion of a particular thrust, through a chain of unpredictable processes of stress accumulation and release, finally transmitted to the sedimentary record through a complex link of geomorphic processes. Drumond & Wilkinson (1996), based on a statistical analysis of hierarchies in thickness and duration of different orders of stratigraphic packages, conclude that 'discrimination of stratigraphic hierarchies and their designation to cycle orders may constitute little more than the arbitrary subdivision of an uninterrupted stratigraphic continuum'. The situation in which 'the use of a particular classification scheme tends to become a framework where in its very application serves to suppress certain lines of inquiry' (Drumond & Wilkinson 1996) should be avoided.

Segmented character of sequence boundaries

From the case studies of Sant Llorenç del Munt and the Ainsa Basin it could be concluded that sequence boundaries are formed by different processes in each of the three parts (subaerial, shallow marine and deeper marine) of the sedimentary profile. Sequence boundaries are therefore segmented, and their continuity interrupted at the coastline, between the alluvial plain and shallow marine, and at the upper slope, between shallow and deeper basin.

Within the sedimentary profile, considered from the fluvial valley down to the basin floor, changes of gradient and a accomodation space differ from one profile segment to the next.

It may be concluded that tectonics and climate (and therefore eustasy), interact to such a degree in sequence generation, so that discrimination between these factors becomes not only difficult, but to some extent irrelevant. What really matters, in terms of predictive potential, is the understanding of sequence upbuilding within the particular basin fill, and the integration of this understanding as an additional piece in the geological knowledge of the area.

'The authors wish to thank Tom Dreyer and Kjell-Owe Hager for the review of the manuscript, and for their sensible and useful suggestions. Thanks are also due to P. Arbues, J. Corregidor, D. Garcia Castellanos, A. Martinez, J. Poblet, J. Verges for allowing us to use previously published figures, some of them not yet published. As the aim of this paper is to give an overview of the results of the IBS Project, and at the same time introduce the papers included in this volume, all the authors and members of the Project are here acknowledged.'

References

AMOROSI, A., CAPORALE, L., PUIGDEFÀBREGAS, C. & SEVERI, P. 1997. Correlation between drainage and Depositional Basins: Late Quaternary Sequences From The Reno Fluvial System (Northern Italy). *In:* ROGERS, J. (ed.) *6th. International Conference on Fluvial Sedimentology,* University of Cape Town. Abstract volume, 7.

ARBUÉS, P., PI, E. & BERÁSTEGUI, X. 1996. Relaciones entre la evolución sedimentaria del Grupo de Areny el cabalgamiento de Bóixols. *Geogaceta,* **20,** 446–449.

BEHRMANN J., BROWN K., MOORE J. C., MASCLE A. & TAYLOR E. 1988. Evolution of structures and fabrics in the Barbados Accretionary Prism. Insights from Leg 110 of the Ocean Drilling Program. *Journal of Structural Geology,* **10,** 577–591.

BENTHAM, P. A. 1992. *The tectonestratigraphic development of the western oblique ramp of the south-central Pyrenean thrust system, northern Spain.* Doctoral Disertation, University of Southern California, Los Angeles.

BERÁSTEGUI, X., LOSANTOS, M., MUÑOZ, J. A. & PUIGDEFÀBREGAS, C. 1993. *Tall geològic del Pirineu Central 1/200.000.* Publ. Inst. Cartogràfic de Catalunya.

CLAVELL, E. 1992. *Geologia del petroli des les conques terciàres de Catalunya.* Tesi Doctoral, Univ. Barcelona.

CRUMEYROLLE, P. 1987. *Stratigraphie physique et sédimentogie des systèmes de dépôt de la Sequence de Santa Liestra (Eocene sud-pyrénéeen, Pyrénées aragonaises, Espagne).* PhD Thesis, University of Bordeaux III.

DECELLES, P. G. & GILES, K. A. 1996. Foreland Basin Systems. *Basin Research,* **8,** 105–123.

DERAMOND, J., SOUQUET, P., FONDECAVE-WALLEZ, M. J. & SPETCH, M. 1993. Relationships between thrust tectonics and sequence stratigraphy surfaces in foredeeps: model and examples from the Pyrenees (Cretaceous–Eocene, France, Spain). *In*: WILLIAMS, G. D. & DOBB, A. (eds) *Tectonics and Seismic Stratigraphy*. Geological Society, Special Publications London, **71**, 193–219.

DREYER, T., WADSWORTH, J., ARBUÉS, P., PUIGDEFÀBREGAS, C. & TAYLOR, A. M. 1994. Sequence Architecture in a Tectonically Active Setting: The Sobrarbe Formation, Ainsa Basin, Spain. *In*: JOHNSON, S. D. (ed.) *High Resolution Sequence Stratigraphy: Innovations and Applications. Abstract Volume*. University of Liverpool.

DRUMOND, C. N. & WILKINSON, B. H. 1996. Stratal Thickness Frequencies and the Prevalence of Orderedness in Stratigraphic Sequences. *Journal of Geology*, **104**, 1–18.

DUVAL, B. C., CRAMEZ, C. & VALDÉS, G. E. 1995. Giant fields of the 80s associated with an 'A' subduction in S. America. *Oil & Gas Journal*, 17 July and 24 July, 67–71 and 61–64.

ECORS-PYRENEES TEAM 1988. The ECORS deep reflection seismic survey across the Pyrenees. *Nature*, **331**, 508–511.

FAUGÈRES J. C., GONTHIER, E., GRIBOULARD R. & MASSE, L. 1993. Quaternary sandy deposits and canyons on the Venezuelan Margin and South Barbados accretionary prism. *Marine Geology*, **110**, 115–142.

FONDECAVE-WALLEZ, M. J., SOUQUET, P. & GOURINARD, Y. 1988. Synchronisme des séquences sédimentaires du comblement fini Crétacé avec les cycles eustatiques dans les Pyrénées Centro-Méridionales (Espagne). *Comptes Rendus de l'Academiedes Sciences, Paris*, **307**(2), 289–293.

FONNESU, F. 1984. *Estratigrafía física y análisis de facies de la secuencia de Fígols entre el río Noguera Pallaresa e Iscles (Provs. de Lérida y Huesca)*. PhD Thesis of the Universitat Autònoma de Barcelona.

FORD, M., WILLIAMS, E. A., ARTONI, A., VERGÉS, J. & HARDY, S. 1996. Progressive evolution of a fault-related fold pair from growth strata geometries, Sant Llorenç de Morunys, SE Pyrenees. *Journal of Structural Geology*, **19**, 413–441.

GARCIA-CASTELLANOS, D., FERNÁNDEZ, M. & TORNÉ, M. Numerical modeling of foreland basin formation: a program relating thrusting. Flexure, sediment geometry and lithosphere rheology. *Paper accepted in Computers & Geoscience*, in press.

GARRIDO-MEGÍAS, A. 1973. *Estudio geológico y relación entre tectónica y sedimentación del Secundario y el Terciarion de la vertiente meridional Pirenaica en su zona central*. PhD Thesis of the University of Granada.

GIESKES, J. M., BLANC, G., VROLIJK AND THE ODP LEG 110 Scientific Party 1989. Hydrogeochemistry in the Barbados Accretionary Complex: Leg 110 ODP. *Palaeogeography, Palaeoclimatology, Palaeoecology*, **71**, 83–96.

GIMÉNEZ, J. 1993. *Análisis de cuenca del Eoceno inferior de la Unidad Cadí (Pirineo oriental): el sistema deltaico y de plataforma carbonática de la Formación de Corones*. Tesis Univ. Barcelona.

GREBER, E., LEU W., BERNOUILLI, D., SCHUMACHER, M. & WYSS, R. 1997. Hydrocarbon provinces in the Swiss Southern Alps-a gas geochemistry and basin modelling study. *Marine and Petroleum Geology*, **14**, 3–25.

HAQ, B. U., HARDENBOL, J. & VAIL, P. R. 1987. Chronology of Fluctuating Sea Levels Since the Triassic. *Science*, **235**, 1156–1166.

HARDY, S. & POBLET, J. 1994. Geometric and numerical models of progressive limb rotation in detachment folds. *Geology*, **22**, 371–374.

HOMEWOOD, P., ALLEN, P. & WILLIAMS, G. 1986. Dynamics of the Molasse Basin of Western Switzerland. *In*: ALLEN, P. A. & HOMEWOOD, P. (eds), *Foreland Basins*. IAS Special publications, **8**, 199–219.

JIN, J., AIGNER, T., LUTERBACHER, H. P., BACHMAN G. H. & MÜLLER, M. 1995. Sequence stratigraphy and depositional history in the south-eastern German Molasse Basin. *Marine and Petroleum Geology*, **12**, 929–940.

JOLIVET, D;, FAUGÈRES J. C., GRIBOULARD, R., DESBRUYÈRES, D. & BLANC, G. 1990. Composition and spatial organisation of a cold seep community on the South Barbados accretionary prism: tectonic, geochemical and sedimentary context. *Programmes in Oceanography*, **24**, 24–45.

LABAUME, P., HENRY, P., RABAUTE, A. & THE ODP LEG 156 SCIENTIFIC PARTY 1995. Circulation et surpression de l'eau interstitielle dans le prisme d'accrétion Nord-Barbade. *Comptes Rendus de l'Académie des Sciences, Paris*, **320**, série IIa, 977–984.

LARROQUE, CH., CALASSOU, S., MALAVIEILLE, J. & CHANIER, F. 1995. Experimental modelling in forearc basin development during accretionary wedge growth. *Basin Research*, **7**, 255–268.

LE VOT, M., BITEAU, J. J. & MASSET, J. M. 1996. The Aquitaine Basin: oil and gas production in the foreland of the Pyrenean fold-and-thrust belt. New exploration perpectives. *In*: ZIEGLER, P. & HORVATH, F. (eds) *Structure and Prospects of the Alpine Basin and Forelands*. Edition du Muséum, Paris, 159–172.

LOPEZ BLANCO, M., MARZO, M., BURBANK, D., VERGES, J., ROCA, E., ANADON, P. & PINA, J. In press. Tectonic and climatic control on the development of large foreland fan deltas: Montserrat and Sant Llorenç del Munt systems (middle Eocene, Ebro basin, NE Spain). *In*: MARZO, M. & STEEL, R. (eds) *Sedimentology and Sequence Stratigraphy of the Sant Llorenç del Munt clastic wedges (SE Ebro Basin, NE Spain)*. Sedimentary Geology special issue.

MARTÍNEZ, A., VERGÉS, J. & MUÑOZ, J. A. 1988. Secuencias de propagación del sistema de cabalgamientos de la terminación oriental del manto del Pedraforca y relación con los conglomerados sinorogénicos. *Acta Geologica Hispania*, **23**, 119–128.

MARTÍNEZ RIUS, A., BERÁSTEGUI, X. & LOSANTOS, M. 1996. Corte geológico en el Pirineo oriental:

emplazamiento en una secuencia de bloque superior de las láminas cabalgantes que forman el manto inferior del Pedraforca. *Geogaceta*, **20**, 450–453.

MASCLE A., ENDIGNOUX L. & CHENNOUF T. 1990. Frontal accretion and piggyback basin development at the southern edge of the Barbados Ridge accretionary complex. *In: Proceedings of ODP Leg 110 Scientific Results*. College Station,TX, 17–29.

——, VIALLY, R., DEVILLE, E., BIJU-DUVAL, B. & ROY, J. P. 1996. The petroleum evaluation of a tectonically complex area: the western margin of the Southeast Basin (France). *Marine and Petroleum Geology*, **13**, 941–961.

MELLERE, D. 1993. Thrust-generated, back-fill stacking of alluvial fan sequences, South Central Pyrenees, Spain (La Pobla de Segur Conglomerates). *In*: FROSTICK, L. & STEEL, R. (eds) *Tectonic Control and Signatures in Sedimentary Successions. Special Publications of the International Association of Sedimentologists*. **20**, 259–276.

MILLÁN, H., DEN BEZEMER, T. VERGÉS, J., MARZO, M., MUÑOZ, J. A., ROCA, E., CIRÉS, J., ZOETEMEIJER, R., CLOETING, S. & PUIGDEFÀBREGAS, C. 1995. Palaeoelevation and effective elastic thickness evolution at mountain ranges: inferences from flexural modeling in the Eastern Pyrenees and Ebro Basin. *Marine and Petroleum Geology*, **12**, 917–928.

MOORE, J. C. & VROLIJK, P. 1992. Fluids in accretionary prisms. *Reviews of Geophysics*, **30**, 113–135.

MUÑOZ, J. A. 1992. Evolution of a continental collision belt: ECORS-Pyrenees crustal balanced cross-section. *In*: MCCLAY, K. (ed.) *Thrust Tectonics*. Chapman & Hall, London, 235–246.

——, MCCLAY, K. & POBLET, J. 1994. Synchronous extension and contraction in frontal thrust sheet of the Spanish Pyrenees. Geology, **22**, 921–924.

MUTTI, E., DATTILO, P., SGAVETTI, M., TEBALDI, E., BUSATTA, C. & MORA, S. 1994. Sequence stratigraphic response to thrust propagation in the upper cretaceous Aren Group, south-central Pyrenees. *In*: MUTTI, E., DAVOLI, G., MORA, S. & SGAVETTI, M. (eds) *Second high-resolution sequence stratigraphy conference, June 1994, Tremp. Spain*. Part III, 25–36.

——, LUTERBACHER, H. P., FERRER, J. & ROSELL, J. 1972. Schema stratigrafico e lineamenti di facies del paleogeno marino della zona centrale sudpirenaica tra Tremp (Catalogna) e Pamplona (Navarra). *Memorie della Società Geologica Italiana*, **11**, 391–416.

——, SEGURET, M. & SGAVETTI, M. 1988. *Sedimentation and deformation in the Tertiary sequences of the southern Pyrenees*. AAPG Mediterranean Basins Conference, Nice, Field Guide 7. Special Publication of the Institute of Geology, University of Parma, Italy.

——, SGAVETTI, M. & REMACHA, E. 1984. Le relazioni tra piataforme deltizie a sistemi torbiditici nel Bacino Eocenico Sudpirenaico di Tremp-Pamplona. *Giornale di Geologia Ser. 3*, **46**(2), 3–32.

NAGTEGAAL, P. J. C., VAN VLIET, A. & BROUWER, J. 1983. Syntectonic Coastal offlap and Concurrent Turbidite Deposition: the Upper Cretaceous Aren sandstone in the South-Central Pyrenees, Spain. *Sedimentary Geology*, **34**, 185–218.

NIJMAN, W. & NIO, S. D. 1975. *The Eocene Montañana Delta (Tremp-Graus Basin, Prov. Lerida and Huesca, Southern Pyrenees, N. Spain)*. IAS 9th International Sedimentology Congress, Nice, Excursion Guidebook, 19, part B.

PAPPON, J. P. 1969. *Etude de la zone sud-Pyrénéenne dans le Massif de Turbón (Prov. de Huesca - Espagne)*. These 3e cycle. Univ. Toulouse.

PERMANYER, A., VALLÉS, D. & DORRONSORO, C. 1988. Source rock potential of an Eocene carbonate slope: the Armàncies Formation of the Southern Pyrenean Basin, Northeast Spain. *American Association of Petroleum Geologists Bulletin*, **72**, 1019.

POBLET, J., MUÑOZ, J. A., TRAVÉ, A. & SERRA-KIEL, J. Quantifying the kinematics of detachment folds using the 3D geometry: application to the Mediano anticline (Pyrenees, Spain). Geological Society of America Bulletin, in press.

POSAMENTIER, H. W., JERVEY, M. T. & VAIL, P. R. 1988. Eustatic controls on clastic deposition I – conceptual framework. *In*: WILGUS, C. K. & HASTINGS, B. S. (eds) *Sea level changes: an integrated approach*. SEPM. Special Publications, **42**, 109–124.

PUIGDEFÀBREGAS, C. & SOUQUET, P. 1986. Tectonosedimentary Cycles and Depositional Sequences of the Mesozoic and Tertiary from the Pyrenees. *Tectonophysics*, **129**, 172–203.

——, MUÑOZ, J. A. & MARZO, M. 1986. Thrust belt development in the Eastern Pyrenees and related depositional sequences in the southern forelan basin. *In*: ALLEN, P. A. & HOMEWOOD, P. (eds) *Foreland Basins*. Special Publications, **8**, 229–245.

—— & VERGÉS, J. 1992. Thrusting and foreland basin evolution in the southern Pyrenees. *In*: MCCLAY, K. (ed.) *Thrust Tectonics*, Chapman & Hall, London, 247–254.

REILLE, J. L. 1971. *Les relations entre tectogénèse et sédimentation sur le versant sud des Pyrénées Centrales*. Thèse, Univ. Montpellier.

RIBA, O. 1964. Estructura sedimentaria del Terciario continental de la Depresión del Ebro en su parte riojana y navarra. *Aportación Esp. al XX Congr. Geogr. Int. 1964*, 127–138.

—— 1976. Syntectonic unconformities of the Alto Cardener, Spanish Pyrenees: a genetic interpretation. *Sedimentary Geology*, **15**, 213–233.

ROBERTS, G. 1990. Structural control on fluid migration through the Rencurel thrust zone, Vercors, French Sub-Alpine chain. *In*: ENGLAND, R. & FLEET, A. J. (eds) *Petroleum Migration*, Geological Society, London, Special Publications, **59**, 245–262.

SASSI, W. & DEVILLE, E. Maturity modelling in thrust belts, methodology and cases histories. *AAPG Bulletin*, in press.

SIMÓ, A. 1985. *Secuencias deposicionales del Cretácico superior de la Unidad del Montsec (Pirineo Central)*. Tesi doctoral. Univ. de Barcelona.

SOLER, M. & PUIGDEFÀBREGAS, C. 1970. Líneas generales de la geologia del Alto Aragón Occidental. *Pirineos*, **96**, 5–20.

TEN HAAF, E. 1966. Le Flysch sud-Pyrénéen le long du rio Ara (Huesca). *Pirineos*, **81–82**, 143–150.

TOURET, J. & VAN HINTE, J. (eds) 1992. Le role des fluides dans les zones de subduction. *Koninklijke Nederlandse Akademie van Wetenschappen Proceedings*, **95**, 293–403.

VAIL, P. R. & MITCHUM, R. M. 1977. Seismic stratigraphy and global changes of sea level, Part 1: Overview. *AAPG Memoirs*, **26**, 51–52.

VERGÉS, J. 1993. *Estudi geològic del vessant sud del Pirineu oriental i central. Evolució cinemàtica en 3D*. Tesi Doctoral. Univ. Barcelona.

—— & BURBANK, D. W. 1996. Eocene–Oligocene thrusting and basin configuration in the eastern and central Pyrenees (Spain). *In*: FRIEND, P. F. & DABRIO, C. J. (eds) *Tertiary Basins of Spain*. Cambridge University Press, 120–133.

——, MILLÁN, H., ROCA, E., MUÑOZ, J. A., MARZO, M., CIRÉS, J., DEN BEZEMER, T., ZOETEMEIJER, R. & CLOETINGH, S. 1995. Eastern Pyrenees and related foreland basins: pre-, syn- and post-collisional crustal-scale cross-sections. *Marine and Petroleum Geology*, **12**, 893–915.

VIALLY, R. 1994. The southern French Alps Paleogene basin: subsidence modelling and geodynamics implications. *In*: MASCLE, A. (ed.) *Hydrocarbons and Petroleum Geology of France*, Springer-Verlag, 281–294.

VROLIJK, P. & SHEPPART, S. 1991. Syntectonic carbonate veins from the Barbados accretionary prism (ODP Leg 110): record of paleohydrology. *Sedimentology*, **38**, 671–690.

ZOETEMEIJER, R. 1993. *Tectonic modeling of foreland basins: thin skinned thrusting, syntectonic sedimentation and lithospheric flexure*. PhD Thesis, Vrije Universiteit, Amsterdam.

Geophysical and geological constraints on the evolution of the Guadalquivir foreland basin, Spain

M. FERNÀNDEZ[1], X. BERÁSTEGUI[2], C. PUIG[2], D. GARCÍA-CASTELLANOS[1], M. J. JURADO[1,4], M. TORNÉ[1] AND C. BANKS[3]

[1] *Institute of Earth Sciences (J. Almera), CSIC. Lluís Solé Sabarís s/n, 08028-Barcelona, Spain*
[2] *Servei Geològic de Catalunya, ICC, Parc Montjuic, 08038-Barcelona, Spain*
[3] *Department of Geology, Royal Holloway University of London, Egham, Surrey TW20 0EX, UK.*
[4] *Present address: Institute of Geophysics, University of Karlsruhe, Hertzstrasse, 16, 76187-Karlsruhe, Germany*

Abstract: This paper presents a compilation and reinterpretation of available geophysical and geological data recently acquired for the ENE–WSW Guadalquivir foreland basin, located on the northern margin of the Betic orogen in southern Iberia. The data include seismic reflection and refraction profiles, well logs, gravity, geoid, surface heat-flow data and field observations. The deep structure of the southern Iberian margin is characterized by large variations in crustal thickness and high heat-flow values, which result in a very low lithospheric rigidity for the whole area. Geoid and gravity data show that deformation affected the crust and the lithospheric mantle differently, producing anomalous mass distributions that could act as subsurface loads. Seismic sequence analysis of the basin infill has permitted the re-assessment of the depositional sequential arrangement of the sediments deposited from Late Langhian–Early Serravallian to Messinian. They are arranged in six sequences and do not show any E–W progradational pattern indicating that during this period the acting loads moved essentially in a NNW direction. A careful analysis of the southern border of the basin shows that the 'so-called olistostromes' correspond to lateral diapirs of squeezed Triassic evaporites and internally imbricated Miocene wedges. We discuss the results obtained in terms of palaeo-geographic environments, time distribution and nature of acting loads, and constraints for future basin modelling approaches.

The development of orogenic belts involves vertical loads being imposed on the lithosphere which reacts by deflecting elastically, forming a foreland basin. Modelling foreland basins requires an understanding of (1) the spatial and temporal distribution of surface and subsurface tectonic loads; (2) the geometry and composition of the basin basement; (3) the thickness, composition and palaeo-environments of the major stratigraphic units of the sedimentary infill; (4) the spatial and temporal variations in the mechanical properties of the lithosphere.

The Guadalquivir Basin is one of the youngest European foreland basins which formed as a consequence of the latest stages (Miocene–Recent) of the Alpine Orogeny. It is located in the southern Iberian Peninsula with its northern, foreland margin being the Iberian Massif and its southern, hinterland margin being the Betic Mountain chain (Fig. 1). In spite of the numerous studies carried out for hydrocarbon exploration and academic research purposes, many

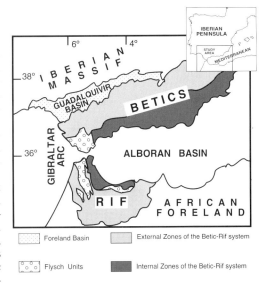

Fig. 1. Geological sketch map of the study area.

FERNÀNDEZ, M., BERÁSTEGUI, X., PUIG, C., GARCÍA-CASTELLANOS, D., JURADO, M. J., TORNÉ, M. & BANKS, C. 1998. Geophysical and geological constraints on the evolution of the Guadalquivir foreland basin, Spain. *In:* MASCLE, A., PUIGDEFÀBREGAS, C., LUTERBACHER, H. P. & FERNÀNDEZ, M. (eds) *Cenozoic Foreland Basins of Western Europe.* Geological Society Special Publications, **134**, 29–48.

aspects concerning the evolution of the Guadalquivir Basin still remain under debate.

Major discussions arise related to: (1) the age of the sediments that unconformably overlie the basement, which could be Late Langhian – Early Serravallian (e.g. Perconig 1960–62; Saavedra 1964; Riaza & Martínez del Olmo 1996) or Mid-Tortonian (e.g. Perconig 1971; Sierro *et al.* 1996); (2) the origin and extent of the main prograding directions of the depositional systems which can be attributed to either eustatic changes or tectonic uplift (e.g. Sierro *et al.* 1996; Riaza & Martínez del Olmo 1996); (3) the nature and mechanisms of emplacement of the so-called 'Olistostromic Unit' that could correspond either to mega-elements emplaced by gravitational sliding (e.g. Perconig 1960–62; Martínez del Olmo *et al.* 1984; Suárez-Alba *et al.* 1989) or to lateral diapirs emplaced by squeezing of Triassic materials, which would have also produced a Miocene frontal wedge (this paper; Berástegui *et al.* this volume); and (4) the composition and geometry of the basement below the Betics, which is poorly constrained (e.g. IGME 1987; Banks & Warburton 1991).

The mechanical properties of the lithosphere are mainly defined by the crustal thickness and the thermal regime (e.g. Ranalli 1994; Burov & Diament 1992). Partial information on the crustal structure and thermal regime in the study area comes from combined seismic and gravity studies and heat-flow surveys carried out in different parts of the southern Iberian margin (e.g. Torné & Banda 1992; Banda *et al.* 1993; ITGE 1993; Polyak *et al.* 1996). However, it is necessary to put together all the available information in order to produce a regional image of the crustal and lithospheric thickness variations over the whole area.

The aim of this paper is to outline some constraints on the present-day structure and Neogene evolution of the Guadalquivir Basin and surrounding areas rather than to develop a self-consistent model, which cannot be achieved from present knowledge. We present a compilation and re-interpretation of geological and geophysical data based on field observations, seismic reflection and refraction profiles, oil-well logs, gravity, geoid and surface heat-flow data. The paper focus on: (1) the deep structure of the whole area including the southern Iberian Massif, the Guadalquivir Basin, the Betic Chain and the Alboran Sea; (2) the architecture of the Guadalquivir Basin, its extension, sequence stratigraphy, and basement geometry; and (3) the nature of the allochthonous, chaotic bodies that are emplaced on the southern border of the basin. A companion paper by Berastegui *et al.* (this volume) deals with the structure and tectonosedimentary evolution of the southern border of the Guadalquivir Basin.

Regional tectonic setting

The main geological units that characterize the southern part of the Iberian Peninsula are: the Betic mountain chain, the Guadalquivir foreland basin, and the Alboran Sea. The Betic Chain is the northern segment of an arcuate orogen that continues westwards across the Gibraltar Arc into the Rif Chain (Fig. 1). The inner part of this orogen is occupied by the Alboran Sea extensional basin. The tectonic evolution of the whole area, which constitutes the westernmost part of the Alpine Chain, was controlled by the relative movement between the African and Eurasian plates since Late Mesozoic times. Plate-motion studies from Dewey *et al.* (1989) suggest that this part of the plate boundary experienced about 200 km of roughly N–S convergence between Mid-Oligocene and Late Miocene times, followed by about 50 km of WNW-directed oblique convergence in Late Miocene to Recent time.

The major palaeogeographic elements forming the Betics–Gibraltar Arc–Rif system belonged to four pre-Miocene domains, which were well delimited at the beginning of the Neogene (Balanyá & García-Dueñas 1988): (1) and (2) are the External Zones of the Betic–Rif chain corresponding to the inverted Mesozoic continental margins of the Iberian and African plates, respectively; (3) the Flysch Units, which are made up of allochthonous sediments; (4) the Internal Zones of the Betic–Rif chain, composed of a polyphase thrust stack that includes three high-pressure–low-temperature metamorphic nappe complexes (e.g. Bakker *et al.* 1989; Tubia & Gil-Ibarguchi 1991).

Late Cretaceous and Palaeogene convergence caused substantial crustal thickening in the Internal Zones and generated an orogen by collisional stacking. Whether this thickening occurred at the present-day Alboran Sea basin or further to the East is still under debate. The Internal Zones represent the disrupted and extended fragments of this pre-Miocene orogen (Balanyá & García-Dueñas 1987, 1988; Platt & Vissers 1989; Monié *et al.* 1991; Vissers *et al.* 1995). The External Zones and the Flysch Units reflect continued crustal shortening during the Miocene, whereas crustal extension occurred in the Internal Zones of the orogen. This shortening began in the lower Aquitanian and continued into the Late Miocene (García-Dueñas *et al.* 1992; Comas *et al.* 1992).

Miocene extensional detachment systems and

fault-bounded sedimentary basins of Miocene age are superimposed upon the continental collision of the Internal Zones (e.g. Galindo Zaldivar et al. 1989; Platt & Vissers 1989; García-Dueñas & Balanyá 1991; García-Dueñas et al. 1992). This Miocene extensional phase was accompanied by a distinctive low pressure–high-temperature metamorphism (Zeck et al. 1992). The crustal thinning over much of the region, both on- and off-shore is likely to be a result of this phase of extension.

Different hypotheses have been formulated to account for the geodynamic evolution of the southern margin of the Iberian Peninsula including models involving a back-arc origin (Zeck et al. 1992; Royden 1993), mantle delamination (Channell & Mareschal 1989; García-Dueñas et al. 1992; Docherty & Banda 1995), extensional collapse (Dewey 1988; Vissers et al. 1995), and rifting (Cloetingh et al. 1992). Up to now, however, none of these models have been successfully tested to fit the available surface and subsurface data.

The Guadalquivir basin formed in an overall environment of plate convergence as the foreland basin to the central and western Betics. This convergence, however, is not directly reflected in the kinematics either of the surrounding mountain chains or in the extension of the Alboran basin. In fact, the present-day Guadalquivir basin correlates only with the later steps (Miocene–Recent) of this tectonic history that developed from Late Cretaceous times.

Regional geophysical data

During the last years several geophysical surveys have been carried out along the southern margin of the Iberian Peninsula. These surveys include deep seismic refraction, wide-angle and reflection profiles, gravity and heat flow. In this section we present a compilation of those datasets that have a predominantly crustal- and lithospheric-scale significance, namely Bouguer gravity anomalies and geoid height maps, crustal thickness and surface heat-flow.

Bouguer gravity anomalies and geoid height maps

Gravity and geoid data primarily reflect variations in the density distribution of the Earth's interior. Gravity anomalies, because of the inverse square law of the gravity field, are particularly sensitive to density variations at crustal levels. In contrast, under the assumption of local isostatic equilibrium, the geoid anomaly is proportional to the density-moment function (Haxby & Turcotte 1978) and therefore is more sensitive to density variations at sub-crustal levels. With the aim of highlighting the major features of crustal and lithospheric structure along the southern margin of the Iberian Peninsula, we have mapped both Bouguer gravity anomalies and geoid height.

The Bouguer anomaly map (Fig. 2) has been constructed using available data from the 'Bureau Gravimétrique International', from the Spanish 'Instituto Geográfico Nacional' and from available ship tracks including a cruise carried out by research vessel *RD Conrad* in the Alboran Sea. Additional data points at sea have been taken from Morelli et al. (1975). These data have been reduced following the same procedure as Torné et al. (1992). Figure 2 shows that the Bouguer anomaly, at the Guadalquivir Basin, is characterized by a gentle South-Southeastward decrease of about 60–70 mGal that reflects the basement topography fairly well. Towards the Betics, the Bouguer anomaly decreases down to -120/-130 mGal associated with the existence of a crustal root, and increases rapidly towards the south to more than +180 mGal in the easternmost Alboran Basin as a response of a prominent crustal thinning. In the western Alboran Basin, the Bouguer anomaly yields negative values, which reflect the combined effect of crustal thinning and large sediment accumulations (up to 6–7 km thick).

The geoid height map has been constructed using available data from the Deutsche Geodätische Kommission (Brennecke et al. 1983). These data include gravimetric geoid determinations off-shore and on-shore with a coverage of 6' × 10' arc. Figure 3 shows the contoured map of the resulting geoid heights. The most striking feature is the high gradient across the Guadalquivir Basin, which amounts to 0.1 m km^{-1}. In this area the geoid anomaly decreases by 6 m towards the Betics. Such anomalies are common in continental margins, and are related to the thinning of continental crust towards the oceanic crust. However, since there are no major changes in crustal thickness across the Guadalquivir Basin (see Fig. 4), this geoid anomaly must be related to deeper structures. A similar trend is observed in North Africa where the minimum geoid height (39 m) is localized on the African–Atlantic margin.

Crustal thickness

The southern Iberian Peninsula has been the objective of numerous deep seismic experiments, which include refraction, wide-angle and reflection. The crustal thickness values compiled

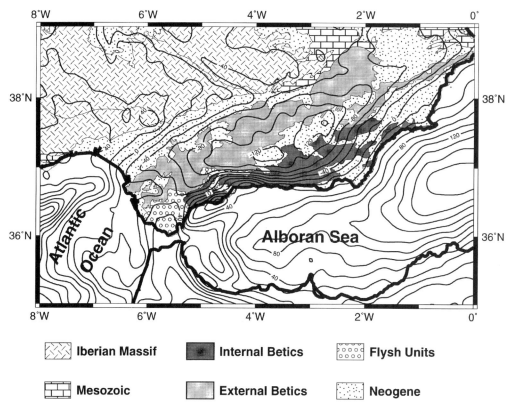

Fig. 2. Bouguer anomaly map. Isolines every 20 mGal. Compiled from 'Bureau Gravimétrique International', Spanish 'Instituto Geográfico Nacional' and *Morelli et al.* (1975).

in this section come primarily from seismic refraction and wide-angle data. The contoured map of the Moho depth (Fig. 4) obtained from the above mentioned data has been filtered and smoothed and therefore shows the main features and trends of the crustal structure in the study area. The Iberian Massif is characterized by a 32–34 km thick crust with a Pn velocity of 8.0–8.1 km s^{-1} (Banda *et al.* 1981, 1993; Banda 1988; Suriñach & Vegas 1988; ILIHA 1993). In the western part of the Iberian Massif and close to the northern border of the Guadalquivir Basin, González *et al.* (1993) propose a crustal thickness of about 29 km increasing up to 31 km towards the Gulf of Cadiz.

In the Central-Eastern Betics seismic data indicate that the Moho deepens up to 38 km under the Internal Betics (Banda & Ansorge 1980; Banda *et al.* 1993). Deep multichannel seismic data show noticeable differences in the reflective character of the crust between the External and the Internal Betics (García-Dueñas *et al.* 1994). The External Betics are characterized by an almost transparent upper crust and a weakly reflective lower crust that seems to thin out towards the SE. In contrast, the Internal Betics show a prominent reflector in the upper crust, which is interpreted as a detachment surface, and a moderate reflectivity in the lower crust with dipping reflectors in its central part. The crustal structure obtained for the westernmost Betics (Gibraltar Arc) is characterized by a Moho depth of about 31 km and a very thick sedimentary cover (up to 10 km towards the Gulf of Cadiz) with large lateral variations (Medialdea *et al.* 1986).

A pronounced crustal thinning is observed from the Betics to the Alboran Sea. Different seismic profiles acquired near the shoreline (Banda & Ansorge 1980; Medialdea *et al.* 1986; Barranco *et al.* 1990) show a crustal thickness of 23–24 km. The crustal structure in the Alboran Sea is poorly constrained, since the only available data come from seismic experiments shot in 1974 (Hatzfeld 1976; Boloix & Hatzfeld 1977) and in 1979 (Suriñach & Vegas 1993) which show that the Moho lies at about 15 km depth in the central part of the basin. Combined seismic and

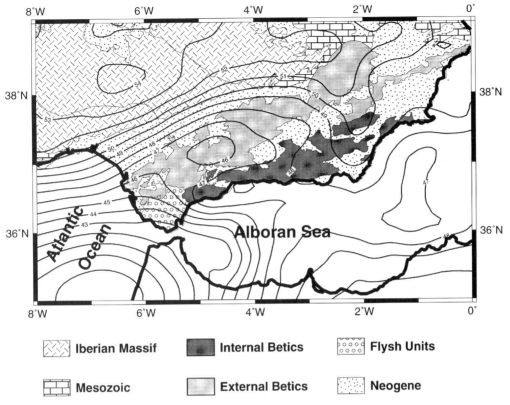

Fig. 3. Absolute geoid height map. Isolines every 1 m. Compiled from Brennecke *et al.* (1983).

gravity studies suggest that the thinning observed from the Betics to the Alboran Sea is mainly produced over a narrow area (30–35 km wide) which resembles the pattern observed in transform margins (Torné & Banda 1992; Torné *et al.* 1992; Watts *et al.* 1993). To the east however this thinning occurs over a much broader zone of about 120 km (Torné & Banda 1992).

Surface heat-flow

The thermal data available in southern Spain come from thermal gradient determinations carried out in oil wells (Albert-Beltran 1979; Banda *et al.* 1991) and water exploration wells (ITGE 1993), and from sea-floor heat-flow measurements in the Alboran Sea (Polyak *et al.* 1996). Thermal gradients obtained in water exploration wells were corrected for topographic and palaeoclimatic effects. A number of thermal conductivity measurements were also performed on rock-samples representative of the main lithologies to calculate the surface heat-flow in water wells (ITGE 1993). Heat flow in oil wells has been calculated by multiplying the mean thermal gradient, deduced from bottom hole temperature (BHT) data, by a constant thermal conductivity of 2.1 W m^{-1} K^{-1} (Albert-Beltran 1979).

Figure 5 shows the heat-flow distribution obtained in the study area. The high scatter observed in heat flow data indicates an active ground-water circulation as is evidenced in some thermometric logs recorded in water wells (Fig. 6). Thermal perturbations due to ground water are particularly noticeable along the northern border of the Guadalquivir Basin and the Betic Chain. Thermal data from oil wells confirm these results and yield temperature gradient values ranging from 20 to 45 mK m^{-1} in the Guadalquivir Basin, and from 13 to 21 mK m^{-1} in the External Betics. The stack of the BHT data (Fig. 7) shows that fluid circulation is not restricted to shallow depths (few hundred metres) but it is also evident at depths up to 5 km. In spite of that, a regional thermal pattern characterized by a surface heat-flow that varies from 70–80 mW m^{-2} in the south of the Iberian

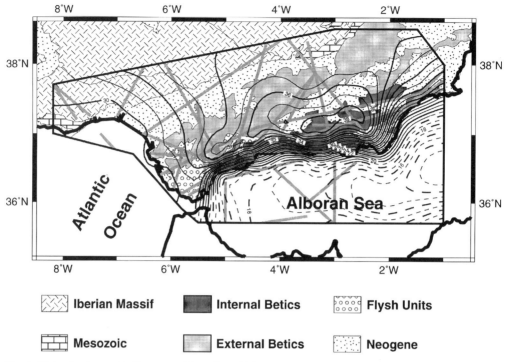

Fig. 4. Moho depth map. Isolines every 1 km. Grey thick lines indicate the location of seismic refraction profiles.

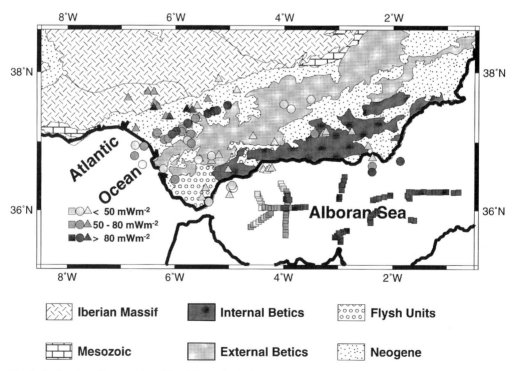

Fig. 5. Surface heat flow map in mW m^{-2}. Triangles indicate measurements from water exploration wells. Circles indicate indicate measurements from oil exploration wells. Squares indicate sea-floor measurements.

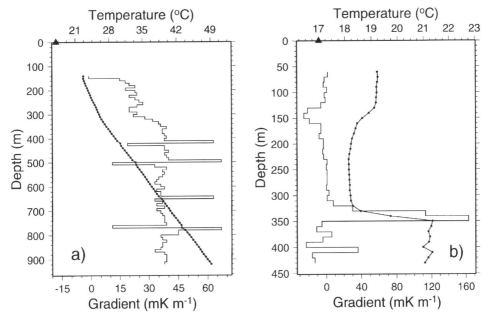

Fig. 6. Thermometric logs recorded in water exploration wells. Dotted line indicates temperature-depth log. Stepped line indicates thermal gradient-depth log. (**a**) Example of linear temperature log measured at the western part of the Guadalquivir Basin. Only perturbations of short wave-length associated with fractures are evident. (**b**) Example of highly perturbed thermometric log measured in a Neogene basin (Almeria) located at the southeastern Betic Cordillera. Circulation of cold water between 130 m and 330 m and perturbations associated with fractures between 350 m and 430 m are evident.

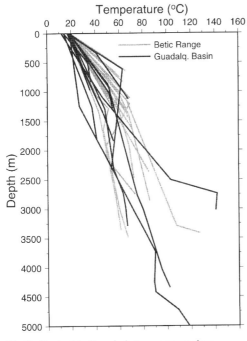

Fig. 7. Stack of bottom-hole temperature data obtained from oil exploration wells.

Massif, to 80 mW m^{-2} in the Guadalquivir Basin, and to 40–50 mW m^{-2} in the External Betics can be inferred. The few measurements carried out in the Internal Betics also show low heat-flow values (< 50 mW m^{-2}). However, further to the south, in the Alboran Sea, the heat-flow shows very high values increasing both in a W–E and N–S direction from about 50 mW m^{-2} to 120 mW m^{-2}.

Geology of the Guadalquivir Basin

Previous interpretations of the basin infill

From early works (see Perconig 1960–62) it is widely accepted that the Neogene infill of the Guadalquivir Basin consists of marine marls including sandy intercalations deposited above a 'basal calcarenite' ('grès a *Heterostegina costata*'), which unconformably overlies a pre-Miocene basement.

Regional interpretations (e.g., Perconig 1960–62; Martínez del Olmo et al. 1984; Roldán García & García Cortés 1988; Suárez-Alba et al. 1989; Sierro et al. 1996; Riaza & Martínez del Olmo 1996) agree that a northern passive margin (the Iberian Massif) provided a clastic infill to the basin, and a southern active margin (the External Zones of the Betic Cordillera) provided a gravitational infill. The clastics derived from the northern margin were re-distributed by turbiditic currents along the major axis of the basin (ENE–WSW) and finally deposited in the deepest areas (close to the southern margin) forming small, relatively sand-rich stacked lobes interbedded with blue marls and clays. According to these authors, during Late Tortonian times, gigantic chaotic masses of rocks were emplaced by gravitational sliding into the basin from the External Zones of the Betic Cordillera as olistostromes, mega-turbidites or mega-elements, responsible for the N–S narrowness of the basin. These rock masses mainly consist of Triassic evaporites, clays and limestones, and blocks of upper Cretaceous to Palaeocene limestones. In spite of the general agreement on the mode of emplacement of these units, other mechanisms can be considered as discussed below.

Riaza & Martínez del Olmo (1996) arranged the sedimentary infill of the basin into five tectono sedimentary units using seismic and well data, and tied them to the emplacement of the olistostromes. In contrast, Sierro et al. (1996), from outcrop observations and micropalaeontological data, defined five offlapping depositional sequences forming a westward progradational set. Both stratigraphic models show strong discrepancies (See Table 1 for a summary).

Although the dating of the Neogene sedimentary infill has been investigated by several authors (e.g. Perconig 1960–62, 1971; Saavedra 1964; Perconig & Granados 1973; Flores & Sierro 1989; Sierro et al. 1993), the results obtained show important discrepancies that span 5 Ma (about 30% of the age of the basin). Perconig (1960–62) proposed that the oldest sediments (the aforementioned 'grès a *Heterostegina costata*') were Helvetian in age (currently Langian–Serravallian). Later, Perconig (1971) attributed a Tortonian age to these sediments. Finally, Sierro et al. (1996) proposed that these 'basal calcarenites' are diachronous.

All the above-mentioned uncertainties led us to propose an alternative, self-consistent geological framework, which includes a re-interpretation of the basin infill using seismic stratigraphy, and an analysis of the nature and mode of emplacement of the materials in the southern margin of the basin.

A reinterpretation of the Guadalquivir Basin infill

A total of about 1400 km of seismic profiles together with data from 35 exploration hydrocarbon wells (Fig. 8) were interpreted following the usual procedures in seismic-sequence analysis (i.e. Mitchum et al. 1977; Vail 1987).

The lithological composition and physical properties of the sediments and their lateral variations were studied on six selected wells by inversion of the geophysical well logs. Additional information from the corresponding company well reports was also used. Two well-to-well correlations were performed at the westernmost and central parts of the Guadalquivir Basin (Figs 9 and 10) to study the transition from the Betics to the foreland basin. The main lithologies in each well together with the gamma-ray and sonic logs are displayed in these figures. Figure 9 outlines the lithologic composition of the 'so-called olistostromes' (hereinafter referred to as the chaotic unit) in the westernmost Guadalquivir Basin and its lateral thickness variations. Betica 14-1 well shows that the chaotic unit overlies undisturbed Miocene sediments, and mainly consists of undercompacted shales and clays with intervals (up to 160 m thick) of Triassic evaporites. Miocene to Pliocene sediments are recognized at the uppermost part of the section. The thickness of the chaotic unit decreases dramatically from 2600 m (Betica 14-1) to 100 m in the Casa Nieves-1 well located about 20 km to the northwest, and is not recognized further north at the Villamanrique-1 well. In all three wells, the lowermost sediments overlying the autochthonous basement correspond to the 'basal calcarenite' previously recognized by Perconig (1960–62). In the central part of the basin (Fig. 10), the Nueva Carteya-1 well, located on the External Betics, drilled more than 3000 m of Mesozoic carbonates overthrusting

Table 1. *Summary of stratigraphic models from Riaza & Martínez del Olmo (1996); Sierro* et al. *(1996) and Berástegui* et al. *(this volume)*

AGE		Riaza & Martínez del Olmo, 1996		Sierro et al., 1996	Berástegui et al., this volume
QUATERNARY 1.65		Odiel sequence	Post-Olisthostromic		
PLIOCENE 5.2		MARISMAS GROUP		Sequence E	
				Sequence D	
Messinian 6.3		ANDALUCIA GROUP		Sequence C	Sequence 6
Tortonian		BETICA GROUP	Sin-	Sequence B	Sequence 5
10.2				Sequence A	Sequence 4
Serravallian		ATLANTIDA GROUP	Pre-Olisthostromic		Sequence 3
					Sequence 2
15.2 Langhian					Sequence 1

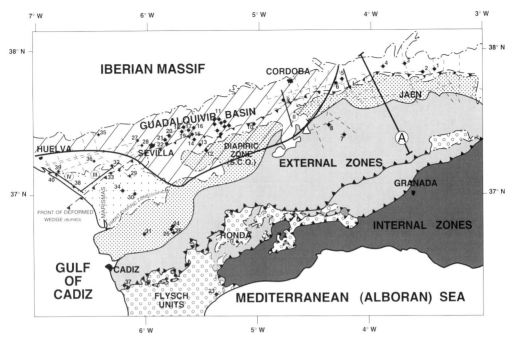

Fig. 8. Location map of subsurface data. Dashed lines indicate the location of seismic profiles used in this study. Solid lines indicate the location of interpreted seismic lines shown in this paper: I (Fig. 12); II (Fig. 16); III (Fig. 15); IV (Fig. 13). Segment 'A' indicate the location of Martos cross section (Fig. 17). Oil-wells: 1, Baeza-1; 2, Baeza-2; 3, Baeza-4 or Bailén; 4, Villanueva de la Reina or Baeza-3; 5, Rio Guadalquivir K-1; 6, Bujalance; 7, Rio Guadalquivir H-1; 8, Nueva Carteya-1; 9, Rio Guadalquivir N-1; 10, Ecija 1 and 2; 11, Córdoba A-1 to A-7, Córdoba B-1 and B-2, and Córdoba C-1; 12, Carmona 6; 13, Carmona-5; 14, Carmona-4; 15, Carmona-3; 16, Carmona-2; 17, Sevilla-3; 18, Carmona-1; 19, Sevilla-1; 20, Ciervo; 21, Sevilla-2; 22, Sevilla-4; 23, Cerro Gordo-3; 24, Bornos-3; 25, Bornos-1; 26, Angostura-Bornos; 27, Salteras-1; 28, Castilleja; 29, Isla Mayor, 30, Bética 14-1; 31, Bética 18-1; 32, Villamanrique; 33, CasaNieves; 34, Sapo-1; 35, Villalba del Alcor-1; 36, Almonte-1; 37, Chiclana; 38, Asperillo; 39, Huelva-1; 40, Moguer-1. Striped area indicates the pre-Mesozoic basement overlain directly by Neogene.

Miocene sediments. Towards the N and NW, at the Rio Guadalquivir N-1 and Rio Guadalquivir K-1 wells, respectively, the drilled sections correspond to autochthonous Miocene shales with poorly compacted sands.

The top of the pre-Neogene basement, as imaged from well data and seismic profiles, dips gently (2–4°) towards the SE beneath the basin, whereas below the thrust belt, where the Iberian crust is more heavily loaded, an increase in deep of as much as 10° is expected. Towards the southern margin of the basin, the seismic character of the top of the basement is lost below the thrust belt — this basin margin is not defined by basement structure, but by the thrust front. Figure 11 shows a smoothed contour map of depth to basement obtained from well data, seismic profiles and cross sections (Banks & Warburton 1991; Berástegui *et al.* this volume).

In most of the basin, the basement consists of Palaeozoic sediments in varying states of metamorphism and, below the central area of the basin, intrusive rocks occur also, similar to those cropping out in the Iberian Meseta. In the southwesternmost areas (Marismas), the Palaeozoic rocks are overlain by a partially eroded cover of Mesozoic strata, ranging in age from Triassic to upper Cretaceous, which have been drilled by the Moguer, Almonte, Asperillo, Casa Nieves and Isla Mayor wells (see Fig. 8 for location). In the easternmost area, a partially eroded Triassic cover was drilled in the Bailen-1 and Baeza-2 wells. Figure 8 also displays a map of the Neogene subcrop on which can be observed the present-day boundaries of the Mesozoic cover.

On seismic profiles the materials forming the pre-Neogene basement are involved in normal faulting (Figs 12 and 13). These faults, striking roughly NE–SW, die out in the lower part of basin fill, in the first sequence that shows thickening into the basin. Fault offsets are generally less than 100 ms (two-way travel time), and there

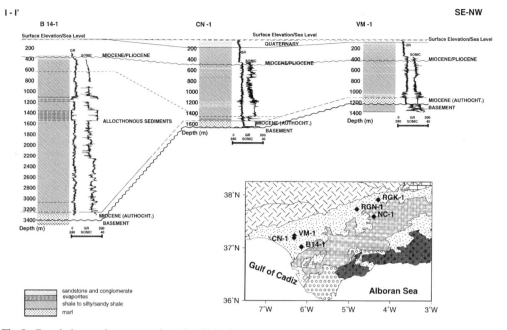

Fig. 9. Correlation section across selected wells in the westernmost Guadalquivir Basin. Allochthonous undercompacted sediments are recognized at the SE tip of the section, in the Betica 14-1 (B14-1) well, which progressively disappear towards the NW in the CasaNieves-1 (CN-1) and Villamanrique-1 (VM-1) wells. Main lithologies, gamma-ray log (GR) expressed in API units, and sonic log (μs/ft) are shown for each well.

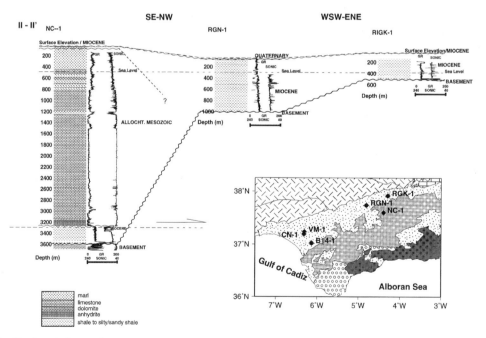

Fig. 10. Correlation section across selected wells from the Central Betic units that overthrust the autochthonous Guadalquivir Basin sediments at the Nueva Carteya-1 (NC-1) well towards the unperturbed basin interior at the RioGuadalquivir N-1 (RGN-1) and RioGuadalquivir K-1 (RGK-1) wells. Main lithologies, gamma-ray log (GR) expressed in API units, and sonic log (μs/ft) are shown for each well.

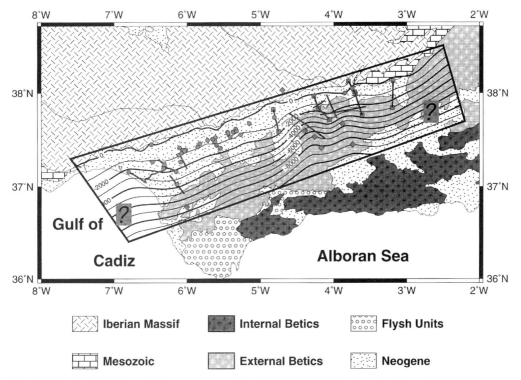

Fig. 11. Depth of the Palaeozoic basement map of the Guadalquivir basin referred to the sea level. Isolines every 1000 m. The data used come from oil wells (circles), seimic lines (squares), and cross-sections (diamonds) from Banks & Warburton (1991) and Berástegui et al. (this volume).

is a preponderance of faults downthrowing to the south. However, some large north-throwing faults are observed (Fig. 12), which have probably resulted from crustal arching when the tectonic load was emplaced during early Mid-Miocene (Langhian) times, thus the set leaves a longitudinal 'central high'. In a few places there is slight evidence of compressive reactivation of the faults (basin inversion), but this is not thought to be a major process in this basin.

The Neogene sedimentary infill is here arranged into six seismic-stratigraphic, depositional sequences bounded by unconformities and their correlative conformable boundaries, recording the time span from Late Langhian to Late Messinian. Because of the aforementioned dating uncertainties, we have tied the identified sequences to the standard Exxon cycle-chart (Haq et al. 1987) (Fig. 14) taking as a datum the well-known pre-Pliocene erosion. The nature of the boundaries, seismic facies, thicknesses, lithologies, sedimentary environments and ages attributed to each sequence are summarized in Table 2. Table 1 shows the correlation of the here-defined sequences to previous works. In contrast to Sierro et al. (1996), from the set of studied seismic profiles (see Fig.8) we cannot recognize any NE–SW progradational arrangement of the Late Langhian to Messinian set of sequences. Conversely, the Late Messinian to Plio-Quaternary deposits show a remarkable westward progradational pattern in the Marismas area (Fig. 15). The absence of the 'central high' in the westernmost part of the basin results in thickness variations of the sedimentary sequences from east to west.

The southern margin of the Guadalquivir Basin

As mentioned above, previous works agree that the most important infill supplied by the southern margin of the basin consisted of gigantic masses of chaotic materials that were emplaced by gravitational sliding.

Seismic profiles acquired recently show that the chaotic unit is formed by two kinds of seismic fabrics. The southern half is seismically noncoherent and corresponds to Triassic evaporites drilled in wells (e.g. Betica 18-1, Bornos-1, Bornos-3). The northern half consists of a set of

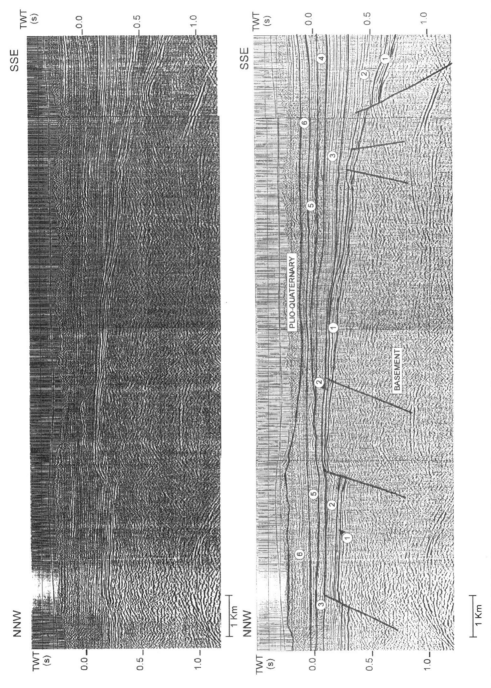

Fig. 12. Profile I (line 89-01). The interpreted version shows extensional structures involving the basement and Neogene sequences 1 and 2 (see text). Numbered sequences correspond to Tables 1 and 2.

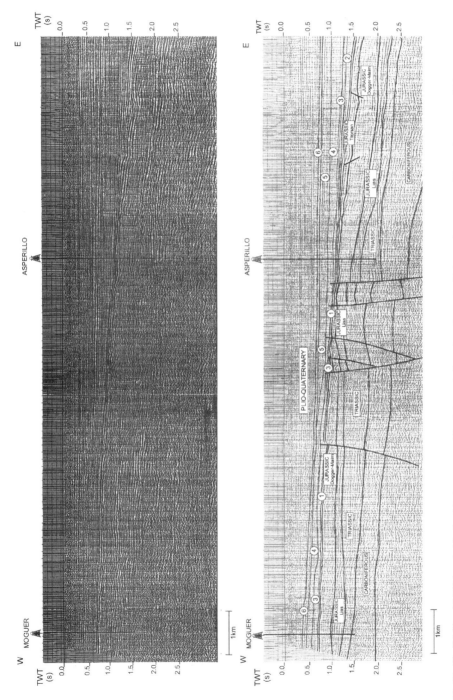

Fig. 13. Profile IV (line MA2) located in the Marismas area. The interpreted version shows the Palaeozoic structure, Mesozoic cover and Neogene sequences. Numbered sequences correspond to Tables 1 and 2.

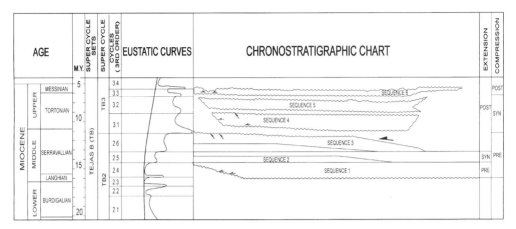

Fig. 14. Cycle chart showing the established sequences and its correlation to the standard cycle chart (Exxon). Sequences are also correlated to the basement extension and compression.

imbricate thrust slices involving Miocene sediments (e.g. Isla Mayor-1, Carmona-4, Ecija-1-2), displaying a wedge-shaped structure (Fig. 16). Moreover, it can be observed that the southernmost, more internal, seismically non-coherent part, is beneath the Mesozoic limestones forming the External Zones of the chain, and that its contact with the frontal Miocene imbricates can be interpreted as a more or less complex reverse fault, or as a diapiric contact. To summarize, from the integration of field, seismic and well data, we propose that the 'so-called olistostromes' are the result of squeezing of Keuper materials from below the External Zones (Fig. 16) that pushed the Miocene sediments ahead, thus forming the frontal deformed wedge. The processes involved include evaporite lateral diapirism, glacier-like flow and cap-rock formation. The chaotic character of this unit in outcrop corresponds to the above-mentioned cap-rock formation, together with small vertical evaporite developments that can involve the overlying Miocene sediments. This interpretation differs from those given by previous authors in that: (1) the 'so-called olistostromes' are not a single body but encompass an internal Triassic part and a frontal Miocene deformed wedge; (2) their emplacement does not correspond to gravitational sliding but to a lateral diapiric mechanism; (3) they do not form part of the basin sedimentary infill. A wider interpretation and discussion about the origin and

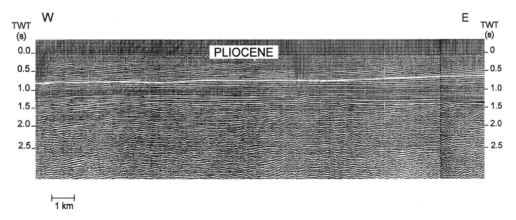

Fig. 15. Profile III (Line MA-4). Pliocene to Quaternary progradational stratal pattern towards the West (Marismas area).

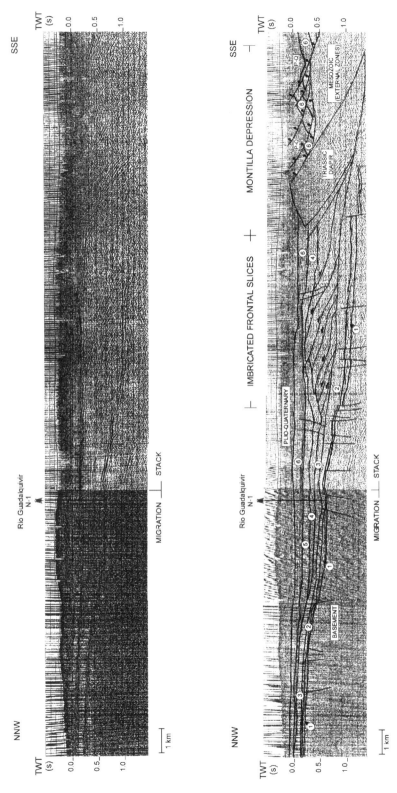

Fig. 16. Profile II (line S84-40 and RGK09110). The interpreted version shows the entire 'chaotic body', which from its contact with the External Betics includes extensional faulting in the Montilla depression, internal Triassic part of the chaotic body itself, external imbricates, frontal structure, and basin infill.

mechanisms of emplacement of this frontal unit is presented by Berástegui et al., (this volume).

The NW–SE cross-section displayed in Fig. 17 (see Fig 8 for location) summarizes the tectonostratigraphic zones of the Guadalquivir Basin and External Betics. Details on the basement dip, structure of the frontal imbricates and lateral diapir, as well as the arrangement of the External Betics are shown.

Discussion

The evolution of the Guadalquivir Basin is closely related to the formation of the Betic Chain and the Alboran Sea. This relation was already pointed out by van der Beek & Cloetingh (1992), who presented the only attempt to model the flexural response of the Guadalquivir Basin. These authors considered the present-day basement deflection under topographic loads associated with the External and Internal Betics and used gravity modelling to constrain the resulting crustal structure. Several important features were derived from this work: (a) very low elastic thickness values equivalent to 10 km in the western Betics and nearly zero in the eastern Betics; (b) the necessity to invoke subsurface loads; (c) the necessity to invoke an Oligocene–Early Miocene extensional event affecting the margin before the main phase of overthrusting; (d) a qualitative relationship between elastic thickness and rheology that was controlled by the thermal regime.

The role of the crustal structure and thermal regime on the rigidity of the continental lithosphere and therefore on its flexural response has been established by different authors (e.g. Beaumont 1979; Bodine et al. 1981; Burov & Diament 1992). The combination of high heat-flow and thick crust results in a low value of the total lithospheric strength. The heat-flow data (Fig. 5) show a thermal pattern characterized by a clear eastwards increase in heat flow (up to 125 mW m^{-2}) in the Alboran Sea (Polyak et al. 1996). The western part of the Guadalquivir Basin is also characterized by rather high heat-flow values (70–85 mW m^{-2} on average), whereas the available data in the Betics show strong perturbations due to groundwater circulation, that completely mask the deep thermal signature. Nevertheless, the regional thermal pattern together with the crustal thickness variations (Fig. 4) suggest that the present-day elastic thickness across the Guadalquivir Basin and the Betics should decrease towards the east as inferred by van der Beek & Cloetingh (1992).

An outstanding feature deduced from the interpretation presented is the non-existence of gigantic olistostromic masses in the southern border of the basin as has been claimed by previous authors. According to Berástegui et al. (this volume) these bodies are, in fact, the result of tectonically squeezed Triassic materials and Miocene deformed frontal wedges. This interpretation has strong implications for the palaeogeography. The presence of olistostromes in the southern margin of the basin has led authors to propose that the pre-Guadalquivir Basin corresponded to a foredeep in order to have the necessary palaeo-relief for their deposition. As a consequence, Early Miocene palaeogeographic schemes have been characterized by deep waters and a frontal trough (e.g. Sanz de Galdeano & Vera 1992). However, it is quite difficult to reconcile this environment with the subsidence of the basement and with the observed stratigraphy. The new interpretation of the southern margin of the Guadalquivir Basin does not require deep waters and is more tectonically consistent with the evolution of the orogenic loads, the geometry of the sedimentary basin-fill, and with the thermomechanical behaviour of the lithosphere.

According to van der Beek & Cloetingh (1992), the origin of subsurface loads acting in the Internal Betics is related to the sharpness of the crustal towards the Alboran Sea. This allowed them to account for the extra load, associated with the replacement of crustal material by denser mantle, making it compatible with observed Bouguer gravity anomaly. However, further gravity models, constrained by deep seismic data, show that the crustal thinning from 34–35 km to 20–22 km is produced over an area 15–30 km wide (Torné & Banda 1992; Torné et al. 1992), which is noticeably narrower than that proposed from flexural modelling. This steep thinning affects the northern margin of the Alboran Sea except its easternmost part as is evident from the Moho depth map (Fig. 4). Therefore, subsurface loads associated with the crust–mantle transition if any, should be of less significance.

The necessity of subsurface loads to fit the basement deflection together with the assumption of deep palaeo-bathymetries in the Betic front led van der Beek & Cloetingh (1992) to propose that the margin should have undergone a rifting event prior to the Betic overthrusting. From the arguments presented above, we think that the existence of such an Oligocene–Early Miocene rifting event is highly questionable, at least in the Central Betics.

To date, no attempts have been made to numerically model the progressive evolution of the Guadalquivir Basin. The geometry of a foreland basin as well as the architecture of its sedimentary fill records the main tectonic features related to thrust sheet displacements, crustal shortening and associated changes in the thermo-mechanical properties of the lithosphere. Most of the work carried out in this project has been directed toward establishing the main features that characterize the evolution of the Guadalquivir Basin which are: (a) initiation of subsidence; (b) time distribution of surface loads; and (c) lithospheric structure and subsurface loads.

Initiation of subsidence. A major problem in understanding the evolution of the Guadalquivir Basin is the determination of the time at which the first sediments were deposited. The results obtained from our seismic stratigraphic interpretations indicate that the oldest sediments overlaying the basement are Late Langhian although radiometric dating is necessary to confirm this age. This sequence extends to the south below, at least, the Miocene wedge and the lateral diapir (formerly called olistostromes). However, the presence of older sediments further to the south cannot be ruled out since compression initiated at Late Cretaceous thus producing a progressive northward migration of the foreland basin. Probably, the sediments filling this pre-Guadalquivir basin were incorporated to the hanging wall of the frontal thrusts. This picture is actually very similar to that imaged on seismic profiles at the most frontal part of the External Zones of the Betic Chain.

Time distribution of surface loads. The main surface loads acting in the Guadalquivir Basin correspond to the External and Internal Betics. The interpretation of the available data has permitted a better definition of the depth to basement towards the SSE (Figs 11 and 17). In these figures it can be seen that the Mesozoic and Palaeozoic basements reach up to 7 and 10 km depth respectively below the External Betics and dip below the Internal Betics. The sedimentary infill of the Guadalquivir Basin indicates that from Late Langhian to Messinian the acting loads were preferentially displaced in a NNW direction since no E–W onlaps and progradations are evident. The situation could have been radically different from Messinian to Recent as eastwards progradations are recognized in the western part of the basin together with a huge Pliocene erosion in its eastern part. These facts indicate that the Guadalquivir Basin probably underwent a differential uplift in its eastern border which could be related to the regional Pliocene uplift that affected most of the Mediterranean Spanish coast (Sanz de Galdeano & Vera 1992; Janssen *et al.* 1993).

Lithospheric structure and subsurface loads. Different studies have shown that the load produced by thrust stacking (surface load) is in some cases too large (Ganga Basin and Himalaya; Lyon-Caen & Molnar 1983, 1985) or too small (Pyrenees and Apennines; Brunet 1986, Royden 1988) to fit the observed basement deflection and gravity anomaly. As a consequence, it is necessary to invoke the existence of subsurface loads or bending moments. A plausible cause for this extra load could be the redistribution of mass at deep crustal and lithospheric levels during tectonic shortening (e.g. Brunet 1986; Royden 1988). Since the density contrast between the crust and the lithospheric mantle and between the mantle and the asthenosphere has an opposite sign, the resulting net load will depend on the lateral and vertical distribution of strain. Deep seismics, gravity and geological data indicate that the southern margin of the Iberian Peninsula underwent a large amount of tectonic deformation that affected different crustal levels. Furthermore, heat flow and geoid data show that deformation is not restricted to the crust but also affected the upper mantle. The mass distribution in the crust differs noticeably from that in the upper mantle as evidenced when comparing gravity and geoid data. It is therefore possible that subsurface loads resulting from mass redistribution over time may have played an important role in determining basement subsidence and the associated sedimentary pattern in the study area.

In summary, the compilation and re-interpretation of available data have permitted the improvement of the present-day knowledge of the Guadalquivir foreland basin and to change radically some of the previous interpretations. The results obtained from the presented work put important constraints on further numerical models of the basin. Clearly, there are still some aspects that remain unresolved, such as the quantification and identification of the acting loads and the resulting flexural rigidity; the role of the formation of the Alboran Sea and the associated unloading; the transition from the Atlantic to the Mediterranean and the continuation of the Guadalquivir Basin into the Gulf of Cadiz; and the nature of the vertical movements that the basin seems to have experienced from Pliocene to Recent. Ongoing studies dealing with some of these topics are being developed by the institutions involved in this project.

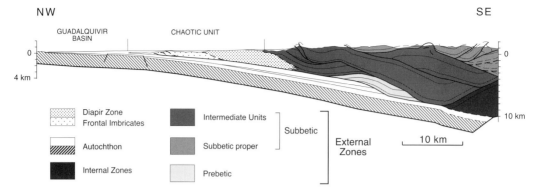

Fig. 17. 'Martos' cross-section showing the tectonostratigraphic units and the extrussion of the diapir zone along the base of the Intermediate Units.

This work has been financially supported by the European Union 'Integrated Basin Studies' project (JOU2-CT92-0110). We are indebted to M.C. Comas and C. Taberner for fruitful discussions during the preparation of this paper. The authors also wish to thank to M. Ford and F. Roure for their thorough and constructive reviews.

References

ALBERT-BELTRAN, J. F. 1979. El mapa español de flujos caloríficos. Intento de correlación entre anomalías geotérmicas y estructura cortical. *Boletin Geológico y Minero*, **90**, 36–48.

BAKKER, H. E., DE JONG, K., HELMERS, H. & BIERMANN, C. 1989. The geodynamic evolution of the Internal Zone of the Betic Cordilleras (SE Spain): a model based on structural analysis and geothermobarometry. *Journal of Metamorphic Geology*, **3**, 359–381.

BALANYÁ, J. C. & GARCÍA-DUEÑAS, V. 1987. Les directions structurales dans le Domaine d'Alboran de part et d'outre du Detroit de Gibraltar. *Comptes rendus de l'Acadèmie des Sciences Paris*, **304**, 929–933.

—— & —— 1988. El cabalgamiento cortical de Gibraltar y la tectónica de Béticas y Rif. *In: Proceedings II Congreso Geológico de España (Simposios)*, 35–44.

BANDA, E. 1988. Crustal parameters in the Iberian Peninsula. *Physics of the Earth and Planetary Interiors*, **51**, 222–225.

—— & ANSORGE, J. 1980. Crustal structure under the central and eastern part of the Betic Cordillera. *Geophysical Journal of the Royal Astronomical Society*, **63**, 515–532.

——, SURIÑACH, E., APARICIO, A., SIERRA, J. & RUIZ DE LA PARTE, E. 1981. Crust and upper mantle structure of the central Iberian Meseta (Spain). *Geophysical Journal of the Royal Astronomical Society*, **67**, 779–789.

——, ALBERT-BELTRAN, J., TORNÉ, M. & FERNÀNDEZ, M. 1991. Regional geothermal gradients and lithospheric structure in Spain. *In*: CERMAK, V. & RYBACH, L. (eds) *Exploration of the Deep Continental Crust. Terrestrial Heat Flow and the Lithosphere Structure*. Springer-Verlag, Berlin, 176–186.

——, GALLART, J., GARCÍA-DUEÑAS, V., DAÑOBEITIA, J. J. & MAKRIS, J. 1993. Lateral variation of the crust in the Iberian peninsula: new evidence from the Betic Cordillera. *Tectonophysics*, **221**, 53–66.

BANKS, C. J. & WARBURTON, J. 1991. Mid-crustal detachment in the Betic system of southeast Spain. *Tectonophysics*, **191**, 275–289.

BARRANCO, L. M., ANSORGE, J. & BANDA, E. 1990. Seismic refraction constraints on the geometry of the Ronda peridotitic massif (Betic Cordillera, Spain). *Tectonophysics*, **184**, 379–392.

BEAUMONT, C. 1979. On rheological zonation of the lithosphere during flexure. *Tectonophysics*, **59**, 347–365.

BERÁSTEGUI, X., BANKS, C. J., PUIG, C., TABERNER, C., WALTHAM, D. & FERNÀNDEZ, M. 1998. Lateral diapiric emplacement of Triassic evaporites at the southern margin of the Guadalquivir Basin, Spain. *This volume*.

BODINE, J. H., STECKLER, M. S. & WATTS, A. B. 1981. Observations of flexure and the rheology of the oceanic lithosphere. *Journal of Geophysical Research*, **86**, 3695–3707.

BOLOIX, M. & HATZFELD, D. 1977. Preliminary results of measurements along seismic profiles in the Albora Sea. *Publications of the Institute of Geophysics, Polish Academy of Sciences*, Warsaw, **A-4 (115)**, 365-368.

BRENNECKE, J., LELGEMANN, D., REINHART, E., TORGE, W., WEBER, G. & WENZEL H. G. 1983. *A European astro-gravimetric geoid*. Deutsche Geodätische Kommission. Verlag des Instituts fur Angewandte Geodasie Frankfurt am Main. Reihe B, **269**, Nr. 169.

BRUNET, M. F. 1986. The influence of the evolution of the Pyrenees on adjavent basins. *Tectonophysics*, **129**, 343–354.

BUROV, E. B. & DIAMENT, M. 1992. Flexure of the continental lithosphere with multilayered rheology. *Geophysical Journal International*, **109**, 449–468.

CHANNEL, J. E. T. & MARESCHAL, J. C. 1989. Delamination and asymmetric lithospheric thickening in the development of the Tyrrhenian Rift. *In*: COWARD, M. P., DIETRICH, D. & PARK, R. G. (eds) *Alpine Tectonics*. Geological Society, London, Special Publications, **45**, 285–302.

CLOETINGH, S., VAN DER BEEK, P. A., VAN REES, D., ROEP, T. B., BIERMANN, C. & STEPHENSON, R. A. 1992. Flexural interaction and the dynamics of Neogene extensional basin formation in the Alboran-Betic region. *Geo-Marine Letters*, **12**, 66–75.

COMAS, M. C, GARCÍA-DUEÑAS, V. & JURADO, M. J. 1992. Neogene tectonic evolution of the Alboran Basin from MCS data. *Geo-Marine Letters*, **12**, 144–149.

DEWEY, J. F. 1988. Extensional collapse of orogens. *Tectonics*, **7**, 1123-1139.

—, HELMAN, M. L., TURCO, E., HUTTON, D. H. W. & KNOTT, S. D. 1989. Kinematics of the western Mediterranean. *In*: COWARD, M. P., DIETRICH, D. & PARK, D. G. (eds.) *Alpine Tectonics*. Geological Society, London, Special Publications, **45**, 265–283.

DOCHERTY, C. & BANDA, E. 1995. Evidence for the eastward migration of the Alboran Sea based on regional subsidence analysis: A case for basin formation by delamination of the subcrustal lithosphere? *Tectonics*, **14**, 804–818.

FLORES, J. A. & SIERRO, F. J. 1989. Calcareous nannoflora and planktonic foraminifera in the Tortonian–Messinian boundary interval of East Atlantic DSDP Sites and their relation to Spanish and Moroccan. *In*: VAN HECK, S. E. & CRUX, J. (eds) *Nannofossils and Their Applications*. British Micropalaeontological Society Series. Ellis Horwood, 249–266.

GALINDO ZALDIVAR, J., GONZÁLEZ LODEIRO, F. & JABALOY, A. 1989. Progressive extensional shear structures in a detachment contact in the western Sierra Nevada (Betic Cordilleras, Spain). *Geodynamica Acta*, **3**, 73–85.

GARCÍA-DUEÑAS, V. & BALANYÁ, J. C. 1991. Fallas normales de bajo ángulo a gran escala en las Béticas Occidentales. *Geogaceta*, **9**, 33–37.

—, — & MARTÍNEZ MARTÍNEZ, J. M. 1992. Miocene extensional detachments in the outcropping basement of the northern Alboran basin (Betics) and their tectonic implications. *Geo-Marine Letters*, **12**, 88–95.

—, BANDA, E., TORNÉ, M., CÓRDOBA, D. & ESCI-BÉTICAS WORKING GROUP 1994. A deep seismic reflection survey across the Betic Chain (southern Spain): first results. *Tectonophysics*, **232**, 77–89.

GONZÁLEZ, A., CÓRDOBA, D., MATÍAS, L. M., VEGAS, R. & TÉLLEZ, J. 1993. A reanalysis of P-wave velocity models in the southwestern Iberian peninsula-Gulf of Cadiz motivated by the ILIHA-DSS experiments. *In*: MEZCUA, J. & CARREÑO, E. (eds) *Iberian Lithosphere Heterogeneity and Anisotropy*. Instituto Geográfico Nacional, Madrid, Monografias, **10**, 215–227.

HAQ, B. U., HARDENBOL, J., VAIL, R. R. & 10 OTHERS 1987. Mesozoic–Cenozoic Cycle Chart. *In*: BALLY, A. W. (ed.) *Atlas of Seismic Stratigraphy*. American Association of Petroleum Geologists, Studies in Geology, **27**.

HATZFELD, D. 1976. Etude sismologique et gravimetrique de la structure profonde de la mer d'Alboran: Mise en évidence d'un manteau anormal. *Compte rendu de l'Acadèmie des Sciences Paris*, **283**, 1021–1024.

HAXBY, W. F. & TURCOTTE, D. L. 1978. On isostatic geoid anomalies. *Journal of Geophysical Research*, **83**, 5473–5478.

IGME 1987. *Contribución de la exploración petrolífera al conocimiento de la geología de España*. Instituto Geológico y Minero de España, Madrid.

ILIHA DSS GROUP 1993. A deep seismic sounding investigation of lithospheric heterogeneity and anisotropy beneath the Iberian Peninsula. *Tectonophysics*, **221**, 35–51.

ITGE 1993. *Trabajos de medición e inventario de datos del flujo de calor en España: Cordilleras Béticas y Suroeste peninsular*. Instituto Tecnológico y Geominero de España, Madrid.

JANSSEN, M. E., TORNÉ, M., CLOETINGH, S. & BANDA, E. 1993. Pliocene uplift of the eastern Iberian margin: Inferences from quantitative modelling of the Valencia trough. *Earth and Planetary Science Letters*, **119**, 585–597.

LYON-CAEN, H. & MOLNAR, P. 1983. Constraints on the structure of the Himalaya from analysis of gravity anomalies and a flexural model of the lithosphere. *Journal of Geophysical Research*, **88**, 8171–8191.

— & — 1985. Gravity anomalies, flexure of the Indian plate, and the structure, support and evolution of the Himalaya and Ganga basin. *Tectonics*, **4**, 513–538.

MARTÍNEZ DEL OLMO, W., GARCÍA-MALLO, J., LERET-VERDÚ, G., SERRANO-OÑATE, A. & SUÁREZ-ALBA, J. 1984. Modelo tectonosedimentario del Bajo Guadalquivir. *In: Proceedings I Congreso Español de Geología*, **I**, 199–213.

MEDIALDEA, T., SURIÑACH, E., VEGAS, R., BANDA, E. & ANSORGE, J. 1986. Crustal structure under the western end of the Betic cordillera (Spain). *Annales Geophysicae*, **4(B4)**, 457–464.

MITCHUM, R. M., VAIL, P. R. & THOMPSON, S. 1977. The depositional sequence as a basic unit for stratigraphic analysis. *In*: PAYTON, C. E. (ed.) *Seismic Stratigraphy. Aplications to Hydrocarbon Exploration*. American Association of Petroleum Geologists Memoirs, **26**, 53–62.

MONIÉ, P., GALINDO-ZALDÍVAR, J., GONZÁLEZ LODEIRO, F., GOFFÉ, B. & JABALOY, A. 1991. $^{40}Ar/^{39}Ar$ geochronology of Alpine tectonism in the Betic Cordilleras (southern Spain). *Journal of the Geological Society, London*, **148**, 289–297.

MORELLI, C., PISANI, M. & CANTAR, C. 1975. Geophysical anomalies and tectonics in the western Mediterranean. *Bolletino di Geofisica*, **67**, 211–449.

PERCONIG, E. 1960-1962. Sur la constitution géologique de l'Andalousie Occidentale en particulier du Bassin du Guadalquivir (Espagne

méridionale). *In: Livre Mémoir du Professor Paul Fallot*. Mémoires hors-Série de la Société géologique de France. 229–256.
—— 1971. Sobre la edad de la transgresión del Terciario Marino en el borde meridional de la Meseta. *In: Proceedings I Congreso Hispano-Luso-Americano de Geología Económica*, **I**, 309–319.
—— & GRANADOS, L. 1973. Límite Mioceno-Plioceno. Corte de la Autopista Km 17. El estratotipo del Andaluciense. La 'caliza Tosca' de Arcos de la Frontera. *In: Proceedings XIII Coloquio Europeo de Micropaleontología*, Madrid.
PLATT, J. P. & VISSERS, R. L. 1989. Extensional collapse of thickened continental lithosphere: a working hypothesis of the Alboran Sea and the Gibraltar Arc. *Geology*, **17**, 540–543.
POLYAK, B. G., FERNÀNDEZ, M. & FLUCALB GROUP 1996. Heat flow in the Alboran Sea (the western Mediterranean). *Tectonophysics*, **263**, 191–218.
RANALLI, G. 1994. Nonlinear flexure and equivalent mechanical thickness of the lithosphere. *Tectonophysics*, **240**, 107–114.
RIAZA, C. & MARTÍNEZ DEL OLMO, W. 1996. Depositional model of the Guadalquivir- Gulf of Cádiz Tertiary basin. *In: FRIEND, P. & DABRIO, C. J. (eds) Tertiary Basins of Spain: The stratigraphic record of crustal kinematics*. Cambridge University Press, 330–338.
ROLDÁN-GARCÍA, F. J. & GARCÍA-CORTÉS, A. 1988. Implicaciones de los materiales triásicos en la Depresión del Guadalquivir (províncias de Córdoba y Jaén). *In: Proceedings II Congreso Geológico de España*, **I**, 189–192.
ROYDEN, L. H. 1988. Flexural behaviour of the continental lithosphere in Italy: constraints imposed by gravity and deflection data. *Journal of Geophysical Research*, **93**, 7747–7766.
—— 1993. Evolution of retreating subduction boundaries formed during continental collision. *Tectonics*, **12**, 629–638.
SAAVEDRA, J. L. 1964. Datos para la interpretación de la estratigrafía del Terciario y Secundario de Andalucía. *Notas y Comunicaciones del Instituto Geológico y Minero de España*, **73**, 5–50.
SANZ DE GALDEANO, C. & VERA, J. A. 1992. Stratigraphic record and paleogeographical context of the Neogene basins in the Betic Cordillera, Spain. *Basin Research*, **4**, 21–36.
SIERRO, F. J., FLORES, J. A., CIVIS, J., GONZÁLEZ-DELGADO, J. A. & FRANCES, G. 1993. Late Miocene globorotaliid event-stratigraphy and biogeography in the NE Atlantic and Mediterranean. *Marine Micropaleontology*, **21**, 143–168.
——, GONZÁLEZ-DELGADO, J. A., DABRIO, C. J.,
FLORES, J. A. & CIVIS, J. 1996. Late Neogene depositional sequences in the foreland basin of Guadalquivir (SW Spain). *In: FRIEND, P. & DABRIO, C. J. (eds) Tertiary Basins of Spain: The stratigraphic record of crustal kinematics*. Cambridge University Press, 339–345.
SUÁREZ-ALBA, J., MARTÍNEZ DEL OLMO, W., SERRANO-OÑATE, A. & LERET-VERDÚ, G. 1989. Estructura del sistema turbidítico de la Formación Arenas del Guadalquivir. Neógeno del valle del Guadalquivir. *In: Libro Homenaje R. Soler*. Asociación de Geólogos y Geofísicos Españoles del Petróleo, Madrid, 123–132.
SURIÑACH, E. & VEGAS, R. 1988. Lateral inhomogeneitiesof the Hercynian crust in central Spain. *Physics of the Earth Planetary Interiors*, **51**, 226–234.
—— & —— 1993. Estructura general de la corteza en una transversal del Mar de Alborán a partir de datos de sísmica de refracción-reflexión de gran ángulo. Interpretación geodinámica. *Geogaceta*, **14**, 126–128.
TORNÉ, M. & BANDA, E. 1992. Crustal thinning from the Betic Cordillera to the Alboran Sea. *Geo-Marine Letters*, **12**, 76–81.
——, ——, GARCÍA-DUEÑAS, V. & BALANYÁ, J. C. 1992. Mantle-lithosphere bodies in the Alboran crustal domain (Ronda peridotites, Betic-Rif orogenic belt). *Earth and Planetary Science Letters*, **110**, 163–171.
TUBIA, J. M. & GIL-IBARGUCHI, J. L. 1991. Eclogites of the Ojen nappe: a record of subduction in the Alpujarride complex (Betic Cordilleras, southern Spain). *Journal of the Geological Society, London*, **148**, 801–804.
VAIL, P. R. 1987. Seismic stratigraphy interpretation procedure. *In: BALLY A. W. (ed.) Atlas of Seismic Stratigraphy*. American Association of Petroleum Geologists, Studies in Geology, **27**, 1–10.
VAN DER BEEK, P. A. & CLOETINGH, S. 1992. Lithospheric flexure and the tectonic evolution of the Betic Cordilleras (SE Spain). *Tectonophysics*, **203**, 325–344.
VISSERS, R. L., PLATT, J. P. & VAN DER WAL, D. 1995. Late orogenic extnesion of the Betic Cordillera and the Alboran Domain: A lithospheric view. *Tectonics*, **14**, 786–803.
WATTS, A. B., PLATT, J. P. & BUHL, P. 1993. Tectonic evolution of the Alboran Sea basin. *Basin Research*, **5**, 153–177.
ZECK, H. P., MONIER, P., VILLA, I. M. & HANSEN, B. T. 1992. Very high rates of cooling and uplift in the Alpine belt of the Betic Cordilleras, southern Spain. *Geology*, **20**, 79–82.

Lateral diapiric emplacement of Triassic evaporites at the southern margin of the Guadalquivir Basin, Spain

X. BERÁSTEGUI[1], C.J. BANKS[2], C. PUIG[1], C. TABERNER[3], D. WALTHAM[2] & M. FERNÀNDEZ[3]

[1]*Servei Geològic de Catalunya, ICC, Parc Montjuic, 08038-Barcelona, Spain*
[2]*Department of Geology, Royal Holloway University of London, Egham, Surrey, TW20 0EX, UK*
[3]*Institute of Earth Sciences (J. Almera), CSIC, Lluís Solé Sabarís s/n, 08028-Barcelona, Spain*

Abstract: The Guadalquivir Basin is the Neogene foreland basin of the central and western Betic thrust belt in southern Spain. At the boundary between the basin and the outcrops of thrust nappes of Mesozoic limestones of the Prebetic and Subbetic is a broad belt of outcrops of Triassic evaporitic sediments with scattered younger rocks: the so-called 'Olistostrome' unit. This is highly deformed, in places chaotic, and its mode of emplacement has been attributed by various authors to olistostromal debris flow, diapirism, or tectonic melange. Studies of outcrop data in conjunction with seismic and well data, integrated using restorable cross-sections lead us to propose the following sequence of emplacement mechanisms. (a) Loading above a Triassic evaporite formation, probably in the Intermediate Units depositional zone, by north vergent thrusting of thick nappes of Mesozoic sediments, causes northward expulsion of evaporitic sediments between a basal thrust and the base of the limestones. (b) Continued thrust loading drives the diapiric body forwards ahead of the thrust belt, into the floor of the deepening Miocene foreland basin. The body includes blocks of Triassic rocks in normal stratigraphic sequence, as well as blocks of younger rocks broken off the leading hanging-wall cutoffs of the nappes. (c) When the diapiric body reaches the sea-floor of the basin, its top becomes subject to modification by sedimentary processes such as dissolution of evaporites leaving a cap rock and debris flow, both submarine and subaerial but rarely, if ever, forming true olistostromes. (d) At the leading edge of the diapir, northward compression of Miocene basin sediments results in thin-skinned thrusting within these sediments, and formation of duplex structures with a north-dipping monoclinal deformation front. Results from analogue and numerical modelling match the main geological features observed in the study area, thus supporting the plausibilty of the proposed lateral diapiric emplacement of the chaotic unit.

Between the frontal thrust zone of the central and western Betics and the Guadalquivir foreland basin (Fig. 1) there is a unit characterized by outcrops of chaotic Triassic gypsum and folded red-bed clastics, with some scattered younger blocks, mostly of Upper Cretaceous limestones. These outcrops have been referred to as 'Olistostromic Zone' by Perconig (1960–62), and many authors have followed his nomenclature and interpreted this unit as a sedimentary melange or olistostrome (Pérez-López & Sanz de Galdeano 1994). Alternative names used are 'Guadalquivir Allochthonous Unit' (Blankenship 1990, 1992), and by our group informally 'SCO' standing for 'So-Called Olistostrome' (chaotic unit). In fact, the most recently published geological maps are based on the olistostrome model, where Tertiary formations have been interpreted as the matrix supporting blocks of gypsum and other rocks.

The unit made up mainly of chaotic Triassic sediments at the front of the thrust belt is a very unusual aspect when compared with 'classical' thrust fronts. It is generally agreed that the unit is derived from the thrust belt, but its composition, geometry, exact origin and mode of emplacement are all contentious.

The aim of this paper is to interpret the mechanism of emplacement of the chaotic allochtonous unit and its geodynamic relationships with both the thrust belt and the foreland basin. Seismic reflection, oil wells and field geological data including those from recent (1980s) oil and gas exploration surveys and exposures from road building have been used in refining the stratigraphical and structural interpretation of the study area. The application of sequence-stratigraphic methodology to the basin-fill has permitted us to time the emplacement of the chaotic unit. The new information presented

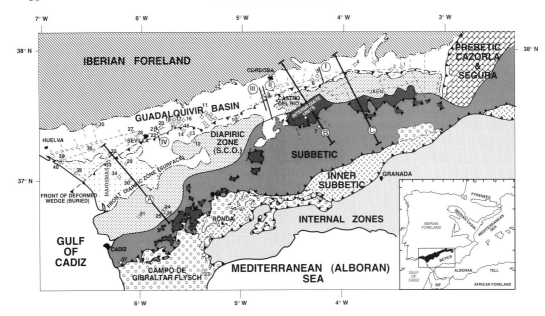

Fig. 1. Structural map of the central and western Betics and Guadalquivir Basin and location map of oil-wells, seismic profiles and cross-sections. Cross-sections: A, Marismas section (Fig. 4); B, Baena section (Fig. 3); C, Martos section (Figs 2 and 5). Seismic profiles: I (Fig. 9); II (Fig. 6); III (Fig. 7); IV (Fig. 10); broken lines correspond to different profiles used in the interpretation. Well number: 1, Baeza-1; 2, Baeza-2; 3, Baeza-4 or Bailén; 4, Villanueva de la Reina or Baeza-3; 5, Rio Guadalquivir K-1; 6, Bujalance; 7, Rio Guadalquivir H-1; 8, Nueva Carteya-1; 9, Rio Guadalquivir N-1; 10, Ecija 1 and 2; 11, Córdoba A-1 to A-7, Córdoba B-1 and B-2, and Córdoba C-1; 12, Carmona 6; 13, Carmona-5; 14, Carmona-4; 15, Carmona-2; 16, Carmona-2; 17, Sevilla-3; 18, Carmona-1; 19, Sevilla-1; 20, Ciervo; 21, Sevilla-2; 22, Sevilla-4; 23, Cerro Gordo-3; 24, Bornos-3; 25, Bornos-1; 26, Angostura-Bornos; 27, Salteras-1; 28, Castilleja; 29, Isla Mayor, 30, Bética 14-1; 31, Bética 18-1; 32, Villamanrique; 33, Casanieves; 34, Sapo-1; 35, Villalba del Alcor-1; 36, Almonte-1; 37, Chiclana; 38, Asperillo; 39, Huelva-1; 40, Moguer-1. Guadalquivir Basin: in white, Miocene to Quaternary; in dots, marine Pliocene to Quaternary (Marismas area).

here on the structure of the External Betics, and the Guadalquivir Basin enables us to propose a diapiric hypothesis for the emplacement of the chaotic unit. Restorable geological cross sections and numerical and physical (sand-box) modelling have been applied to test the reliability of the proposed geodynamic hypothesis.

Geological setting

The Guadalquivir Basin is the Neogene foreland basin of the central and western half of the Betic Cordillera (Fig. 1). At its eastern termination it is bounded by the NNE-striking thrust imbricates of the Cazorla Zone and the big thrust anticlines of the Sierra de Segura. The eastern half of the Cordillera has no foreland basin because deformation of the thick sedimentary cover extends far beyond the limits of load-induced flexural subsidence of the basement. East of Córdoba the Guadalquivir Basin is very narrow and shallow, but it widens and deepens to the west. The basin floor dips fairly uniformly at a shallow angle to the SSE, with basement (Hercynian Iberian Massif) emerging at a remarkably straight but unfaulted basin boundary to the NNW.

The Betic Cordillera comprises the Internal and External Zones. The Internal Zones consist of metamorphic basement and Palaeozoic–Triassic sediments of varying metamorphic grade in big domal culminations elongated more-or-less E–W, including some outcrops of peridotites in the Ronda area. The emplacement of the dense rocks of the Internal Zones onto the former passive margin of southern Iberia was responsible for both the load-induced subsidence of the margin and the compressive deformation of the Betics (van der Beek & Cloetingh 1992; Banks & Warburton 1991). Any obliquity of transport direction (Leblanc & Olivier 1984; Frizon de Lamotte et al. 1991), and any non-compressive deformation, were resolved during deformation of the External Zones (for example by vertical pole rotation of individual thrust slices; Allerton et al. 1993) so that the only process affecting the thrust front was NNW-directed compression. The first compressive phase that affected the southern margin of the External Zones took place in the Eocene. The main phase of compression was in the Mid-Miocene (Late Langhian–Serravallian), but compression did not finally die out until the Tortonian.

The External Zones

Stratigraphy

The stratigraphy and palaeogeography of the External Zones of the Betic Cordillera have been the subject of many works since the 1920s; key papers are Fontboté (1965); García Hernández et al. (1980); Baena & Jerez (1982); Vera et al. (1982) and Fontboté & Vera (1984). In summary, the External Zones consist of a tectonically detached cover of Mesozoic sediments ranging from Middle Triassic to Upper Cretaceous–Palaeocene pre-tectonic units, and Eocene to Lower Miocene syn-tectonic deposits. From palaeogeographic and tectonic considerations, the External Zones are divided into Prebetic and Subbetic (Fig. 1). The Prebetic is made up of shallow-marine and continental Mesozoic carbonates and capped by continental Palaeogene sediments. It does not crop out in the central Betics. The Subbetic is characterized by Mesozoic dominance of pelagic facies in the strata of Upper Liassic to Upper Cretaceous–Palaeocene age. From thickness variations and tectonic position, Subbetic is usually divided into Intermediate Units and Subbetic proper. Among the significant characteristics of the Intermediate Units are the thickness of the Mesozoic strata (more than 2000 m of Jurassic and Lower Cretaceous, mostly slope to basinal sediments). Well data show that the Triassic in the Intermediate Units comprises only a thin section of 'Supra-Keuper' dolomites and anhydrites of uncertain Jurassic to Triassic age lying immediately above a regional detachment.

The Mesozoic series in the Subbetic proper are much thinner than in the Intermediate Units, and include radiolarites and ammonitico rosso facies as well as carbonate turbidites at different levels in the Jurassic succession. In places, there are also pillow lavas defining a 'Median Subbetic' unit. The Triassic in the External Subbetic crops out in very extensive areas and has been the subject of many studies, from Blumenthal (1927) to Pérez López (1991). The latter author defined, from bottom to top, the Majanillos Formation (Muschelkalk), Ladinian in age, which consists of dolomites and limestones; the Jaén Group (Keuper), Karnian in age, which consists of multicoloured mudstones including evaporites, carbonates and clastics (K1), red sandstones and mudstones (K2), red mudstones and dolomites ('carniolas', K3) and evaporites (K4–5); and the Ocres Rojos or Zamoranos Formation, Norian in age, consisting of dolomites and limestones.

Structure

Between Córdoba and Granada, structures are well exposed, and seismic and well data are

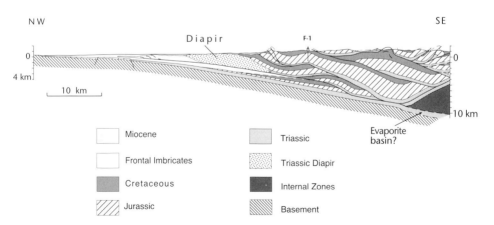

Fig. 2. 'Martos' geological cross-section based on outcrop data, seismic lines RGKO-89-01, 82-31 & 32 and Fuensanta-1 well, and speculative extrapolation.

BAENA CROSS SECTION

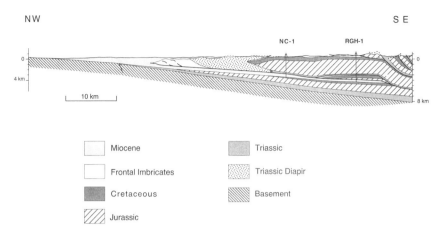

Fig. 3. 'Baena' geological cross-section, based on outcrop data, seismic lines BT-8A, S-83-34, S-84-40, S-83-44 and AES lines, and Nueva Carteya-1 and Rio Guadalquivir H-1 wells. It shows variation of structure about 40 km down strike.

available. The data have been compiled and interpreted using conventional thrust-belt techniques. The three structural cross-sections, 'Martos', 'Baena' and 'Marismas', (see Fig. 1 for location and Figs 2, 3 and 4) are drawn at true scale with the available data accurately placed, and although they are in places unconstrained, and not unique solutions, they are restorable. The only previous attempt to interpret the region using this methodology was by Blankenship (1992). The interpretation proposed in this paper is based in part on the interpretation of the eastern Betics of Banks & Warburton (1991), and involves a relatively large amount of shortening: at least 80 km shortening of the Jurassic on the part of the Betics shown in Fig. 2.

Figure 5 shows an interpretation of the Martos section (see Fig. 1 for location) in terms of the tectonostratigraphic zones discussed above and demonstrates the volumetric importance of the Intermediate Units (proven by the two wells on the Baena section, Fig. 3) and the insignificance of the Prebetic in this area (in this paper we consider the 'Prebetic of Jaén' and similar units as a part of the Intermediate Units). The 'frontal thrust' of the Intermediate Units is in fact a subtractive contact with the chaotic unit. The Subbetic can be deduced to be a duplex, with basal detachment consistently within the Middle Triassic. The frontal Subbetic thrust carries it over the Intermediate Units duplex, in which the basal detachment is in Supra-Keuper and the roof detachment is in the Lower Cretaceous.

The proposed general relationship between the Intermediate Units, the Subbetic proper, and the Internal Zones is best seen on the Martos structural section (Fig. 2). A wedge of Internal Zones basement is driven in at the basal (Triassic) detachment, so that the Subbetic proper is uplifted in a synform. The Subbetic leading edge is thrust over the Intermediate Units in a shallow flat-lying nappe, with windows and klippen, but otherwise the Subbetic proper is mainly internally deformed into a backthrust duplex thrusting top-to-south back onto the Internal Zones (all of the actual displacements, of course, being northwards the Iberian Massif).

The Triassic at the base of the frontal nappe of the Subbetic proper forms extensive areas of outcrop and has obviously moved in part diapirically (Pérez-López & Sanz de Galdeano 1994). For example, in the low area between the Martos and Baena cross-sections, the Triassic appears to overlie completely the Intermediate Units so as to merge with the Triassic of the chaotic unit (whereas the Jurassic limestones, that are carried by it are overturned, suggesting a possible 'tank-track' mode of emplacement). However, the bedded Subbetic Triassic can be distinguished in the field from the Triassic forming the chaotic unit, in that the former contains well-bedded sequences of mudstones to sandstones, with relatively little gypsum. These

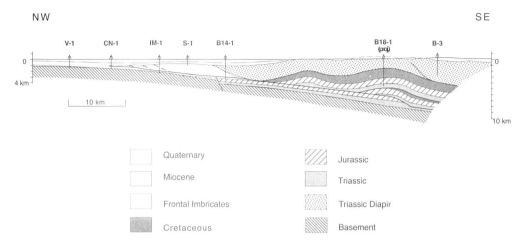

Fig. 4. 'Marismas' geological cross-section, based on a few outcrop data, seismic line MA-3 and others, and 7 wells (V, Villamanrique-1; CN, Casa Nieves-1; IM, Isla Mayor-1; S, Sapo-1; B14-1 and B18-1 are Esso Betica wells; B-3, Bornos-3). In Betica-18-1 the diapir is mostly halite.

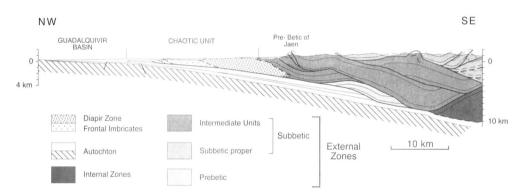

Fig. 5. 'Martos' cross-section displaying the tectonostratigraphic zones. It shows how diapir zone extruded along the base of Intermediate Units. Prebetic of Jaen is represented structurally as a part of Intermediate Units.

bedded outcrops are mainly bright red, as opposed to the very mixed purple-red, green and other colours of the chaotic unit.

In the central Betics, there is some evidence indicated on both cross-sections for deformation below the base of the Intermediate Units thrust sheet, arching it upwards. This must be the youngest thrust deformation, and we consider it to be parautochthonous Prebetic (Fig. 4). In fact, seismic data in the vicinity of well Bética 18-1 show a broad antiformal structure that may indicate a Prebetic structure below the 'chaotic unit'.

Composition and structure of the chaotic unit

The aforementioned 'Martos', 'Baena' and 'Marismas' structural sections (Figs 2, 3 and 4) show how the deeper SE half (or more) of the Guadalquivir foreland basin is occupied by the allochthonous chaotic unit that is the subject of this paper. From seismic and well data (see below), the chaotic unit has two parts. The most external part, which is not exposed, comprises imbricated Miocene sediments of the basin fill. It has a basal thrust in the early Mid-Miocene, leading to a blind deformation front. Seismic stratigraphy (see also Fernández et al. this volume) shows that only the Serravallian and Tortonian are deformed, with the youngest Tortonian, Messinian and Pliocene onlapping the top surface of the deformed Miocene strata, which is an unconformity. In some places there is a possible passive backthrust in the upper part of the deformed unit. The inner part of the chaotic unit appears as a chaotic mass on seismic profiles. Outcrops indicates that it consists of Triassic gypsum and clays with the characteristic colours including purple, red, green and grey, with no coherent bedding or structure at regional scale. This mixture of lithologies encloses blocks, tens to hundreds of metres in size, of various formations including Upper Cretaceous to Palaeocene carbonates (Capas Rojas Formation), Triassic carbonates (Zamoranos and Majanillos Formations) and Triassic basic volcanic rocks (ophites); Jurassic carbonates apparently do not occur as blocks. The contacts, where seen, are typically tectonic or diapiric, with brecciation of the Keuper and the block material extending over a few decimetres. The chaotic unit is typically 2–3 km thick, and 10 km wide in the central Betics. However, further west (see 'Marismas' section, Fig. 4) the width increases to at least 50 km, and the wells show that it includes a significant proportion of halite. In the Montilla area, the top of the chaotic unit is depressed, and seismic shows that this is due to an array of shallow listric normal faults of probably Pliocene age (Fig. 6).

In most of the outcrop area the chaotic unit lies directly below Quaternary soils and fluvial terraces, but in around Castro del Rio it is overlain by Serravallian clastics and marls (turbidites of Castro del Rio area; see also Roldán et al. 1992) which very commonly show steep diapiric contacts.

The seismic image of the inner part of the chaotic unit consists of non-coherent reflections (Fig. 6) within a package defined by south-dipping boundaries so that the Intermediate Units overlie the internal edge of the chaotic unit. The external sub-unit is seismically much more coherent than the inner part (see Figs 6 and 7), and consists of sub-parallel sets of reflections displaying fold geometries cut by faults, which make the whole interpretable as a small north-verging imbricated thrust system. The contact between the two parts of the chaotic unit could be either a thrust or a primary diapiric contact (Figs 2, 3, 6 and 7). On the third cross-section, Marismas (Fig. 4), located 160 km further west, the chaotic unit is considerably larger and its limit to the SE is not constrained.

The external sub-unit is clearly seen on seismic and in wells; Isla Mayor-1 penetrated the sub-unit completely and showed that it consists in deformed and repeated Miocene. The wells Betica 18-1 and Bornos-3 drilled thick sequences of the internal sub-unit and showed that it consists mainly of halite (north of Jaén, halite is produced commercially by solution from the internal sub-unit).

Since the early works by Perconig (1960–62) the chaotic unit has been interpreted by many authors (e.g. Pérez López 1991; Riaza & Martínez del Olmo 1995) as a chaotic mass formed by a matrix of Triassic sediments and including younger blocks, which flowed into the basin during Tortonian times in the form of gigantic olistostromic masses derived from the Subbetic. Therefore, the chaotic unit has been considered as a mega-turbidite forming a part of the sedimentary infilling of the basin (e.g. Suárez-Alba et al. 1989).

However, the term olistostrome denotes 'a sedimentary deposit which consists of a chaotic mass of rock and contains large clasts composed of material older than the enclosing sedimentary sequence. The clasts may be gigantic and are then called "olistoliths". Such deposits are generally formed by gravity sliding of material, sometimes into oceanic trenches. Olistostromes have also been called "sedimentary melanges"' (Allaby & Allaby 1990).

Keuper gypsum, as seen in the outcrops of the chaotic unit, is not a plausible matrix lithology for a deep-water sediment and is different from the observed Triassic of the Subbetic. Furthermore, a Triassic olistostrome matrix cannot enclose younger (Upper Cretaceous) olistoliths. We interpret the subtractive contacts at the front of the Intermediate Units referred to above (Figs 2, 3 and 5) as diapiric contacts, and therefore we propose that the evaporites were derived from below the Intermediate Units, where the Keuper is now absent, and emplaced below the seismically recognized south-dipping tectonic contact. This is a diapiric process.

The Guadalquivir Basin

The basement of the basin consists of Palaeozoic and Mesozoic rocks, its top dipping some 2–4° towards the SE. It holds a Neogene to Quaternary sedimentary infilling, with a major megasequence boundary between them. Previous works focusing on the stratigraphy of the infilling considered it as formed by two different types of deposits, both from lithological and genetic points of view: (1) massive, allochthonous olistostromes dropping into the basin from

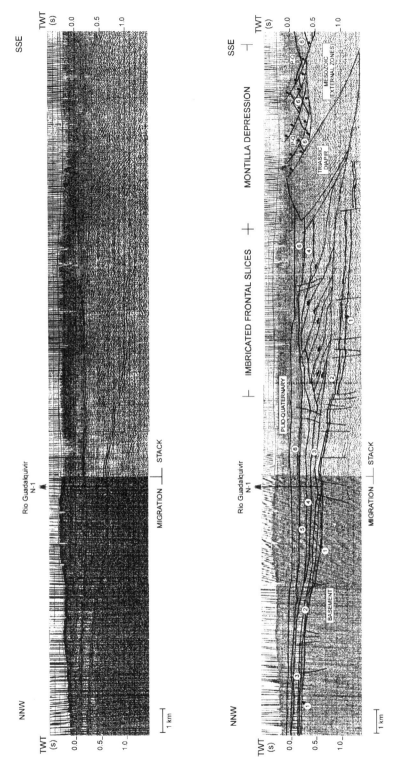

Fig. 6. Profile II (line S84-40 and RGK09110). Uninterpreted and interpreted versions. It shows the entire 'chaotic unit' from its contact with the External Zones. There are displayed extensional faulting in the Montilla depression, internal Triassic part of chaotic unit, external imbricates, frontal structure and basin infill.

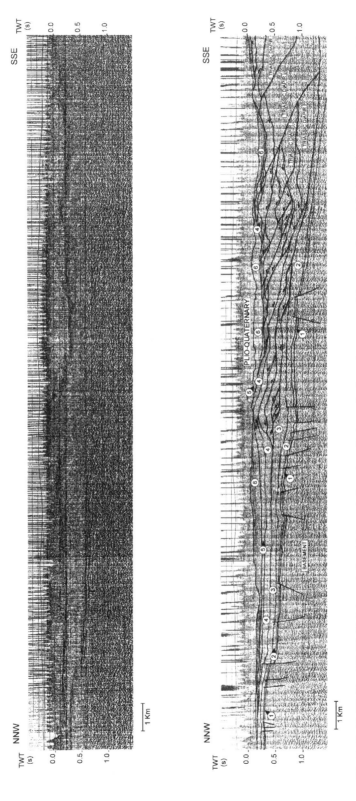

Fig. 7. Profile III (line S83-44). Uninterpreted and interpreted versions. It shows the frontal structure of the external part of the 'chaotic unit' and its relation with the infill.

the External Zones of the chain, and (2) 'autochthonous' sedimentation (see Fernàndez et al. this volume).

The interpretation of seismic stratigraphy from more than 1400 km of seismic lines and 35 exploration wells (Lanaja et al. 1987; see Fig. 1 for location) has enabled us to divide the Miocene sedimentary infill of the basin into six seismic-stratigraphic, depositional sequences (Mitchum et al. 1977; Vail 1987) numbered, from bottom to top, sequences 1 to 6 (Fig. 8). They are widely recognizable in the whole basin. Because of problems in dating the coastal and nearshore sediments forming the sequence 1 in outcrop and in the available wells (see details in Fernández et al. this volume), the age attributions have been made using seismic-stratigraphic techniques. Thus, taking as a datum the regionally well-known Late Messinian Unconformity, the defined sequences have been tied to third-order cycles 2.4 (Late Tortonian–Early Serravallian) to 3.3 (Messinian) in supercycles TB2 and TB3 of the Standard Cycle Chart (Haq et al. 1987) (Fig. 8). The nature of the boundaries, seismic facies, thicknesses, lithologies and attributed sedimentary environments are summarized in Table 1.

In relation to the extensional faulting involving the basement, sequence 1 clearly predates the extensional structures; the thickness of sequence 2 is controlled by normal fault offsets, showing important N–S variations, increasing towards the internal zones of the basin, and pinching out towards the structural highs of the basement developed in the northern half of the basin (Córdoba area). From outcrop data, sequence 2 can be correlated with the Castro del Rio sands and marls, where it unconformably overlies the Triassic, internal part of the chaotic unit, and is involved in minor diapiric structures. The structural highs also control the arrangement of the sequences 3 and 4 (Figs 9 and 10), which post-date the normal faults.

Sequence 3 is involved in the frontal imbricates, sequences 4 and 5 forming its frontal monocline; also some remains of these sequences can be found on the hanging wall of the structure (Fig. 7). Sequence 6 post-dates all the structural features, and the bottom of the Pliocene deposits, especially in the Córdoba area, is deeply incised in the previous sediments (Fig. 10).

Structural modelling of lateral diapirism

In the last few years, big advances have been made in the understanding of diapiric processes, and their relationship with other structural and stratigraphic processes. The physical modelling work of the Applied Geodynamics Laboratory (AGL) in Austin, Texas, using sand and polymers, has made a particularly important contribution to the understanding of lateral movement of salt in the Gulf of Mexico (Jackson & Talbot 1989). Further advances in this passive continental margin setting have been made using cross-section balancing techniques (McGuinness & Hossack 1993; Hossack 1995). Application of these techniques to seismic data from the Gulf Coast has led to the recognition that major lateral flow of diapiric material often takes place both within sediments as intrusions,

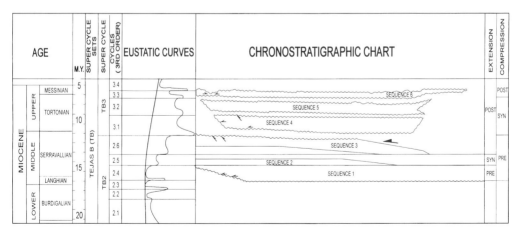

Fig. 8. Cycle chart showing the established sequences and its correlation to the standard cycle chart (Exxon). Also there are two columns in which the sequences are correlated to the basement extension and to the chaotic unit emplacement (compression), pre-syn-post in each.

Table 1. *Summary of the seismic stratigraphic interpretation and correlation with the Standard Cycles of Haq et al. (1987)*

This work	Standard cycle chart Haq et al. (1987)		Age	Boundaries	Thickness	Lithology	Internal seismic pattern	Depositional environment
	Supercycles	Ma						
6	3.3		Messinian	Lower boundary: erosive unconformity (toplap terminations below it). Very late Tortonian unconformity. Upper boundary: very important erosive surface with deeply incised channels and valleys. Very late Messinian unconformity.	Very irregular in the whole basin. It ranges from 0 to 400 ms (TWT). The higher values are in Marismas area.	Marls and sandy marls with sand intercalations.	Sub-parallel, continuous reflections with areas of strong amplitude (specially near the bottom) and slight amplitude (in the upper half of the section), of low apparent frequency.	Basinal/turbiditic
		6.3						
5	3.2		Mid- and late Tortonian	Lower boundary: Mid-Tortonian unconformity. Upper boundary: Very late Tortonian unconformity.	From 20 ms to 80 ms (TWT). Its maximum value is 200 ms (TWT) in Córdoba area.	Marls, silts and sands.	Strong amplitude, high frequency and good continuity reflections which show mounted, channelled and progadationals geometries.	Turbiditic
		8.2						
4	3.1		Early Tortonian	Lower boundary: Very late Serravallian unconformity. Upper boundary: erosive unconformity. Mid-Tortonian unconformity.	It is in infilling the erosion features from Tortonian erosive surface. Its thickness is very irregular ranging from 300 ms to 0 ms (TWT).	Sands and marls.	Strong to moderate amplitude, low apparent frequency and good continuity reflections, showing sub-parallel, mounded and channelled geometries in strike sections.	Turbiditic
		10.5						
3	2.6		Late Serravallian	Lower boundary: erosive unconformity (toplap terminations of the sequence 2). Mid-Serravallian unconformity. Upper boundary: very important erosive unconformity defined by toplap reflection terminations. Very late Serravallian unconformity.	From 30 ms to 80 ms (TWT). Its maximum value is 200 ms in Córdoba area.	Marls with sandy intercalations.	Sub-parallel to oblique, apparent low frequency, slight amplitude and poor continuity in the deepest areas.	Basin/turbiditic
		12.5						
			Early Serravallian	Lower boundary: Onlap surface. Early Serravallian	Controlled by normal faults. Its maximum	Marls with intercalations	Low amplitude and apparent poor continuity, sub-parallel reflections	Basin/turbiditic

Seq	Age (Ma)	Age	Boundaries	Thickness	Lithology	Seismic facies	Environment
2	2.5 – 13.8		unconformity. Upper boundary: unconformity including an erosional toplap below the sequence 3. Mid-Serravallian unconformity.	value is 300 ms (TWT).	of sands.	in the shallower parts of the basin, and high amplitude, low apparent frequency and good continuity, sub-parallel, oblique, mounded and channel-shaped reflections in the deepest areas.	
1	13.8 – 15.5	Very late Langhian to very early Serravallian	Lower boundary: Very late Langhian unconformity. Upper boundary: Early Serravallian	From 20 to 40 m. The maximum values are found in Sevilla area.	Calcarenites, conglomerates and sandstones.	Not imaged on the available profiles.	Coastal and near shore.

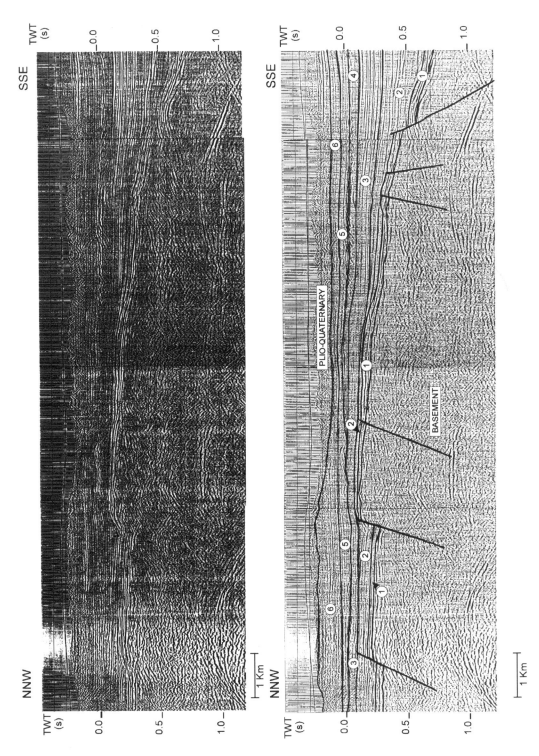

Fig. 9. Profile I (line 89-01). Uninterpreted and interpreted versions. It shows extensional structures involving basement and Neogene sequences No. 1 and 2.

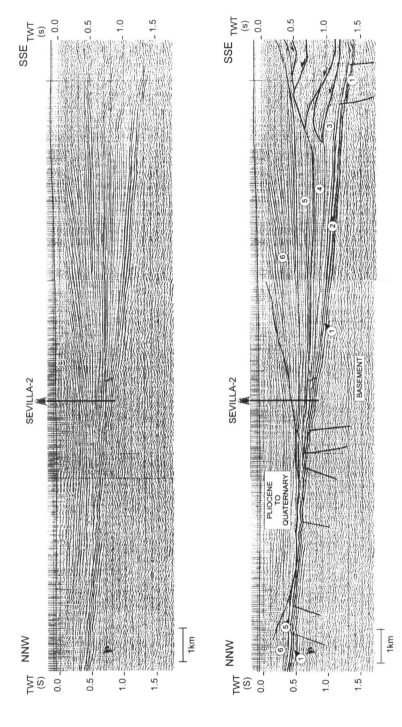

Fig. 10. Profile IV (line 81-58). Uninterpreted and interpreted versions. It shows the pinching out of sequences 2 and younger over the basement highs (Córdoba–Sevilla area). Also shows S–N progradational pattern of Pliocene sequences and paleo-Guadalquivir incision. Includes well Sevilla-2.

and at the sea bed by processes analogous to the flow of glaciers. Salt 'glaciers' have also been studied where diapirs emerge at surface in the semi-arid conditions of the Zagros fold-belt, Iran. Here the term 'namakier' (from the Farsi word for salt, *namak*) has been given to them (Talbot 1981). When they form at sea-bed they partly dissolve, until they acquire a carapace of insoluble residue and marine sediments. So long as the supply of diapiric material from below is sufficient to maintain some surface or seabed topography, the namakier will continue to flow (Fletcher *et al.* 1993).

Analogue sand-box physical model

A provisional attempt was made to model the structure as interpreted on the Martos and Baena cross-sections (Figs 2 and 3), on the assumption that it is a diapiric body emplaced laterally under compression (most of the AGL published models involve extension). The experiment was carried out in the Fault Dynamics Project laboratory at Royal Holloway University of London. Figure 11 shows the starting configuration of the experiment. It includes an unstable asymmetric 'basin' of SGM polymer to simulate the original basin of Keuper evaporite, overlain by a wedge of sand (dark grey and white) thinning towards the foreland, represented by a block of wood with a 25° ramp. The model was tilted about 3°, then compressed from the 'internal' end.

We had expected deformation to start with the formation of an imbricate stack of thrust slices propagating from the moving end wall (right), simulating the Betic thrust belt, but this did not occur. Instead, the sand layers detached as a single block on a thrust that ramped to surface above the basement step. This meant that the tectonic load necessary to drive out the polymer did not develop. At this point we stopped the compression, and introduced an artificial load in the form of metal weights. This made the polymer flow from below the weights and up into a pillow above the basement ramp and into the initially formed thrust. Then the metal weights were removed, and the space filled with light grey and thin white sand layers (which can be regarded as early syn-compressive beds). During subsequent compression, 'pop-up' thrust structures developed in the internal part of the model, but still not a good analogue for the external Betics.

As compression continued, the diapir enlarged, and was pushed by the large block of pre- and syn-tectonic material. Its base moved on a flat thrust over the basal (white) layer of pre-tectonic sediment. At the top of the diapir, the thin pre-tectonic layers soon stretched to negligible thickness, and the polymer emerged, flowing down the slope into the foreland basin and overriding deformed basin-fill sediments (white and thin light grey layers). Polymer also intruded the sand layers on both sides of the diapir.

The final result (Fig. 12) shows striking similarities with the interpreted cross-sections, especially the 'Baena' section, (Fig. 3) although it must be admitted that in this preliminary experiment the procedure was somewhat contrived. Figure 13 shows a schematic model for the emplacement of the diapiric body into the southern margin of the Guadalquivir Basin, showing also how blocks of younger limestones can be broken off the tip of the overthrusting Jaen Zone and incorporated into the diapir.

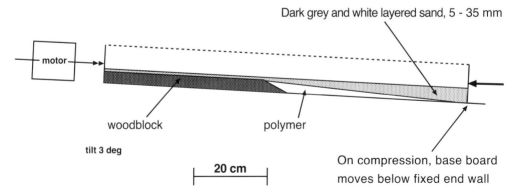

Fig. 11. Model set-up for sandbox experiment. The box measures 100 cm long, 30 cm wide, 10 cm deep. On compression, the base-board moves below the fixed end-wall at the right. This means that deformation starts at this end (internal) and propagates to the left (external).

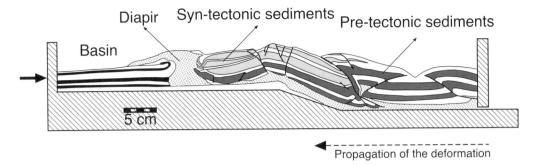

Fig. 12. Sandbox model – final structure redrawn from a cut slice. Compare especially with 'Baena' geological cross-section (Fig. 3). Note that the light grey and white layers in the centre of the structure were added during the diapiric phase of the experiment induced by adding metal weights.

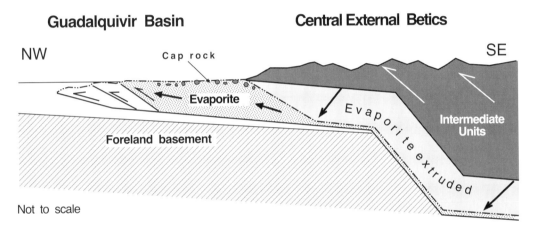

Fig. 13. Schematic model for emplacement of diapiric evaporite–sediment body into the southern margin of the Guadalquivir Basin.

Numerical model

A 2D computer simulation was generated with the aim of improving understanding of the basic mechanisms for lateral diapiric emplacement and to reproduce the early stages of deformation that were not modelled satisfactorilly in the sand-box. The model assumed vertical simple shear deformation along a thrust plane. The hangingwall to this thrust included a Coulomb wedge which corresponds to a thrust sheet stacking piled formely. The footwall consisted of the basement which remained undeformed and which supported an evaporitic basin. For simplicity, we considered only the dynamics of the salt layer rather than using a fully self-consistent dynamic model for the salt and the overburden. The salt layer, which is considered to behave as a viscous newtonian fluid, was overlain by sediments with no strength which deformed in response to the salt-layer dynamics. As the overburden moved with respect to the salt it produced drag of the top of the salt layer. Hence, the numerical approach modelled the response of a thin viscous layer to buoyancy, differential loading and drag forces.

Figure 14 shows the geometric set up considered in this approach which, for consistency, was the same as in the sand-box model. A salt body, of thickness $H(x)$, was acted upon by a pressure $P(x)$ and the salt top was dragged along at a horizontal velocity $u(x)$. The salt body had a density ρ_s and viscosity μ, and the overburden had a density ρ_0. The vertical coordinate y increased upwards from $y = y_0$ at the salt base.

With this set up, and assuming thin film theory, the horizontal salt flux, Q, is given by

$$Q = -(H^3/12\mu)\partial P/\partial x + uH/2. \quad (1)$$

The pressure derivative in equation (1) is controlled by the physical properties of the overburden. Then, for a weak overburden

$$\partial P/\partial x = \rho_0 g\, \partial S/\partial x + (\rho_s - \rho_0)\, g\, \partial H/\partial x \quad (2)$$

where $S(x)$ is the height of the top of the overburden. The first term on the right accounts for differential loading whilst the second term models buoyancy effects. A more complete analysis incorporating the effects of overburden strength and in-plane stresses has been derived in Waltham (1997). Under the assumption of cross-sectional area conservation, the rate of change of salt thickness is given by

$$\partial H/\partial t = -\partial Q/\partial x. \quad (3)$$

Vertical simple shear implies that the hanging-wall horizontal velocities are constant and equal to the compression rate whilst, in the footwall, the horizontal velocities are zero except within the salt body. In addition, the vertical rate of movement, v, within the hanging-wall only depends on x and hence, knowledge of u and $\partial H/\partial t$ allows v to be calculated by (e.g. Waltham & Hardy 1995):

$$\partial H/\partial x = v - u\partial f/\partial x \quad (4)$$

where $f(x,t)$ is the bottom of the overburden or the detachment surface. Once velocities are specified throughout the overburden, any horizon given by $h(x)$ will evolve according to

$$\partial h/\partial t = v - u\partial h/\partial x. \quad (5)$$

Equations (1) to (5) are solved iteratively with time using an explicit finite difference scheme. Boundary conditions allow for flow material associated with the moving overburden through lateral boundaries. The basement remains at rest and undeformed during the process while the top of the model acts as a free surface.

Figure 15 shows the results from the computer model for different time steps and using reasonable values of rate of thrusting, evaporite viscosity, initial evaporite thickness, densities and surface slopes for the study area (Table 2). After 2.5 Ma (Fig. 15a) a thrust, off the right of the model, has generated a Coloumb wedge acting as a differential load on the evaporite. The load affects only the thin part of the evaporite which has therefore not yet moved significantly. After 4.5 Ma (Fig. 15b) the Coloumb wedge has advanced over the thick part of the evaporite and is now producing a diapir ahead of the wedge toe. A new thrust, formed after 5 Ma, detaches in the evaporites and breaks surface off

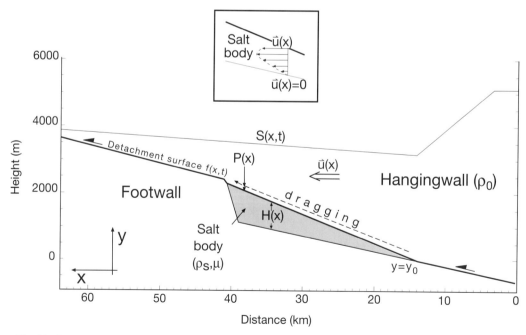

Fig. 14. Geometric set up considered in the numerical model of the salt emplacement. Upper in set: schematic velocity distribution with depth within the salt body. $S(x,t)$ = free surface; $f(x,t)$ = detachment surface, note that this surface change in time.

Table 2. *Computer model parameters*

Coloumb wedge slope	10°
Maximum wedge height	6100 m
Overburden density	2400 kg m^{-3}
Evaporite density	2200 kg m^{-3}
Rate of thrusting	5 m ka^{-1}
Evaporite viscosity	10^{18} Pa s

the left of the model (Fig. 15c). This results in shearing and partial transport leftwards of the evaporite. The final model configuration after 8 Ma is showed in Fig. 15d. Therefore, the differential load on the evaporitic layer associated with the thrusting progression together with the drag from the overburden and buoyancy forces results in the squeezing and further extrussion of the evaporites. These results show that the mechanism proposed for lateral diapiric emplacement is dynamically feasible and matches the main time and spatial features of the evaporite emplacement as observed in the Betics.

The final element of the model is the Coulomb wedge. This is simply imposed as a constant, specified slope which progresses across the model at a specified rate jointly with the overburden. The effect of this is to impose a slowly increasing load on the salt in addition to the load generated by the moving overburden. Note that thin film theory only applies for early stages of salt movement during which concordant (in the sense of Waltham 1997) low amplitude structures are formed. Therefore, the final stages of this process, during which the evaporite diapir breaks through to the surface, has not been modelled numerically. This final process was, however, well reproduced by the physical model.

Discussion

The anomalous chaotic unit in the southern margin of the Guadalquivir Basin was not emplaced as an olistostrome. We interpret this body as emplaced through lateral diapiric processes, partly comparable to those interpreted in the Gulf of Mexico (McGuinness & Hossack 1993). Physical and numerical models can achieve a result that is a good match with the structure of the central Betics as explained and illustrated on Figs 12, 13 and 15.

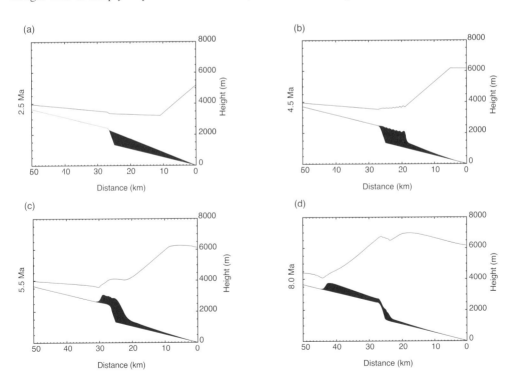

Fig. 15. Numerical modelling results for the evaporite emplacement. (**a**) After 2.5 Ma; (**b**) after 4.5 Ma; (**c**) after 5.5 Ma; (**d**) after 8 Ma. Final model configuration.

In the eastern Betics, evaporites in the Prebetic are responsible for all the diapiric structures (Martínez del Olmo et al. 1986; De Ruig 1992). However this evaporite basin dies out some 100 km to the east of the study area. In the central Betics, the Prebetic is much reduced in thickness and extent, and would seem an unlikely place for thick evaporites to have originated (as was concluded by Blankenship 1992 based on 1:200 000 scale geological maps). In the study area, the diapiric Triassic forming the chaotic unit lies immediately north of the Intermediate Units where the Keuper has been tectonically removed, and it seems reasonable to propose that the diapiric body that is now in the Guadalquivir Basin originated in a Triassic basin below the known Jurassic–Cretaceous basin of the Intermediate Unit trough. The approximate northern edge of the original basin is indicated on the 'Martos' cross-section (Fig. 2): its position based on restoration of the small Prebetic thrust slice coincides with the estimated position of the leading edge of the overthrust basement block (admittedly there is little constraint on either of these positions!). Mobile evaporites deposited in such a basin would certainly have become diapirically displaced in some way by loading, first by the Intermediate Units limestones and then, by the denser basement block. If this is the true origin of the material, it implies a lateral displacement of the order of 40 km.

The path along which this expulsion took place, and the mechanism by which the ductile Keuper was pushed ahead of the overlying thrust sheets, is evident from the cross-sections. We consider the most likely expulsion path is above the basal thrust which ramps from the base of ductile Keuper, first to the Prebetic Cretaceous, and then to the Miocene, and below the basal detachment of the Intermediate Units, i.e. at the base of Supra-Keuper. This has become a 'salt weld'. The ductile body could have picked up blocks of Cretaceous and, less easily, Jurassic material by breaking it away from the roof of the diapir as it passed below the hanging-wall cut-off of the Intermediate Units. Any originally coherent-bedded units within the Keuper would have become disrupted as random blocks during this long-distance diapiric transport. The final events of the emplacement are recorded in the most external unit, where sediments originally deposited in the Guadalquivir Basin during Serravallian to Early Tortonian are involved in deformation caused by lateral movement of the diapir, which ceased in Mid-Tortonian times.

Conclusions

(a) The interpretation of geological and geophysical data has allowed us to construct three true-scale, structural cross-sections, which indicates that compression in External Betics was driven by northward emplacement of basement culminations onto a passive continental margin. Compression started probably in the Eocene and propagated through the External Betics during the Miocene, dying out in Tortonian times.

(b) The presence in the central and western Betics of the large chaotic unit composed mainly of Triassic evaporites, emplaced into the southern margin of the Guadalquivir foreland basin, is interpreted as a body extruded from an evaporitic formation originally below the Intermediate Units. It moved by lateral diapiric flow in front of an advancing thrust sheet, and was eventually emplaced as a sea-floor diapir.

(c) Seismic stratigraphy has been used successfully to define sequential arrangement of the sediments filling the basin, and to attribute accurate ages for the emplacement of the diapir. Submarine exposure of the diapir took place initially in the Serravallian, allowing marine sediments of that age to be deposited on top of the Triassic in places, but lateral emplacement of the diapir continued, compressing the basin-floor sediments ahead of it, until at least Mid-Tortonian.

(d) Lateral diapiric emplacement in a compressional setting has been successfully reproduced using both physical and numerical modelling thus demonstrating the plausibility of this process.

(e) From these results, we recommend that this zone should from now on be called the 'Guadalquivir lateral diapir', instead of the word 'olistostrome'. Evaporites are involved in many of the world's thrust belts, and lateral diapirism in some form is probably a widespread phenomenon.

We thank BP Exploration for allowing the use of unpublished data and reports, and J. Hossack for discussions on salt tectonics. B. Colleta and M. Jackson reviews are gratefully acknowledged. We thank M. Keep, M. James and N. Deeks of RHUL's Fault Dynamics Group for providing facilities and assistance in running the sand-box models. Repsol Exploracion supplied seismic profiles and well data. A. Pérez-López and J. Fernandez of Granada University are thanked for useful discussions in the field area. We also thank M. Losantos, for her help in organizing the first manuscript and constructive criticism. This work has been supported financially by the European-Union 'Integrated Basin Studies' project (JOU2-CT92-0110).

References

ALLABY, A. & ALLABY, M. (eds) 1990. *The concise Oxford Dictionary of Earth Sciences*. Oxford University Press, Oxford.

ALLERTON, S., LONERGAN, L., PLATT, J. P., PLATZMAN, E. S. & MCCLELLAND, E. 1993. Palaeomagnetic rotations in the eastern Betic Cordillera, southern Spain. *Earth and Planetary Science Letters*, **119**, 225–241.

BAENA, J. & JEREZ, L. 1982. *Síntesis para un ensayo paleogeográfico entre la Meseta y la Zona Bética s. str.* Colección Informe, I.G.M.E.

BANKS, C. J. & WARBURTON, J. 1991. Mid-crustal detachment in the Betic system of southeast Spain. *Tectonophysics*, **191**, 275–289.

BLANKENSHIP, C. L. 1990. *Structural evolution of the central External Betic Cordillera, southern Spain*. MA Thesis, Rice University, Houston.

—— 1992. Structure and palaeogeography of the External Betic Cordillera, southern Spain. *Marine and Petroleum Geology*, **9**, 256–264.

BLUMENTHAL, M. M. 1927. Versuch einer tektonischen Gliederung der betischen Cordilleren von central und Sudwest Andalusien. *Eclogae Geologicae Helvetiae*, **20**, 487–532.

DE RUIG, M. J. 1992. *Tectono-sedimentary evolution of the Prebetic fold belt of Alicante (SE Spain)*. PhD Thesis, Vrije Universiteit Amsterdam.

FERNÁNDEZ, M., BERÁSTEGUI, X., PUIG, C., GARCÍA, D., JURADO, M. J., TORNÉ, M. & BANKS, C. 1998. Geological and geophysical constraints on the evolution of the Guadalquivir Foreland Basin (Spain). *This volume*.

FLETCHER, R. C., HUDEC, M. R. & WATSON, I. A. 1993. Salt glacier model for the emplacement of an allochthonous salt sheet. *In: American Association of Petroleum Geologists, Hedberg Research Conference on Salt Tectonics abstracts*, Bath, UK, 13–17 September 1993.

FONTBOTÉ, J. M. 1965. *Las Cordilleras Béticas. La Depresión del Guadalquivir*. Mapa Geológico de España y Portugal. Nota explicativa. Ed. Paraninfo. Madrid.

—— & VERA, J. A. 1984. La Cordillera Bética. Introducción. *In: Libro Jubilar J.M. Rios, Geologia de España*, **2**. IGME, 343–349.

FRIZON DE LAMOTTE, D., ANDRIEUX, J. & GUÉZOU, J.-C. 1991. Cinématique des chevauchements néogènes dans l'Arc bético-rifain: discussion sur les modèles géodynamiques. *Bulletin de la Société Géologique de France*, **162**, **4**, 611–626.

GARCÍA-HERNANDEZ, M., LÓPEZ-GARRIDO, A. C., RIVAS, P., SANZ DE GALDEANO, C. & VERA, J. A. 1980. Mesozoic palaeogeographic evolution of the External Zones of the Betic Cordillera. *Geologie en Mijnbouw*, **59**, 155–168.

HAQ, B. U., HARDENBOL, J., VAIL, P. R., AND 10 OTHERS 1987. Mesozoic–Cenozoic Cycle Chart. *In*: BALLY, A. W. (ed.) *Atlas of Seismic Stratigraphy*. American Association of Petroleum Geologists, Studies in Geology, **27**.

HOSSACK, J. R. 1995. Geometrical rules of section balancing for salt structures. *In:* JACKSON, M. P. A., ROBERTS, D. G. & SNELSON, S. (ed.) *Salt tectonics: a global perspective*. AAPG Memoir, **65**, 29–40.

JACKSON, M. P. A. & TALBOT, C. J. 1989. Salt canopies. *In: Gulf Coast Section, Society of Economic Paleontologists and Mineralogists Foundation, 10th Annual Research Conference, Extended Abstracts*, 72–78.

LANAJA, J. M., NAVARRO, A., MARTÍNEZ ABAD, J. L., DEL VALLE, J., RIOS, L. M., PLAZA, J., DEL POTRO, R. & RODRIGUEZ DE PEDRO, J. 1987. *Contribución de la exploración petrolífera al conocimiento de la geologia de España*. Instituto Geológico y Minero de España, Madrid.

LEBLANC, D. & OLIVER, P. 1984. Role of stike-slip faults in the Betic–Rifian Orogeny. *Tectonophysics*, **101**, 345–355.

MCGUINNESS, D. B. & HOSSACK, J. R. 1993. The development of allochthonous salt sheets as controlled by the rates of extension, sedimentation, and salt supply. *In: Gulf Coast Section of the Society of Economic Paleontologists and Mineralogists Foundation 14th Annual Research Conference*, Extended Abstracts, 127–139.

MARTÍNEZ DEL OLMO, W., LERET VERDÚ, G. & SUÁREZ-ALBA, J. 1986. La estructuración diapírica del Sector Prebético. *Geogaceta*, **1**, 43–44.

MITCHUM, R. M., VAIL, P. R. & THOMSON, S. III. 1977. The depositional sequence as a basic unit for stratigraphic analysis. *In*: PAYTON, C. E. (ed.) *Seismic Stratigraphy. Applications to Hydrocarbon Exploration*. American Association of Petroleum Geologists, Memoirs, **26**, 53–62.

PERCONIG, E. 1960–1962. Sur la constitution geologique de l'Andalousie Occidentale, en particulier du bassin du Guadalquivir (Espagne Meridionale). *In: Livre Mémoire du Professeur Paul Fallot*. Mémoires hors-Série de la Société géologique de France. 229–256.

PÉREZ LÓPEZ, A. D. 1991. *Trias de facies Germánica del sector central de la Cordillera Bética*. PhD Thesis, Univ. Granada.

—— & SANZ DE GALDEANO, C. 1994. Tectónica de los materiales triásicos en el sector central de la Zona Subbética (Cordillera Bética). *Revista de la Sociedad Geológica de España*, **7**, 141–153.

RIAZA, C. & MARTÍNEZ DEL OLMO, W. 1996. Depositional Model of the Guadalquivir- Gulf of Cadiz Tertiary Basin. *In*: FRIEND, P. & DABRIO, C. J. (eds) *Tertiary Basins of Spain: The stratigraphic record of crustal kinematics*. Cambridge University Press, 330–338.

ROLDÁN, F. J., LUPIANI, E. & VILLALOBOS, M. 1992. *Cartografía y Memoria de las Hojas Geológicas números 927 (Baeza), 945 (Castro del Rio) y 947 (Jaén) del Mapa Geológico Nacional a Escala 1/50.000*. Instituto Tecnológico Geo-Minero de España. Madrid.

SUAREZ-ALBA, J., MARTÍNEZ DEL OLMO, W., SERRANO OÑATE, A. & LERET VERDU, G. 1989. Estructura del sistema turbidítico de la Formación Arenas del Guadalquivir, Neógeno del Valle del Guadalquivir. *In: Libro Homenaje R. Soler*.

Asociación de Geólogos y Geofísicos Españoles del Petróleo, Madrid, 123–132.

TALBOT, C. J. 1981. Sliding and other deformation mechanisms in a glacier of salt, S. Iran. *In*: MCCLAY, K. R. & PRICE, N. J. (eds). *Thrust and nappe tectonics*. Geological Society, London, Special Publications, **9,** 173–183.

VAIL, P. R. 1987. Seismic stratigraphy interpretation procedure. *In*: BALLY, A. W. (ed.) *Atlas of Seismic Stratigraphy*. American Association of Petroleum Geologists, Studies in Geology, **27**, 1–10.

VAN DER BEEK, P. A. & CLOETINGH, S. 1992. Lithospheric flexure and the tectonic evolution of the Betic Cordilleras (SE Spain). *Tectonophysics*, **203**, 325–344.

VERA, J. A., GARCÍA HERNANDEZ, M., LÓPEZ GARRIDO, A. C., COMAS, M. C., RUIZ ORTIZ, P. A. & MARTIN ALGARRA, A. 1982. El Cretácico de las Cordilleras Béticas. *In*: *El Cretácico de España*, Madrid, Universidad Complutense, 515–630.

WALTHAM, D. 1997. Why does salt start to move? *Tectonophysics*, **282**, 117–128.

—— & HARDY, S. 1995. The velocity description of deformation. Paper 1: Theory. *Marine and Petroleum Geology*, **12**, 153–165.

Alluvial gravel sedimentation in a contractional growth fold setting, Sant Llorenç de Morunys, southeastern Pyrenees

EDWARD A. WILLIAMS[1], MARY FORD[1], JAUME VERGÉS[2] & ANDREA ARTONI[1]

[1]*Geologisches Institut, ETH-Zentrum, CH-8092 Zürich, Switzerland (e-mail eaw@erdw.ethz.ch)*
[2]*Institute of Earth Sciences "Jaume Almera", Consejo Superior de Investigaciones Científicas, Lluís Solé i Sabarís s/n, 08028 Barcelona, Spain*

Abstract: New data are presented on the classic growth structure at Sant Llorenç de Morunys (NE Ebro Basin, Spain). During the late Eocene to Oligocene thick alluvial-fan gravel sediments accumulated principally by repetitive sub-aerial mass flow (cohesionless debris flow and fluidal sediment flow) events, with smaller volumes of fan-stream flows. Subaerial, high-viscosity (cohesive) debris flows contributed comparatively small volumes of sediment to the successsion. These sediments constructed a complex architecture of conglomeratic and sandstone-bearing lithosomes that were affected by stratal thickening and erosion across a growth fold pair and genetically related internal unconformities, which formed a long-lived thrust-related structure in the immediate footwall of the SE Pyrenean mountain front. Four periods of evolution for the Sant Llorenç growth structure are defined on the basis of distinctive stratigraphical architecture. These describe a gross evolution from onlapping to overlapping growth strata, related to the ultimate demise of growth folding. In detail complex erosional and offlapping events punctuated the growth history, which shows extreme variation parallel to the axis of the structure. Patterns of palaeoflow were highly complex, showing distinct axial and transverse directions relatable to growth fold evolutionary periods. Palaeocurrents are considered to have been deflected and diverted by surficial differential subsidence and areas of relative uplift and erosion generated by fold growth. The complexity of sediment dispersal is compounded by variables intrinsic to alluvial fan environments. The Sant Llorenç de Morunys growth strata provide information on how sediments are reorganized by syndepositionally-growing structures and on the nature of sediment distribution between external fold-and-thrust belts and foreland basins.

Gravelly sediments shed from the erosion of collisional orogens are trapped in the proximal zones of flexural foreland basins in several distinct depositional environments. Commonly, where continental (alluvial) settings are developed over a wide zone up to the mountain front, proximal conglomerates may accumulate in bajadas comprising steep, small-radius alluvial fans, but also, synchronously, in the channel belts of gravelly braided rivers which transfer large volumes of coarse and fine sediment across foreland basins. Where continental settings are restricted due to marine or lacustrine flooding of the foreland basin, gravels accumulate in fan-deltas (Nemec & Steel 1988) or as river-dominated braid-deltas (McPherson *et al.* 1987) supplying clastic material to coarse-grained coastal environments (Nemec & Steel 1984). Where continental environments predominate, the gravel reaches of permanent river channels, at spaced exits from the mountain belt (Hovius 1996), coincide with the coarse-grained proximal-medial zones of steep fans, posing problems of distinction between environments (e.g. Hirst & Nichols 1986; Nichols 1987*a*; see also Haughton 1989; Nemec & Postma 1993; Blair & McPherson 1994*a*). As well as providing signatures of thrusting, uplift, erosion and drainage reorganization in the source area (Blair & Bilodeau 1988; Burbank *et al.* 1988; Colombo 1994; Gupta 1997) and subsidence regimes within the basin (Paola 1988, 1989; Paola *et al.* 1992; Whipple & Trayler 1996), gravel sediments in proximal foreland basin settings are themselves deformed by thrusting and folding during their accumulation. Active deformation of the aggradational area may episodically or quasi-continuously affect the topography with resultant effects on water–sediment dispersal, erosion (with the creation of new drainage routeways) and gross stratigraphical development. Tectonic deformation of the sediments accumulating at mountain fronts gives rise to so-called growth structures, with their associated

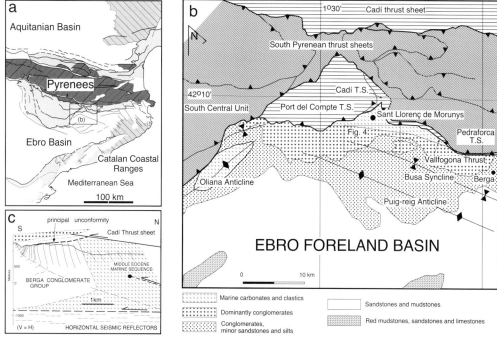

Fig. 1. (**a**) Tectonic map of the eastern Pyrenees and Ebro Basin, showing the location of (b). The Pyrenean axial zone is shown by the dark-shaded hatching. (**b**) Map showing the distribution of Late Eocene–Oligocene continental clastic facies (based on the 1:250 000 geological map of Catalunya), and major structures of the foreland basin in the Sant Llorenç de Morunys area. Studied area is boxed. (**c**) Local N–S cross section showing the deep structural context of the Sant Llorenç de Morunys growth fold (central part), and its relationship to the frontal emergent thrust of the SE Pyrenees which is sealed by the upper units of the Berga Conglomerate Group (coarse stipple).

suite of progressive unconformities (Riba 1976b; Anadón et al. 1986) and growth strata geometries (Artoni & Casero 1997; Ford et al. 1997; Suppe et al. 1997). Evolving topography will distort classical radial dispersal patterns of steep alluvial fans, and potentially deflect or divert more permanent river courses, unless conditions favouring antecedent drainage across structures are met (Burbank et al. 1996). Drainage basins associated with actively deforming zones at mountain fronts are commonly multiple in origin (Friend 1989), and the material they yield will partially accumulate within growth structures developed in this zone.

Growth structures, their associated stratal geometries, sedimentology and petrographic evolution have been widely studied (e.g. Rockwell et al. 1985; DeCelles et al. 1991a, b, 1995; Mellere 1993) and particularly around the margins of the Eo-Miocene Ebro Basin, Spain (Riba 1967, 1976a, b; Anadón et al. 1986; Nichols 1987b; Vergés & Riba 1991; Colombo & Vergés 1992; Burbank et al. 1992a, b, 1996; Millán et al. 1994; Burbank & Vergés 1994; Vergés et al. 1996). Some recent key studies of the interaction between alluvial systems and developing structures in foreland basin settings have concentrated on theoretical explanations of drainage diversion by folds (Burbank et al. 1996), control due to break-back sequences of thrusts associated with detachment folds (Burbank & Vergés 1994) and drainage basin reorganization and megafan location due to frontal thrust propagation and fold growth (Gupta 1997). In this paper, we focus on the growth structure near Sant Llorenç de Morunys (see Riba 1973, 1976a, b), which preserves complex fold and stratal geometries developed in the immediate footwall of the late Eocene–Oligocene SE Pyrenean mountain front (Fig. 1a). This fold pair developed over a long period below the surface of the foreland basin, providing localized accommodation space, but otherwise having a more subtle influence on sedimentary processes than previously described growth folds. The study aims to document and explain the interaction of the coeval fold growth, gravel sedimentation and dispersal, surface erosion and stratigraphical development by integrating interpreted sedimentary processes and environments with a series of palaeogeographical–tectonic maps of the evolving basin margin. The latter have been constructed from sequential restorations of serial cross sections through the growth structure (Ford *et al.* 1997). Sediment dispersal has been investigated by the measurement of a suite of palaeocurrent indicators across the structure.

Geological setting

The Sant Llorenç de Morunys growth structure affects Eocene–Oligocene conglomerates (and minor interbedded sandstones) up to 5 km south of the east–west-trending Vallfogona Thrust (Fig. 1b), and persists laterally for approximately 25 km. A variable width fringe of conglomerates characterizes this part of the NE Ebro Basin, from the northern flank of the Oliana growth fold at the SE margin of the south-central unit (Burbank & Vergés 1994) to the coarse-grained alluvial dispersal systems in the Berga area (Mató *et al.* 1994; Fig. 1b). The very low-angle (7°) frontal emergent (Vallfogona) thrust emplaces the Port del Compte and Cadí thrust sheets (consisting mainly of thin Mesozoic and thick lower to middle Eocene shallow-water limestones and marls) towards the SSW (Vergés 1993), over mid Eocene shallow marine clastic-carbonate shoreline and coastal plain sediments (equivalent to the Banyoles and Igualada Marl Formations, Vergés 1993), and the alluvial Berga Conglomerate Group. These rocks define a large-scale growth fold pair which developed as a fault propagation fold ahead of a footwall splay of the Vallfogona thrust (Fig. 1c; Ford *et al.* 1997, fig. 3a). On a seismic profile horizontal reflectors indicate that the fold pair is detached on a thrust flat at depth (Fig. 1c). This thrust accommodates *c.* 8 km of southward displacement, and the minimum displacement accommodated by the Vallfogona thrust is estimated to be 8.4 km (Vergés 1993). The true displacement is difficult to define as no clear stratigraphical correlation can be made across the thrust. The Vallfogona thrust is, however, buried by conglomerates of the upper Berga Group to the west of Sant Llorenç, and would appear to be the principal mountain-front structure which influenced sedimentation in the Sant Llorenç de Morunys area. Traced to the south, the proximal Berga Group conglomerates become finer-grained and more thinly bedded, replaced by finer grained facies (sandstones and siltstones) in the succession (Fig. 1c). Conglomerates extend *c.* 5 km beyond the trace of the Puig-reig growth ramp anticline (Vergés 1993), where they are replaced by sandstones and mudstones (Fig. 1b).

On a larger scale the Port del Compte and Cadí thrust sheets are part of a stack of thrust sheets that are preserved above foreland basin deposits. These are bounded to the north by the axial zone of the Pyrenees (Fig. 1a) which comprises Upper Palaeozoic low-grade metasediments and Variscan plutons. Regional balanced sections constructed across the SE Pyrenean thrust belt reveal a shortening of *c.* 150 km in the central part (Muñoz 1992) and 125 km in the eastern part (Vergés 1993). The major period of emplacement of the thrust sheets in this region of the orogen (latest Cretaceous to Oligocene), and the Eocene–Oligocene age of the sediments that seal the Vallfogona thrust indicates that the sediment transport path for the Berga Group from known source areas in the axial zone (Riba 1976*b*) was approximately the present-day distance (Fig. 1a). The Berga Conglomerate Group represents the waning stages of foreland basin sedimentation in the NE Ebro Basin, although Miocene alluvium accumulated in western-central areas. The late Eocene closure of the Ebro Basin to become an area of centripetal drainage (Riba *et al.* 1983) led Coney *et al.* (1996) to suggest that the Berga Group was continuous with conglomerates which 'backfilled' the southern Pyrenean fold and thrust belt, the evidence for which has been challenged by González *et al.* (1997).

Local stratigraphical-structural setting of the conglomerates

An internal lithostratigraphy of the Berga Conglomerate Group at Sant Llorenç de Morunys is presented in Fig. 2, which is based on field mapping and sequence logging of distinguishable units of different scale (see Ford *et al.* 1997). The base of the Berga Group is an abrupt conformable transition, occurring over some 10–15 m, from interbedded marine marls, sandstones, rare nummulitic limestones and essentially non-reddened pebble–cobble conglomerates, to stacked, thick conglomerates associated with finer-grained reddened sediments. The mid-Eocene biostratigraphy of the upper marine sediments, suggests a late Eocene age for the lower Berga Conglomerates (Riba 1973), although no more accurate dating is currently available.

Previous stratigraphical work (Riba 1976a, b) divided the Berga conglomerates into four members based on several 'conglomeratic key beds' that were traced from aerial photographs. These key beds, on examination, comprise metre-scale bedsets of conglomerates and sandstones, and are thus multiple event deposits, often found to form part of new mapping divisions. The new scheme (Fig. 2) defines a series of geometrically complex conglomeratic and sandstone-rich formational units. A simplified measured section through some of these units (Fig. 2) is shown in Fig. 3, emphasizing the differentiation of the succession and constituent lithosomes by bulk grain size. Overall, all formations are dominated by conglomerate lithologies (Fig. 3), whereas those units designated sandstone-rich (Fig. 2) contain c. 20–30% of sandstone or pebbly sandstone lithologies. Several instances of bulk facies changes occur in depositional-strike parallel directions (WNW–ESE to E–W), which result in the pinching out or sharper termination of conglomerate-dominant formations (Figs 2 & 4). Some units have subtle triangular cross-sections on a scale of >2 km, suggesting that they have wedge- or lobate-forms in three-dimensions, similar to geometries of gravelly alluvial fan deposits described by Koltermann & Gorelick (1992, fig. 3). These geometrical attributes indicate that the formations comprise large-scale lithosomes in the sense of Wheeler & Mallory (1956). Geometrical complexity also occurs within units designated as formations. In the dip-parallel direction, more marked thickness changes occur, which are clearly relatable to the axial surfaces of the growth folds (see below and Fig. 5).

The growth structure (Fig. 4) can be divided into (1) a northern 3 km thick, overturned to sub-vertical panel comprising the marine clastic–carbonate succession and lower alluvial conglomerates (Casa Blanca, El Castell and Sobirana Formations, Fig. 2) and (2), in the south, a growth fold pair, comprising the upper alluvial conglomerates (Camps de Vall-llonga, Pont de les Cases to Mirador Formations). The steep panel and fold structures to the south are considered (Ford et al. 1997) to be part of a single large growth fold pair as shown in Fig. 1c that grew continuously during deposition of the Berga Conglomerate Group (see Suppe et al. 1997 for an alternative interpretation). The steep panel can be traced for over 25 km eastward from Coll de Jou (Fig. 4). Three cross sections through the area (Ford et al. 1997; Fig. 5) show that the geometry of the growth fold varies along strike. On the western flank of Vall de Lord (profile 1; Figs 5a & 6a) these two structural zones are separated by a spectacular

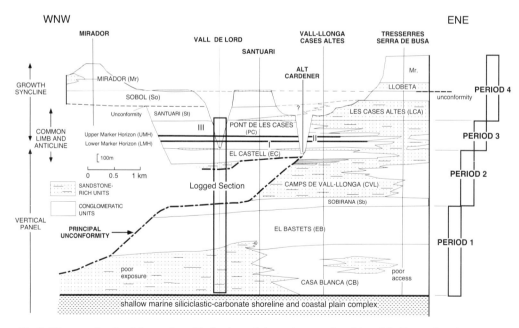

Fig. 2. Diagram showing lithostratigraphical units and the main unconformities of the Berga Conglomerate Group near Sant Llorenç de Morunys, and the relationship to Periods 1–4 defined by stratigraphical architecture. The general location of the logged section in the Vall de Lord is boxed.

angular unconformity (Fig. 6a) which becomes bedding-parallel toward the south. This is the principal growth unconformity (see also Fig. 2 and Riba 1976b). The growth fold pair is clearly visible in younger strata to the south (Fig. 6b) and can be traced eastward into the Tossal de Vall-llonga on profile 2 as can the principal growth unconformity (Fig. 5b). East of Vall-llonga (the next valley to the east), only the growth syncline is clearly visible (profile 3, Fig. 5c) and there is a gradual transition from overturned beds southward into flat-lying beds. The significant along-strike variations in growth strata history partly correlate with an eastward increase in fold size.

Structural orientation data for the whole region (Fig. 4) give a mean sub-horizontal fold axis trending 1°–284°; minor local variations in fold plunge are detailed in the three-dimensional structural analysis of Ford et al. (1997). The asymmetrical fold pair comprises an anticline whose axial surface can be continuous or fragmented into en-echelon segments and which overall shows either a shallow hinterland, or sub-horizontal sheet-dip, and a syncline whose curved axial surface has a moderate hinterland dip (Figs 5 & 6b). The generalized fold axial surfaces converge upwards. However they do not meet at a point, but become parallel or die out in the highest strata (Fig. 6b). Principal and subsidiary angular unconformities are developed in the growth anticline and upper parts of the common limb, while composite offlap-onlap is observed in the growth syncline. Major thickness changes occur across the anticlinal and synclinal axial surfaces and across the common limb. Dips shallow upward into younger beds within the common limb. An anticlockwise axially transecting cleavage exists in sandstones of the growth syncline hinge region. Minor thrusts and strike-slip faults record brittle downdip and along-strike extension in steep strata.

Based on these observations and sequential restoration of three cross sections (Fig. 5), it was concluded by Ford et al. (1997) that the fold pair grew principally by progressive rotation of the common limb and that folding was ongoing at all levels of the structure. Deposition of wedge-shaped units of sediment across the growing fold pair forced the limb to lengthen with time. The fold is considered to have developed ahead of a propagating low-angle blind thrust (i.e. a fault propagation fold; Fig. 1c), the position of the tip of which is unknown. Suppe et al. (1997) present an alternative interpretation for the upper 500 m

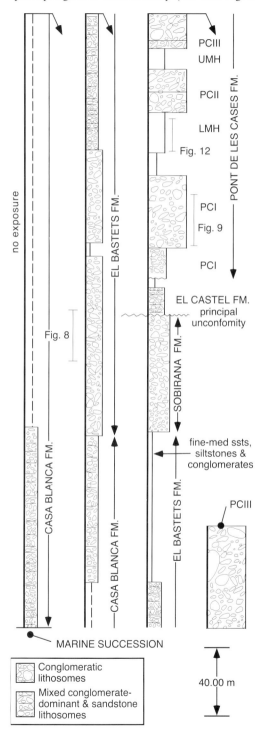

Fig. 3. Simplified vertical stratigraphic section, exposed along the Cami del Santuari (Vall de Lord, see Figs 2 and 4 for location), from the marine-continental boundary to base of the Les Cases Altes Formation.

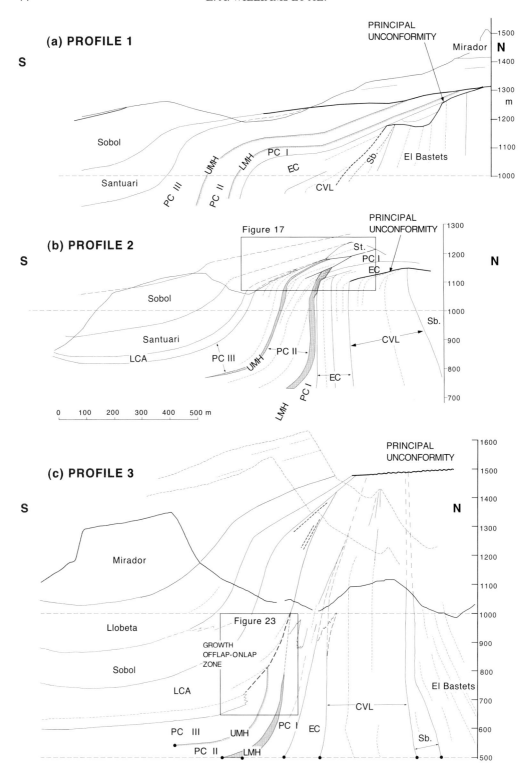

Fig. 5. Three dip-parallel cross-sections across the Sant Llorenç growth structure, located in Fig. 4 (modified slightly from Ford *et al.* 1997, fig. 19). Abbreviations for stratigraphical units as in Fig. 2

Fig. 6. (**a**) Photograph (view to the west) of the principal unconformity as depicted on profile 1, in an approximately dip-parallel section. Vertical to overturned conglomerate beds of the El Bastets and Sobirana Formations showing a continuous spectrum of dips until truncated by the unconformity. The highest, gently-dipping beds are of the Sobol and Mirador Conglomerate Formations. (**b**) Photograph of an oblique view of the section depicted in profile 1 on the west side of Vall de Lord, illustrating the thickening of stratigraphical units across the growth anticline, and the rounded hinge forms of the anticline and syncline. In the middle ground are flat-lying conglomerates above the principal unconformity up to the Mirador Formation. In the background to the right are the Eocene carbonates of the Port del Compte thrust sheet.

of the growth fold, arguing that it developed by curved-hinge kink-band migration predicted by the fault-bend fold mechanism, which caused lengthening of the common limb but no limb rotation.

Sedimentology

The purpose of this section is to establish the *basic* emplacement processes of the volumetrically dominant (modal) conglomerate and sandstone facies that make up the growth strata at Sant Llorenç de Morunys, in order to understand the relationship between surface processes and deformation. The sedimentological analysis is centred on a well exposed road section in Vall de Lord (Fig. 4) which cuts obliquely across the growth fold pair, including conglomerates of the vertical panel, and thus samples several geometrically complex lithosomes. With only one large exposure gap (Fig. 3) the section extends from the conformable contact with marine strata to the base of the Les Cases Altes Formation (Figs 2 & 3). Approximately 700 m of this section were logged on a bed-by-bed basis at a scale of 1:25; this database was supplemented by observations and measurements from other well exposed regions of the growth structure. Data collected for each bed included size/frequency of framework clast modes, framework clast shape and roundness, percentage and type of matrix, clast support system and fabric type. Details of the bed geometry, structures and stratification were routinely recorded. Clast size-bed thickness and palaeocurrent analyses were also carried out, the details of which are given below.

Formation and member divisions of the Berga Conglomerate Group have been principally established on the basis of proportions of three broad lithologies: granule to small boulder conglomerates, lithic arenites and pebbly (lithic) sandstones. The latter pair tend to vary inversely in occurrence with the more varied conglomerate facies. Siltstone lithologies are comparatively rare, tending to occur in overall finer-grained formations. Detailed information regarding the clast populations of the conglomerates will be given elsewhere. In general, however, typical beds are polymict, with varied carbonate rock types predominating. As noted by Riba (1976*b*) clasts of acid–intermediate plutonic rocks, Permo-Triassic red beds, Carboniferous micro-conglomerates, rare metasedimentary rocks and basic igneous rocks indicate that drainage basins extended (or were linked) to the Pyrenean axial zone where basement rocks outcrop. The large volumes of varied, generally fine-grained, carbonate rock types reflect significant source areas in the Mesozoic–Cenozoic sequences of the south Pyrenean thrust sheets (Fig. 1b). No major difference in roundness or mean shape is noticeable between clast types of clear axial zone origin or those from the less distant thrust sheets, although roundness is greater for axial zone clasts. In general, more petromict assemblages appear at the top of the Casa Blanca Formation, and persist with significant internal variations until mid levels of the Mirador Formation (Fig. 2), where assemblages are very enriched in limestone and sandstone clast types. Signatures of radically different provenances within the succession, however, have not been detected.

Conglomerate facies

The conglomerates (Fig. 7) are split into two broad categories: (1) unstratified, coarse-grained, matrix-rich conglomerates, which form the modal conglomerate type and (2) variably stratified conglomerates, with varied matrix type and proportion. All facies categories have been analysed in terms of their bed thickness (B.Th) and clast size characteristics, following the methods of Bluck (1967) and Nemec & Steel (1984). The mean maximum particle size (MMPS) for individual beds was obtained by measurement of the *a* (long) axis of the ten largest clasts restricted to a 2 m wide zone about the point of observation. In the case of very thick (>1 m) beds, the ten largest clasts were recorded for uniform divisions of the bed in order to eliminate underestimates of the largest clasts, and to establish or verify any coarse tail grading profile. Any clear outsize clasts were noted, and separate MMPS plots prepared. The scatter plots presented are in a similar form to those of Nemec & Steel (1984, fig. 19) in order to facilitate comparisons with the flow competence–thickness models they discuss.

Unstratified, coarse matrix-rich conglomerates. This category forms the volumetric bulk of the conglomerates, and has been analysed initially without further subdivision on the basis of grading type (see below). Characteristically, unstratified conglomerates occur in beds up to 1.75 m thick, though exceptionally up to 6 m. Beds are dominantly unordered, with modal framework grains of rounded to well-rounded large pebble to small cobble size, often characterized by distinct polymodal distributions (Fig. 7a). The matrix is typically coarse-grained sand to small pebbles, with occasional prominent admixtures of orange-red fine–medium silt, or rarer medium grey calcareous mud. The matrix is frequently rich enough to locally support the framework clasts, most commonly allowing point contacts, and coarse matrix-supported facies are commonly noted (Fig. 8). The polymodal nature frequently makes difficult the distinction between matrix and framework. Clast fabrics, where developed, are most commonly $a(p)a(i)$ imbrication, which tends to be a 'dispersed' rather than a pervasive fabric (Fig. 7b), or $a(p)a(i)$ clast nests (Allen 1981). Well developed sub-horizontal, bedding-parallel alignment of *ab* planes is rare; unstratified conglomerates are much more likely to be chaotically-organized with frequent examples of large cobbles or

(a)

(b)

Fig. 7. Photographs showing the texture and fabric of the main conglomerate and sandstone facies. (**a**) Typical unstratified, large cobble to boulder grade conglomerate. Scale ruler is 0.4 m long. (**b**) Moderately-developed $a(p)a(i)$ imbrication fabric of the unstratified conglomerate facies. Ruler scaled in cm, visible length is 0.22 m. (**c**) Fine matrix-supported medium-large pebble conglomerate (cohesive debris flow), in which orange silt matrix becomes dominant upwards. Top is truncated by a sharp erosional basal surface of the overlying pebble-cobble conglomerate. Member I of the Pont de les Cases Formation. (**d**) Sequence of generally small pebble-grade sheet-like, stratified clast-supported and matrix-supported conglomerates from the lower marker horizon (see Fig. 12), from c. 84–87 m. Notebook is 0.2 m long. (**e**) Draping coarse-grained, granule-very small pebble bearing sandstone, overlying a clast-supported, well-imbricated (probably $a(t)b(i)$ type) pebble- to large cobble-grade conglomerate.

(e)

small boulders with steep or vertical *ab* planes. Fabric development is not thought to be controlled by the availability of discoidal or bladed clasts; such shapes are readily found in all of the coarse conglomerate facies, and are not preferentially sorted as part of lateral facies assemblages of the type generated by processes operating on gravel river bars (e.g. Bluck 1976; Haughton 1989). Rare outsized clasts are frequently concentrated at the top or upper part of beds. A small proportion of unstratified conglomerates display poorly–moderately developed $a(t)b(i)$ fabrics and clast clusters (Brayshaw 1984), and tend to show better clast support systems (Figs 7e & 9).

Unstratified conglomerate beds have flat–sharp and sharp–irregular erosional bases (Fig. 7c), and tend to be sheet-like at the scale of exposure. Basal surfaces are also characterized by erosional longitudinal scours. Conglomerate sheets frequently show steep to vertical erosional margins, less often low-angle (concave-type) channel margins, and equally rarely transitional–gradational contacts to other (finer-grained) facies. Relief on steep margins is frequently of the order of the bed thickness, but may be several metres in rare cases.

A plot of MMPS against B.Th for unstratified conglomerates shows moderate positive linear regression (Fig. 10a), with increasing scatter of the data at higher values of both parameters. The distribution indicates a low value positive y axis intercept, which closely approaches the origin. Plots of the *maximum* clast size versus B.Th give similar results. The very high correlation of MMPS and maximum clast size in beds ($r = 0.94–0.952$, Fig. 10b) points to a lack of outsized clasts; a nearly identical relationship for similar alluvial conglomerate facies was interpreted by Nemec *et al.* (1984) to indicate efficient sorting of the framework clast population. Sorting of the coarse tail component, however, appears to decrease with increasing grain size (Fig. 10c).

Interpretation. The conjunction of significant positive correlations between bed thickness and both MMPS and maximum clast size, with matrix-supported textures, occasional parallel imbrication but largely unordered fabrics suggests a mass flow origin for the bulk of these conglomerates (Bluck 1967; Nemec & Steel 1984). The absence of silt- or clay-rich material in the matrix of the majority of unstratified conglomerates and the trend for data points to pass through the origin of a MMPS:B.Th plot further suggests that the deposits were emplaced as density-modified grain flows (Lowe 1976), equivalent to the cohesionless debris flows of Nemec & Steel (1984). Separate analyses of the small number of clast-supported unstratified conglomerates also show similar positive correlations, and thus imply comparable emplacement mechanisms operated. The highly

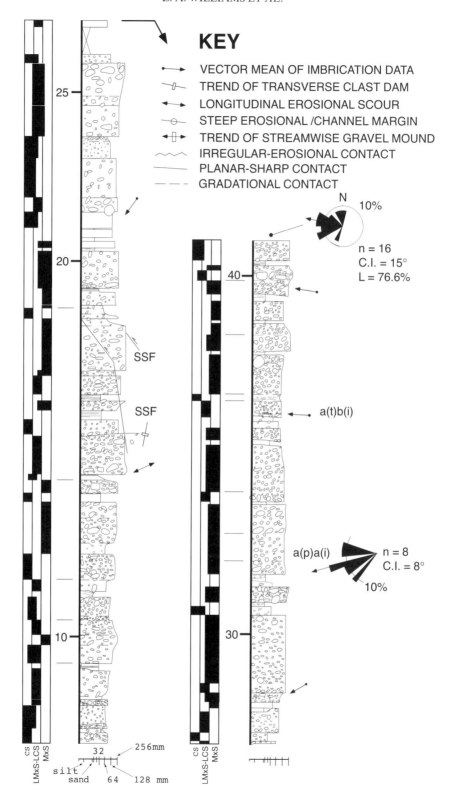

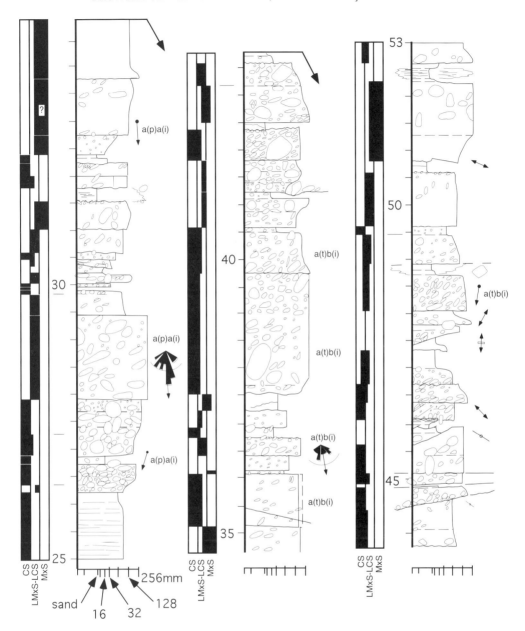

Fig. 9. Graphic log detailing conglomerate facies in member I of the Pont de les Cases Formation. This coarse-grained section above the principal unconformity of the growth structure, contains unstratified, polymodal, coarse matrix-rich conglomerates, which show a suite of grading profiles. Several examples of matrix- to local matrix-supported $a(p)a(i)$ imbricated large cobble to small boulder grade conglomerates. Rose diagrams for imbrication data show vector mean arrow, and the radius of the arc indicates 10% of points.

Fig. 8. Detailed graphic log of characteristic conglomerate facies from the El Bastets Conglomerate Formation (of Period 1). See Fig. 3 for location. Vertical scale for all logs in metres. Filled areas of the left hand columns indicate the clast support system for each bed, where CS is clast-supported, LMxS-LCS is local matrix support to local clast support, and MxS is matrix-supported. The matrix in this section is almost invariably in the range from medium-grained sand to small pebbles. North for palaeocurrents is to the top of the page. Note the pair of minor syn-sedimentary contraction faults (SSF) cutting conglomerate beds at 15–18 m.

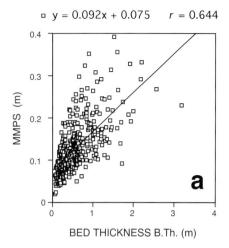
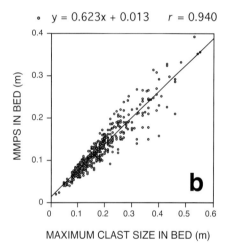
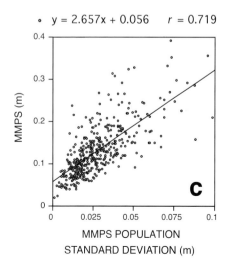

extensive, frequently erosionally based and fine-matrix-poor nature of the gravels imply that these were during single, variable magnitude flood events, rather than typical sediment gravity flows (Blair & McPherson 1994a).

Sub-division of unstratified, matrix-rich conglomerates into grading types (e.g. Nemec *et al.* 1984) has been carried out, and the results plotted in Fig. 11 using the maximum particle-size parameter. This can be used in place of MMPS because of the high correlation between the two parameters (Fig. 10b). Ungraded conglomerates form the largest sub-group ($n = 174$), and show a moderate positive linear correlation (Fig. 11). Separate analyses of unstratified conglomerate categories of different grading profile (top-only normally-graded, inversely-graded, base-only inversely graded, and inversely to normally graded types) all give significant positive linear correlations with r values ranging from 0.64 to 0.73 for different ranges of grain size. These types are generally considered the result of cohesionless mass flows (cf. Allen 1981; Nemec & Steel 1984). The MMPS:B.Th relationship of unstratified normally graded conglomerates ($n = 112$) is considerably less well correlated. This probably reflects the inclusion of polymodal clast-supported (relatively matrix poor) beds, with dispersed type $a(t)b(i)$ fabrics, in the analysis. These are likely to result from bed load transport in stream flows (e.g. Allen 1981; Steel & Thompson 1983). A relatively small population of silt–sand matrix-supported, massive conglomerates have been included in the data analysis summarized in Fig. 11. The poorly sorted texture, chaotic fabric tendency (Fig. 7c) and capability of supporting outsize clasts (e.g. Fig. 12 at 93 m) suggest that these are cohesive (viscous) debris flows.

The particle size–bed thickness data presented in Fig. 11 is broken down into information (a) from Period 1 (Fig. 11a) of the growth structure evolution (structurally beneath the principal unconformity in the measured section, Figs 2 & 4a), and (b) from Periods 2 and 3 (Fig. 11b), from rocks associated with the growth fold

Fig. 10. (**a**) Scatter plot of mean maximum particle size (MMPS) versus bed thickness (B.Th) for the unstratified conglomerate facies up to bed thicknesses of *c.* 3 m, showing regression equation and correlation coefficient (r). Rare cases of thicker beds (up to *c.* 6 m) have not been included but give a similar correlation ($y = 0.099x + 0.071$, $r = 0.729$). (**b**) Plot of maximum particle size against mean maximum particle size in all unstratified conglomerate beds. (**c**) Plot of mean maximum particle size against population standard deviation of the MMPS for the same facies.

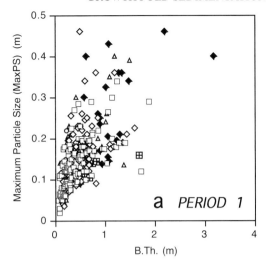

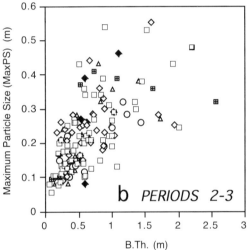

Fig. 11. Plots of maximum particle size (MaxPS) against bed thickness (B.Th.) for the unstratified conglomerate facies sub-divided by grading profile for (**a**) beds of Period 1 within the vertical panel beneath the principal unconformity ($n = 275$), and (**b**) beds of Periods 2 and 3 from the common limb and anticline of the growth structure ($n = 145$). Both show well-developed positive correlations, suggesting similar emplacement mechanisms operated for conglomerates of the vertical panel, and those above the principal unconformity.

pair. This is in order to test whether apparently similar conglomerate facies had similar flow competence-thickness relationships (and magnitudes) when deposited in different regions of the palaeosurface affected by the growth structure, possibly under the influence of contrasting syn-depositional surface deformation. The similar distributions revealed (Fig. 11a, b) suggest that the same emplacement mechanisms operated during Period 1 as in later periods for which there is data, arguing that there was a continuity of processes (and probably environmental conditions) throughout the history of the structure.

Stratified and sand-rich conglomerates. Of this category, the most important representatives are horizontally stratified, and (low-angle) inclined stratified conglomerates, both of which tend to be clast-supported and normally- to un-graded (Figs 7d, 9 & 12). The horizontally stratified facies tend to show basal lags of large pebbles and cobbles above irregular-sharp erosive surfaces, in beds of typical thickness range 0.2–1.5 m. Low-angle (inclined) stratified conglomerates are often heterolithic, in containing lenses or layers of pebbly sandstone, and also show *ab* planes of bladed clasts sub-parallel to the inclined layering. Evidence of 'lateral accretion', from the parallel directions of basal erosional structures and the strike of inclined layering is evident in a number of examples. Both facies are transitional to sand-rich (pebbly) conglomerates, which may be flat-bedded, contain inclined strata or structureless. Large-scale cross-stratified conglomerate facies are very rare throughout the succession.

Interpretation. Horizontally stratified conglomerates are interpreted as the product of rolling of bedload beneath stream flows. The sand-rich gradational version of this facies probably involved the simultaneous transport in suspension of relatively fine gravel and coarse sand. Conglomerates with inclined stratification indicate the existence of significant depositional relief in local sub-environments, and probably represent the flanks of longitudinal bars. A similar designation can be made to the horizontally stratified conglomerates, which would constitute the low-angle bar tail region (Hein & Walker 1977; Nemec & Postma 1993), although lateral facies transitions do not reveal bar head conglomerate facies.

Sandstone facies

Sandstones occur in three associations: (i) as thin (<0.6 m), typically laterally impersistent

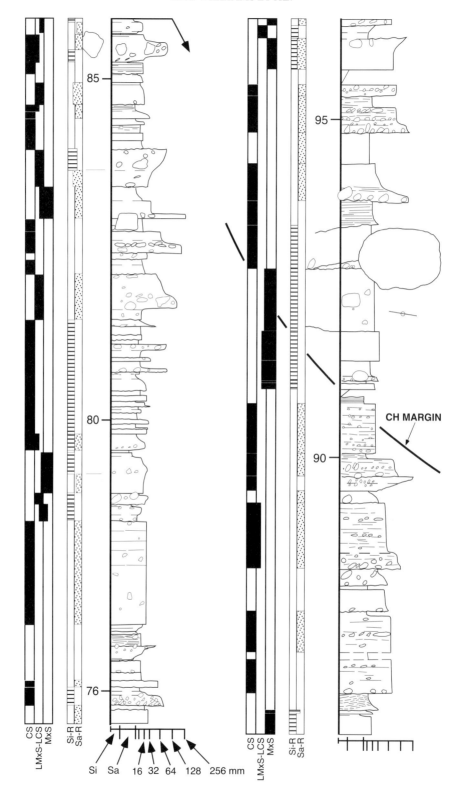

gradational caps on conglomerate beds, (ii) as decimetre- to metre-scale bedsets, associated with siltstone and sandy conglomerate facies, and (iii) reddened silt-rich sandstones laterally-equivalent to fine matrix-supported conglomerates.

Sandstones are typically medium to very coarse grained, and frequently contain dispersed ('floating') granules and small-medium pebbles, which may also be concentrated in impersistent layers. Those occurring as conglomerate caps (Figs 8 & 9) tend to be coarser grained and fine-gravel bearing (Fig. 7e). This type is frequently horizontally parallel-laminated, but forms a continuum to three-dimensional non-parallel undulatory–sinusoidal stratification, observed in rare cases as lateral transitions. Sinusoidal–undulatory sandstones may incorporate solitary outsize (cobble-grade) clasts, and thin lenses of fine pebbly gravel. Occasionally, primary convex-upward tops, and isolated domes or hummocks above conglomerates are preserved. This sub-facies clear connection with often normally graded, clast-supported conglomerates suggests emplacement from standing wave-antidune type bedforms in supercritical flows, and shallowing high velocity flows, to give parallel-laminated and massive beds. In this connection it is notable that very few of the sandstones transitional from conglomerates throughout the succession are (large-scale) cross-stratified as has been commonly reported in other successions (e.g. Allen 1981; Todd 1989).

Medium–fine-grained sandstones occurring in thick packages are massive to horizontally planar parallel laminated. They are commonly highly reddened, and show evidence of strong bioturbation by ?*planolites*- and *cylindricum*-type burrowers, and unidentified grazers. These sandstones are interpreted as the product of shallow unconfined flood events, which accumulated in regions bypassed by major gravel input. The third association of distinctive silt-rich sandstone facies contains floating, unordered gravel clasts, and has highly irregular, gradational contacts to conglomerate beds, and clearly suggests cohesive sandy ('clast-poor') debris flows.

Siltstone facies

This volumetrically rare facies occurs as two contrasting sub-facies: (i) structureless or blocky red-orange siltstones, in thick (metre-scale units) or very thin 'drapes' and (ii) poorly-sorted sand- (and more rarely granule-) bearing siltstones, associated with other fine matrix-bearing or matrix-supported sandstones and conglomerates. Sub-facies (i) is interpreted as repeated suspension-deposit events, subsequently weakly pedogenically modified. This is in frequent association with medium–fine-grained parallel-laminated sandstones, organized in thick (metre-scale) units of alternating facies. Sub-facies (ii) is considered to represent viscous mud flows, the dominant silt matrix having the strength to support sand and very fine gravel.

Palaeocurrents

Sediment dispersal within the system was investigated by the measurement of a suite of primary structures considered to represent flood-stage palaeoflow. The most common structures for which palaeocurrent directions were determined were longitudinal erosional scours on conglomerate bases, steep margins of conglomerate beds, incised channel margins, conglomerate clast imbrication, clast *a*-axis alignment, obstacle scours, cross-stratification, and primary current lineation in sandstones. Because of the complex and systematically varying bedding orientations intrinsic to the Sant Llorenç de Morunys growth structure, data were individually stereographically restored to horizontal using locally collected bedding data. Corrections for the variably oriented fold axes were not performed because of the low (1–5°) plunge values for sub-areas of the structure (Ford *et al.* 1997). Derivation of palaeocurrents from conglomerate fabrics involved direct measurement of *ab* planes and *a* axes of representative bladed–discoidal clasts in single beds. Following structural restoration, the inferred palaeoflow direction for each clast was determined by the dip-azimuth of *ab* + 180° and the vector mean and magnitude of the sample calculated. Rayleigh tests were routinely applied, and vector means rejected if the test was failed at the $p = 0.05$ level of significance (Curray 1956). The spatial relationship of *ab* and *a* was stereographically assessed, following restoration, to distinguish $a(p)a(i)$ and $a(t)b(i)$ type fabrics. Palaeoflow from the preferred orientation of clast *a*-axes, was

Fig. 12. Detailed graphic log of the lower-mid section of the lower marker horizon (Pont de les Cases Formation). This horizon contains a mixture of (i) small-medium pebble, thinly-bedded horizontally-stratified conglomerates with sub-angular framework clasts, and (ii) fine-grained granule to pebble, silt matrix-rich to matrix-supported conglomerates. Filled areas of the far left hand columns indicate the clast support system (abbreviations as for Fig. 8), and ornamented areas of the central columns specify the predominant matrix type, Si-R, red-orange coloured silt/mud-rich; Sa-R, sand- (plus granule–very small pebble gravel-) rich.

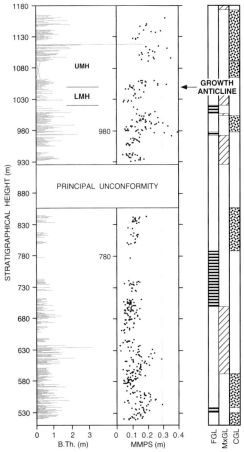

Busa Syncline northwards to the El Bastets Formation outcrop (Fig. 4). Information is thus common, in the west of the structure, in association with the steep panel and growth anticline, whereas in the east it applies mainly to the growth syncline and to the steep common limb (Fig. 5c). It is thus currently not viable to compare palaeoflow in different limbs of the growth fold pair *directly* across strike, but this can be done for the structure as a whole.

Examples of the local variability of palaeoflow in relation to facies units and structures are shown in Figs 8 and 9. Palaeoflow in these cases remained consistent over vertical intervals of 10–25 m. No systematic switching of palaeocurrents (e.g. Haughton 1989, fig. 8) is clearly evident between stacked facies units, or in rare gradationally bounded conglomerate packages. The texture, facies structure and organization of the conglomerates (see above) are unlike those of low-sinuosity, gravel-bed permanent rivers in which palaeocurrents are frequently at high angles to the channel direction (regional palaeoslope; Bluck 1976, figs 19 & 20), suggesting that differing palaeocurrent modes (see below) do reflect differing palaeoslopes within the growth structure.

Grain-size variations

Details of the vertical variation in conglomerate bed thickness and corresponding data on the mean maximum clast size are shown in Fig. 13. Vertical cycles based on grain size (MMPS) are not clearly developed on the scale of succession illustrated (El Bastets to Pont de les Cases Formations). Intervals of c. 100 m show neutral trends in MMPS, and are not always affected by boundaries of formational (or lower-order) lithosomes of contrasting grain size. However the obliquity of the measured section with respect to the geometrically complex lithosomes partly accounts for the complexity of the vertical trends. A major increase in MMPS and a corresponding increase in bed thickness occurs approximately 50 m above the principal unconformity (Fig. 13). This abrupt increase is not compatible with definition of a gradational coarsening-upward megasequence; indeed organized sequences of this scale are not immediately clear from the successions preserved in the growth structure. The effects of the principal unconformity on laterally equivalent and overlying conglomerates were limited due to the restricted time span over which it operated within the growth structure. It is therefore not thought to be responsible for recycling

Fig. 13. Combined bar graph of bed thickness (B.Th) and plot of mean maximum grain size (MMPS) against stratigraphical height for the principal logged section in Vall de Lord (Fig. 3). Broken lines in the bar graph denote individual bed thicknesses at similar stratigraphical levels not resolvable at this scale. Also shown are positions of the principal unconformity, position of the growth anticline axial surface in the section, and the Upper and Lower Marker Horizons. Right hand columns characterize the bulk grain size of the lithosomes represented in the section: FGL, fine-grained; MxGL, mixed grain size, and CGL, coarse-grained (conglomeratic) lithosome.

determined using measurements made in the bedding plane.

Data were collected from a large area of the structure. Much information is derived from the principal logged section, spanning the region between profiles 1 and 2 (Fig. 4), whereas a more areally spread dataset exists in the region between profiles 2 and 3 from the trace of the proximal facies to build medial and distal parts of lithosomes on a significant scale. Analysis of the grain size data on a smaller scale reveals neutral, upward-coarsening and upward-fining cycles of variable thicknesses ranging from 6.5 to 25 m. These are broadly comparable in scale to the sequences, small-scale cycles and fifth-order lithosomes respectively of Heward (1978, fig. 11), Gloppen & Steel (1981, fig. 17) and DeCelles *et al.* (1991*a*, figs 2 & 12), although not in detailed facies organization. These are interpreted as alluvial fan-trench (incised fan-channel) deposits (Heward 1978; DeCelles *et al.* 1991*a*) and progradational fan-lobe deposits (Gloppen & Steel 1981).

The restricted horizontal extent of the measured section in growth strata above the principal unconformity does not allow specification of trends in lateral grain size variation. Therefore it is not currently possible to make inferences on fan scale from rates of grain-size change.

General environmental model

The modal conglomerate making up, in particular, the conglomerate-dominated lithosomes were emplaced during moderate- to high-magnitude flood events charged with high concentrations of gravel and sand (cohesionless debris flows). Associated fine-grained facies, conglomerate textures and matrix distribution imply that the majority of this conglomerate group was emplaced under sub-aerial conditions. In only a few cases where conglomerate beds show distinct upward increases in matrix volume, considered by Nemec & Steel (1984, p. 16) to indicate sub-aqueous deposition, and are associated with ephemeral lacustrine facies is this tendency excepted. The sheet-like bed geometry, laterally extensive in depositional dip and strike directions for distances of the order of 10^2 m in recent cliff and valley walls (e.g. Fig. 6a, b), suggests a frequently unconfined (sheetflood) mode of deposition. Evidence of laterally equivalent steep to sub-vertical conglomerate margins, almost certainly underestimated during data collection, indicate that the flood events had considerable erosional potential, and were in many cases emplaced as wide, but laterally confined bodies. The larger relief instances of these wide rectangular-profile channels resemble incised fan head channels (Blair & McPherson 1994*b*), and there is evidence (see below) that some existed as stable features, presumably supplying fan lobes, and were subsequently filled/buried by later sedimentation events. Inferred rectangular-profile channels scaling approximately with bed thickness are more likely to have formed simultaneously with low-frequency–high-magnitude floods, rather than have been cut during normal conditions by sediment-poor water flows. The latter are secondary fan processes (Blair & McPherson 1994*a, b*) forming braided-channel networks, likely to produce lower-angle erosional surfaces. Subaerial high viscosity (cohesive) debris flows contributed comparatively small volumes of sediment to the growth strata, and occurred more frequently during later periods of growth history. This facies is commonly associated with proximal–medial zones of steep alluvial fans (Bull 1972; Rust & Koster 1984; Blair & McPherson 1994*a*). Sandstone facies associated with the modal conglomerate type were typically deposited under upper flow regime to supercritical flow conditions. This is compatible with typical hydraulic conditions operating over alluvial fan surfaces (Blair & McPherson 1994*a*).

Stream-flow-related conglomerates are distributed throughout the succession of the growth structure, and are considered to represent a volume of the sediment that was due to relatively high-frequency–low-magnitude flow events, which may have been localized in wide-shallow channels. The sequential and textural disorganization of the several facies attributed to traction currents, and particularly the absence of large-scale cross-strata due to mesoscale bedforms or bars (*sensu* Bluck 1979) indicates that permanent, low-gradient rivers were not responsible for the alluvium accumulated in the growth structure.

The assemblage of sub-environments and processes involved suggest a type II alluvial fan setting (terminology of Blair & McPherson 1994*a, b*) for the sediments of the Sant Llorenç de Morunys growth structure. This fan type is defined by dominance of unconfined fluid-gravity sedimentation events, in this case characterized by cohesionless mass flows of the range discussed by Nemec & Steel (1984, fig. 15), including cohesionless debris flows and fluidal gravelly-sediment flows. Mean radial slopes of type II fans are between 2 and 8° (Blair & McPherson 1994*b*, p. 394), although a clear majority of alluvial fan types have surface inclinations of 2–5° (Blair & McPherson 1994*a*, fig. 4). Such fans (e.g. Heward 1978) are designated as 'steep', in order to distinguish them in terms of nomenclature from gravelly 'wet-type' alluvial fans (McPherson *et al.* 1987) or 'megafans', in which conglomerates are deposited in permanent (and possibly antecedent) river channel belts (Haughton 1989; Burbank *et al.* 1996). The essentially sub-horizontal to very low-angle fan surfaces have implications for (1) the potential for deflection of palaeoflow and (2) sequential structural restoration of the growth structure. The very low-angle surfaces inferred for this environment (angles usual for depositional growth strata, Burbank *et al.* 1996) favour the deflection of palaeoflow, and in particular axial deflection, by surface perturbation caused by growth folding. Considerably greater deformation would be required to disturb transverse flows crossing steeper surfaces. As it is not possible to specify the true palaeo-depositional dip of the growth strata, it was considered valid to use the horizontal in the restorations (Ford *et al.* 1997), knowing that surficial dips were probably ≪5°.

Although affected by topographic diversion due to fold growth, this proximal zone allowed large volumes of gravel to be dispersed southwards beyond the axial trace of the Puig-reig anticline (Fig. 1b). In the simplest case, involving sediment dispersal approximately perpendicular to the mountain front, this across-basin transport of gravel indicates a fan (or fans) which had a radial length of approximately 16 to 20 km. This is compatible with typical alluvial fan lengths collated by Heward (1978, fig. 3) of between 5 and 20 km, and also Blair & McPherson (1994*a*, fig. 2) who indicate a range of ≤10–15 km). A value at the upper limit of *typical* fans is qualitatively in keeping with the regional-scale drainage basin which supplied this system.

Three-dimensional evolution of the growth fold

The sequential restoration of three cross sections (Fig. 5) in up to nine steps presented in Ford *et al.* (1997) provides a model for the development through time of the Sant Llorenç de Morunys growth fold. Restoration is based on classical section balancing techniques (Dahlstrom 1969; Hossack 1979), driven principally by line length balancing, but with constant volume control. Successively older stratigraphical boundaries that approximate bedding are restored to horizontal about a hinterland pin-line. The pin-line was positioned such that negligible distortion was generated across the complex geometries of the anticlinal crests. This was dictated by the observation that in this hinge zone major and minor unconformities record no displacement or distortion. Based on these restorations we attempt here to reconstruct the three-dimensional palaeogeographical environment and lateral variations in fold growth history at Sant Llorenç de Morunys. Based on field observations (summarized above), sequential section restoration and numerical modelling it was concluded in Ford *et al.* (1997) that the Sant Llorenç de Morunys folds grew predominantly by limb rotation but that limb lengthening also occurred, induced principally by the deposition of dip-parallel wedge-shaped sedimentary bodies (formational lithosomes).

Using stratigraphical architecture (e.g. offlap–onlap–overlap), principally recorded in the crestal regions of the growth anticline, four periods of fold growth have been recognized. In this study we depend entirely on growth geometries as, currently, no control on absolute time intervals is available. Within the growth syncline we have to assume ongoing sedimentation, as periods of non-deposition or even erosion generally cannot be recognized. Stratigraphical geometries are controlled by the interaction of sedimentary processes and growth of the fold pair (Burbank & Vergés 1994; Artoni & Casero 1997). We attempt to track this relationship on a very large scale by a graph of time against the relative rates of sedimentation (R_s) and fold growth (uplift, R_u) for each of the profiles (Fig. 14a). Vertical movements due to fold growth are relative to the sedimentation base level. As the fold developed at the Earth's surface R_u is the rate of surface uplift as defined by England & Molnar (1990). Increased R_u relative to R_s can lift the crest of the anticline above base level inducing erosion. In the syncline the same change will simply reduce the amount of sedimentation. Thus the anticlinal crest is the most sensitive area of the growth fold to changes in R_s/R_u. On a short time-scale sedimentation and tectonic activity are periodic in these environments (e.g. Wolman & Miller 1960; Heward 1978; Suppe *et al.* 1997); however these fluctuations are not visible on the longer time scale implied in Fig. 14a. When the anticlinal closure is considered in more detail (see below) the episodic nature of both sedimentation and fold growth can be detected with packages of sediment separated by unconformities that are subsidiary to the principal unconformity. The gross evolution of the Sant Llorenç de Morunys growth fold which is visible and analysed in this paper involves a general decrease in fold growth with respect to sedimentation, as previously detected by Riba (1973). In other words fold activity was dying out. This process was diachronous so that folding was still active to the east while to the west the inactive part of the structure was being buried. An idealized growth fold showing four stages of fold demise (A to D) is represented in Fig. 14b. Although these four

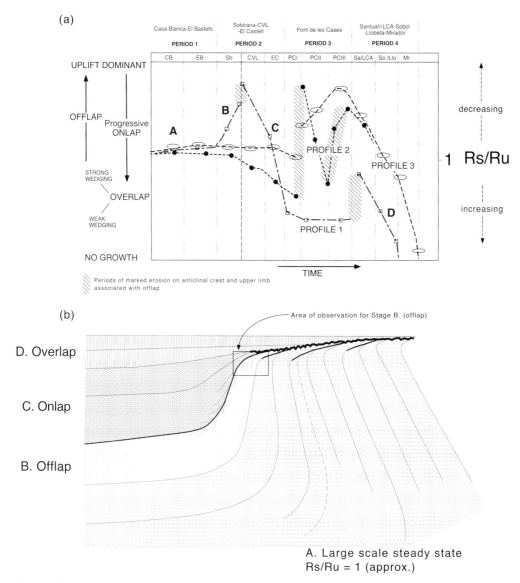

Fig. 14. (a) Graph showing the ratio of rate of uplift (R_u) to rate of sedimentation (R_s) for the three cross sections through the Sant Llorenç structure. (b) Schematic growth fold showing four stages of fold demise (A–B). These stages are seen in all three cross sections of the Sant Llorenç de Morunys growth fold but at different times due to the diachroneity of fold demise.

stages of fold demise can be recognized on each profile, they do not correspond to the four periods of evolution of the Sant Llorenç de Morunys fold pair as a whole because fold demise was diachronous along strike. In our idealized fold, the system is relatively stable during stage A with R_s and R_u approximately equal but fluctuating. The anticlinal axial plane would therefore lie either directly above or below (as in Fig. 14b) the principal unconformity. During stage B a prolonged decrease in sedimentation rate with respect to uplift rate causes a major period of growth offlap followed in stage C by an increase in R_s relative to R_u causing growth onlap. Finally in stage D, overlapping strata show gradually weaker wedging indicating a decrease in tectonic activity. Variations on this general model which represent variation in sedimentation and tectonic processes during fold development across the region will be discussed below.

Block diagrams representing palaeogeography during periods 2 to 4 have been constructed to illustrate the three-dimensional variations in sedimentation, erosion and fold development. The diagrams are based on stages in the sequential restoration of the three cross-sections. The diagrams show the Cadí thrust sheet, comprising Eocene limestones, advancing toward the foreland and a transfer zone (Schumm 1981) between the thrust (mountain) front and the fall line. We show a hypothetically located bedrock exit canyon in the Cadí thrust sheet to emphasize that large volumes of exotic sediment were supplied to this region over the time scale of an epoch. The nature of the detritus suggests linkage to a large-scale drainage basin, and therefore probably a sub-permanent exit canyon of the type related to the half width-scale of the orogen (Hovius 1996). To the south of the fall line the depositional surface is underlain by the growing fold pair. There are several uncertainties involved in these models, in particular in the character and size of the sediment transfer area between the fall line and the tip line of the Cadí thrust sheet to the north which controlled the palaeotopographic mountain front. The width of the transfer zone must however have been sufficiently large, and persisted over the entire history of the growth structure, to have excluded any highly proximal alluvial fan facies, such as avalanche, rock slide, rock fall deposits, talus cones and colluvial slides (Blair & McPherson 1994b), from the depositional area. While we cannot constrain on what time scale deposition and erosion occurred across this zone, it is likely that these parameters were in long-term equilibrium, for the following reasons. (1) Incised upper fan channels supplying the Berga Conglomerates must have traversed this region and switched regularly to judge by the contrasting lithosomes and their palaeoflow (see below). If sediment did not completely by-pass this zone, aggradation would have proceeded over wide areas. (2) No petrographic signature of the Middle Eocene marine rocks, incorporated into the steep panel beneath the transfer zone, is known in the Berga Group, favouring net aggradation in this zone.

Sediment dispersal within the depositional system as recorded by palaeocurrent data for each period is integrated with the palaeogeographical models. These data provide information on the interaction between the growing fold and the depositional system.

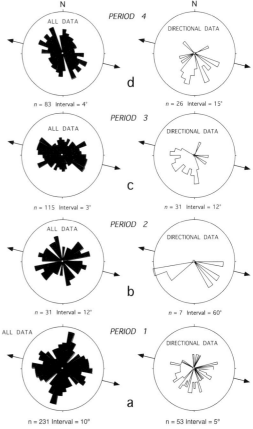

Fig. 15. Rose diagrams of combined directional and non-directional palaeocurrents for (**a**) Period 1 (Casa Blanca to Sobirana Formation times), (**b**) Period 2 (Camps de Vall-llonga–El Castell Formations), (**c**) Period 3 (Pont de les Cases Formation) and (**d**) Period 4 (Les Cases Altes–Mirador Formations). n refers to the number of readings, Interval is the class interval of the histogram, and the radius of the circle is 10% of points in all cases. Arrows outside the circles indicate the trend of the growth fold axes (104-284°). The vector mean for the aggregate data for Period 2 (b) is 085°, although the very low Von Mises concentration parameter k (= 0.39) indicates that this is not statistically significant.

Period 1 (Casa Blanca–El Bastets Formations)

In this paper we redefine this period to include only the Casa Blanca and El Bastets Formations (cf. Ford et al. 1997), because the Sobirana Formation records a clear decrease in R_s/R_u (Fig. 14a) heralding a change in growth fold dynamics. The period 1 succession forms part of the steep panel (Fig. 4) and thus reveals little about growth geometries. As the fold closures are not preserved (or exposed) there are few data

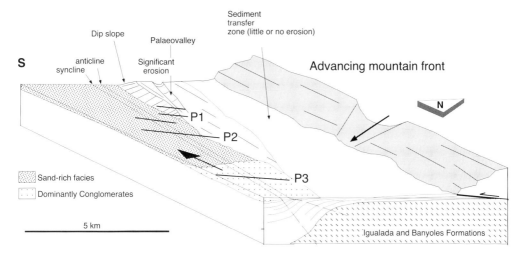

Fig. 16. Block diagram of the palaeogeography during the Camps de Vall-llonga Formation (Period 2). The position of the Cadí thrust sheet is unconstrained. In all reconstructions we show some palaeotopography on this limestone-dominated thrust sheet, including a hypothetical bedrock canyon providing a conduit for erosional material from the drainage basin to the north to the foreland basin to the south. The actual position of this conduit is unknown. This block diagram is not constrained by restored cross sections and is therefore the most schematic of those presented. The mean palaeocurrent direction is obliquely axial as shown by the large arrow. Positions of the three profiles are for reference. All reconstructions are to scale.

available to indicate the character of the fold growth before deposition of the Sobirana Formation and therefore no reconstruction has been attempted for this period. However the down-dip thickening of parts of these sequences (e.g. El Bastets Formation) are interpreted as sedimentary and indicative of growth during earlier evolution of the fold (Ford *et al.* 1997). We envisage that the growth fold developed steadily throughout this period and thus the anticlinal hinge may have described a sub-horizontal array of en echelon segments which lies above the present day erosion surface, (Fig. 14a; as in stage A of the idealized fold development in Fig. 14b).

The Casa Blanca Formation comprises a major conglomeratic lithosome in the eastern sections of the growth structure, which expanded to the WNW with time (Fig. 2). These conglomerates are laterally equivalent to a sandstone-rich unit from which the palaeoflow data was measured (see below). On a larger scale, the Casa Blanca and El Bastets Formations show a mutual thickening–thinning relationship (Fig. 4), in which the El Bastets Formation thickens to the ESE. The large-scale wedge-shaped units may represent individual fan units, bounded by surfaces which are apparently gradational and diachronous.

Sediment dispersal. Data for period 1 (Fig. 15a) indicate that two distinct directions of palaeoflow operated: (i) a transverse mode with a mean vector to the SSW and (ii) an axial mode of greater dispersion sub-parallel to the orientation of the growth fold axes. The latter mode suggests that older lithosomes responded to controls on dispersion similar to those which operated in the younger units where the growth folds are exposed. These steeply dipping strata would have been deposited to the south of the axial trace of the growth anticline. Directional palaeoflow data for period 1 (Fig. 15a) shows that the axial mode was west- and WSW-directed, though small numbers of data indicate a minor opposite flow. The transverse mode was SSW to SSE.

The bulk of the data for period 1 are restricted to the west in the Casa Blanca and El Bastets Formations. Examination of these data stratigraphically (Fig. 19) shows an upward-increasing tendency towards axially oriented flows. The upper El Bastets Formation is the first unit to show this dominance. The Casa Blanca Formation palaeoflow indicates a weak WSW–ENE mode, slightly oblique to the statistical fold axis orientation. Despite the tendency to axial flow dominance later in period 1, the palaeocurrent distributions in *c.* 50 m vertical sections suggest that individual transverse and axial mass flow and other flow type events occurred approximately in grouped packages, superimposed by fewer variant directions.

Period 2 (Sobirana–Camps de Vall-llonga–El Castell Formations)

The Sobirana Formation (Fig. 2) is an extensive sheet-like unit, approximately 100 m thick, traceable across the steep panel (Fig. 4). In the west (Vall de Lord) the Sobirana Formation shows offlap/apical wedge geometries (Ford *et al.* 1997) indicating that here R_s/R_u was decreasing (Fig. 14a). Erosion was probably ongoing to the north. As shown on profile 1 (Fig. 5a) the principal growth unconformity appears to merge southward with bedding at the top of this formation. Strata in similar positions have been eroded on profiles 2 and 3 (Fig. 5b, c).

After the deposition of the Sobirana Formation, the Camps de Vall-llonga Formation, a sand-rich sequence with limited exposure in the Cardener river bed, appears to record a significant and abrupt offlap on profile 1 followed by gradual onlap. Prolonged erosion, probably accompanying the offlap event (i.e. a period of non-deposition) and deposition of this unit, occurred to the north, further developing the principal unconformity. The 50 m deep strike-parallel palaeo-valley (Fig. 5a) seen on profile 1 was eroded around this time. The importance of this period of non-deposition and erosion further north can be appreciated when the level of the anticlinal closure on either side are compared. Above the unconformity the anticlinal closure lies at or just below river level in the Camps de Vall-llonga Formation (Fig. 5a) while the anticlinal closure for the older Sobirana Formation must have lain above the present position of the principal unconformity. On profile 2 anticlinal closures are preserved just below the principal unconformity in the Camps de Vall-llonga Formation, indicating that the growing fold pair evolved steadily with the level of erosion just above the level of the anticlinal closure (Fig. 5b) and the fall line tracked southward some distance north of the migrating anticlinal hinge. This implies that R_s was close to R_u (Fig. 14a, b). The anticlinal closure for this stratigraphical level has been eroded from profile 3 (Fig. 5c), however the Camps de Vall-llonga Formation is dominated by conglomerates in the east (Tresserres region, Fig. 4). Figure 16 schematically represents the palaeogeography for the Camps de Vall-llonga Formation showing the predominance of erosional features, offlap and sand-rich facies in the west, slight overlap in sand-rich facies in the centre and the incoming of conglomeratic facies in the east. The distribution of these features indicate that R_s/R_u increased eastward, either due to increasing sediment supply or decreasing tectonic activity. The facies change from west to east suggests that the western part of the basin may have been starved of sediment at this time while the main influx of conglomerates was further east.

On profile 1 onlapping geometries in the El Castell Formation record an increasing R_s/R_u (Figs 14a, 17 & 18). The formation eventually overlapped the principal unconformity and filled the palaeovalley along with a 25–30 m thick breccio-conglomerate unit while the growth fold pair to the south controlled wedging geometries. The El Castell Formation on profile 2 abruptly overlapped the anticlinal closure and was deposited horizontally across the upper limb (Fig. 17). This formation shows considerable thickening and numerous internal unconformities across the anticlinal closure (Figs 17 & 18c) suggesting that the anticlinal hinge zone was the focus of periodic, local erosional events. The El Castell conglomerates die out to the east (Fig. 4) and are replaced by sand-rich facies. Figure 18b schematically represents the palaeogeography at the end of El Castell times showing that, by a steady increase in R_s in the west and centre, topographies created during the earlier sediment-starved phase were infilled and overlapped to produce a smoothed depositional surface. Note that the polarity of the conglomerate to sand-rich facies change has changed in this formation indicating that the main input point for the El Castell Conglomerate lay around profile 1 or further west.

Sediment dispersal. Sediment dispersal during period 2 (Fig. 15b) was both sub-parallel to the growth structure and obliquely transverse to it. Palaeoflow data for the early part of period 2 is derived from the sandstone-rich lithosome of the Camps de Vall-llonga Formation, on the steep fold limb (i.e. originally south of the growth anticline) located in the central to eastern-central part of the structure (Fig. 19). Flow was obliquely axially-directed (Fig. 19), with the principal component to the WSW and a minor component to the ESE, although this is based on limited data. This apparent switch of flow direction may in this case be a function of flow into an interfan (sand-rich) low. The presence of a *c.* 50 m-deep valley on the principal growth unconformity incised into steep beds of the El Bastets Formation on the western side of the Vall de Lord (Fig. 5a) is evidence of a sub-axial supply of sediment during Camps de Vall-llonga time. The palaeovalley is exposed in an approximately transverse section, and suggests that in degradational regions north of the growth anticline flow was sub-parallel to the

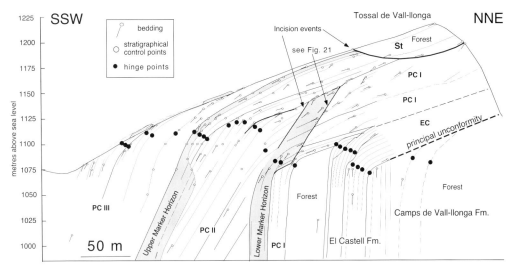

Fig. 17. Detailed cross-section of the growth anticline hinge region on the eastern side of Tossal de Vall-llonga. This shows the stratigraphical-structural relationships generated across the anticline principally during Pont de les Cases Formation times (Period 3 – post-El Castell Formation). Note the interaction of incision and sedimentation events with deformation of the topographic surface controlled by the location of the axial surface trace of the growth anticline. The principal unconformity terminates basinward within the El Castell Formation.

structural grain. During late period 2, palaeoflow in the western El Castell conglomeratic lithosome was across the trend of the growth structure, but not strictly transverse (Fig. 19). These measurements were taken in the back limb of the growth anticline, i.e. in sediments deposited to the north of the growth fold pair. Distinctly axial modes are not developed, instead a mode with a vector mean to the SW is prominent. Stronger transverse flow many have been facilitated by the burial of earlier palaeotopography by the overlapping El Castell Formation.

Period 3 (Pont de les Cases Formation)

The growth fold geometries of the Pont de les Cases Formation are the best preserved of the study area and hence a more detailed description of fold development can be presented for this period. These strata record clear variations in growth geometries and facies across the region (Fig. 5) and represent the most tectonically active period in the centre and east of the structure.

In the west (profile 1), the growth fold pair was overlapped and all units show weak wedging across the fold hinges during period 3 indicating relatively high (and increasing?) R_s/R_u (Figs 14a & 20a). Units appear to have been deposited horizontally and with constant thickness across the upper limb of the growth anticline indicating little rotation of this limb.

Further east (profile 2), accelerated fold growth provided more accommodation space in which a thicker Pont de les Cases Formation accumulated. Probably one of the best exposed growth anticlines in the world is found on Tossal de Vall-llonga (Fig. 4; profile 2, Fig. 5b), the details of which are presented in Fig. 17. Here the lowest member, PC I, overlapped across the upper limb of the anticline but was subsequently largely incised and eroded from the anticlinal crest (Figs 17 & 21). When restored, the strike-parallel incision has a maximum dip of 35° south, a minimum height of 40 m but does not extend beyond 200 m to the west. A 50 cm thick, reddened, matrix-supported breccia lies along the erosion surface. This palaeoriver margin was probably formed by flow parallel to the anticlinal crest by a channel localized on the hinge. The southern margin of the channel appears to lie some 150 m down dip in a poorly exposed area of the steep limb where the boundary of the lower marker horizon steps southward. This large channel relief was infilled and buried by the lower marker horizon, a limestone rich breccio-conglomerate (Fig. 21) which extends beyond the channel margins. The areal distribution of this unit suggests that it was deposited as a small fan lobe-like body (Fig. 22), which locally infilled this channel feature, reaching a

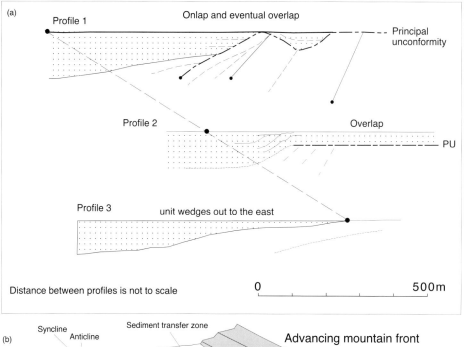
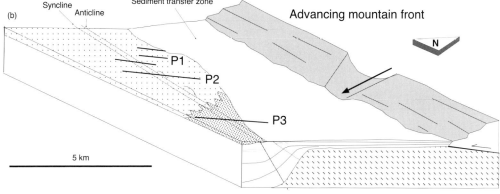
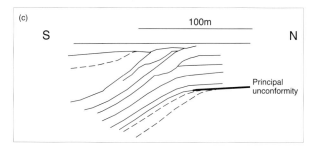

Fig. 18. (**a**) Restored geometry of the El Castell Formation from profiles 1, 2, and 3, linked by pin line. (**b**) Block diagram of the palaeogeography during the El Castell Formation (Period 2) showing on profile 1 onlap of previously formed topography and eventual overlap to fill the palaeovalley to the north. On profile 2 the formation overlapped the anticline continuously. The unit wedges out to the east. See Fig. 16 for key to patterns. (**c**) Detail of restored geometry of the El Castell Formation on the hinge of the anticline on profile 2.

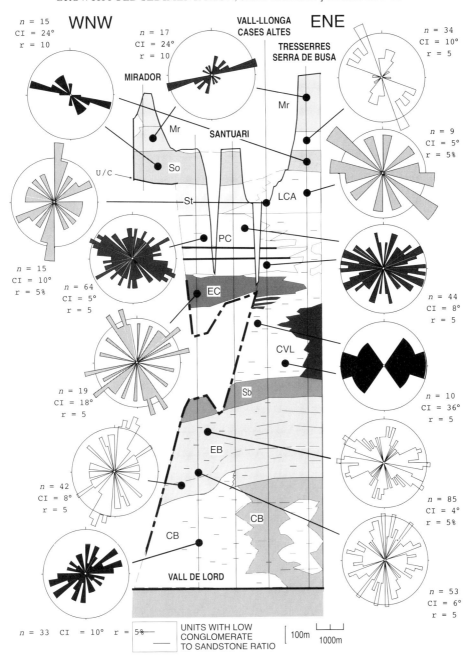

Fig. 19. Palaeoflow data subdivided stratigraphically and spatially through the growth structure. All rose diagrams display palaeocurrents in a non-directional form; north in all cases is to the top of the page. n refers to the number of data, CI is the class interval of rose diagram. The radius (r) of the scale circles is either 5% of points or the specified value of r times the normal distribution. Refer to Figs 2 and 3 for more details of the Berga Conglomerate Group lithostratigraphy.

thickness of 40 m. Elsewhere the unit is only 5–10 m thick. Palaeoflow in the upper beds of the lower marker horizon adjacent to the palaeo-river margin show transverse flow. The lower beds of member PC II record a later period of incision and infill localized on the

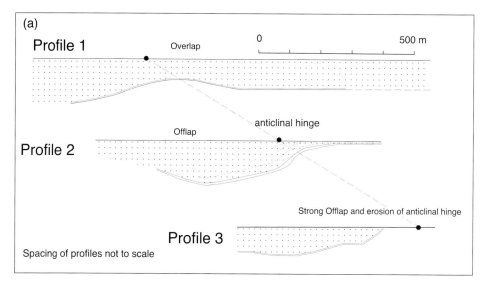
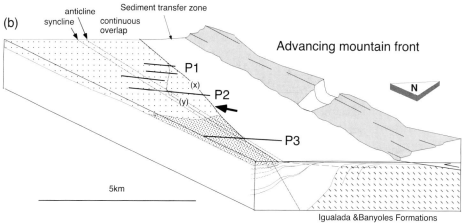
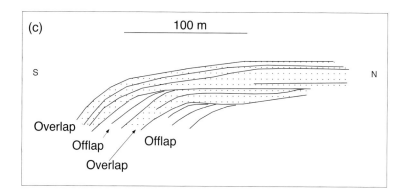

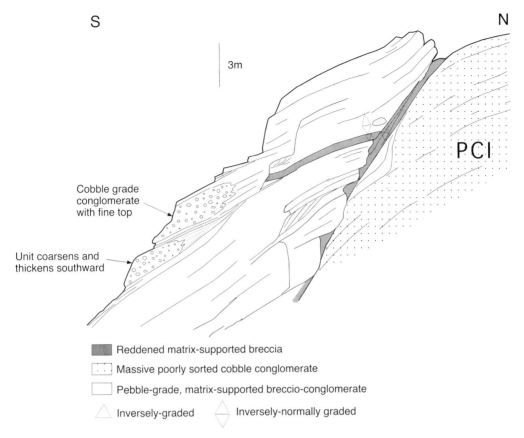

Fig. 21. Field sketch of northern margin of an incised palaeochannel (see Fig. 17) exposed on the eastern side of Tossal de Vall-llonga. This exposure is approximately 5 m deep, climbing away from the viewer. The palaeocliff surface, defined by a 0.5 m thick unit of reddened matrix-supported breccia is continuous but hidden in places by projecting rock units in this view. To the north conglomerates belong to member I of the Pont de les Cases Formation (PCI). The lower units of the Lower Marker Horizon infill the palaeotopography. The depositional dip of the fill abutting the margin is very low, and does not represent angle of repose stratification of a talus cone or scree.

growth anticline hinge (Fig. 17) that cut into the lower marker horizon. The upper units of PCII very clearly record alternating phases of offlap and overlap (Fig. 20c). These short periodicity events were superimposed on a longer timescale increase in R_s/R_u in the middle levels of this formation (Fig. 14a). Continuous growth offlap dominated in the upper member PC III (R_s/R_u decreased; Figs 14a & 20). A 25 m deep incision on the peak of Tossal de Vall-llonga (Fig. 17) may have occurred synchronously or following this period of growth offlap. The (concave) form of the erosion surface implies the base of a sub-axially oriented valley, possibly favoured by location over the northern limb of the growth anticline, where subsidence was more likely to be attenuated, and entrenchment promoted.

To the east (profile 3, Fig. 5c) the Pont de les Cases Formation is significantly thinner in the core of the syncline than on profile 2 and

Fig. 20. (a) Restored geometries of the upper member of the Pont de les Cases Formation (PCIII) on the profiles 1, 2 and 3, linked by a pin-line. (b) Palaeogeographical block diagram of the growth structure at the end of Period 3 (Pont de les Cases Formation). The heavy arrow indicates the approximate entrance point of a probable incised fan channel. As shown in (a) the formation changes from overlap geometries in the west to strong offlap and erosion in the east. On profile 2 the geometries varied from overlap (x position of fall line) to offlap (y position of fall line) through time. (c) Detail of growth geometries within the middle member of the Pont de les Cases Formation (PCII) on profile 2, showing alternating offlap and overlap.

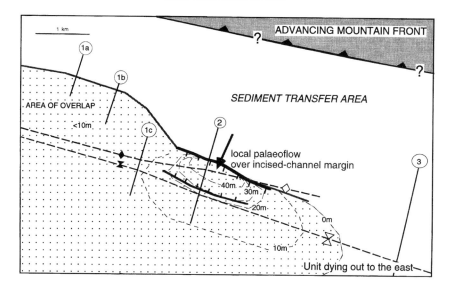

Fig. 22. Palaeogeographical map of the Lower Marker Horizon (within the Pont de les Cases Formation) showing infill by this unit of an erosional-topographical feature on Tossal de Vall-llonga (see Figs 17 and 21). The position of the advancing Cadí thrust sheet is unconstrained. Palaeoflow outside this feature was axial.

sand-rich facies are dominant (Figs 2 & 20). The main input point for conglomerates at this time lay therefore to the west. The lower member PC I thins gradually upward along the steep limb (Fig. 5c) and may have overlapped the anticline as in profile 2. The upper two members appear to offlap but their history is difficult to establish with certainty because of the presence of an unusual local unconformity at the base of the overlying Les Cases Altes Formation. In the steep limb this surface is bedding parallel, separating PC I and Les Cases Altes Formation. However toward the hinge of the syncline the surface cuts across the PC II and PC III members. The details of this structure is shown in Fig. 23. The boundary is complicated by a lateral (N-S) facies change in the upper part of PC III (sand to the south). Because it truncates PC II and III this unconformity may record a phase of erosion post-dating deposition of PC III. However several observations suggest a more subtle structure. At several points just below the unconformity, dips in the underlying conglomerates appear to fan rapidly from steep values upward to parallelism with overlying beds suggesting evolution as a progressive unconformity associated with growth offlap. However the surface becomes bedding parallel to the north and to the south and can only be traced laterally for over 1 km. This feature is important (1) because it demonstrates that significant erosion (whether progressive or abrupt) can occur close to or even in the core of the growth syncline and (2) because it has important implications for the nature of the folding mechanism. The trace of the growth syncline, continuous on profiles 1 and 2, is clearly displaced across this unconformity (Fig. 5c). Numerical modelling (Ford et al. 1997) has shown that this cannot be reproduced by simple models of kink band migration but can be reproduced by a limb rotation (trishear) model by incorporating a significant break in sedimentation.

Figure 20b shows a block model for the area at the end the Pont de les Cases Formation. Conglomerates were fed into the western part of the structure, while a sandstone-rich lithosome was deposited in the east. R_s/R_u is thus interpreted to have decreased to the east (Fig. 14a). In addition, the fold clearly amplified to the east at this time, contributing further to an eastward decrease in R_s/R_u. This is reflected in the continuous overlap on profile 1 (Fig. 20a), changing to the more complex history on profile 2 (Fig. 20c) which itself records a decrease in R_s/R_u with time through the formation. The, as yet unresolved, structures of profile 3 which record a complex interrelationship between offlap, erosion and a break in sedimentation, all indicate a decreasing R_s/R_u with time, but perhaps not as gradual as in the region of profile 2.

Sediment dispersal. Aggregate data for period 3 indicates that axial-parallel palaeoflow dominated over transverse dispersal (Fig. 15c). This

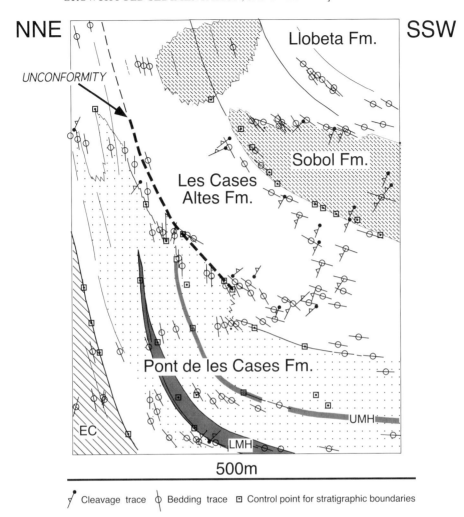

Fig. 23. Detailed cross-section through zone of composite off-onlap ('progressive unconformity') on profile 3. (see Fig. 5c for location) showing progressive changes in dip through several levels across the unconformity, and the geometries of the associated lithosomes. White units are sand-rich.

period shows the most strongly developed axial flow in the history of the structure, and appears to have been closely related to the axial trace of the growth folds. Data collected from the complex, dominantly conglomeratic lithosomes in the west and centre of the structure, show west- and WNW-directed flow (Fig. 19). The directional dataset (Fig. 15c) also indicates south-directed palaeoflow modes. The gradual ESE termination of these lithosomes, and replacement by finer-grained, sandy equivalents suggest that a major transverse/high-angle entry point, which supplied dominantly gravel, was located between profiles 2 and 3, as indicated by the arrow in Fig. 20b. This may have been controlled by the relative uplift (lower rates of subsidence) of the structure in the east, as indicated by the offlapping stratigraphy during this period.

The predominant axial palaeoflow is also reflected in the series of large-scale erosional structures revealed in the central regions of the growth structure (Fig. 17). The axial surface trace of the growth anticline is associated with major incision surfaces immediately pre-dating (Fig. 21) and post-dating the lower marker horizon, the latter an *apparently* 'one-sided' incision, possibly representing a palaeocliff (Fig. 17). The planar, strike-parallel form of this, and the steeply-incised channels, suggests alluvial down-cutting by water flows parallel to the anticline trace. However, although hanging valleys (or canyons) are not observed incised into period 2 gravelly alluvium in the region of profile 2, the fan-lobate form of the lower marker horizon (Fig. 22) suggests an apical entry point at (and crossing) the northern margin of the palaeochannel. This implies a transverse drainage across the surface trace of the anticline which acted episodically as a local fall line during period 3, e.g. during lower marker horizon aggradation. The characteristics of this and the upper marker horizon (restricted sedimentary clast assemblages, higher angularity and immature shapes) suggest either (1) a short transport path from a relatively small drainage basin in the frontal ranges of the mountain front or, perhaps less likely, (2) reworking of local material from the erosional/non-depositional region to the north of the growth anticline (Fig. 22). The less well defined limits of the overlapping upper marker horizon, though similar to those of the lower marker horizon in the east, suggests a similar interval of transverse drainage across the anticlinal crest, possibly preceded by axial erosional flows (Fig. 17). No palaeocurrent data yet exists for the Pont de les Cases Formation on the back limb of the growth anticline to compare the dispersal with that in the common limb and syncline. The dominantly conglomeratic members PCI to III showing W and WNW palaeodispersal thus alternated with probably shorter intervals of localized transverse dispersal represented by the marker horizons.

Period 4 (Santuari–Les Cases Altes–Mirador Formations)

Above the Pont de les Cases Formation the entire growth structure is dominated by growth overlap (Fig. 5), which followed a phase of early onlap in the east, as folding died out and the structure became buried. Thus sedimentation rates everywhere became greater relative to fold growth rate (Fig. 14a; equivalent to phase D in Fig.14b). In the west (profile 1), the upper limb of the growth anticline was tilted and eroded (R_s/R_u decreased) during deposition of the Santuari Formation, which tapers strongly across the fold pair. The resultant erosion surface steps up proximally across a 100 m high strike-parallel cliff (Fig. 24a, b; see also Ford et al. 1997, fig. 7). The Santuari Formation forms a rapidly eastward-tapering conglomeratic lithosome, interfingering with the laterally equivalent sand-rich Les Cases Altes Formation (Fig. 24b). The overlying Sobol and Mirador Formations overlapped the entire western area and record an upward decrease in growth. The Sobol Formation banked against the cliff, which was itself probably only a few hundred metres south of the mountain front.

On profile 2 the Santuari Formation overlapped the growth anticline and infilled a 50 m incision into older beds of the upper limb on Tossal de Vall-llonga (Fig. 17). The geometries of younger units across the growth anticline are not preserved, however the syncline is clearly developed in the Sobol Formation (Fig. 5b). On profile 3 the lower beds of the Les Cases Altes Formation show growth onlap (Fig. 5c) across the top of the Pont de les Cases Formation in the hinge of the growth syncline, while upper beds formed an apical wedge. The growth anticline is preserved in overlying formations (Fig. 5c) which show growth overlap continuing into the uppermost levels of the Mirador Formation (Fig. 24c). These relationships suggest that in this eastern area sedimentation rates were high and that fold growth continued while further west folding died out and local development of accommodation space was reduced.

Sediment dispersal. Data for period 4 (Figs 15d & 19) indicate dominant oblique-transverse palaeoflow to the SSW. The sandstone-rich Les Cases Altes Formation had differing dispersal in the west (dominantly transverse) to in the east (oblique-axial; Fig. 19). The similarly sandy Llobeta Formation, restricted to the eastern parts, dispersed sediment to the SSE (Fig. 19), the data extending to the south limb of the Busa syncline (Fig. 4). Only the Sobol conglomeratic sheet-like lithosome dispersed sediment in an approximately axial parallel direction during period 4, possibly influenced by the cliff feature in the NW of the structure and a WNW–ESE fall line in the east (Fig. 24b). Palaeodispersal within the overlapping Mirador Formation, although based on few data, shows (i) an oblique transverse SW mode in both the west and east of the structure, and (ii) modes directed toward the south quadrant.

Fig. 24. (**a**) Restored geometry of Sobol Formation in each of the three profiles linked by a pinline. (**b**) Palinspastic block diagram detailing the palaeogeography during Sobol Formation time (Period 4) showing the cliff in the west. Note that the folds are dying out to the west. (**c**) Palinspastic block diagram of palaeogeography during Mirador Formation time showing the continued fold activity in the east while the Vallfogona thrust is emergent in the west. See Fig. 18 for key to patterns.

GROWTH FOLD SEDIMENTATION, SANT LLORENÇ DE MORUNYS

(a) Restored geometries of Sobol Formation

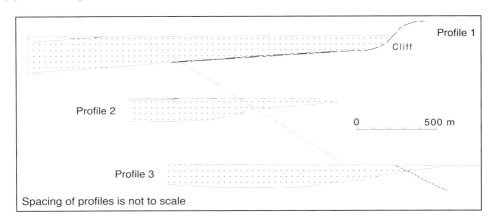

(b) Sobol Formation reconstruction

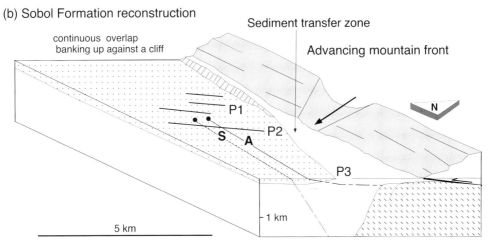

(c) Mirador Formation reconstruction

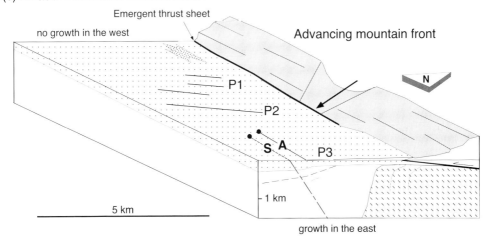

While distribution of sand-rich facies was restricted to the east during periods 3 and 4, during early period 1 and period 2 sand-rich facies were aggraded to the west of their conglomeratic equivalents. This variation is thought to be related to switching of proximal fanhead trenches (incised fan channels; cf. Denny 1967) between stable mountain front exit points and the fall line controlled by the growth structure. The contrasting along-strike evolution of the growth structure may also have had an important influence on facies distribution. In summary, growth folding died out diachronously across the area, continuing in the east after it had ceased in the west.

Discussion and conclusions

The sediments

The alluvial stratigraphical pile preserved in the Sant Llorenç de Morunys growth structure can be considered as multiple discrete event deposits covering very extensive areas, unlike the alluvium of established perennial channels and their enclosing floodplains. The nature of the emplacement of this sheet-like alluvium (related to steep alluvial fans) precludes to a large extent consideration of factors relevant to how antecedent river channels are affected by growth structures (Burbank et al. 1996), and also subsidence, interconnectedness and deposit density relationships of alluvial architecture models (Bridge & Leeder 1979). However, palaeodispersal patterns through the history of this structure suggest that the prevailing high concentration–high magnitude cohesionless debris flows and fluidal sediment flows were deflected by surface topography associated with, in particular, the common limb to growth syncline, and also the growth anticline. However, for sequences of tens to hundreds of metres, transverse as well as axial flows operated across the actively deforming surface. A classical radial palaeoflow pattern is not immediately obvious from the existing database. The high dispersion of much of the palaeoflow data (Figs 15 & 19) is likely to be the product of fan form and scale, fan head channel switching, fan lobe expansion angle, evolving surface slope (controlled by growth deformation) and vertical grouping of the data (up section changes with time). The mixture of structural types synthesized in the rose diagrams is thought to be less significant than the primary factors.

The bulk of the sediment volume was supplied externally to the region, and although significantly diverted at certain times (e.g. Period 3), appears to have been ultimately transferred through the growth structure to areas to the south. The length of axial (mountain front-parallel) diversion of incoming sediment is presently unknown (see below).

The effect of the growth fold on sedimentation

In comparison to most growth folds affecting foreland basin deposits the Sant Llorenç de Morunys fold pair has a quite different geometry and thus had a very different effect on the accumulating sediments. Its highly asymmetrical geometry and essentially flat-lying anticlinal back limb, which seems not to have significantly rotated throughout fold development, (1) facilitated largely unhindered sediment transfer from the mountain front and (2) resulted in growth strata being developed almost exclusively on the basinward side of the anticlinal axial surface. The back limb of the structure was cut by the frontal (Vallfogona) thrust which controlled the mountain front. Units overlapping the back limb of the anticline are horizontally-bedded, of constant thickness and extend far to the north of the anticlinal crest (e.g. the El Castell Formation on profile 2, the Pont de les Cases Formation on profile 1). Thus no simple anticnal uplift capable of acting as a barrier (e.g. Burbank et al. 1996) was generated by folding; subsiding areas, until late in the history of the structure, were therefore demarcated principally by the rotating and lengthening common limb. This implies that the axial surface trace of the growth syncline was the favoured site for axial diversions of palaeocurrents.

An 'intrabasinal' fall-line at the axial surface trace of the growth anticline is thought to have developed due to the change in subsidence regime from the static to uplifting anticlinal backlimb to the subsiding common limb and growth syncline. A clearer intrabasinal fall-line existed at the time of the principal unconformity. Sedimentation rates appear to have been sufficiently high throughout most of the fold history to have suppressed *major relief* potentially generated across the growth fold pair, and at the fall line generated by the principal unconformity. However, sufficient surficial subsidence occurred to have influenced palaeoflow, particularly in the earlier history of the structure (periods 1–3).

Regional context

The southern margin of the south Pyrenean thrust sheets is a complex segmentation of

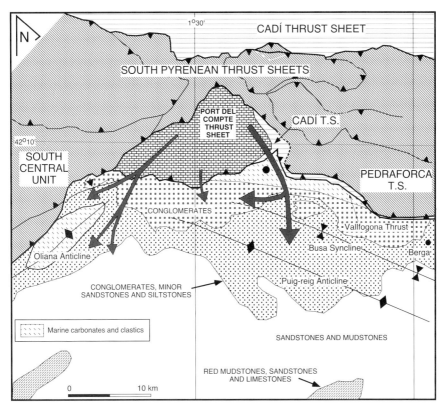

Fig. 25. Tectonic map of the SE Pyrenean Mesozoic-Cenozoic thrust sheets, and the present distribution of Late Eocene-Oligocene continental clastic facies (as in Fig. 1b), superimposed by the currently known major dispersal pathways. The longitudinal system in the Oliana area is after Burbank & Vergés (1994), and the bimodal south- and west-directed dispersal is simplified from this study. Note that both systems emerge from the re-entrant in the lower Pedraforca thrust sheet, the area presently occupied by the Port del Compte thrust sheet.

frontal and oblique ramps comprising the external thrust of the system (Fig. 25). Synchronous with the motion of the frontal thrust, the Puig-reig and Oliana anticlines developed within the foreland basin, sub-parallel to their respective thrust fronts (Fig. 25). The Oliana structure grew over a period of several million years as documented by associated growth strata (Burbank et al. 1992a; Vergés 1993).

The evolving landscape during late Eocene and Oligocene thrusting was dominated by drainage basins eroding marginal thrust sheets and the high axial Pyrenean zone, supplying sediment to the centre of the Ebro Basin. This central region was characterized by prograding subaerial terminal fans and playa lakes, for example on the northern limb of the Súria anticline, 75 km south of Sant Llorenç de Morunys (Sáez 1987; Fig. 1). Dispersal through the Sant Llorenç growth structure had distinct southward and westward modes (Fig. 25), whereas a large-scale longitudinal fluvial dispersal system was oriented parallel to the growing Oliana Anticline (Burbank & Vergés 1994), parallel to the local mountain front. Taken together these systems apparently diverge from the re-entrant displayed by the lower Pedraforca thrust, at present occupied by the Port del Compte thrust sheet (Fig. 25). This thrust sheet was emplaced at a late stage of Pyrenean shortening (Vergés 1993). The region defined by the lower Pedraforca re-entrant is interpreted as a relatively low topographic region prior to emplacement of the Port del Compte thrust sheet (Vergés 1993), at the same time as conglomerate deposition in the Oliana and Sant Llorenç de Morunys areas. It may therefore have represented a primary mountain front re-entrant or embayment. Late-stage dispersal in alluvial fan settings west of Sant Llorenç, thought to be sourced by the Port del Compte thrust sheet, is also southward-directed.

Sedimentation rates and growth strata evolution

Geomorphic evolution of the drainage basin (e.g. Schumm 1981) forced by the main phase of Pyrenean shortening, basin-scale flexural subsidence generated by thrust sheet loading (Doglioni & Prosser 1997), and potential climatic cyclicity suggest that sedimentation is unlikely to have maintained a steady long term rate. This is confirmed by several studies of successions of similar age in the Spanish Pyrenees. For example, accumulation rates varied from 0.15 to 0.5 mm a^{-1} for uppermost Eocene to upper Oligocene fluvial and lacustrine deposits of the southeastern South Central Unit at Artesa del Segre (Meigs et al. 1996; Fig. 1a), and conglomeratic growth strata associated with the Oliana growth anticline (Fig. 25) had rates of sediment accumulation from 0.14 to 0.25mm a^{-1}, with a low rate period of 0.06 mm a^{-1} (Burbank & Vergés 1994; Vergés et al. 1996). As there is no current absolute time frame for the Sant Llorenç sediments, we estimate that the minimum thickness of 2.5 km of Berga Conglomerates at Sant Llorenç were deposited in >10 Ma, extrapolating the mean accumulation rate of 0.25 mm a^{-1} calculated for the stratigraphically equivalent conglomerates at Oliana (Fig. 25; Burbank & Vergés 1994). The establishment of a reliable chronometric framework for these sediments is clearly required to specify controlling parameters by quantifying R_s and R_u. Superimposed on the above suite of controls are local processes such as the temporally and spatially variable rate of fold uplift (and subsidence), and the rate of advance and erosion of the palaeomountain front. Despite the likelihood of non-steady accumulation rates on a megasequence scale, the patterns of stratigraphical architecture and aspects of conglomerate sedimentology described in this paper suggest that variations in structurally controlled uplift and subsidence generated by the mountain front tectonic regime was a most significant variable that influenced the growth strata at Sant Llorenç for much of its history.

This work is part of a multidisciplinary study into growth fold evolution in foreland basins, as part of the EU JOULE II Geosciences project 'Integrated Basin Studies', directly funded by the Swiss Bundesamt für Bildung und Wissenschaft (project 94.0153). J.V. has been partially supported by DGICYT project PB94-0908 and by the Comissionat per Universitats i Recerca de la Generalitat de Catalunya, Quality Group GRQ94-1048. Aspects of this study were presented at the BSRG annual meeting in Dublin 1996 and at EUG 9 at Strasbourg 1997. Many thanks are due to P. D. W. Haughton and D. W. Burbank for their constructive, thought-provoking reviews, and to A. Mascle for his editorial assistance.

References

ALLEN, P. A. 1981. Sediments and processes on a small stream-flow dominated, Devonian alluvial fan, Shetland Islands. *Sedimentary Geology*, **29**, 31–66.

ANADÓN, P., CABRERA, L., COLOMBO, F., MARZO, M. & RIBA, O. 1986. Syntectonic intraformational unconformities in alluvial fan deposits, eastern Ebro Basin margins (NE Spain). *In*: ALLEN, P.A. & HOMEWOOD, P. (eds) *Foreland Basins*. International Association of Sedimentologists, Special Publications, **8**, 259–271.

ARTONI, A. & CASERO, P. 1997. Sequential balancing of growth structures, the late Tertiary example from the central Apennine. *Bulletin de la Société Géologique de France*, **168**, 35–49.

BLAIR, T. C. & BILODEAU, W. L. 1988. Development of tectonic cyclothems in rift, pull-apart, and foreland basins: sedimentary response to episodic tectonism. *Geology*, **16**, 517–520.

—— & MCPHERSON, J. G. 1994a. Alluvial fans and their natural distinction from rivers based on morphology, hydraulic processes, sedimentary processes, and facies assemblages. *Journal of Sedimentary Research*, **A64**, 450–489.

—— & —— 1994b. Alluvial Fan Processes and Forms. *In*: ABRAHAMS, A. D. & PARSONS, A. J. (eds) *Geomorphology of Desert Environments*. Chapman & Hall, London, 354–402.

BLUCK, B. J. 1967. Deposition of some Upper Old Red Sandstone conglomerates in the Clyde area: A study in the significance of bedding. *Scottish Journal of Geology*, **3**, 139–167.

—— 1976. Sedimentation in some Scottish rivers of low sinuosity. *Transactions of the Royal Society of Edinburgh: Earth Sciences*, **69**, 425–456.

—— 1979. Structure of coarse grained braided stream alluvium. *Transactions of the Royal Society of Edinburgh: Earth Sciences*, **70**, 181–221.

BRAYSHAW, A. C. 1984. Characteristics and origin of cluster bedforms in coarse-grained alluvial channels. *In*: KOSTER, E. H. & STEEL, R. J. (eds) *Sedimentology of Gravels and Conglomerates*. Canadian Society of Petroleum Geologists, Memoirs, **10**, 77–85.

BRIDGE, J. S. & LEEDER, M .R. 1979. A simulation model of alluvial stratigraphy. *Sedimentology*, **26**, 617–644.

BULL, W .B. 1972. Recognition of alluvial fan deposits in the stratigraphic record. *In*: RIGBY, J.K. & HAMBLIN, W. K. (eds) *Recognition of Ancient Sedimentary Environments*. Society of Economic Paleontologists and Mineralogists Special Publications, **16**, 63–83.

BURBANK, D. W. & VERGÉS, J. 1994. Reconstruction of topography and related depositional systems during active thrusting. *Journal of Geophysical Research*, **99**, 20281–20297.

——, BECK, R. A., RAYNOLDS, R. G. H., HOBBS, R. & TAHIRKHELI, R. A. K. 1988. Thrusting and gravel progradation in foreland basins: a test of post-thrusting gravel dispersal. *Geology,* **16,** 1143–1146.

——, MEIGS, A. & BROZOVIĆ, N. 1996. Interactions of growing folds and coeval depositional systems. *Basin Research,* **8,** 199–223.

——, PUIGDEFÀBREGAS, C. & MUÑOZ, J. A. 1992b. The chronology of the Eocene tectonic and stratigraphic development of the eastern Pyrenean foreland basin, northeast Spain. *Geological Society of America Bulletin,* **104,** 1101–1120.

——, VERGÉS, J., MUÑOZ, J. A. & BENTHAM, P. 1992a. Coeval hindward- and forward-imbricating thrusting in the south-central Pyrenees, Spain: timing and rates of shortening and deposition. *Geological Society of America Bulletin,* **104,** 3–17.

COLOMBO, F. 1994. Normal and reverse unroofing sequences in syntectonic conglomerates as evidence of progressive basinward deformation. *Geology,* **22,** 235–238.

—— & VERGÉS, J. 1992. Geometria del margen S.E. de la Cuenca del Ebro: discordancias progresivas en el Grupo Scala Dei, Serra de La Llena (Tarragona). *Acta Geológica Hispànica,* **27,** 33–53.

CONEY, P. J., MUÑOZ, J. A., MCCLAY, K. R. & EVENCHICK, C. A. 1996. Syntectonic burial and post-tectonic exhumation of the southern Pyrenees foreland fold-thrust belt. *Journal of the Geological Society, London,* **153,** 9–16.

CURRAY, J. R. 1956. The analysis of two-dimensional orientation data. *Journal of Geology,* **64,** 117–131.

DAHLSTROM, C. D. A. 1969. Balanced cross sections. *Canadian Journal of Earth Sciences,* **6,** 743–757.

DECELLES, P. G., GRAY, M. B., RIDGWAY, K. D., COLE, R. B., PIVNIK, D. A., PEQUERA, N. & SRIVASTAVA, P. 1991a. Controls on synorogenic alluvial-fan architecture, Beartooth Conglomerate (Palaeocene), Wyoming and Montana. *Sedimentology,* **38,** 567–590.

——, ——, ——, ——, SRIVASTAVA, P., PEQUERA, N. & PIVNIK, D. A. 1991b. Kinematic history of a foreland uplift from Paleocene synorogenic conglomerate, Beartooth Range, Wyoming and Montana. *Geological Society of America Bulletin,* **103,** 1458–1475.

——, LAWTON, T. F. & MITRA, G. 1995. Thrust timing, growth of structural culminations, and synorogenic sedimentation in the type Sevier orogenic belt, western United States. *Geology,* **23,** 699–702.

DENNY, C. S. 1967. Fans and pediments. *American Journal of Science,* **265,** 81–105.

DOGLIONI, C. & PROSSER, G. 1997. Fold uplift versus regional subsidence and sedimentation rate. *Marine and Petroleum Geology,* **14,** 179–190.

ENGLAND, P. & MOLNAR, P. 1990. Surface uplift, uplift of rocks and exhumation of rocks. *Geology,* **18,** 1173–1177.

FORD, M., WILLIAMS, E. A., ARTONI, A., VERGÉS, J. & HARDY, S. 1997. Progressive evolution of a fault-related fold pair from growth strata geometries, Sant Llorenç de Morunys, SE Pyrenees. *Journal of Structural Geology,* **19,** 413–441.

FRIEND, P. F. 1989. Space and time analysis of river systems, illustrated by Miocene systems of the northern Ebro Basin in Aragon, Spain. *Revista de la Sociedad Geologica España,* **2,** 55–64.

GLOPPEN, T. G. & STEEL, R. J. 1981. The deposits, internal structure and geometry in six alluvial fan-fan delta bodies (Devonian – Norway) – a study in the significance of bedding sequence in conglomerates. *In*: ETHRIDGE, F. G. & FLORES, R. M. (eds) *Recent and Ancient Nonmarine Depositional Environments: Models for Exploration.* Society of Economic Paleontologists and Mineralogists, Special Publications, **31,** 49–69.

GONZÁLEZ, A., ARENAS, C. & PARDO, G. 1997. Discussion on syntectonic burial and post-tectonic exhumation of the southern Pyrenees foreland fold-thrust belt. *Journal of the Geological Society, London,* **154,** 361–365.

GUPTA, S. 1997. Himalayan drainage patterns and the origin of fluvial megafans in the Ganges foreland basin. *Geology,* **25,** 11–14.

HAUGHTON, P. D. W. 1989. Structure of some Lower Old Red Sandstone conglomerates, Kincardineshire, Scotland: deposition from late orogenic antecedent streams? *Journal of the Geological Society, London,* **146,** 509–525.

HEIN, F. J. & WALKER, R. G. 1977. Bar evolution and development of stratification in the gravelly braided Kicking Horse River, British Columbia. *Canadian Journal of Earth Sciences,* **14,** 562–570.

HEWARD, A. P. 1978. Alluvial fan sequence and megasequence models: with examples from Westphalian D-Stephanian B coalfields, northern Spain. *In*: MIALL, A.D. (ed.) *Fluvial Sedimentology.* Canadian Society of Petroleum Geologists, Memoirs, **5,** 669–702.

HIRST, J. P. P. & NICHOLS, G. J. 1986. Thrust tectonic controls on Miocene alluvial distribution patterns, southern Pyrenees. *In*: ALLEN, P. A. & HOMEWOOD, P. (eds) *Foreland Basins.* International Association of Sedimentologists, Special Publications, **8,** 247–258.

HOSSACK, J. R. 1979. The use of balanced cross-sections in the calculation of orogenic contraction. *Journal of the Geological Society, London,* **136,** 705–711.

HOVIUS, N. 1996. Regular spacing of drainage outlets from linear mountain belts. *Basin Research,* **8,** 29–44.

KOLTERMANN, C. E. & GORELICK, S. M. 1992. Paleoclimatic signature in terrestrial flood deposits. *Science,* **256,** 1775–1782.

LOWE, D. R. 1976. Grain flow and grain flow deposits. *Journal of Sedimentary Petrology,* **46,** 188–199.

MCPHERSON, J. G., SHANMUGAM, G. & MOIOLA, R. J. 1987. Fan-deltas and braid deltas: Varieties of coarse-grained deltas. *Geological Society of America Bulletin,* **99,** 331–340.

MATÓ, E., SAULA, E., MARTÍNEZ, A., MUÑOZ, J.A., VERGÉS, J. & ESCUER, J. 1994. *Memoria explicativa y cartografía geológica de la Hoja 293 (Berga) del mapa geológico de España Plan Magna a escala 1:50,000.* ITGE.

MEIGS, A.J., VERGÉS, J. & BURBANK, D.W. 1996. Ten-million-year history of a thrust sheet. *Geological Society of America Bulletin,* **108,** 1608–1625.

MELLERE, D. 1993. Thrust-generated, back-fill stacking of alluvial fan sequences, south-central Pyrenees, Spain (La Pobla de Segur Conglomerates).

In: FROSTICK, L.E. & STEEL, R.J. (eds) *Tectonic Controls and Signatures in Sedimentary Successions*. International Association of Sedimentologists, Special Publications, **20**, 259–276.

MILLÁN, H., AURELL, M. & MELENDEZ, A. 1994. Synchronous detachment folds and coeval sedimentation in the Prepyrenean External Sierras (Spain): a case study for tectonic origin of sequences and systems tracts. *Sedimentology*, **41**, 1001–1024.

MUÑOZ, J. A. 1992. Evolution of a continental collision belt: ECORS-Pyrenees crustal balanced cross-section. *In*: MCCLAY, K.R. (ed.) *Thrust Tectonics*. Chapman & Hall, London, 235–246.

NEMEC, W. & POSTMA, G. 1993. Quaternary alluvial fans in southwestern Crete: sedimentation processes and geomorphic evolution. *In*: MARZO, M. & PUIGDEFÀBREGAS, C. (eds) *Alluvial Sedimentation*. International Association of Sedimentologists, Special Publications, **17**, 235–276.

—— & STEEL, R. J. 1984. Alluvial and coastal conglomerates: their significant features and some comments on gravelly mass-flow deposits. *In*: KOSTER, E. H. & STEEL, R. J. (eds) *Sedimentology of Gravels and Conglomerates*. Canadian Society of Petroleum Geologists, Memoirs, **10**, 1–31.

—— & —— (eds) 1988. *Fan Deltas: Sedimentology and Tectonic Settings*. Blackie and Son Ltd.

——, ——, POREBSKI, S. J. & SPINNANGR, Å. 1984. Domba Conglomerate, Devonian, Norway: process and lateral variability in a mass flow-dominated, lacustrine delta. *In*: KOSTER, E. H. & STEEL, R. J. (eds) *Sedimentology of Gravels and Conglomerates*. Canadian Society of Petroleum Geologists, Memoirs, **10**, 295–320.

NICHOLS, G. J. 1987*a*. Structural controls on fluvial distributary systems- the Luna System, northern Spain. *In*: ETHRIDGE, F .G., FLORES, R. M. & HARVEY, M. D. (eds) *Recent Developments in Fluvial Sedimentology*. Society of Economic Paleontologists and Mineralogists, Special Publications, **39**, 269–277.

—— 1987*b*. Syntectonic alluvial fan sedimentation, southern Pyrenees. *Geological Magazine*, **124**, 121–133.

PAOLA, C. 1988. Subsidence and gravel transport in alluvial basins. *In*: KLEINSPEHN, K. L. & PAOLA, C. (eds) *New Perspectives in Basin Analysis*. Springer-Verlag, New York, 231–244.

—— 1989. A simple basin-filling model for coarse-grained alluvial systems. *In*: CROSS, T.A. (ed.) *Quantitative Dynamic Stratigraphy*. Prentice Hall, Englewood Cliffs, N.J., 363–374.

——, HELLER, P. L. & ANGEVINE, C. L. 1992. The large-scale dynamics of grain-size variation in alluvial basins, 1: Theory. *Basin Research*, **4**, 73–90.

RIBA, O. 1967. Resultados de un estudio sobre el Terciario continental de la parte este de la depresión central catalana. *Acta Geológica Hispànica*, **2**, 1–6.

—— 1973. Las discordancias sintectónicas del Alto Cardener (Prepirineo catalán), ensayo de interpretación evolutiva. *Acta Geológica Hispànica*, **8**, 90–99.

—— 1976*a*. Tectogenèse et Sedimentation: deux modèles de discordances syntectoniques pyrénéenes. *Bulletin du Bureau des Recherches Géologique et Minières 2ème sér Sect I*, **4**, 387–405.

—— 1976*b*. Syntectonic unconformities of the Alto Cardener, Spanish Pyrenees: a genetic interpretation. *Sedimentary Geology*, **15**, 213–233.

——, REGUANT, S. & VILLENA, J. 1983. Ensayo de síntesis estratigráfica y evolutiva de la cuenca terciaria del Ebro. *In*: COUMBA, J. A. (ed.) *Libro Jubilar J.M. Ríos. Geología de España*, **II.**, IGME, Madrid, 131–159.

ROCKWELL, T. K., KELLER, E. A. & JOHNSON, D. L. 1985. Tectonic geomorphology of alluvial fans and mountain fronts near Ventura, California. *In*: MORISAWA, M. & HACK, J. T. (eds) *Tectonic Geomorphology*. Allen & Unwin, Boston, 183–207.

RUST, B. R. & KOSTER, E. H. 1984. Coarse Alluvial Deposits. *In*: WALKER, R. G. (ed.) *Facies Models*. Geological Association of Canada, Toronto, 53–69.

SÁEZ, A. 1987. *Estratigrafía y sedimentología de las formaciones lacustres del tránsito Eoceno-Oligoceno del NE de la cuenca del Ebro*. PhD thesis, Universitat de Barcelona.

SCHUMM, S.A. 1981. Evolution and response of the fluvial system, sedimentologic implications. *In*: ETHRIDGE, F.G. & FLORES, R.M. (eds) *Recent and Ancient Nonmarine Depositional Environments: Models for Exploration*. Society of Economic Paleontologists and Mineralogists, Special Publications, **31**, 19–29.

STEEL, R. J. & THOMPSON, D. B. 1983. Structures and textures in Triassic braided stream conglomerates ('Bunter' Pebble Beds) in the Sherwood Sandstone Group, North Staffordshire, England. *Sedimentology*, **30**, 341–367.

SUPPE, J., SÀBAT, F., MUÑOZ, J.A., POBLET, J., ROCA, E. & VERGÉS, J. 1997. Bed-by bed fold growth by kink-band migration: Sant Llorenç de Morunys, eastern Pyrenees. *Journal of Structural Geology*, **19**, 443–461.

TODD, S. P. 1989. Stream-driven, high-density gravelly traction carpets: possible deposits in the Trabeg Conglomerate Formation, SW Ireland and some theoretical considerations of their origin. *Sedimentology*, **36**, 513–530.

VERGÉS, J. 1993. *Estudi geològic del vessant sud del Pirineu oriental i central. Evolució cinemàtica en 3D*. PhD thesis, Universitat de Barcelona.

—— & RIBA, O. 1991. Discordancias sintectónicas ligadas a cabalgamientos: modelo kinemático. *Comunicaciones Congreso del Grupo Español del Terciario*, 341–345.

——, BURBANK, D. W. & MEIGS, A. 1996. Unfolding: An inverse approach to fold kinematics. *Geology*, **24**, 175–178.

WHEELER, H. E. & MALLORY, V. S. 1956. Factors in lithostratigraphy. *American Association of Petroleum Geologists Bulletin*, **40**, 2711–2723.

WHIPPLE, K. X. & TRAYLER, C. R. 1996. Tectonic control of fan size: the importance of spatially variable subsidence rates. *Basin Research*, **8**, 351–366.

WOLMAN, M. G. & MILLER, J. P. 1960. Magnitude and frequency of forces in geomorphic processes. *Journal of Geology*, **68**, 54–74.

Quantified vertical motions and tectonic evolution of the SE Pyrenean foreland basin

J. VERGÉS[1,2], M. MARZO[1], T. SANTAEULÀRIA[1], J. SERRA-KIEL[1], D. W. BURBANK[3], J.A. MUÑOZ[1] & J. GIMÉNEZ-MONTSANT[4]

[1]*Grup de Geodinàmica i Anàlisi de Conques, Departament de Geologia Dinàmica, Geofísica i Paleontologia, University of Barcelona, Martí i Franquès s/n, 08071 Barcelona, Spain*
[2]*Present address: Institute of Earth Sciences 'Jaume Almera' Solé i Sabarís, s/n 08071 Barcelona, Spain (e-mail: jverges@ija.csic.es)*
[3]*Department of Earth Sciences, University of Southern California, Los Angeles, CA 90089-0740, USA*
[4]*Department of Geology, Royal Holloway, University of London, Egham, Surrey TW20 0EX, UK*

Abstract: Local isostatic backstripping analysis is performed across the eastern part of the Ebro foreland basin between the Pyrenees and the Catalan Coastal Ranges. The subsidence analysis is based on two well-dated field-based sections and four oil-wells aligned parallel to the tectonic transport direction of the eastern Pyrenean orogen. The marine infill of the foreland basin is separated into four, third-order, transgressive–regressive depositional cycles. The first and second depositional cycles are located in the Ripoll piggy-back basin and the third and fourth ones are located south of the syn-depositional emergent Vallfogona thrust. Subsidence curves display a typical convex-up shape with inflection points recording the onset of rapid tectonic subsidence. Inflection points coincide roughly with the base of depositional cycles. Rates of tectonic subsidence are less than 0.1 mm a^{-1} in distal parts of the basin and up to 0.53 mm a^{-1} in proximal parts during second depositional cycle. Younger depositional cycles show maximum rates of tectonic subsidence of 0.26 mm a^{-1}. The locus of subsidence within the basin migrated southward at a rate of c. 10 mm a^{-1}. This flexural wave crossed the complete Ebro foreland basin in 10–11 Ma. The intraplate Catalan Coastal Ranges at the southeastern margin of the Ebro foreland basin produced an increase of tectonic subsidence rate at 41.5 Ma. Maximum rates of tectonic subsidence coincide with deep-marine infill of the basin, maximum rates of shortening and thrust front advance, and low topographic relief orogenic wedge. Transgressive–regressive depositional cycles can be controlled partly by reductions of available space within the basin during tectonic thickening of the sedimentary pile by layer parallel shortening, folding and thrusting.

Although much less constrained, an approximation of post-thrusting exhumation and isostatic and tectonic uplift, as well as a first determination of possible amounts of eroded material of parts of the Ebro basin illustrate the impact of post-depositional erosion and uplift on the foreland.

The geometry and distribution of the infill of a flexural foreland basin register the interplay between different surficial and deep geological processes and are dependent on the ratio between rates of sediment supply and space available in the basin. Rates of tectonic growth and denudation control sediment supply, mainly in the orogenic wedge, whereas lithospheric strength, advancing tectonics and sea-level variations control the accommodation space in the foreland basin. All these factors are linked to define a unique linkage between orogenic evolution and the foreland basin during several tens of millions of years. However, most foreland basins evolve in a similar way, from marine and largely underfilled to continental, overfilled and largely bypassing (e.g. Allen *et al.* 1986; Fig. 1). The Alpine Molasse basin (Homewood *et al.* 1986; Sinclair & Allen 1992) and the Pyrenees (Puigdefàbregas *et al.* 1986, 1992) as part of the western European foreland basins show good examples of this general evolutionary trend. The internal distribution of sediments within the basin is characterized by several depositional cycles with a duration of a few millions of years (third-order depositional cycles).

VERGÉS, J., MARZO, M., SANTAEULÀRIA, T., SERRA-KIEL, J., BURBANK, D. W., MUÑOZ, J. A. & GIMÉNEZ-MONTSANT, J. 1998. Quantified vertical motions and tectonic evolution of the SE Pyrenean foreland basin. *In:* MASCLE, A., PUIGDEFÀBREGAS, C., LUTERBACHER, H. P. & FERNÀNDEZ, M. (eds) *Cenozoic Foreland Basins of Western Europe.* Geological Society Special Publications, **134**, 107–134.

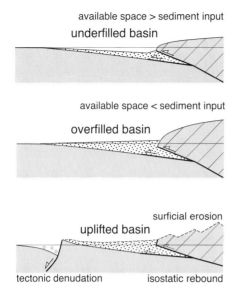

Fig. 1. Vertical motions in a foreland basin during thrusting are governed by the downward flexure of the plate produced by tectonic loads. The large-scale infill distribution on the foreland basin is dependent on the ratio between available accommodation space within the basin and the volume of sediment influx. For low sediment input the basin will be underfilled (Flysch stage). For large sediment influx, the basin will be overfilled (Molasse stage). Post-orogenic vertical motions are mainly controlled by the interplay between tectonic and surficial denudation with concomitant isostatic rebound.

In this paper we present the depositional organization of the Palaeogene marine infill of the southeastern Pyrenean foreland basin together with a quantitative analysis of vertical motions within the basin. This work has been possible by the integration of surface data, oil-well data and non-published commercial seismic lines. An important effort has been made in refining the chronostratigraphic framework of this part of the Pyrenean basin by combining magnetostratigraphic and new palaeontological data. Furthermore, a backstripping analysis, based on local isostatic compensation, has been made across a N–S transect throughout the complete foreland basin. This approach allows us to calculate rates of total and tectonic subsidence in single localities as well as their migration across the basin, ahead of the forelandward-moving Pyrenean deformational front. Moreover, this analysis illustrates how an active distal margin (the Catalan Coastal Ranges) influenced subsidence distribution and subsidence rates across the basin.

Geological setting

The Pyrenees correspond to the western termination of an orogenic belt formed during the Tertiary closure of the Tethyan Sea located between the converging African and European plates (Fig. 2). The relatively straight and E–W-trending Pyrenean orogen developed by the northern subduction of the Iberian lithosphere beneath Europe as imaged by the deep seismic reflection ECORS-Pyrenees profile (Choukroune et al. 1989). The Pyrenees merge eastward with the highly arcuate Alps, which formed above a south-dipping European lithospheric subduction underneath Adria, the northern promontory of the African plate, (ECORS-CROP Deep Seismic Sounding Group 1989). Both orogens are doubly sided with a major foreland basin on top of the lower plate (e.g. Muñoz 1992; Pfiffner 1992). The Ebro basin and the Molasse basin represent the latest evolutionary stage of the flexural foreland basin. Earlier foreland basin strata are preserved in piggyback basins on top of thrust sheets. The Aquitaine basin developed on the northwestern side of the Pyrenees and represents a retro-foreland basin.

The eastern side of the Ebro basin displays an irregular shape bounded by the Pyrenees to the north and the Catalan Coastal Ranges to the southeast (Fig. 3). This irregular geometry is due to both the oblique trend of the Catalan Coastal Ranges with respect to the Pyrenean chain and the succession of frontal and oblique segments of the Pyrenean front, inherited from the Mesozoic extensional basin geometry (Puigdefàbregas et al. 1992; Vergés & Burbank 1996). The southeastern Pyrenean foreland basin developed almost entirely over pre-Mesozoic basement and stands in contrast to the more westerly Jaca Basin which developed on top of a detached Mesozoic section (e.g. Séguret 1972; Teixell 1996).

The Vallfogona thrust represents the major thrust boundary between the southeastern Pyrenean thrust sheets and the deformed foreland strata (Fig. 3). The thick Tertiary succession cropping out in the Cadí thrust sheet is folded by the Ripoll syncline (Muñoz et al. 1986; Puigdefàbregas et al. 1986). The lower segment of this stratigraphic succession represents the northern part of the former south Pyrenean foreland basin whereas the upper segment represents the individualization of this part of the trough as a piggyback basin, the Ripoll basin, after the

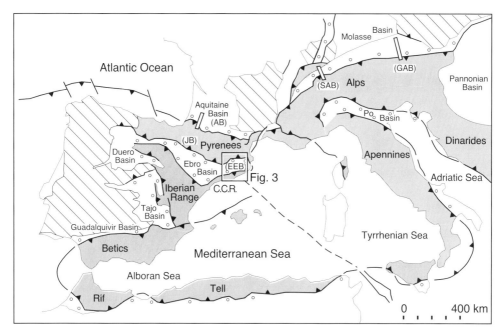

Fig. 2. Tectonic map of western Mediterranean with location of Tertiary orogenic belts and adjacent foreland basins. Thick black lines with black triangles bound shaded orogenic regions. Open circles highlight foreland basins. Oblique lines define major outcrops of basement. Narrow boxes indicate location of compared subsidence studies in the Jaca basin in the western Pyrenees (JB), Aquitaine foreland basin (AB), Swiss Alpine and German Alpine basins (SAB and GAB).

inception of the Vallfogona thrust. A large part of the Ebro foreland basin is deformed by a system of folds and thrusts, in part coeval to deposition (Puigdefàbregas et al. 1986; Burbank et al. 1992a), detached above a suite of foreland evaporitic levels (Vergés et al. 1992; Sans et al. 1996).

The subsidence analysis presented in this paper has been made along a N–S transect (for location see in Fig. 3), parallel to the tectonic transport direction in this segment of the Pyrenees. The location of the studied stratigraphic sections and oil-wells have been projected into a regional cross-section, presented in Fig. 8 and described in Vergés (1993) and Vergés & Burbank (1996). This projection into the regional section is nevertheless problematic due to the rapid lateral changes in both the stratigraphy and the tectonic style and has to be viewed as an approximate couple between subsidence results and regional fold-and-thrust belt geometry. For instance, the Puig-reig oil-well is projected parallel to the Puig-reig anticlinal structure from the east where the anticline ends and displays a much reduced shortening (Vergés & Burbank 1996).

Depositional cycles and chronostratigraphy

The Palaeogene marine succession of the SE Pyrenean basin has been divided into four major transgressive–regressive depositional cycles across the southeastern Pyrenean foreland basins. First and second cycles are defined within the Ripoll syncline (Cadí thrust sheet in Fig. 4), whereas third and fourth cycles are contained in the Ebro basin (Figs 3 & 4). The lateral continuity of these four cycles throughout the foreland basin is not straightforward. Difficulties arise for the following reasons: (1) the stratigraphic succession is partitioned in different tectonic units and basins as discussed above; (2) the interplay between tectonics and sedimentation occurs at different scales; (3) the Catalan Coastal Ranges represent an active, rather than passive, southern basin margin (Figs 2 & 3); (4) the sediments vary and comprise carbonates, clastics and evaporites; and (5) precise chronostratigraphic control is lacking in the oil-wells of the central part of the basin.

In this situation, an accurate chronostratigraphy is especially needed to determine reliable

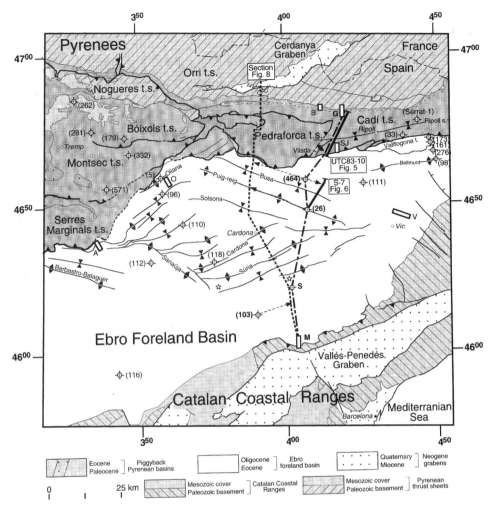

Fig. 3. Structural map of the Eastern part of the Ebro basin between the Pyrenean thrust belt to the North and the Catalan Coastal Ranges to the South. Dashed line shows the analysed transect from G to M. Short thick lines represent the location of seismic lines (UTC83-10 crossing the Cadí thrust sheet and S-7 in the northern side of the Ebro basin). Open circles with crosses show oil-wells with a number referred to an oil-well index published in Lanaja (1987). Thick dashed N–S line shows the position of a regional balanced and restored cross-section (Fig. 8) used in the construction of a crustal-scale section (Vergés et al. 1995). Subsidence results have been projected, parallel to the tectonic structures to both the analysed section to the east and to the regional cross-section to the west (see black dashed arrows). Sections and oil-wells used in this study are represented by G-Gombrèn section; (464), Jabalí oil-well; (26), Puig-reig oil-well; S, Santpedor oil-well; (103), Castellfollit oil-well; and M, Montserrat section. Additional locations are represented by V, Vic section; SJ, Sant Jaume de Frontanyà section; B, Bagà section; O, Oliana section; and A, Artesa del Segre section. Little stars show location of vitrinite samples used in this study (Santpedor and Calaf localities).

depositional cycles boundaries (Fig. 4). Chronostratigraphy is based on biostratigraphic and magnetostratigraphic data compilation from published works and their correlation to the magnetic polarity time scale. Biostratigraphy is based mainly on larger foraminifera, especially *Nummulites* and *Alveolina* (Hottinger 1960; Ferrer 1971; Schaub 1981; Serra-Kiel 1984; Tosquella 1995). Magnetostratigraphic data come from seven different sections within the studied region (Fig. 3; Burbank et al. 1992a,b; López-Blanco et al. in press) as well as from the central Pyrenees where the Palaeocene–Eocene boundary is well defined in the Campo section

(Serra-Kiel et al. 1994). The correlation between magnetic sections and larger foraminifera corresponding to Shallow Benthic Zones (SB), mainly based on south Pyrenean specimens (Serra-Kiel et al. in press) is shown in Fig. 4. The magnetic polarity time scale of Cande & Kent (1995) is used here.

Ages corresponding to either transgressive–regressive boundaries or lithological formations are calculated by interpolation assuming constant rates of sediment accumulation during well-defined magnetic intervals. The use of lithostratigraphic formations minimize the possible changes of sediment accumulation rates within analyzed intervals. Ages of the successions derived from oil-wells have been calculated from biostratigraphic data, correlation with surface sections, and the support of the magnetostratigraphic sections of Vic (Burbank et al. 1992a) and Oliana (Burbank et al. 1992b), corresponding to the basin centre (Fig. 3). The age for the uppermost part of the Bellmunt deposits in the Gombrèn section has been calculated by extrapolating the mean sediment accumulation rate determined for the interval 45.82–43.78 Ma. The age of the uppermost deposits of the Solsona Fm in the Jabalí, Puigreig, Santpedor and Castellfollit sections have been determined assuming mean sedimentation rates of 0.23 mm a^{-1} determined from roughly coeval deposits in Oliana (Burbank et al. 1992b) and in Artesa del Segre (Meigs et al. 1996) (see Fig. 3).

During the Palaeogene, the inception of marine conditions on the SE Pyrenean foreland basin occurred above a regional unconformity dated as 55.9 Ma old in the Campo section (chron 24.3r; Serra-Kiel et al. 1994) as well as farther west in the Urbasa section, Basque country (Pujalte et al. 1994). First depositional cycle is Ilerdian to early Cuisian in age comprising SB5 to SB10 based on magnetostratigraphy (Serra-Kiel et al. 1994; Bentham & Burbank 1996) correlated with biostratigraphic data (Serra-Kiel et al. 1994). This first transgressive–regressive depositional cycle started with shallow-marine Alveolina limestones (Cadí Fm), Ilerdian and Cuisian in age. The age of the base of the Cadí Fm decreases toward the distal margin of the basin: early Ilerdian in the Pyrenean thrust sheets, middle Ilerdian in the centre of the Ebro basin and middle Ilerdian 2 in the southern and distal margin (Orpí Fm). The shallow-marine carbonate platform represented by both the Cadí and Orpí Formations merged northward into deeper water calcareous mudstones of the Sagnari Fm (Luterbacher et al. 1991; Giménez-Montsant 1993). The regressive hemicycle is represented by a prograding, deltaic, clastic wedge capped by fluvio-lacustrine facies of the Corones Fm. (Giménez-Montsant 1993), early Cuisian in age (Tosquella 1995). Based on the magnetostratigraphic studies, the total duration of this transgressive–regressive cycle is 5.1 Ma (from 55.9 to 50.8 MaBP).

In the Pyrenean thrust sheets, the second cycle is mid-Cuisian to early Lutetian in age containing SB11 to SB13 based on magnetostratigraphy (Bentham & Burbank 1996) correlated with biostratigraphy (Samsó et al. 1994; Tosquella 1995). This cycle starts with shallow-marine limestones (uppermost part of the Corones Fm), which rapidly grade upward into a c. 1000 m thick, slope facies association mostly represented by carbonates and calcareous mudstones, often slumped (Armàncies Fm). This slope association includes up to seven, decametric, megabreccia units interpreted as the result of slope resedimentation of co-existing carbonate platforms which constitute the Penya Fm (Giménez-Montsant 1993). Above the slope marls, over 700 m of siliciclastic turbidites (Campdevànol Fm), linked to southward-prograding fluvio-deltaic and fan-delta systems (lowermost part of the Bellmunt Fm), are present in the northern margin of the Cadí thrust sheet. These turbidites filled a trough adjacent to the advancing Pyrenean thrust sheets and was rimmed southward by the above mentioned carbonate platform (Puigdefàbregas et al. 1986). The uppermost part of the turbiditic succession could be interstratified with evaporitic deposits (Martínez et al. 1989), although the thick evaporitic alternation was deposited southward of the turbiditic depocentre (Fig. 4). The thickness of this evaporitic unit (Beuda Fm) greatly increases southwards, as evidenced by the Serrat-1 oil well (Union Texas Inc. 1987), drilled in the northern flank of the Ripoll syncline (Cadí thrust sheet; Fig. 3). In the footwall of the Vallfogona thrust, the well cuts c. 2000 m of gypsum, and alternating gypsum, sandstones and marls, with a 200 m thick intercalation of salt (Martínez et al. 1989). The 100 m thick level of gypsum cropping out along the northern flank of the Ripoll syncline represents the northern margin of the Beuda depocentre. This second depositional cycle comprises the mid-Cuisian and the latest part of the early Lutetian, spanning a total of 4.54 Ma (from 50.8 to 46.26 MaBP).

The third and fourth marine depositional cycles are exposed south of the Vallfogona thrust, in the Jabalí and Puig-reig oil-wells and in the southern margin of the Ebro basin.

The third cycle is early Lutetian to uppermost Lutetian in age. It contains SB14, SB15 and

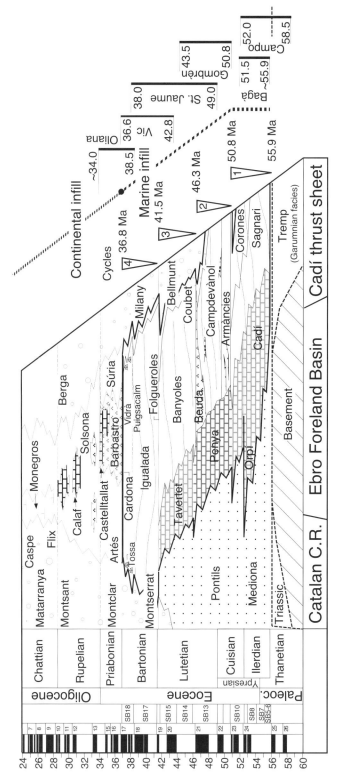

Fig. 4. Stratigraphic panel across the Ripoll basin (Cadí thrust sheet), Ebro foreland basin and Catalan margin. The marine infill of the basin (limited from the continental deposits by a thick continuous line) is separated into four transgressive-regressive third-order depositional cycles of about 5 Ma duration. Biostratigraphic data combined with magnetostratigraphic information define the chronostratigraphic framework of the study area. Magnetostratigraphy comes from Bagà, Gombrèn, Sant Jaume de Frontanyà, Vic and Oliana sections (Burbank et al. 1992a,b; Vergés & Burbank 1996) and from the Campo section (Serra-Kiel et al. 1994). Global polarity time scale from Cande & Kent (1995). Larger foraminifera shallow bentic zones (SBZ) from Serra-Kiel et al. (in press).

SB16 zones based on magnetostratigraphy (Burbank et al. 1992a; Bentham & Burbank 1996) correlated to biostratigraphic data (Serra-Kiel & Travé 1995; Samsó et al. 1994). This third cycle starts with a renewed transgression recorded at the northern basin margin by a glauconite-rich, mostly sandy, nearshore-to-deltaic facies association (Coubet Fm) that unconformably overlies previous continental deposits (lowermost part of the Bellmunt Fm). These sandy facies are overlain by a thick, southward-prograding, terrigenous fluvio-deltaic wedge (Mató et al. 1994), comprising conglomerates, sandstones and marls (Bellmunt and Banyoles Formations). The thickness of this fluvio-deltaic wedge is around 700 m east of the studied transect (Saula et al. 1994), c. 600 m in the Jabalí oil-well and only c. 200 m in the Puig-reig oil-well. To the south, these fluvio-deltaic clastics overlap a southern carbonate platform (Tavertet Fm). The age of this cycle is well-constrained by both biostratigraphy and magnetostratigraphy; it covers a time span of 4.76 Ma (from early Lutetian at 46.26 to late Lutetian at 41.50 MaBP).

The fourth cycle is Bartonian in age. It comprises SB17 and SB18 zones based on magnetostratigraphy (Burbank et al. 1992a) integrated with biostratigraphic data (Serra-Kiel & Travé 1995). This fourth depositional cycle is characterized by the absence of well-developed platform carbonate deposits, because of the simultaneous progradation from both basin margins of Montserrat Fm fan-delta deposits in the south and Milany Fm fluvio-deltaic sandstones and conglomerates in the north grading basinward to prodeltaic-offshore marls of the Igualada Fm. The overall offlapping trends of these terrigenous wedges are locally punctuated by minor onlapping transgressive pulses leading to the local and temporary development of carbonate platforms and reef build-ups which become more frequent toward the top of the cycle (Tossa Fm). These mixed terrigenous–carbonate wedges overlie a transgressive, glauconite-rich, shelf sandstone complex (Folgueroles Fm) and are capped in turn by an evaporitic plug (Cardona Fm) that fills up the marine basin. The thickness of this fourth cycle is about 1560 m in the Jabalí oil-well and 1760 m in the Puig-reig oil-well, although a thrust fault located at c. 1600 m depth produced tectonic duplication of part of the section that could be up to 450 m determined from vitrinite reflectance data from the Puig-reig oil-well (Clavell 1992). If these determinations are correct, there would be an overestimation of both total sedimentary thickness and calculated value and rate of subsidence for this interval.

However, this tectonic duplication is not represented in the seismic line S-7 projected from the eastern termination of the Puig-reig anticline (see Fig. 6).

South of the Vallfogona thrust, the fourth sedimentary cycle shows two internal angular unconformities located at the base and within the cycle. These two unconformities fossilize older folds and thrusts and are subsequently folded and thrusted (Mató et al. 1994). These cross-cutting relationships suggest continuous deformation in the south Pyrenean frontal thrust zone (at the time slices we are considering) at least during the total duration of the last marine cycle, which is estimated to be 4.7 Ma (Bartonian, from 41.5 to 36.8 MaBP).

The above described depositional cycles are not recognizable, except the fourth one, at the southern margin of the Ebro foreland basin near the contact with the Catalan Coastal Ranges (Fig. 4). In this sector the carbonate platform deposits (Cadí-Orpí, Penya, and Tavertet Formations) are substituted by continental red beds (Pontils Group, Mediona and Cairat Fms). The precise chronostratigraphic attribution of these continental units is not well-constrained, and the possible existence of several major stratigraphic gaps, although suspected, can not be proved.

Two seismic lines have been selected to show the geometry of the Ripoll piggyback basin, the position of the underlying Vallfogona thrust (seismic line UTC-83-10; Fig. 5), and the geometry of the southward prograding shallow-water carbonate platforms in the central northern part of the Ebro foreland basin (seismic line S-7; Fig. 6).

Cadí thrust sheet structure: seismic line UTC-83-10

Line UTC-83-10 (Union Texas España Inc. 1983, migrated) shows the geometry of the Cadí thrust sheet and underlying tectonic units (Fig. 5). The line is NNE–SSW oriented, crosses the Cadí thrust sheet but remains parallel to the southern flank of the Vilada anticline in the southern part of the line. This anticline plunges toward the NNE as a result of a folding interference with the Ripoll syncline (Vergés et al. 1994). The Vilada anticline is parallel to the eastern oblique termination of the lower Pedraforca thrust sheet and synchronous to its emplacement and further development (Martínez et al. 1988; Burbank et al. 1992a). The southern end of the seismic line is parallel to and located 1 km northward of the present outcrop of the Vallfogona thrust (Fig. 3).

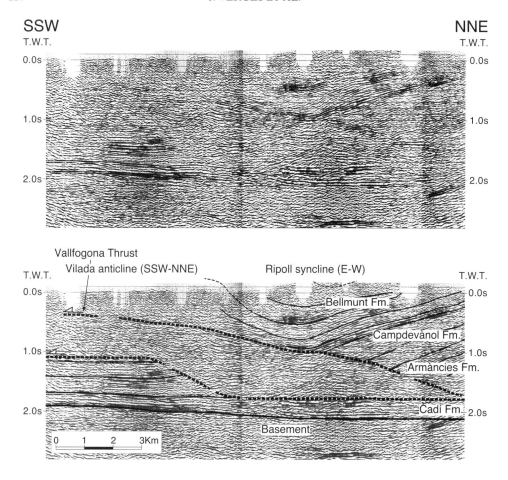

Fig. 5. Seismic line UTC-83-10 (Union Texas España Inc., 1983, migrated) and interpreted line drawing. The line parallels the eastern oblique termination of the Pedraforca thrust sheet and the Vilada anticline in its southern part as shown in Fig. 3. The line shows the geometry of asymmetric Ripoll syncline on top of the Vallfogona thrust that define the Cadí thrust sheet. The thrust showing a ramp geometry between two flat segments in the footwall of the Vallfogona thrust represents the main detachment level within the deformed Ebro basin. Vertical and horizontal scales are similar.

An almost parallel and gently north-dipping group of reflectors, interpreted as the lower Eocene carbonate platforms (Cadí Fm), characterizes the bottom part of the line between 1.75 and 1.9 s (TWT) to the south. Above these reflectors, the line shows three different panels of reflectors that, from north to south, define (a) the Ripoll syncline (as part of the Cadí thrust sheet) above the Vallfogona thrust, (b) an intermediate unit in the footwall of the Vallfogona thrust, and (c) nearly flat-dipping and parallel reflectors in the southern part of the line between 1.1 and 1.75 s.

The Ripoll syncline displays a highly asymmetric geometry characterized by a complete and thick south-dipping Palaeocene–Eocene stratigraphic section in the northern flank whereas only the uppermost part of these deposits crop out in the southern flank. The cut-off relationships between the Ripoll syncline strata and the underlying Vallfogona thrust can be expressed as follows: the northern flank of the syncline represents a high angle hanging wall ramp with respect the Vallfogona thrust whereas the hinge zone and the southern flank of the syncline represent a very low angle hanging wall ramp. The Ripoll syncline shows a northward migration of its axis through time. The footwall of the Vallfogona thrust is characterized by short and irregular reflectors interpreted as a set of folds and thrusts that crop out to the south of the Vallfogona thrust (Muñoz et al. 1994; Mató et al.

1994). Well-imaged subhorizontal reflectors located in the southernmost part of the seismic line between 1.1 and 1.75 s display an abrupt northern termination interpreted as the footwall ramp of a north-dipping and south-directed thrust related to the foreland fold and thrust system. This thrust can be followed in the regional cross-section, although, the geometry of the Cadí thrust sheet west of the seismic line shows an important southern tilt due to the emplacement of basement tectonic units in the northern part of the section (Vergés et al. 1995; see Fig. 8).

Ebro foreland basin structure: seismic line S-7

Line S-7 (UERT, S.A. 1977, fold stack) shows the geometry and distribution of shallow-marine prograding platforms in the central-northern segment of the Ebro foreland basin (Fig. 6). The line with a NNE-SSW direction, crosses the gentle Busa syncline and ends at the Puig-reig oil-well in the northern flank of the Puig-reig anticline. The Jabalí oil-well has been projected into the S-7 line from 6 km to the WNW, parallel to the northern flank of the Busa syncline (Fig. 3).

The seismic line shows different vertical segments according to their seismic facies reflectivity (Fig. 6). The lower segment is composed of subhorizontal, parallel and non-deformed group of reflectors between 1.3 and 1.0 s to the south of the line. The southern ends of these reflectors display a progressive southward onlap over a less reflective level interpreted as the basement (upward arrows in Fig. 6). The northern ends of individual reflectors grade laterally and northward into the overlying less reflective segment, corresponding to more fine grained sediments (downward arrows in Fig. 6). The lowermost reflectors correspond to the lower and middle Eocene platform limestones (Fig. 4) as confirmed by well information (Jabalí and Puig-reig oil-wells). The length of individual reflectors is on the order of few tens of kilometres. The less reflective segment, located between 1.05 and 0.85 s (at the southern part of the line) shows increasing reflectivity toward the north and has been interpreted as corresponding to marls and sandstones of the Banyoles Fm. Between 0.85 and 0.5 s, there is a reflective segment delineating the Busa syncline and corresponding to the upper part of the Igualada Fm. The uppermost segment, above 0.5 s is much less reflective although individual reflectors outline the syncline geometry. These reflectors correspond to the fluvial pelites, sandstones and conglomerates of the Solsona Fm.

Sedimentary model and palaeobathymetries

Figure 7 shows the interpretation of the studied part of the foreland basin. The shallow-water platform limestones migrate and become thinner toward the south. The northern Jabalí oil-well cuts through c. 820 m of limestones (Cadí to Tavertet Formations) and about 600 m of overlying terrigenous deposits (Banyoles Fm), whereas the southern Puig-reig oil-well cuts across c. 480 m of limestones and c. 200 m of marly deposits. The fluvio-deltaic Banyoles Fm passes southward and overlaps the platform deposits. Although the ages of these platforms are not well constrained across the whole basin, seismic information shows that the uppermost well-imaged reflector cut in the Jabalí oil-well corresponds to the middle part of the group of reflectors cut in the Puig-reig oil-well, indicating therefore a fairly rapid southward migration of the shallow marine platform systems, as well as a younger age for the base and top of these carbonate platforms in the same direction (Fig. 7).

The south-dipping geometry of some of the reflectors corresponding to the Banyoles Fm can be interpreted as depositional slopes. Nevertheless, the regional tectonic framework indicates that there is a regional detachment level that follows approximately the contact between limestones and marls (Vergés et al. 1992; Sans et al. 1996). This detachment climbs up from the limestone-marl contact to the Cardona salt horizon and produces the Puig-reig anticline above the thrust ramp.

Facies distribution within the basin shows an inner, northern part characterized by thick marine terrigenous wedges grading southward to shallow-water carbonate platforms, overlapping thin continental deposits located on the unflexed part of the foreland basin (Fig. 4). Foreland basin carbonate platforms on the southern margin migrated toward the foreland side of the basin although several progradational episodes toward the inner part of the basin are recorded by at least seven megabreccia units during the second depositional cycle. These megabreccias are intercalated with talus marls cropping out in the northern flank of the Ripoll syncline (Puigdefàbregas et al. 1986; Barnolas 1992; Giménez-Montsant 1993). Above the talus deposits, important north-derived turbiditic wedges (Campdevànol Fm) produced the shift of depositional facies belts to the south. The end

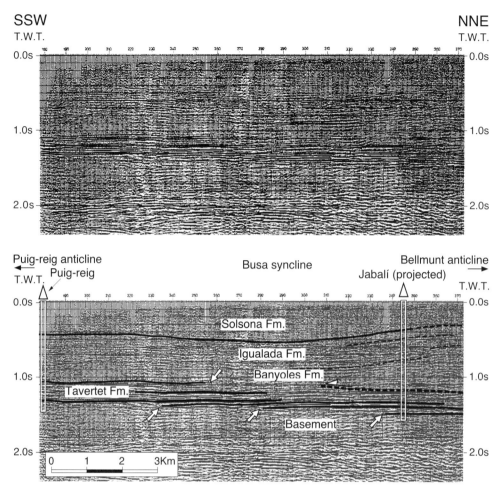

Fig. 6. Seismic line Seismic line S-7 (UERT, S.A., 1977, fold stack) and interpreted line drawing. The line is NNE-SSW and crosses the Busa syncline to the north and the Puig-reig anticline to the south as shown in Fig. 3. The line shows the geometry of the northern part of the Ebro basin, especially the southward migration of the lower and middle Eocene shallow marine platform limestones (constrained by Jabalí and Puig-reig boreholes), represented by the lower set of subhorizontal reflectors of the line. Arrows point to the northern and southern ends of individual reflectors. Upward-pointing-arrows indicate the southward migration of platforms. Vertical and horizontal scales are similar.

of carbonate platform development during the fourth cycle coincides with the onset of generalized clastic deposition by large fan-delta and fluvio-deltaic complexes around the Ebro basin (Fig. 4).

These relationships between carbonate platforms and clastic wedges are well exposed along the N–S, forelandward direction of facies migration in the Jaca piggyback basin (JB in Fig. 2). Shallow-water carbonate platforms developed in the distal margin of the Jaca basin, whereas a thick unit of northerly derived siliciclastic turbidites (Hecho Group) filled the basin during a 9–10 Ma long interval spanning from Ypresian to late Lutetian (e.g. Labaume et al. 1985; Barnolas & Teixell 1994). Eight megabreccia units are recognized within the turbidites, interpreted as single depositional events produced by either episodes of thrusting in the northern side of the trough (Johns et al. 1981; Labaume et al. 1985) or episodes of drowning and collapse of the southern carbonate platform during the migration of the tectonic flexure toward the foreland side of the basin (Barnolas & Teixell 1994). The southern depocenter migration during this period of time was 3–5 mm a^{-1} (Labaume et al. 1985; Barnolas & Teixell 1994).

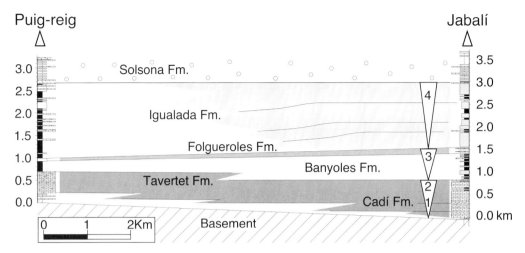

Fig. 7. Interpretative stratigraphic panel from Jabalí to Puig-reig oil-wells supported by seismic data. The lower and middle Eocene limestones migrate toward the foreland coevally with the migration of the foreland flexure. Triangles represent depositional cycles.

Palaeobathymetry is an important factor in subsidence analysis. For this reason we utilize multiple approaches to estimate palaeowater depths. Palaeobathymetries of southeastern Pyrenean foreland platform and shallow-water deposits are well constrained by larger foraminifera fossil record. At present, *Alveolina* develop between 10 and 60 m depth and are especially abundant between 25 and 35 m whereas the range of *Nummulites* is between 30 and 130 m with their maximum abundance at 45–85 m (Hottinger 1977; Reiss & Hottinger 1984). The palaeobathymetric values for the Armàncies and Campdevànol Fms presents more uncertainties because they lack good palaeobathymetric indicators. Planktonic foraminifera, are almost absent in these sediments. Within the Armàncies Fm and in the front of carbonate platforms, the megabreccias extend for around 20 km along the strike of the basin centre (Vergés et al. 1994; Muñoz et al. 1994), as other modern analogs in active tectonic settings, that extend along 30 km downslope and 16 km along the axis of the basin (Hine et al. 1992). This modern-analogue massive deposition occurred above slopes of 2.4°, reaching depths of up to 300 m. Although it is difficult to compare different tectonic scenarios, we think that the second depositional cycle follows a deepening–shallowing pattern, starting at very shallow depth of c. 15 m above the upper layer of the Corones Fm and increasing the depth to the base of the Campdevànol turbidites at a maximum depth of c. 300 m. From this point, there is a shallowing sense of the bathymetry, very well constrained by a *Turritella* horizon, located on the lowermost part of the Beuda Fm (deposited around 50 m below sea level), to the final very shallow-water or even subaerial exposure at the top of the Beuda evaporites (Ortí et al. 1988). Although a more precise palaeobathymetric analysis would be desirable, especially for the Armàncies and Campdevànol Formations, the values discussed above give us adequate lower and upper palaeobathymetric boundaries for the subsidence analysis presented below (Table 1).

Palaeobathymetry is less constrained for continental deposits. We assumed an average altitude of 50 m (0 to –100 m) above sea level for Bellmunt and Solsona fluvial deposits. Fluvial sediments of the Bellmunt Fm were deposited, at least partially, after the Ripoll basin become a piggyback basin on top of the Vallfogona thrust. However, the Ripoll basin was not completely disconnected from the main foreland basin to the south where marine sedimentation represented by the Coubet and Banyoles Fms prevailed during this interval (Muñoz et al. 1994; Saula et al. 1994). Alluvial and fluvial Solsona deposits in the basin centre were deposited after the closure of the south Pyrenean foreland basins toward the Atlantic Ocean at the Bartonian–Priabonian boundary. These deposits produced an extensive backfilling of the basin and external parts of the Pyrenean thrust belt (Coney et al. 1996) that could be concurrent with a continuous increase of the palaeobathymetry within the basin. However, we assumed a fairly low altitude between 0 and 100 m above sea level

Table 1. *List of initial parameters used in this subsidence analysis for each lithology (rock densities, initial porosities and porosity coefficients)*

Lithology	Shales	Siltstones	Sandst.	Cong.	Marls	Limest.	Gypsum	Anhydrite
Initial Porosity ϕ_o	0.63	0.59	0.49	0.5	0.58	0.5	0.5	0.5
Constant c (m^{-1})	0.00051	0.00045	0.00027	0.0003	0.00059	0.0007	0.0009	0.0009
Density (g cm^3)								
General	2.50	2.53	2.60	2.60	2.50	2.67	2.31	2.85
Garumnian		2.51	2.54			2.67		
Cadi Fm						2.69		
Orpi Fm						2.65		
Sagnari Fm					2.58			
Corones Fm			2.62		2.56	2.66		
Amàncies Fm					2.50	2.60		
Campdevànol Fm		2.55	2.69		2.48			
Beuda Fm								2.85
Igualada Fm					2.35			
Solsona Fm			2.53					

for the Solsona Fm deposits (Table 2). Finally, above sea-level palaeobathymetries for Montserrat fan-delta deposits are deduced from the fan-delta palaeotopographic reconstructions (López-Blanco *et al.* in press; Table 2).

Subsidence analysis

Subsidence analysis is made along a N–S profile, parallel to the direction of the Pyrenean tectonic transport, along the southeastern Pyrenean foreland basin from the Ripoll piggyback basin in the north to the Catalan Coastal margin of the Ebro basin in the south (Figs 3, 8). Local isostatic backstripping analysis is carried out on six detailed stratigraphic sections along this N–S transect. The northernmost, outcrop-based, Gombrèn stratigraphic section spans the northern flank of the Ripoll syncline (Cadí thrust sheet). The presented section is based on a detailed unpublished section from UERT, S.A. (1977) shown in Fig. 8. The section records the continental Palaeocene section that underlies the first and second depositional cycles and is combined with good magnetic time-control for the Eocene sequence (Bagà and Gombrèn magnetostratigraphic sections; Burbank *et al.* 1992a; Fig. 3). South of the Vallfogona thrust, in the northern crops out of the Ebro basin, the Jabalí and Puig-reig oil-wells cut the complete Eocene marine depositional sequence (first to fourth cycles). The Santpedor and Castellfollit oil-wells characterize the central and southern margins of the Ebro basin. These two sections, as well as the outcrop-based Montserrat stratigraphic column represent mostly the fourth depositional cycle.

The Montserrat section and magnetostratigraphy are based on López-Blanco *et al.* (in press).

The stratigraphic intervals used in this subsidence analysis correspond to lithostratigraphic formations that show fairly homogeneous sedimentary facies. We assume a constant rate of sedimentation within well-constrained intervals of time. The lithological percentages for each interval have been resolved from calculations based on detailed 1:500 sections, except for the Santpedor section in which information was less reliable and the section was incomplete in its upper part corresponding to the Solsona Fm.

This subsidence analysis, assuming Airy isostasy, is made from decompaction calculations and uses estimates of palaeobathymetry. Decompaction calculations are based on van Hinte (1978) using a simple exponential relationship for changes in porosity with depth. Initial porosities and constant c are based on Sclater & Christie's (1980) values for detritic and carbonatic rocks and on Sonnenfeld's (1984) values for evaporitic rocks. To determine possible errors during decompaction we calculate decompacted thickness of the section using a more accurate algorithm (Allen & Allen 1990; Angevine *et al.* 1990). The formal errors have been omitted as they are insignificant (a minimum of 1.1% in the Montserrat section to a maximum of 4.6% in the Gombrèn section). The errors associated with the specific choices of porosities and compaction exponent c are unknown. They are probably considerably greater than the formal errors cited above, but they are unquantified and are ignored here. Table 1 lists initial porosities and constant c for each lithology.

The backstripping technique that allows us to calculate tectonic subsidence is based on Steckler & Watts (1978). We applied specific final densities for both particular lithologies and formations based on regional studies by Rivero (1993) as well as evaporitic rock densities based on Sonnenfeld (1984), that are listed in Table 1. We used 3.3 g cm^{-3} for mantle and 1.0 g cm^{-3} for water densities. Equations for continental basins, in which tectonic subsidence is calculated for an air-loaded basin, rather than a water-loaded basin, are applied to the Montserrat section and decrease the calculated tectonic subsidence by a factor of 30% with respect to marine equations. Nevertheless, we also analysed this section using only marine equations to compare the results.

Conservative estimations of palaeobathymetry for marine and continental deposits have been discussed in the preceding section and are listed in Table 2.

Because the uncertainties on the correlation between sea-level curves and South Pyrenean foreland basin infill, especially for short-term events like the Cardona salt deposition, we do not include them in the presented backstripping analysis. First-order cycle sea level decreases regularly from c. 220 m at the base of the Eocene to c. 190 m at the top of the Eocene (Vail et al. 1977; Haq et al. 1987). The Beuda and particularly the Cardona evaporitic intervals approximately correspond to marked sea-level falls (−100 m) in the published curve (as noted in Burbank et al. 1992a), and may introduce additional uncertainties to the subsidence analysis. The Beuda and Cardona evaporitic successions are characterized by shallowing-upward sequences culminating in subaerial exposure at the top of the unit in the case of the Beuda succession (Ortí et al. 1988). The decreasing palaeobathymetry related to these units might reflect the variation of the ratio between tectonic subsidence and variations in sea-level. If we hold the tectonic subsidence constant for these short intervals of time, then the shallowing-upward sequence would be produced only by a sea-level drop. However, if the shallowing-upward palaeobathymetry observed in the Beuda unit is related to a sea-level fall, then the tectonic uplift associated with this evaporitic unit would be apparent (see Gombrèn subsidence curve in Fig. 8).

Apparently abrupt slope changes in the subsidence curves are primarily artifacts of the calculation technique in which constant rates are calculated between dated endpoints that are widely separated in time. In addition use of local Airy isostasy tends to amplify apparent rate changes that would be more subdued if lateral flexural strength were taken into account. These abrupt changes have no geological significance as can be deduced by the widespread gradational contacts between different depositional units. The uppermost interval of most of the analysed sections corresponds to continental deposits that have been partially eroded, such as the Bellmunt Fm in the Gombrèn section and the Solsona Fm south of the Vallfogona thrust. Subsidence curves for these uppermost segments of the sections have been determined using marine equations for water-loaded crust and assuming the extrapolated age of the top of each section represents its final stage of deposition (Figs 8 & 9). Table 2 lists the input parameters (age of the top of the selected interval, present thickness, final density of the whole interval, and bathymetry) and results (total and tectonic subsidence and decompacted thickness) for each of the analysed sections.

Subsidence and tectonics: geohistory

The subsidence curves display a typical foreland basin convex-up shape with an inflection point dating the onset of the most intense period of tectonic subsidence (Figs 8 & 9). The Gombrèn section displays the first inflection point at 55.9 Ma coinciding with the onset of marine conditions within the basin after a long period of time characterized by widespread continental Palaeocene deposition (Figs 4 & 8). The second inflection point, at 50.7 Ma coincides with the basal part of the second depositional cycle represented by marls and calcareous marls of the Armàncies Fm. During deposition of siliciclastic turbidites of the Campdevànol Fm, there is a marked decrease in tectonic subsidence values (Figs 8 & 9). During Beuda deposition, the tectonic subsidence pattern starts to show an upward vector of motion that is distinctive of the latest evolution of the Ripoll basin as a piggyback basin during the Bellmunt Fm fluvial deposition. Nevertheless, as has been noted previously, a sea-level drop of c. 100 m would be sufficient to invert the calculated upward disposition of the total subsidence curve during the Beuda interval.

The Jabalí and Puig-reig subsidence curves show the southward migration of their first inflection point. In the Jabalí section the first inflection point corresponding to a considerable increase in tectonic subsidence is located around 46.26 Ma, several millions of years later than the Gombrèn inflection points. However, the most important inflection point in this curve corresponds to 43 Ma as well as in the Puig-reig

Table 2. *List of stratigraphic unit age, present thickness, final density and bathymetry for each of the analysed intervals. Total and tectonic subsidence as well as decompacted thickness for each sedimentary interval of the analysed sections are given.*

Gombrén

Unit	Age (Ma)	Present thickness (m)	Density (g cm⁻³)	Bathymetry min (m)	Bathymetry max (m)	Total subsidence (m)	Tectonic (m)	Decompacted thickness (m)							
Ma:								55.90	52.05	50.69	49.36	46.85	46.18	45.82	42.65
Bellmunt	42.65	1203.4	2.56	−100	0	4430.6	2053.9								1203.4
Coubet	45.82	126.8	2.52	0	30	3741.7	1829.7							195.7	126.8
Beuda	46.18	170.0	2.39	0	100	3702.9	1844.7						278.8	244.6	170.0
Campdevànol	46.85	723.2	2.54	100	300	3721.7	1945.5					1064.3	940.1	895.4	723.2
Armàncies	49.36	1076.6	2.52	15	300	3097.4	1704.2				1624.0	1212.9	1197.7	1168.5	1076.6
Corones	50.69	225.6	2.60	−5	15	1757.8	995.7			423.6	259.4	241.4	238.6	235.9	225.6
Sagnari	52.05	465.0	2.59	0	75	1555.0	934.8		857.1	705.0	520.4	491.1	485.7	480.3	465.0
Paleocene	55.90	490.0	2.59	−100	0	793.9	478.4	843.9	660.4	624.2	536.1	512.0	512.0	506.3	490.0
Basement	65.00	4480.6				0.0	0.0	843.9	1517.5	1752.8	2939.9	3251.7	3652.9	3726.7	4480.6

Jabalí

Unit	Age (Ma)	Present thickness (m)	Density (g cm⁻³)	Bathymetry min (m)	Bathymetry max (m)	Total subsidence (m)	Tectonic (m)	Decompacted thickness (m)						
Ma:								49.0	46.26	43.0	41.5	40.0	36.81	33.77
Solsona	33.77	698.0	2.52	−100	0	3650.0	1767.8							698.0
Igualada-Tossa	36.81	1410.0	2.53	30	150	3364.3	1727.1						1621.5	1410.0
Folgueroles	40.0	155.0	2.59	100	150	2173.1	1184.7					227.9	166.1	155.0
Banyoles	41.5	580.0	2.55	50	100	2004.7	1090.2				914.6	834.4	609.7	580.0
Tavertet	43.0	417.0	2.65	10	30	1286.8	697.5			699.9	515.8	496.2	430.8	417.0
Penya	46.26	143.0	2.72	10	15	726.6	419.2		259.0	190.7	163.4	161.5	146.0	143.0
Cadí	49.0	297.0	2.68	0	30	528.0	322.8	513.0	455.1	376.2	335.9	328.1	300.2	297.0
Basement	55.90	3700.0				0.0	0.0	513.0	714.1	1266.8	1929.7	2048.1	3274.3	3700.0

Puig-Reig

Unit	Age (Ma)	Present thickness (m)	Density (g cm⁻³)	Bathymetry min (m)	Bathymetry max (m)	Total subsidence (m)	Tectonic (m)	Decompacted thickness (m)						
Ma:								49.0	46.26	43.0	41.5	40.0	36.81	34.36
Solsona	34.36	563.2	2.54	−100	0	3175.7	1604.9							563.2
Igualada-Tossa	36.81	1662.8	2.44	30	150	3013.3	1640.6						1896.6	1662.8
Folgueroles	40.0	97.5	2.54	100	150	1556.7	940.4					175.9	101.1	97.5
Banyoles	41.5	218.0	2.52	50	100	1401.4	841.9				402.8	363.3	226.0	218.0
Tavertet	43.0	127.0	2.64	10	30	1078.6	641.3			226.6	183.4	175.1	129.9	127.0
Penya	46.26	96.7	2.67	10	15	925.6	572.3		177.9	156.1	132.8	127.1	98.9	96.7
Cadí	49.0	460.5	2.59	0	30	820.9	530.1	805.9	735.2	675.9	607.3	580.2	470.8	460.5
Basement	55.90	3225.7				0.0	0.0	805.9	913.1	1058.6	1326.3	1431.6	2923.3	3225.7

QUANTIFIED VERTICAL MOTIONS AND TECTONIC EVOLUTION

Santpedor							Ma:	52.7	41.5	40.0	36.81	36.15
Solsona	36.15	150.0	2.53	-100	0	1690.0	963.7				–	150.0
Igualada-Tossa	36.81	920.0	2.44	30	150	1742.4	1067.1				967.6	920.0
Collbas	40.0	160.0	2.58	0	50	926.9	601.4			212.0	162.5	160.0
Pontils	41.5	360.0	2.52	-10	0	762.8	509.1		580.9	511.2	370.4	360.0
Orpí	52.7	150.0	2.61	0	30	262.9	177.8	–	186.9	178.7	151.9	150.0
Basement	55.90	1740.0				0.0	0.0	247.9	767.8	901.9	1652.4	1740.0

Castellfollit							Ma:	52.7	41.5	40.0	36.81	34.31
Solsona	34.31	574.0	2.52	-100	0	1885.0	1077.4				–	574.0
Igualada–Cardona	36.81	683.0	2.41	30	150	1687.7	1062.5				860.1	683.0
Collbas	40.0	108.0	2.52	0	50	1016.2	675.7			177.6	119.3	108.0
Pontils	41.5	521.0	2.53	-10	0	891.5	603.1		838.1	756.0	567.0	521.0
Orpí	52.7	49.0	2.65	0	30	98.6	68.1	83.6	58.4	57.6	51.3	49.0
Basement	53.90	1935.0				0.0	0.0	83.6	896.5	991.2	1597.7	1935.0

Montserrat							Ma:	41.52	41.25	40.13	38.42	38.11	37.60	37.47	37.2
(17.1n)	37.2	370.0	2.60	-130	-60	1796.8	661.7							–	370.0
	37.47	50.0	2.60	-100	-60	1552.4	592.7							56.0	50.0
(17.2.3n)	37.60	432.5	2.58	-100	30	1560.7	622.6						482.4	473.3	432.5
	38.11	28.0	2.55	20	30	1268.5	556.3					–	30.9	30.4	28.0
(1.8n)	38.42	399.5	2.57	-100	20	1168.9	476.6				–	38.4	433.4	433.4	399.5
	40.13	68.0	2.57	-100	0	752.6	305.1			–	501.3	501.3	72.4	71.2	68.0
(1.9n)	41.25	120.0	2.58	-100	0	684.3	279.3		166.5	95.5	81.6	80.1	127.5	125.5	120.0
Mediona-La-Salut	41.52	423.8	2.57	-100	0	551.8	225.4	–	166.5	160.0	140.7	138.3			
Basement	55.90	1891.8				0.0	0.0	601.8	567.7	547.1	485.3	485.3	449.1	442.5	423.8
								601.8	734.2	802.6	1208.9	1243.4	1595.7	1623.3	1891.8

Montserrat section calculated with continental and marine equations (shaded intervals). The ages of the continental Orpí and Pontils Fms are fixed although they could become younger to the south as discussed in text.

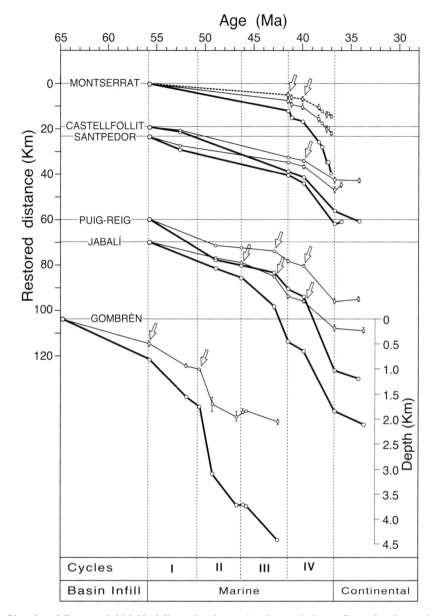

Fig. 9. Plot of total (lower and thick black lines of each curve) and tectonic (upper lines of each curve) subsidence versus time. The position of each profile is restored to its pre-thrusting position according to the restored version of the cross-section shown in Fig. 8 (Vergés 1993; Vergés & Burbank 1996). As in Fig. 8, tectonic subsidence curves are calculated for marine equations using the age of the uppermost strata preserved in the section as a final age-value of the analysis. The Montserrat curve shows results using both continental (dashed thin black line) and marine equations (thin black line) for the sake of comparison. Arrows show the position of the inflection points related to the initiation of subsidence events. From Gombrèn to Montserrat, the migration of the flexural wave occurred at a rate of c. 10 mm a^{-1} throughout 104 km of the southern Pyrenean foreland basin.

section. This period of strong subsidence corresponds to deposition of marly deposits of Banyoles Fm. The younger segments of these curves display smaller amounts and rates of tectonic subsidence during deposition of alluvial and fluvial deposits corresponding to the Solsona

Fm. To the south of these two wells, the Santpedor and Castellfollit subsidence curves show their first and unique inflection point at 40.0 Ma as well as the Montserrat curve. However, the Montserrat section shows a previous and very rapid period of tectonic subsidence at 41.5 Ma at the base of the conglomeratic Montserrat Fm that contains growth strata geometries proving its unambiguous syntectonic character (López-Blanco 1993). Although time control in the Santpedor and Castellfollit sections is not so well constrained as it is in the Montserrat section, the short-term high-rate of tectonic subsidence determined at 41.5 Ma suggests that Catalan Coastal Ranges influenced subsidence in the southern segment of the Ebro foreland basin. The product of this tectonic activity in both margins of the Ebro basin during middle Eocene time resulted in a double flexure as observed in cross-section (Fig. 8) and two opposite prograding deltaic clastic wedges related to the active margins of the foreland basin (Fig. 4).

Based on the combination of balanced and restored cross-sections and palinspastic maps (Vergés 1993; Vergés & Burbank 1996), Fig. 9 shows the tectonic subsidence curves in a restored position with the Montserrat section used as a pin point. This restoration permits calculation of the rate of southward migration of the position of the first inflection point and thus the onset of the tectonic subsidence within different segments of the Ebro foreland basin. The rate of flexural wave migration across 104 km of basin width (from Gombrèn to Montserrat sections) is c. 10 mm a^{-1}. At the same time along the southern margin of the basin, the shallow-water carbonate platforms (as imaged in seismic line S-7, Fig. 6) migrate at the same velocity to maintain the particular bathymetric conditions required for their growth (backstepping foreland carbonate ramps; Dorobek 1995).

Subsidence in the foreland basins is coupled to the emplacement of tectonic loads within the orogen (Price 1973; Beaumont 1981; Jordan 1981). The parallel trend of the subsidence transect and the Pyrenean tectonic transport permits investigation of a direct relationship between tectonics and subsidence. The ages of the emplacement of southeastern Pyrenean thrust sheets have been previously documented based on cross-cutting relationships between thrusts and syn- and post-thrusting sediments dated by classical methods (Puigdefàbregas et al. 1986; Vergés & Martínez 1988). A more complete approach on the timing of thrusting was presented by Burbank et al. (1992a,b), using both magnetostratigraphy and subsidence analysis. Thrust emplacement ages were recalibrated for the Cande & Kent (1992) global polarity time scale in Vergés & Burbank (1996), whereas the links between cover and basement thrust sheets as well as a compilation of ages of thrusting along a complete N–S crustal transect crossing the Pyrenees was presented in Vergés et al. (1995). This subsidence analysis represents a complementary and independent constraint to date thrust development and will allow us to determine better the onset of different events of thrust, the geometry and rates of tectonic subsidence, and the rate of its propagation across the foreland (Figs 9 & 10).

The strong subsidence events determined in the northernmost Gombrèn section, cropping out in the northern flank of the Ripoll syncline, reflect the emplacement of the Pedraforca thrust sheets (the upper southern Pyrenean thrust sheets: Muñoz et al. 1986). Tectonic subsidence starts at the base of the first depositional cycle (55.9 Ma) and strongly increases at the base of the second cycle (50.69 Ma). The parallel increase of the rate of tectonic subsidence, to 0.53 mm a^{-1}, is related to the southward emplacement of the Pedraforca cover thrust sheets and Nogueres basement unit. The submarine and rapid emplacement of the front of these tectonic units is demonstrated by syntectonic sediments containing blocks and olistolites derived from the front of the overriding thrust sheets (Vergés & Martínez 1988) and palinspastic maps (Vergés & Burbank 1996). At basin-scale, relatively deep marine sedimentation (Armàncies and Campdevànol Fms), recording the underfilled stage of the foreland basin, coincides with the period marked by highest rates of tectonic shortening (> 4 mm a^{-1}), rapid thrust-front advance and a first stage of Pyrenean orogenic growth characterized by a relatively low topographic relief (Vergés et al. 1995). This parallelism of underfilled basin conditions and high rates of tectonic advance has been observed in the Swiss Molasse basin (Allen et al. 1991) and simulated with foreland basin modelling (Flemings & Jordan 1989; Sinclair et al. 1991).

Decreasing tectonic subsidence values and rates (0.10 mm a^{-1}) recorded during the turbiditic deposition of the Campdevànol Fm between 49.36 and 46.85 Ma, are related in this work to the initial deformation within the south Pyrenean foreland basin before its partition into a transported Ripoll piggyback basin to the north and a major foreland basin to the south. This deformation was achieved by a combination of layer parallel shortening, folding and thrusting probably related to the forelandward propagation of the Vallfogona thrust across foreland deposits that triggered the large

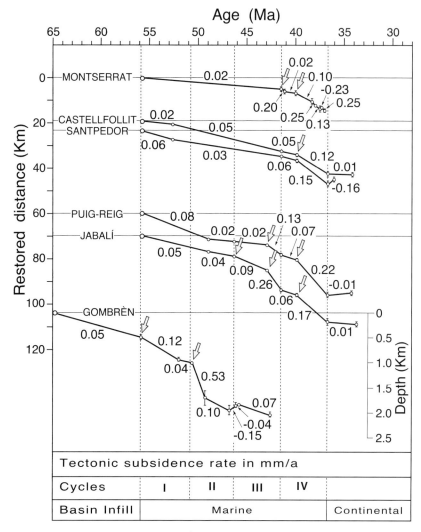

Fig. 10. Plot of tectonic subsidence rates in mm a^{-1} using reference time slices. The position of each of the profiles is restored to the pre-thrusting position as in Fig. 9. Maximum subsidence rates are as high as 0.53 mm a^{-1} (from 50.69–49.36 Ma) in the Ripoll basin. These rates decrease, migrate and become younger toward the foreland ranging from 0.26 mm a^{-1} (43–41.5 Ma at 34 km away of the Vallfogona thrust) to 0.22 mm a^{-1} (40.0–36.8 Ma at 44 km away of the same position) to 0.15 mm a^{-1} (same age at 80 km away of the Vallfogona thrust).

number of slumped deposits found within the turbidites of the Campdevànol Fm, after 49.36 Ma. Following this interpretation, the deformation within the Ripoll basin started before the final emplacement and fossilization of the lower Pedraforca thrust sheets at around 47 Ma, as determined by cross-cutting relationships (Martínez et al. 1988; Vergés et al. 1994) and middle Lutetian basin reconstruction (Vergés et al. 1995).

The general decrease of tectonic subsidence (0.07 mm a^{-1}) related to the fluvial deposition of the Bellmunt Fm is attributed to the onset of upward motion of the Ripoll basin on top of the Vallfogona thrust ramp at 42.65 Ma. To the south, the Jabalí section records a first pulse of tectonic subsidence with rates of almost 0.1 mm a^{-1} at 46.26 Ma and a stronger pulse with rates of 0.26 mm a^{-1} at 43 Ma that is also recorded in the Puig-reig section (0.13 mm a^{-1}). Tectonic subsidence registered in Jabalí and Puig-reig sections was persistent reaching values of 0.17 and 0.22 mm a^{-1} from 40.0 to 36.81 Ma, although the last value could be overestimated because tectonic duplication is inferred underneath the Puig-reig anticline. These events of subsidence are interpreted to result from the emplacement of the Cadí thrust sheet and attached Orri basement unit, coincident with deformation of syntectonic sediments located in the footwall of the Vallfogona thrust, that took place from at least 42 Ma to the late Eocene–early Oligocene time (Burbank et al. 1992a; Mató et al. 1994). Deformation within the footwall of the Vallfogona thrust is widely recognized in both geological maps and seismic lines. The geological map of Berga shows different stratigraphic units bounded by unconformities and affected by different sets of folds and thrusts. This is exemplified by unconformities that seal older sets of folds and are folded by a new group of tectonic structures (Mató et al. 1994). According to Mató et al. (1994), these growth strata units correspond to the fourth depositional cycle. South of the Vallfogona thrust, the western termination of the Bellmunt anticline is also coeval with the fourth depositional cycle, starting after 41.5 Ma (Fig. 7). Deformation of the foreland-basin strata above a regional detachment level is associated with emplacement of the basement Rialp unit after 37 Ma (Vergés et al. 1995) and to the activity as an out-of-sequence fault of the Ribes–Camprodon thrust (Burbank et al. 1992a).

Abrupt increase of tectonic subsidence at 41.5 Ma in the southernmost margin of the Ebro basin is linked to the onset of tectonic activity related to the front of the Catalan Coastal Ranges involving basement rocks (López-Blanco et al. in press). It should be noted that this increase of tectonic subsidence is slightly recorded in Castellfollit and Santpedor sections, but not in the more northern Puig-reig and Jabalí ones. This differential subsidence response agrees with the doubly flexed geometry of the basin determined in the regional cross-section (Vergés & Burbank 1996; Fig. 8). The onset of strong deformation along the southern margin of the basin is at least partly coeval with the forelandward propagation of deformation in the Ebro basin, suggesting that the initial tectonic activity of the deep-seated thrust of the front of the Catalan Coastal Ranges could be triggered by the transmission of stresses throughout the upper crustal levels of the Iberian plate from its northern margin underneath the Pyrenees.

All the curves within the Ebro basin show either decreasing rates of tectonic subsidence or tectonic uplift during the continental deposition of the Solsona Fm following the marine Cardona salt deposition at 36.8 Ma. Post-thrusting evolution of the Ebro basin is important to complete the history of vertical motions of any of the sections that are located south of the Vallfogona thrust and which are affected by tectonic uplift related to the extensional fault system developed during the opening of the València Trough starting at c. 25 Ma (uppermost Oligocene–lower Miocene; Bartrina et al. 1992) but probably effective at 16 Ma coinciding with increasing sediment accumulations in the València Trough continental platform (Roca 1992).

Geohistory of Gombrèn and Jabalí sections across the Vallfogona thrust shows distinctive vertical motions, especially after deposition of relative deep-marine, marly Armàncies and turbiditic Campdevànol Fms (Fig. 11). Development of the Ripoll syncline was coeval with fluvial Bellmunt Fm deposition. Uppermost flat-lying conglomerates display an angular unconformity with the lower Eocene marine series cropping out in the northern flank of the Ripoll syncline and permit quantification of synthrusting rock uplift and denudation achieved during Oligocene times (Vergés et al. 1995). Using a linear geothermal gradient of $30°C\ km^{-1}$ that is slightly high for foreland basins, it is possible to estimate the palaeotemperature of the section during its burial history (Fig. 11). This simple correlation indicates that the lower part of the Armàncies Fm was buried to 3250 m reaching a maximum temperature of c. $100°C$. These results are in agreement with determinations of Tmax of the lower part of the Armàncies Fm source rock indicating that it moved through the upper part of the oil window area (Permanyer et al. 1988; Clavell 1992). The comparison of geohistories across the Vallfogona thrust shows opposite vertical motions. The Gombrèn geohistory indicates a general uplift event during the southern transport of the Cadí thrust sheet on top of the Vallfogona thrust whereas to the south of this thrust, the Jabalí geohistory shows pervasive subsidence to the end of Solsona Fm deposition. Using the same geothermal gradient of $30°C\ km^{-1}$ the base of the section reached 3650 m of palaeoburial depth corresponding to $110°C$ of palaeotemperature.

Discussion

Subsidence analysis

The subsidence analysis presented in this study is carried out along a N–S transect, parallel to the tectonic transport direction. Although we do not incorporate the flexural component of the plate, the calculated amounts and rates at which the foreland basin has subsided due to the

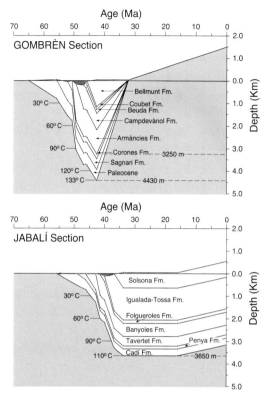

Fig. 11. Geohistory of the Gombrèn and Jabalí sections located in both sides of the Vallfogona thrust. An assumed (slightly high) geothermal gradient of 30°C km^{-1} is plotted to show palaeotemperatures during burial. Geochemical data from the lower part of the source rock Armàncies Fm indicate that it passed through the upper part of the oil window whereas the presented geohistory suggests that it did not. The most convincing explanation is that a part of the Bellmunt Fm has been eroded after the end of deposition in mid- Eocene times. Uplift of the northern flank of the Ripoll syncline occurred during thrusting and before the deposition of continental Tertiary flat-lying conglomerates that unconformably overlie the south dipping lower and middle Eocene series. The Jabalí geohistory shows a younger age for the onset of subsidence, a renewed subsidence event during uplift in the Gombrèn section and widespread subsidence to the end of deposition.

Pyrenean tectonic loading are adequate to show the relative vertical motions throughout the basin. Each of the studied sections shows a convex-up subsidence curve with one or two intervals of rapid tectonic subsidence marked by inflection points. The almost flat segment of the curve before the first inflection point represents a period of slow sedimentation, hiatus or even erosion that has not been corrected due to the lack of accurate information. If our interpretation of seismic line S-7 is correct (Fig. 6), in which the shallow marine platforms (Penya and Tavertet Fms) migrate to the south with basal and top boundaries also becoming younger to the south, then the nearly flat segments of the subsidence curve would show a different geometry than that indicated by the subsidence curve: the initial long segment of the curve would be shorter and would start later.

In the southernmost sections analysed in this work, the thicknesses have not been corrected for overburden due to post-thrusting eroded continental deposits. Additional depositional loads in these sections can decrease or even reverse the upward tendency of the upper segment of the tectonic subsidence curve as shown in the next section.

Post-thrusting Ebro foreland basin history

The Post-thrusting history of the Ebro basin is more difficult to unravel because it has been characterized by erosion, especially in its southeastern margin. Uplift and erosion of the southern part of the basin is related to the development of onshore normal faults oriented roughly parallel to the coast (Fig. 3) and linked to the development of the València Trough (Lewis et al. 1996), starting in latest Oligocene–Early Miocene time. In order to estimate the amount of denudation related to the southeastern segment of the Ebro basin, two coals (five specimens) from Santpedor and Calaf (see location in Fig. 3) have been studied to determine the vitrinite reflectance expressed as mean values of random reflectance (R_m). These samples can be projected to near the top of the Castellfollit oil-well in cross-section. The two coals furnished R_m values between 0.44% and 0.48% with deviations of 0.05. However, in this study we compare these results with R_m values provided by oil-wells more to the north. In the studied transect, the Puig-reig oil-well shows R_m values of 0.45% from 1.45 to 1.9 km of depth (Clavell 1992) and the Jabalí one at 1.65 km. However, the best data come from the Basella oil-well, which cuts a thick and complete Solsona Fm continental succession (located westward of the study transect; see oil-well 96 in Fig. 3). This oil-well shows R_m values of 0.45% at a depth of 2.9 km (Clavell 1992). The correspondence between R_m values and burial depths observed, especially in the Basella oil-well, is consistent with estimations of paleotemperatures and palaeodepths calculated using standard conversion methods and thus corroborating the interpretation that observed depths of 2.9 km

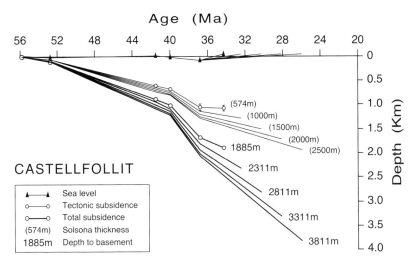

Fig. 12. Interpretative subsidence analysis for the Castellfollit section assuming extra amounts of sediments on top of the section to account for potential erosion since the end of deposition. These extra amounts are slices of 500 m up to a total Solsona Fm of 2500 m (see discussion in the text). The subsidence curves are calculated using marine equations, the age of the uppermost strata preserved in the section as a final age-value of the analysis and a mean palaeobathymetry for the extra continental deposits of 50 m above sea level.

correspond to maximum burial depths. According to this correlation between R_m values and observed burial depths, the calculated R_m of 0.44–0.48% located at the surface strongly suggests a palaeoburial depth of 2–3 km for these sediments. The correlation of R_m values along the studied N–S transect implies differential denudation south of the Vallfogona thrust that increases to a maximum close to the Catalan Coastal margin.

To test the effects in the subsidence analysis of these inferred extra amounts of material in the southern side of the Ebro foreland basin, we added slices of 500 m of strata on top of the Castellfollit section to up to 2000 m (Fig. 12). The subsidence analysis is made using marine equations. These calculations assume a constant negative bathymetry above sea-level between 0 and −100 m for the upper slice, post-sedimentation uplift to the present position with concomitant erosion, and ages of the topmost strata as estimated by extrapolation. The age of the topmost part of each of these slices has been calculated using mean values of sediment accumulation rate of 0.23 mm a^{-1} as described previously. The age of the top of the total 2500 m of Solsona Fm continental fluvial strata would correspond to the Late Oligocene (c. 26 Ma). This extrapolated age agrees with the well-determined upper Oligocene age reported from equivalent strata attached to the Pyrenean front, cropping out 40 km to the northwest in Artesa del Segre (Meigs et al. 1996; see location in Fig. 3).

The results of these analysis show that independent of the amount of material added in the top of the Solsona strata, the tectonic subsidence curves for the underlying intervals show little variation. Tectonic subsidence curve corresponding to 574 m (the present Solsona Fm thickness) show almost a flat trend whereas curves corresponding to 1000, 1500, 2000 and 2500 m show downward trend during Solsona deposition. The total subsidence increases as the thickness of the upper interval increases from 1885 m for the top of basement at the end of deposition of 574 m of Solsona Fm to 3811 m at the end of the deposition of 2500 m of Solsona Fm and material located on top of the present section (at 574 m), cropping out today at 729 m above sea-level would be located at a minimum depth of 1.9 km at the end of Solsona Fm deposition (Fig. 12).

Since the palaeoaltitude of the top of the additional Solsona Fm deposits has been fixed at 50 m above sea-level, the removal of almost 2000 m of material would not increase this altitude to the present 729 m by simple isostatic rebound. Only with palaeoaltitudes of slightly over 1000 m would be possible to reach the present altitude due to erosionally driven isostasy. Because it is unlikely that the basin was ever filled to this altitude, the present elevation of the outcrops of Solsona strata suggests significant tectonically driven uplift following deposition. This analysis shows the potential behavior of the foreland basin filled with additional material that presumably has been denuded since the end of deposition in the Ebro basin.

Subsidence and third order depositional cycles

Rapid progradations of continental clastic deposits covering a large part of foreland basins have been associated with periods of tectonic quiescence (Heller *et al.* 1988). In this model, during periods of thrusting, clastic sediments are trapped in rapidly subsiding troughs close to the thrust front. During tectonic quiescence, isostatic flexural rebound triggers erosion in the mountain range and the products prograde into the uplifting basin and thus depositing thin sheets of coarse material unconformably overlying syntectonic series. Using a dynamic model, Jordan & Flemings (1991) reproduce transgressive–regressive third-order marine–continental sedimentary packages by using repeated periods of tectonic activity and quiescence. They reproduce a complete depositional cycle with 4 Ma of thrusting followed by 2 Ma of tectonic quiescence, a result that can not be attained with sea-level variations alone. However, these models also utilize a simple deformation front defined by a single thrust.

According to the above mentioned models and using the second depositional cycle as an example, the talus marls (Armàncies Fm) and siliciclastic turbidites (Campdevànol Fm) would be related to the emplacement of hinterland tectonic loads, whereas the deposition of evaporites (Beuda Fm) would be related to tectonic quiescence. However, widespread syndepositional deformation together with intercalated olistostromes and slumps indicate a roughly continuous forelandward displacement of the Pedraforca–Nogueres thrust sheet during this entire interval. The final emplacement of this large thrust sheet took place on top of the evaporites of the Beuda Fm. According to these data it seems that the transgressive–regressive depositional cycle could be linked to a different mechanism of formation than that proposed in the activity–quiescence models.

The fact that folding and thrusting occurs during the foreland infill has been widely recognized (e.g. Ricci Luchi 1986). The effect of this syn-depositional faulting in the sedimentary infill of a foreland basin has been kinematically modeled using simple cases (e.g. Zoetemeijer *et al.* 1992; Peper 1993). However, the frontal regions of the eastern Pyrenees foreland basin show distributed deformation affecting large areas of syn-tectonic strata. This deformation was produced above detachment levels that could transfer deformation very efficiently. The thickening of the sedimentary pile by layer-parallel shortening, thrusting and folding could be responsible for simultaneous decrease of the available space in the basin (Fig. 13). In such a case, the rapidity of this thickening process would depend on the efficiency with which the detachment level propagates the deformation far away from the previous deformation front and the velocity at which this process could take place.

Furthermore, overthickened depositional units give overestimated amounts and rates of subsidence, depending on the lithologies. Several locations of the apparently non-deformed eastern Ebro foreland basin display tectonic thickening above 20% (Casas *et al.* 1996). We agree with DeCelles & Giles (1996) that distinction of different segments of foreland basins as well as the controls on particle paths during foreland basin evolution are important to understanding and interpreting subsidence analysis.

Comparison with other European foreland basins

Tertiary Western European foreland basins are relatively complicated, commonly partitioned,

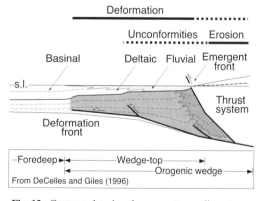

Fig. 13. Cartoon showing the geometry, sedimentary facies and deformation distribution within a stratigraphic unit affected by an advancing buried deformation front. Tectonic thickening of the strata located above a detachment level (shaded) produces a decrease of available space on this segment of the basin. Efficient detachment levels within the basin and concomitant overthickening of large segments of the basin could be responsible for the transgressive part of a third order depositional cycles (at least partially).

and limited in size due to the complex interaction between small plates and rigid basement blocks that controls the geometry of both foreland basins and mountain ranges (Fig. 2). These foreland basins are located between the convergent African and European large-scale plates and developed after the collision of continental blocks moving at relatively low rates. The history of these basins was similar and partially synchronous, with depocentres migrating at the front of forelandward propagating thrusts. The post-foreland basin histories are different depending on the final geodynamic evolution of the basin, and they vary from uplifted basins like the southeastern Pyrenees and Swiss Molasse basins to more or less preserved basins like the northwestern Pyrenean and German Molasse basins.

The available information on transects similarly located with respect to the front of thrusting permits the comparison of subsidence results among our study area and the Aquitaine basin, the Swiss Molasse basin and the German Molasse basin (Table 3).

The Tertiary development of the Aquitaine basin (Fig. 2), formed on top of Mesozoic rifted European plate, was produced by the inversion of Mesozoic extensional structures toward the north (e.g. Daignieres et al. 1994). A local isostatic backstripping analysis from ten oil-wells has been determined by Dubarry (1988). Her results show an increasing subsidence rate starting at around 60 Ma (Thanetian) and culminating at 54–53 Ma (early Ypresian) with rates of tectonic subsidence ranging 0.16–0.2 mm a^{-1} (Dubarry 1988). This rapid phase of subsidence coincided with the emplacement of northern Pyrenean thrust sheets. The end of the rapid subsidence was marked by a progressive decrease of bathymetry at the beginning of continental deposition a few million years later (Table 3). More regional studies suggest that the end of subsidence migrated toward the west, where it was greatest during Oligocene times in positions close to the present boundary of the Bay of Biscay (Desegaulx & Brunet 1990). The control of previous rifted margins also influenced the subsidence results in the Aquitaine basin (Desegaulx et al. 1991).

Located in front of the emergent Alpine thrust front, the Swiss Molasse basin is in part located on top of the detached fold-and-thrust system of the Jura Mountains, although older foreland basin deposits are incorporated in the Alpine thrust system. Analysing the results from the Swiss Molasse basin deposits adjacent to the Alps, we see a comparable subsidence signature to that of the southern Pyrenean foreland.

Homewood et al. (1986), using an Airy model of compensation, determined tectonic subsidence rates similar to the studied Spanish example. These were 0.1 mm a^{-1} at the front of the mountains and 0.06 mm a^{-1} in the distal parts of the basin. The rate of migration of depositional depocentres was 9 mm a^{-1} during Oligocene times but decreased to 2 mm a^{-1} during Miocene times (Homewood et al. 1986).

To the east, the German Molasse basin seems to be largely undisturbed, but is slightly deformed close to the Alpine front. In the German Molasse basin a flexural backstripping analysis defines rates of tectonic subsidence between 43 and 23 Ma. The rates were as high as 0.2–0.6 mm a^{-1} in adjacent to the front (the highest is 0.9 mm a^{-1} during 27–26 Ma), and 0.08–0.3 mm a^{-1}, 60 km away from the front positions, using an effective elastic thickness of 48 km (Jin 1995).

Conclusions

The general pattern of subsidence curves along a N–S transect crossing the South Pyrenean foreland basin shows a coupling between thrusting and flexural response in the basin which produces propagation of depocenters and sedimentary cycles toward the foreland side of the underthrust Iberian plate.

The subsidence curves of the sections analysed in this work show the typical convex-up signature of foreland basins. The Gombrèn section represents an early foreland basin (55.9 Ma to at least 42.65 Ma) that has been transported as a piggyback basin by the Vallfogona thrust starting at around 47 Ma although internal deformation could start 2 Ma earlier (first and second depositional cycles). Maximum rates of tectonic subsidence are 0.53 mm a^{-1} in the Gombrèn section. This relatively short interval of rapid subsidence (50.69–49.36 Ma) was characterized by the deposition of relatively deep marine marls intercalated with slope megabreccia deposits (Armàncies Fm), linked to several northward progradations of shallow-marine carbonate platforms. South of the Vallfogona thrust, maximum rates of tectonic subsidence are around 0.26 mm a^{-1} in the center and southern margin of the Ebro basin. The flexural wave propagated at c. 10 mm a^{-1} across the basin. At this rate, the propagation of the flexural wave took only 10–11 Ma to affect the whole foreland basin, an interval much shorter than the duration of thrusting activity (Vergés et al. 1995).

The activity of the southern margin of the foreland basin, starting at 41.5 Ma, influenced

Table 3. *Comparison of subsidence results among different European foreland basins together with an estimation of the duration of the deep-marine stage and the shallow-marine and continental stage during the foreland basin infill. Maximum rates of tectonic subsidence are shown for localities in both proximal and distal positions with respect to the thrust front*

Locality	(EPD) Eastern Pyrenean basin	(JB) Jaca basin	(AB) Aquitaine basin	(SAB) Swiss Alpine basin	(GAB) German Alpine basin
Backstripping	Local		Local	Local	Flexural (EET = 48Km)
Reference	This work	Labaume et al. (1985)	Brunet et al. (1984)	Homewood et al. (1986)	Jin (1995)
		Barnolas & Teixell (1994)	Dubarry (1988)	Pfiner (1992)	
			Desegaulx et al. (1991)		
Total duration	56–25 MaBP (31 Ma)	56–25 MaBP (31 Ma)	60–25 MaBP (35 Ma)	46–15 MaBP (31 Ma)	43–8 MaBP (35 Ma)
Marine stage	56–37 MaBP (19 Ma)	56–37 MaBP (19 Ma)	60–49 MaBP (11 Ma)	46–34 MaBP (12 Ma)	43–28 MaBP (15 Ma)
Continental stage	37–25 MaBP (12 Ma)	37–<25 MaBP (12 Ma)	49–>25 MaBP (24 Ma)	34–15 MaBP (19 Ma)	28–8 MaBP (20 Ma)

Maximum rates of tectonic subsidence (in mm a^{-1}): frontal-distal locality, at (x km of the front)

Maximum rates	0.53–0.15 (at 60 km)		0.2–0.13 (at ≤15 km)	0.1–0.06	0.9–0.3 (at 60 km)
Flexure migration	c. 10 mm a^{-1}	5 mm a^{-1}	0.3–0.7 mm a^{-1}	9–2 mm a^{-1}	

the subsidence geometry of the basin, especially during the fourth depositional cycle as confirmed by the double flexure of the basin, represented in the geological cross-section. The end of important carbonate production during this last marine depositional cycle, coeval with the progradation of significant siliciclastic systems in the margins of the basin, marks the initial overfilled conditions of the whole basin although tectonic subsidence is still important.

Western European foreland basins developed during Tertiary times by the collision of continental plates flexed downward at a similar maximum tectonic rates of 0.6–0.9 mm a^{-1} close to the tectonic front and $c.$ 0.1 mm/a^{-1} far away from it. Maximum rates of tectonic subsidence are about an order of magnitude smaller than rates of thrust front advance.

This local isostatic backstripping analysis represents a first step in the quantification of vertical motions in the south Pyrenean foreland basin, and it has to be improved at different scales. At a large scale, flexural rigidity of the Iberian crust during thrusting (Millán et al. 1995) has to be taken in account. For areas with irregular shapes like the Ebro basin where other margins affected the basin evolution, a 3D flexural backstripping analysis is necessary to determine the contributions of such margins to the tectonic subsidence within the basin. At basin scale the refinement of the chronostratigraphic framework of the Ebro foreland basin infill as well as the analysis of palaeobathymetric indicators is needed. Also a good correlation with refined sea-level curves would be necessary, especially at smaller scales of third order depositional cycles. The geometry of the deforming thrust front and its relation with syn-thrusting deposits are important to understand foreland basin depositional geometries during thrusting. It is important to discriminate tectonic thickening of the sedimentary succession due to layer parallel shortening, folding and thrusting in subsidence analysis. Indeed, determination of the relative position of successive analysed intervals with respect to the advancing thrust wedge deformation front is important to properly relate changes in subsidence trend with tectonic events. This only can be accomplished by the integration of detailed structural, stratigraphic and chronologic studies. Combination of subsidence and geochemical results is necessary to corroborate geohistories of sedimentary basins and especially to determine syn-thrusting and post-thrusting exhumation events.

We thank S. Crews who wrote the backstripping analysis program we used for our calculations, C. L. Angevine for useful comments about the program, and A. Permanyer for vitrinite reflectance analysis. We also thank to H.-P. Luterbacher and M. Séguret for constructive and critical reviews. This work was founded by IBS Project, Joule II Programme (JOU2-CT92-110), the 'Comissionat per Universitats i Recerca de la Generalitat de Catalunya', Quality Group GRQ94-1048, Total Exploration Production France and by grants from the National Science Foundation (EAR8816181, 9018951) and the Petroleum Research Fund (ACS-PRF 20591, 17625, 23881) to D.W.B.

References

ANGEVINE, C. L., P. L. HELLER, & PAOLA, C. 1990. *Quantitative Sedimentary Basin Modeling.* Continuing Education Course Note Series 32, American Association of Petroleum Geologists, Tulsa.

ALLEN, P. A. & ALLEN, J. R. 1990. *Basin Analysis. Principles and Applications.* Blackwell Scientific Publications, Oxford.

——, CRAMPTON, S. L. & SINCLAIR, H. D. 1991. The inception and early evolution of the North Alpine Foreland Basin, Switzerland. *Basin Research*, **3**, 143-163.

——, HOMEWOOD, P. & WILLIAMS, G. D. 1986. Foreland basins: an introduction. *In:* ALLEN, P. A. & HOMEWOOD, P. (eds) *Foreland Basins.* Special Publications of the International Association of Sedimentologists, **8**, 3–12.

BARNOLAS, A. 1992. Evolución sedimentaria de la Cuenca Surpirenaica Oriental durante el Eoceno. *Acta Geologica Hispànica*, **27**, 15–31.

BARNOLAS, A. & TEIXELL, A. 1994. Platform sedimentation and collapse in a carbonate-dominated margin of a foreland basin (Jaca basin, Eocene, southern Pyrenees). *Geology*, **22**, 1107–1110.

BARTRINA, M. T., CABRERA, L., JURADO, M. J., GUIMERÀ, J. & ROCA, E. 1992. Evolution of the central Catalan margin of the Valencia trough (western Mediterranean). *Tectonophysics*, **203**, 219–47.

BEAUMONT, C. 1981. Foreland basins. *Geophysical Journal of the Royal Astronomical Society*, **65**, 291–329.

BENTHAM, P. & BURBANK, D. W. 1996. Chronology of Eocene foreland basin evolution along the western oblique margin of the South-Central Pyrenees. *In:* FRIEND, P. F. & DABRIO, C. J. (eds) *Tertiary Basins of Spain.* Cambridge University Press, World and Regional Geology, **E11**, 144–152.

BURBANK, D. W., PUIGDEFÀBREGAS, C. & MUÑOZ, J. A. 1992*a*. The chronology of the Eocene tectonic and stratigraphic development of the eastern Pyrenean Foreland Basin, NE Spain. *Geological Society of America Bulletin*, **104**, 1101–1120.

——, VERGÉS, J., MUÑOZ, J. A. & BENTHAM, P. 1992*b*. Coeval hindward- and forward-imbricating thrusting in the Central Southern Pyrenees, Spain: Timing and rates of shortening and deposition. *Geological Society of America Bulletin*, **104**, 3–17.

CANDE, S. C. & KENT, D. V. 1995. Revised calibration

of the geomagnetic polarity timescale for the Late Cretaceous and Cenozoic. *Journal of Geophysical Research*, **100**, 6093–6095.

CASAS, J. M., DURNEY, D., FERRET, J. & MUÑOZ, J. A. 1996. Determinación de la deformación finita en la vertiente sur del Pirineo oriental a lo largo de la transversal del río Ter. *Geogaceta*, **20**, 803–805.

CHOUKROUNE, P. & ECORS TEAM 1989. The ECORS Pyrenean deep seismic profile reflection data and the overall structure of an orogenic belt. *Tectonics*, **8**, 23–39.

CLAVELL, E. 1992. *Geologia del petroli de les conques terciàries de Catalunya*. PhD Thesis, University of Barcelona.

CONEY, P. J., MUÑOZ, J. A., MCCLAY, K. R. & EVENCHICK, C. A. 1996. Syntectonic burial and post-tectonic exhumation of the southern Pyrenees foreland fold-thrust belt. *Journal of Geological Society, London*, **153**, 9–16.

DAIGNIÈRES, M., SÉGURET, M., SPETCH, M. & ECORS TEAM 1994. The Arzacq-Western Pyrenees ECORS Deep Seismic Profile. *In*: MASCLE, A. (ed.) *Hydrocarbon and Petroleum Geology of France*. Special Publication of the European Association of Petroleum Geoscientists, **4**, Springer-Verlag, 199–208.

DECELLES, P. G. & GILES, K. A. 1996. Foreland basin systems. *Basin Research*, **8**, 105–123.

DESEGAULX, P. & BRUNET, M.-F. 1990. Tectonic subsidence of the Aquitaine basin since Cretaceous times. *Bulletin de la Societe Geologique de France*, **8**, VI, n. 2, 295–306.

——, KOOI, H. & CLOETINGH, S. 1991. Consequences of foreland basin development on thinned continental lithosphere: application to the Aquitaine basin (SW France). *Earth and Planetary Science Letters*, **106**, 116–132.

DOROBEK, S. L. 1995. Synorogenic carbonate platforms and reefs in foreland basins: Controls on stratigraphic evolution and platform/reef morphology. *In*: DOROBEK, S. L. & ROSS, G. M. (eds) *Stratigraphic Evolution of Foreland Basins*. SEPM Special Publications, **52**, 127–147.

DUBARRY, R. 1988. *Interprétation dynamique du Paléocène et de l'Éocène inférieur et moyen de la région de Pau-Tarbes (avant-pays Nord des Pyrénées occidentales; SW France): sedimentologie, correlations diagraphiques, decomposition et calcul de subsidence*. PhD Thesis, University of Pau.

ECORS-CROP DEEP SEISMIC SOUNDING GROUP 1989. A new picture of the Moho under the western Alps. *Nature*, **337**, 249–251.

FERRER, J. 1971. *Le Paléocène et l'Eocène des Cordillères Cotières de la Catalogne (Espagne)*. Mémoires Suisses de Paléontologie, **90**.

FLEMINGS, P. B. & JORDAN, T. E. 1989. Stratigraphic modeling of foreland basins: Interpreting thrust deformation and lithosphere rheology. *Geology*, **18**, 430–434.

GIMÉNEZ-MONTSANT, J. 1993. *Analisis de cuenca del Eoceno inferior de la Unidad Cadí (Pirineo oriental): El sistema deltaico y de plataforma carbonatica de la Formación de Corones*. PhD Thesis, University of Barcelona.

HAQ, B. U., HARDENBOL, J. & VAIL, P. R. 1987. Chronology of Fluctuating Sea Levels Since the Triassic. *Science*, **235**, 1156–1166.

HELLER, P. L., ANGEVINE, C. L. & WINSLOW, N. S. 1988. Two-phase stratigraphic model of foreland-basin sequences. *Geology*, **16**, 501–504.

HINE, A., LOCKER, S. D., TEDESCO, L. P., MULLINS, H. T., HALLOCK, P., BELKNAP, D. F., GONZALEZ, J. L., NEUMANN, A. C. & SNYDER, S. W. 1992. Megabreccias shedding from modern, low-relief carbonate platforms, Nicaraguan Rise. *Geological Society of America Bulletin*, **104**, 8, 928–943.

HOMEWOOD, P., ALLEN, P. A. & WILLIAMS, G. D. 1986. Dynamics of the Molasse Basin of western Switzerland. *In*: ALLEN, P. A. & HOMEWOOD, P. (eds) *Foreland Basins*. Special Publications of the International Association of Sedimentologists, **8**, 199–217.

HOTTINGER, L. 1960. *Recherches sur les Alvéolines du Paléocéne et de l'Eocéne*. Memoires Suisses de Paléontologie, **75–76**.

—— 1977. Distribution of larger foraminifera Peneroplidae, Borelis, and Nummulitidae in the Gulf of Elat, Red Sea. *In*: DROOGER, C. W. (ed.) *Depth-Relations of Recent Larger Foraminifera in the Gulf of Aqaba-Elat*. Utrecht Micropaleontological Bulletin, **15**, 35–109.

JIN, J. 1995. Dynamic stratigraphic analysis and modeling in the South-Eastern German Molasse Basin. *Tübingen Geowissenschaftliche Arbeiten*, **24**, 1–153.

JOHNS, D. R., MUTTI, E., ROSELL, J. & SÉGURET, M. 1981. Origin of a thick, redeposited carbonate bed in the Eocene turbidites of the Hecho Group, South-Central Pyrenees, Spain. *Geology*, **9**, 161–164.

JORDAN, T. E. 1981. Thrust loads and foreland basin evolution, Cretaceous, Western United States. *AAPG Bulletin*, **65**, 2506–2520.

—— & FLEMINGS, P. B. 1991. Large-scale stratigraphic arquitecture, eustatic variation, and unsteady tectonism: A theorical evaluation. *Journal of Geophysical Research*, **96**, 6681–6699.

LABAUME, P., SÉGURET, M. & SEYVE, C. 1985. Evolution of a turbiditic foreland basin and analogy with an accretionary prism: Example of the Eocene South-Pyrenean basin. *Tectonics*, **4**, 661–685.

LANAJA, J.M. 1987. *Contribución de la exploración petrolífera al conocimiento de la geología de España*. Instituto Geológico de España.

LEWIS, C. J., VERGÉS, J., MARZO, M. & HELLER, P. L. 1996. Youtfhful topography indicating active surface uplift in NE Iberia: Mantle upwelling along a leakly transform fault? *Annales Geophysicae*, Supplement I, **14**, C-204.

LÓPEZ-BLANCO, M. 1993. Stratigraphy and sedimentary development of the Sant Llorenç del Munt fan-delta complex (Eocene, southern Pyrenean foreland basin, northeast Spain). *In*: FROSTICK, L. E. & STEEL, R. J. (eds) *Tectonic Controls and Signatures in Sedimentary Successions*. Special Publications of the Association of Sedimentologists, **20**, 67–88.

LÓPEZ-BLANCO, M., MARZO, M., BURBANK, D.,

VERGÉS, J., ROCA, E., ANADÓN, P. & PIÑA, J. In press. Tectonic and climatic controls on the development of large, foreland fan deltas: Montserrat and Sant Llorenç del Munt systems (Middle Eocene, Ebro basin, NE Spain). *In*: MARZO, M. & STEEL, R. (eds) *Sedimentology and sequence stratigraphy of the Sant Llorenç del Munt clastic wedges (SE Ebro basin, NE Spain)*. Sedimentary Geology Special Issue,

LUTERBACHER, H. P., EICHENSEER, H., BETZLER, CH., & VAN DEN HURK, A. M. 1991. Carbonate-siliciclastic depositional systems in the Paleogene of the South Pyrenean foreland basin: a sequence-stratigraphic approach. *In*: MACDONALD, D. I. M. (ed.) *Sedimentation, Tectonics and Eustasy*. Special Publications of the International Association of Sedimentologists, **12**, 391–408.

MARTÍNEZ, A., VERGÉS, J., CLAVELL, E. & KENNEDY, J. 1989. Stratigraphic framework of the thrust geometry and structural inversion in the southeastern Pyrenees: La Garrotxa area. *Geodinamica Acta*, **3**, 185–194.

——, —— & MUÑOZ, J. A. 1988. Secuencias de propagación del sistema de cabalgamientos de la terminación oriental del manto del Pedraforca y relación con los conglomerados sinorogénicos. *Acta Geologica Hispànica*, **23**, 119–128.

MATÓ, E., SAULA, E., VERGÉS, J., MARTÍNEZ-RÍUS, A., ESCUER, J. & BARBERÀ, M. 1994. *Mapa Geológico de España. Plan Magna a escala 1:50.000. Hoja de Berga (293)*. Instituto Tecnológico Geominero de España, Madrid.

MEIGS, A., VERGÉS, J. & BURBANK, D. W. 1996. Ten-million-year history of a thrust sheet. *Geological Society of America Bulletin*.

MILLÁN, H., DEN BEZEMER, T., VERGÉS, J., ZOETEMEIJER, R., CLOETINGH, S., MARZO, M, MUÑOZ, J. A., PUIGDEFÀBREGAS, C., ROCA, E. & CIRÉS, J. 1995. Present-day and Middle Lutetian flexural modelling in the Eastern Pyrenees and Ebro Basin. *Marine and Petroleum Geology*, **12**, 917–928.

MUÑOZ, J.A. 1992. Evolution of a Continental Collision Belt: ECORS-Pyrenees Crustal Balanced Cross-section. *In*: MCCLAY, K. (ed.) *Thrust Tectonics*. Chapman & Hall, London, 235–246.

——, MARTÍNEZ, A. & VERGÉS, J. 1986. Thrust sequences in the eastern Spanish Pyrenees. *Journal of Structural Geology*, **8**, 399–405.

——, VERGÉS, J., MARTÍNEZ-RÍUS, A., FLETA, J., CIRÉS, J., CASAS, J. M. & SÀBAT, F. 1994. *Mapa Geológico de España. Plan Magna a escala 1:50.000. Hoja de Ripoll (256)*. Instituto Tecnológico Geominero de España, Madrid.

ORTÍ, F., BUSQUETS, P., ROSELL, L., TABERNER, C., UTRILLA, R. & QUADRAS, M. 1988. La fase evaporítica del Eoceno medio (Luteciense) en la cuenca surpirenaica catalana. Nuevas aportaciones. *Revista d'Investigacions Geològiques*, **44-45**, 281–302.

PEPER, T. 1993. *Tectonic control on the sedimentary record in foreland basins: inferences from quantitative subsidence analyses and stratigraphic modelling*. PhD. Thesis Vrije Univesitet, Amsterdam.

PERMANYER, A., VALLÉS, D. & DORRONSORO, C. 1988. Source Rock potential of an Eocene carbonate slope: the Armàncies Formation of the Southern Pyrenean Basin, Northeast Spain, *AAPG Bulletin*, **72**, 1019.

PFIFFNER, A. 1992. Alpine orogeny. *In*: BLUNDELL, D., FREEMAN, R. & MUELLER, S. (eds) *A continent revealed. The European Geotraverse*. Cambridge University press, 180–190.

PRICE, R. A. 1973. Large-scale gravitational flow of supr-crustal rocks, southern Canadian Rockies *In*: DE JONG, K. A. & SCHOLTEN, R. A. (eds) *Gravity and Tectonics*. Wiley, New York, 491–502.

PUIGDEFÀBREGAS, C., MUÑOZ, J. A. & MARZO, M. 1986. Thrust belt development in the Eastern Pyrenees and related depositional sequences in the southern foreland basin. *In*: ALLEN, P. A. & HOMEWOOD, P. (eds.) *Foreland Basins*. Special Publications of the International Association of Sedimentologists, **8**, 229–246.

——, —— & VERGÉS, J. 1992. Thrusting and Foreland Basin Evolution in the Southern Pyrenees. *In*: MCCLAY, K. (ed.) *Thrust Tectonics*. Chapman & Hall, London, 247–254.

PUJALTE, V., BACETA, J. I., PAYROS, A., ORUE-ETXEBARRIA, S & SERRA-KIEL, J. 1994. *Late Cretaceous–Middle Eocene Sequence Stratigraphy and Biostratigraphy of the SW and W Pyrenees (Pamplona and Basque Basins, Spain)*. G.E.P. and IGCP Project 286, Field Seminar.

ROCA, E. 1992. *L'estructura de la conca Catalano-Balear: paper de la compressió i de la distensió en la seva gènesi*. PhD. Thesis, University of Barcelona.

REISS, Z. & HOTTINGER, L. 1984. *The Gulf of Aqaba. Ecological micropaleontology*. Ecological studies **50**. Springer Verlag.

RICCI LUCCHI, F. 1986. The Oligocene to Recent foreland basins of the northern Apennines. *In*: ALLEN, P. A. & HOMEWOOD, P. (eds) *Foreland Basins*. Special Publications of the International Association of Sedimentologists, **8**, 105–139.

RIVERO, L. 1993. *Estudio Gravimétrico del Pirineo Oriental*. PhD Thesis, University of Barcelona.

SAMSÓ, J. M., SERRA-KIEL, J., TOSQUELLA, J. & TRAVÉ, A. 1994. Cronostratigrafía de las plataformas lutecienses de la zona central de la cuenca surpirenaica. *In*: MUÑOZ, A., GONZALEZ, A. & PÉREZ, A. (eds) *II Congreso Grupo Español del Terciario, Comunicaciones*, Jaca, 205–208.

SANS, M., MUÑOZ, J.A. & VERGÉS, J. 1996. Thrust wedge geometries related to evaporitic horizons (Southern Pyrenees). *Bulletin of Canadian Petroleum Geology*, **44**, 375–384.

SAULA, E., MATÓ, E., ESCUER, J., MUÑOZ, J. A. & BARNOLAS, A. 1994. *Mapa Geologico de España. Plan Magna a escala 1:50.000, Hoja de Manlleu (294)*. Instituto Tecnológico Geominero de España, Madrid.

SCHAUB, H. 1981. *Nummulites et Assilines de la Téthys Paleogène. Taxinomie, phylogenese et biostratigraphie*. Memoires Suisses Paléontologie, **104-106**.

SCLATER, J. G. & CHRISTIE, P. A. F. 1980. Continental

stretching: an explanation of the post Mid-Cretaceous subsidence of the Central North Sea basin. *Journal of Geophysical Research*, **85**, 3711–3739.

SÉGURET, M. 1972. *Étude tectonique des nappes et séries décollées de la partie centrale du versant sud des Pyrénées*. Publication USTELA, série Geologie structurale, **2**. Montpellier.

SERRA-KIEL, J. 1984. *Estudi dels Nummulites del grup N. perforatus (Monfort)*. Treballs de la Institut Catalan d'Historia Natural, **11**.

—— & TRAVÉ, A. 1995. Lithostratigraphic and chronostratigraphic framework of the Bartonian sediments in the Vic and Igualada areas. *In*: PEREJÓN, J. A. & BUSQUESTS, P. (eds) *VII International Symposium Fossil Cnidaria and Porifera*, Madrid, 11–14.

——, CANUDO, J. I., DINARÈS, J., MOLINA, E., ORTÍZ, N., PASCUAL, J. O., SAMSÓ, J. M. & TOSQUELLA, J. 1994. Cronoestratigrafía de los sedimentos marinos del Terciario inferior de la Cuenca de Graus-Tremp (zona Central Surpirenaica). *Revista de la Soceologel Geologica España*, **7**, 273–297.

——, HOTTINGER, L., DROBNE, K., FERRÁNDEZ, C., LESS, G., JAUHRI, A. K., PIGNATTI, J., SAMSÓ, J. M., SCHAUB, H., SIREL, E., TAMBAREAU, Y., TOSQUELLA, J. & ZAKREVSKAYA, E. in press. Benthic Foraminifera from Paleocene and Eocene. *In*: DE GRACIANSKY, P. C., HARDENBOL, J., JACQUIN, T. & VAIL, P. R. (eds) *Mezosoic-Cenozoic Sequence Stratigraphy of Western European Basins*. SEPM Special Publications, **00**, 00–00.

SINCLAIR, H. D. & ALLEN, P. 1992. Vertical versus horizontal motions in the Alpine orogenic wedge: stratigraphic response in the foreland basin. *Basin Research*, **4**, 215–232.

——, COAKLEY, B. J., ALLEN, P. A. & WATTS, A. B. 1991. Simulation of foreland basin stratigraphy using a difusion model of mountain belt uplift and erosion: an example from Central Alps, Switzerland. *Tectonics*, **10**, 599–620.

SONNENFELD, P. 1984. *Brines and evaporites*. Academic Press Inc., Orlando, Florida.

STECKLER, M. S. & WATTS, A. B. 1978. Subsidence of the Atlantic type continental margin of New York, *Earth and Planetary Science Letters*, **42**, 1–13.

TEIXELL, A. 1996. The Ansó transect of the southern Pyrenees: basement and cover thrust geometries. *Journal of the Geological Society*, London, **153**, 301–310.

TOSQUELLA, J. 1995. *Els Nummulitinae del Paleocè-Eocè inferior de la conca sudpirinenca*. PhD Thesis, University of Barcelona.

VAIL, P. R., MITCHUM, R. M. & THOMPSON, S. 1977. Seismic stratigraphy and global changes of sea level, part 3: relative change of sea level from coastal onlap. *In*: PAYTON, C. A. (ed.) *Seismic Stratigraphy - applications to hydrocarbon exploration*, AAPG Memoirs, **26**, 63–81.

VAN HINTE, J. E. 1978. Geohistory analysis-application of micropalaeontology in exploration geology. *AAPG. Bulletin*, **62**, 201–222.

VERGÉS, J. 1993. *Estudi geològic del vessant sud del Pirineu oriental i central. Evolució cinemàtica en 3D*. PhD Thesis, University of Barcelona.

—— & BURBANK, D. W. 1996. Eocene-Oligocene thrusting and basin configuration in the eastern and central Pyrenees (Spain). *In*: FRIEND, P .F. & DABRIO, C. J. (eds) *Tertiary Basins of Spain*. Cambridge University Press, World and Regional Geology, **E11**, 120–133.

—— & MARTÍNEZ, A. 1988. Corte compensado del Pirineo oriental: geometria de las cuencas de antepaís y edades de emplazamiento de los mantos de corrimiento. *Acta Geologica Hispànica*, **23**, 95–106.

——, MARTÍNEZ-RÍUS, A., DOMINGO, F., MUÑOZ, J.A., LOSANTOS, M., FLETA, J. & GISBERT, J. 1994. *Mapa Geológico de España. Plan Magna a escala 1:50.000. Hoja de La Pobla de Lillet (255)*. Instituto Tecnológico Geominero de España, Madrid.

——, MILLÁN, H., ROCA, E., MUÑOZ, J. A., MARZO, M., CIRÉS, J., DEN BEZEMER, T., ZOETEMEIJER, R. & CLOETINGH, S. 1995. Eastern Pyrenees and related foreland basins: pre-, syn- and post-collisional crustal-scale cross-sections. *Marine and Petroleum Geology*, **12**, 903–916.

——, MUÑOZ, J. A. & MARTÍNEZ, A. 1992. South Pyrenean fold-and-thrust belt: Role of foreland evaporitic levels in thrust geometry. *In*: MCCLAY, K. (ed.) *Thrust Tectonics*. Chapman & Hall, London, 255–264.

ZOETEMEIJER, R., SASSI, W., ROURE, F. & CLOETINGH, S. 1992. Stratigraphic and kinematic modelling of thrusts evolution, northern Apennines, Italy. *Geology*, **20**, 1035–1038.

Cyclicity and basin axis shift in a piggyback basin: towards modelling of the Eocene Tremp–Ager Basin, South Pyrenees, Spain

WOUTER NIJMAN

*Netherlands Research School of Sedimentary Geology
Institute of Earth Sciences, Utrecht University, Postbus 80021, 3508 TA Utrecht, The
Netherlands (e-mail: wnijman@earth.ruu.nl)*

Abstract: A new database of the Eocene Montanyana delta in the Tremp–Ager piggyback basin of the South Pyrenees is exemplified and discussed. It consists of a grid of nine transverse and five longitudinal stratigraphic cross-sections and structural profiles, some examples of which are shown. Detailed facies maps and a large number of digitized sedimentary logs formed the basis for the stratigraphic correlation of the traverses with a resolution at the scale of 5 m, i.e. the size of minor architectural elements. The correlation was calibrated chronostratigraphically with biostratigraphic and recently published palaeomag data. The database so provides a detailed picture of the geometry, internal architecture and stacking patterns of the increments of basin infilling.

The database allowed recognition and definition of eight megasequences, between 148 and 404 m thick, that are aperiodic, spanning time intervals between 400 and 1400 Ma. Although most of the megasequence boundaries can easily be related to third-order sea-level fluctuations, predominant structural control is indicated by their correlation with sharp reversals in the pattern of basin axis shift in transverse cross-sections of the basin, and with flank unconformities. Several modes of structural control by thrust sheet displacement are proposed. Sea-level fluctuation modified the megasequential architecture accounting, for instance, for extreme progradation as observed in the Castissent Sandstone.

The megasequences are subdivided into a large number of basin-wide cycles, on an average 44 m thick, with a calculated average periodicity of 124 ka approximating the 100 ka of orbital forcing. Aggradational, amalgamated sheet, fan progradational and fluvial expansion cycles are distinguished and correlated with episodes of specific structural and sea-level control. The pattern of stacking of the cycles conforms to the structurally controlled megasequential basin-axis shift on which it is superposed. Climatic fluctuations appear to have played a prominent role only in generating minor subcycles observed in some parts of the basin fill.

This basin analysis is meant to be a step towards 3D numerical tectono-sedimentary modelling of the Tremp–Ager piggyback basin.

In the Tremp–Ager area (Fig. 1), the application of sequence-stratigraphic concepts was introduced with the AAPG excursion guide on *Sedimentation and Deformation in the Tertiary Sequences of the Southern Pyrenees* by Mutti *et al.* (1988). At the time, many of the unconformities in the area were already known to separate depositional systems and sequences (tectono-stratigraphic units of Garrido-Megias & Rios 1972; Mutti *et al.* 1972; Nijman & Nio 1975), basin scale tectono-sedimentary cycles (Puigdefàbregas & Souquet 1986), and intrabasin-scale megasequences (Nijman 1981; Cuevas Gozalo 1989). A straightforward relationship of conformable stratigraphic boundaries with flank unconformities and, as a consequence, with thrust tectonics is often evident (Riba 1976). Mutti *et al.* (1988, p. 85) stressed that the unconformities between their allogroups, although clearly bearing a tectonic imprint, correspond quite well with drops in the sea-level curve of Haq *et al.* (1987), suggesting a relationship between deformation phases and global sea level fluctuations.

Whether the cyclicity in the piggyback and foreland basins is primarily tectonically, eustatically or climatically controlled, is being debated (for the South Pyrenees e.g. Mutti & Sgavetti 1987; Séguret 1991; Peper & De Boer 1995; Weltje *et al.* 1996). In general, questions have been raised about the applicability of the sequence-stratigraphic concepts in mobile belts, the continuity and traceability of the unconformities over thrust-sheet boundaries, and the segmentation of sequence boundaries downdip along the slope of deposystem tracts (Puigdefàbregas & Mascle, this volume).

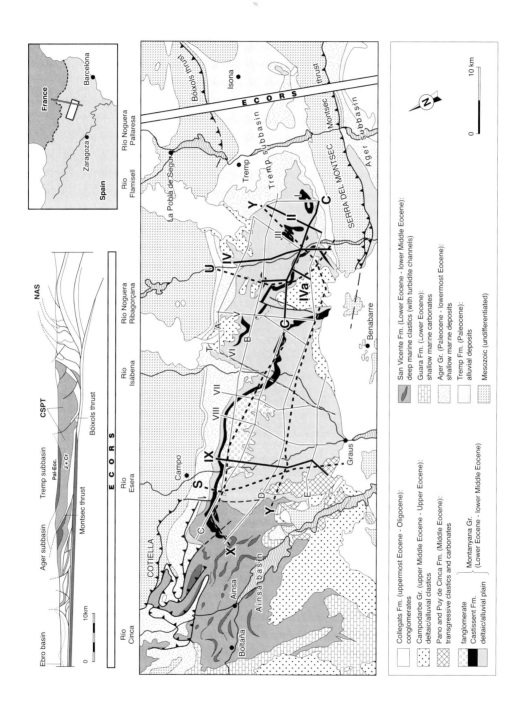

A new database

From 1971 onwards, the rock sequence, composition, and dispersal patterns of the Tremp–Ager basin fill have been recorded by students and staff of the Sedimentology Group at Utrecht University. In 1992, digitization of carefully selected data from the available descriptive information was started in order to construct a large database.

The database has been designed and constructed on Macintosh PCs (Nijman & Van Oosterhout 1994). Sedimentary log data formed the input for Excel spreadsheets from which graphic logs have been generated. A total of 30 km of selected vertical logs have been digitized with a resolution of about 5 m thickness, i.e. the size of individual architectural elements, such as fluvial channels, minor delta lobes, and palaeosols. Subsequently, the logs have been correlated along a grid (Fig. 1) of nine transverse (traverses I–IX) and five longitudinal (traverses A–E) stratigraphic cross-sections, examples of which are shown and discussed in this paper (Figs 2–4). The facies distribution along the traverses is based on direct field observation and detailed mapping throughout the basin. Because of this control, the cross-sectional surfaces of facies units satisfactorily approximate reality, although the minor interdigitations have been drafted arbitrarily. The database also comprises several structural cross-sections (profiles S–U and X–Y) (Figs 2a–4a), and will be completed by processing of detailed facies outcrop maps.

The database and the correlations it enabled to be made substantially modify and elaborate the basin-wide correlation and sequence architecture of allogroups of Mutti et al. (1988). The examples given below show a detailed internal architecture of megasequences and cycles, which allows tracing of the position of the basin axis in time, and provides tools for the distinction between structural, sea level, and climatic control.

Recently, the database has been made operative in an UNIX environment for 3D model studies on an Intergraph Work Station with Voxel Analyst software (Van Hilten 1995), and we plan to incorporate volumetry of architectural elements in successive basin fill increments in sedimentary–tectonic model studies of the Netherlands Research School of Sedimentary Geology (Zoetemeijer 1993; Peper 1993; Den Bezemer et al. this volume). We thereby aim at 3D modelling of the basin architecture in relation to thrust tectonics.

Thrust sheet configuration and basin dynamics

The ECORS cross-section (Fig. 1) disclosed the deep crustal structure of the central sector of the Pyrenees (Roure et al. 1989; Muñoz 1992; Vergés et al. 1995). The Axial Zone might be considered a huge stack of basement thrust sheets formed, almost exclusively composed of mid-to-upper crustal rock (<15 km). Shortening over the entire orogen amounts to about 125 km (Vergés et al. 1995) of which 50% is taken up by the pervasive deformation of the basement units. The Central South Pyrenean Thrust system (CSPT) belongs to the upper thrust sheets, composed of the sedimentary cover of the Axial Zone of the Pyrenees. At the rear, it overlies the Nogueras Zone (NAS in Fig. 1), a duplex of basement-involved lower thrust sheets, that developed into a pronounced antiformal stack at the rear of the cover sheets.

Thrusts were generated by inversion from pre-existing north-facing normal faults in the Cretaceous shelf (Roure et al. 1989). With the development of the Bóixols thrust (Fig. 1), the first to have occurred in the sedimentary cover of this segment of the Pyrenees, the characteristic drainage and provenance pattern for the post-inversion thrust sheet-top basins was initiated. The deeper marine fill of the Vallcarga basin (Fig. 1) preserves the earliest record of longitudinal (i.e. northwestward, subparallel to the orogenic axis) clastic sediment transport (Van Hoorn 1970; Simó 1986, 1989; Souquet & Déramond 1989; Nijman 1989). Quartz-rich turbidity currents followed the basin axis to the northwest, while interfering with marly debris flows from adjacent active (intra)basinal slopes. The shallow-water equivalent of the longitudinal transport component is found in the estuarine and fluvial quartzose sandstones of, respectively, the Maastrichtian Arén Formation north of Tremp, and the Palaeocene Reptile Sandstone near Ager, while in the mean time

---

Fig. 1. Geological map of the Tremp–Ager Basin (Tremp and Ager subbasins) with database grid of stratigraphic traverses (roman numbers) and structural profiles (letters). Inset map gives location of the map area with the ECORS deep seismic profile. Cross-section corresponds to southern part of ECORS profile through Central South Pyrenean thrust sheet (CSPT) and Nogueras antiformal stack (NAS). Bold grid lines refer to traverses and structural profiles of Figs 2–4, 9 & 10.

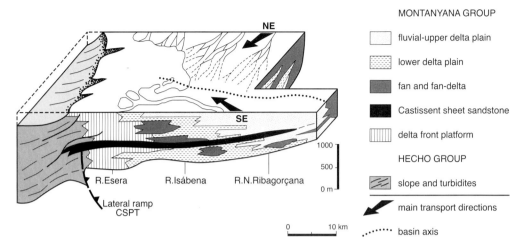

Fig. 5. Block diagram of Montanyana Delta, showing alluvial fan and fluvial feeder systems combining in one deltafront with break-in-slope above lateral ramp of underlying thrust sheet (after Marzo et al. 1988)

palaeocurrent bearings turn to north and northwest (Puigdefàbregas et al. 1991).

In the CSPT, the first flank-derived polymict alluvial fan supply from a northerly, extrabasinal source is recorded from the Tremp Group, north of the type locality (Cuevas in Puigdefàbregas et al. 1991). In course of time, alluvial fans from the Pyrenean Axial Zone became increasingly important, thus implementing the sort of drainage system illustrated in the block diagram of the Eocene Montanyana delta of Fig. 5. Morphologically, the post-inversion piggyback basins are northwest-opening embayments situated in first or second-order synclinal structures (Nagtegaal et al. 1983; Nijman 1989).

From the late Cretaceous onwards, outward depocentre migration shows that basin formation followed the rules of a growing orogen with increasing crustal shortening and upper crustal detachment (Puigdefàbregas & Souquet 1986). The depocentre migrated stepwise (Fig. 6) from an *in situ* early Cretaceous shelf into a succession of sedimentary basins from piggyback basins, first of late Cretaceous, then of Palaeocene–Eocene age, to the Ebro foreland basin during the Oligocene–Miocene.

The process of depocentre migration was accompanied by an increase of clastic supply and by coarsening-up grain size trends at basin-fill scale (Fig. 6). During the late stages, from the upper Eocene onwards, break-back thrusting complicates this pattern (Burbank et al. 1992b).

The fluvial systems responsible for the infilling of the basins show a wide variety of architectural elements. Source areas were situated in the rising Axial Zone of the Pyrenees, probably on the Ebro High before its inversion into a foreland basin (Julivert 1978), and in the Catalonian Coast Ranges (Vergés et al. 1995). Transport routes of these sediments differed in length and gradient and were themselves controlled by the embryonic structures related to forward propagation of the thrust system (Nijman 1989) and accompanied by progressive unconformities. The earlier-formed basins became involved in the thrust process as piggyback basins, while later ones retained a position in front of the leading thrusts.

Another characteristic feature of the upper Cretaceous and Eocene basin-fills is their *northwestward* change from terrestrial to shallow-water (molasse) into deeper-water slope and base-of-slope (flysch) facies (Fig. 5) (Mutti et al. 1972). It reflects a northwestward plunge of the Pyrenean orogen. The relationship of the piggyback basins in the CSPT with the eastern Pyrenean basins is still controversial. According to recent correlations of Vergés et al. (1995; see also Vergés & Burbank 1996), the east Pyrenean Palaeocene–Eocene Ripoll Basin is now considered foreland basin related to the same thrust phase which in the CSPT is represented by either Montsec thrust or frontal thrusts. However, the northwest facing of the Tremp–Ager basin with due north- to northwestward, thrustfold-crossing influx of quartzose sandstone of the Palaeocene–Eocene Tremp–Ager Basin still cannot satisfactorily be matched with the marine foreland basin fill of the Ripoll Basin. It has been suggested (Nijman 1989), that emplacement of

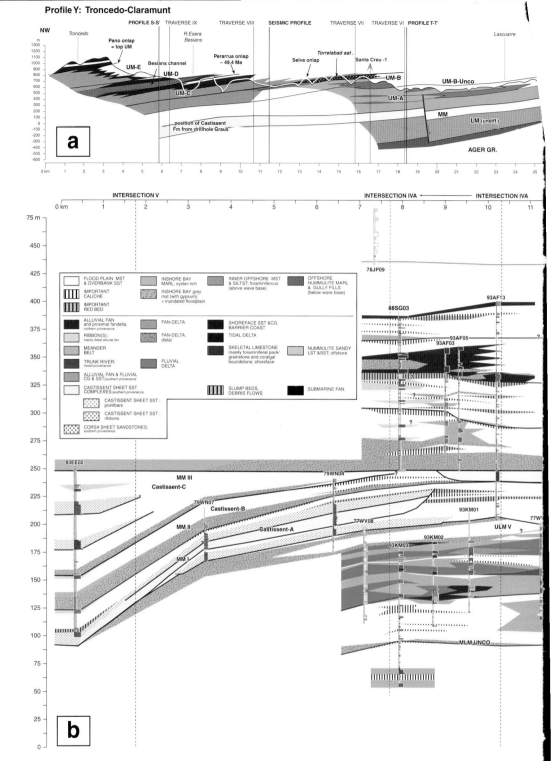

General information on Figs 2–4.
(1) Traces of profiles and traverses are given in Fig. 1. (2) Codes and data of megasequences and cycles in Table 1. (3) Tim[...]
(4) Intersections with stratigraphic traverses and other structural profiles, and seismic lines and boreholes (ENPASA 1966)[...]
mapping data along the section line proper. Traverse II and profile X are shown respectively as Figs 9 and 10. (5) Legend b[...]
respectively, are identical throughout the series of coloured figures and therefore cover a wider range of items than shown [...]
Legend of sedimentary logs is different from those of the facies units in the traverses and is not further explained. Encircle[...]
indicate MPS (mean max. pebble size) in cm. (7) Explanation of the figures has been largely included in the text.

Fig. 2. (a) Longitudinal structural profile of the Montanyana Delta in the Tremp subbasin. The easternmost part is orient[...]
Luzas thrust fold forms part of NNE-striking hinge fault system, in particular influencing the late Eocene Campodarbe Gro[...]
stratigraphic cross-section, transverse C (= C3/4), through the upper deltaic plain of the Montanyana Delta in the eastern se[...]
intersects with traverses IV/IVa of Fig. 3, and traverse II of Fig. 9. The MLM-unconformity seals a NNW-oriented block faul[...]
thrust. The faults have been later reactivated (dashed fault lines), in places with reverse throw: compare traverse II.

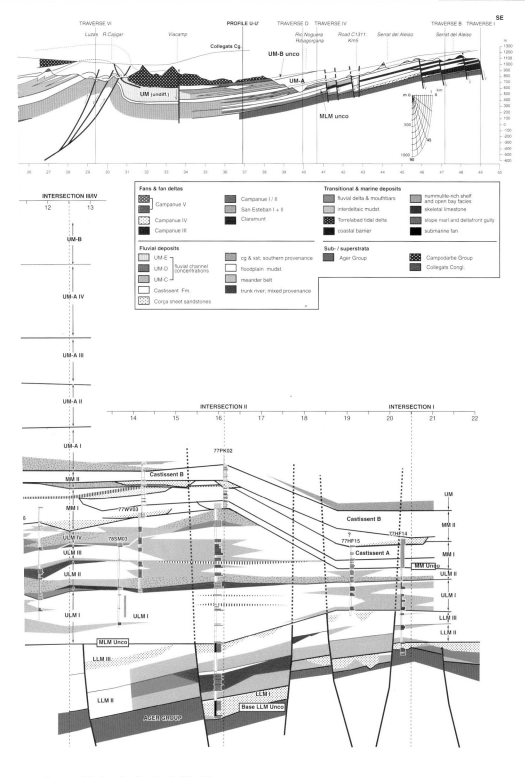

g and general facies distribution in Fig. 13.
ndicate sources of data other than
cks, for profiles and traverses
the single corresponding figures. (6)
numbers, e.g. ⑧, along some of the logs

d NE and oblique to the basin axis. The
p. (**b**) Eastern part of longitudinal
tor of the Tremp subbasin. The traverse
pattern at right angles to the Montsec

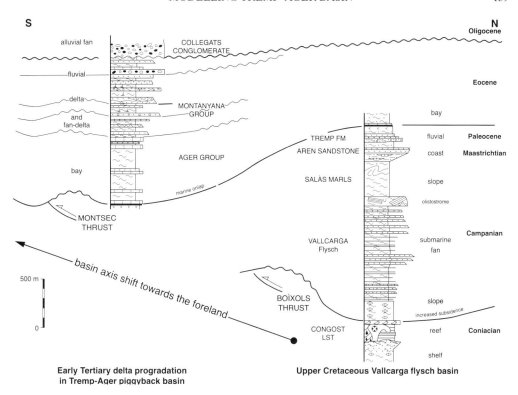

Fig. 6. Basin fill logs and outward depocentre shift of Cretaceous Vallcarga and Eocene Tremp–Ager-thrust sheet top basins. Note that outward depocentre migration relates to thrust migration (Bóixols thrust–Montsec thrust).

thrust sheets was accompanied by rotation to account for Ypresian misfit of the South Pyrenean orogenic basins. Rotational emplacement has now been observed in several places (Bates 1989; Dinarès et al. 1992; Keller 1992; Burbank et al. 1992a; Martinez-Peña et al. 1995), but is still insufficiently evaluated for its effect on the entire basin configuration.

The Montanyana delta in the Tremp–Ager Basin

Basin shape and delta geometry

The upper Ypresian to upper Lutetian Montanyana* deltaic complex is exposed in the Tremp and Ager basins (Fig. 1), separated by the thrust wedge of the Montsec Range. Though visible in the basin architecture, the Montsec thrust was far from obvious during the Eocene and sometimes not expressed at the surface at all. For that reason the *Tremp–Ager Basin* is referred to as one sedimentary basin, the Tremp and Ager *subbasins* (Fig. 7) as component synclinal structural basins. The Tremp subbasin is subdivided in three sectors coinciding with and named after present river valleys (Fig. 1) that cross the subbasin: the eastern or Ribagorçana sector, the central or Isábena sector, and the western or Esera sector.

In longitudinal section (Profile Y, Fig. 2a), the Montanyana Group shows the characteristic deltaic wedge geometry (Fig. 5), while in cross-section (Profile S, Fig. 4a; and profile U, Fig. 3a) the basin fill reflects the shape of the synsedimentarily active, synclinal thrust sheet structure; subordinate undulations correspond to parasitic thrust structures.

Deposition resulted from the combined action of three dispersal mechanisms (Figs. 5 and 8). Apart from the marine processes acting on the delta platform, two major clastic feeder systems have been distinguished (Nijman & Nio 1975): a fluvial system in the southern and

---

*Catalan spelling; Castilian spelling is Montañana.

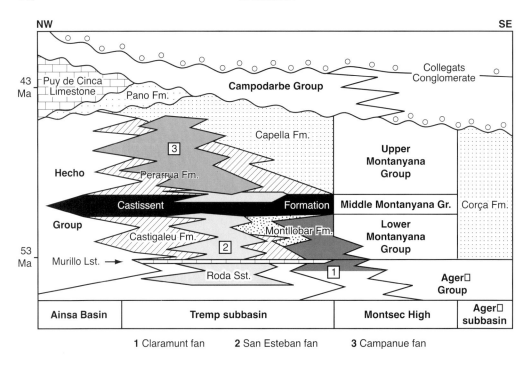

Fig. 7. Scheme of stratigraphic nomenclature of the Tremp–Ager Basin. Stratigraphic names in italics refer to units outside Montanyana Group. Within it, greys refer to alluvial fans, coarse stippling to fluvial and upper deltaic plain; oblique hatching to lower deltaic plain and deltafront.

central parts of the basin, that supplied sand and mud from the southeast, and a complex of alluvial fans and fan deltas rimming the northeastern flanks of the basin. The latter consists of sediment with grain size ranging from mud to boulders derived from the rear of the CSPT and the Pyrenean Axial Zone. Pronounced westward deflections in both drainage patterns have been observed in the interfluvial/interdeltaic area where both supplying systems are laterally juxtaposed. This area is generally occupied by a zone of mud accumulation and of low interconnectedness of the fluvial sandstone bodies it contains (Nijman 1981). Because of its mappability, the conjunction of the two alluvial/deltaic systems, situated in this mud zone or, in its absence, at the confluence of the two systems, is defined as the basin axis (Figs 5 and 8). Its position within a mud zone (see Figs 2–4 and 6) is determined by the palaeocurrent pattern of encased sandstone bodies and may be asymmetric with respect to the centre of the mud zone. In cross-sections of the Tremp subbasin, the basin axis often coincides with the depocentre.

Architectural elements

The architecture of the Montanyana delta is composed of many elements in response to climate, sediment supply, base level changes, and tectonics. In the database, these are grouped in nine categories (Table 1). For practical reasons, all mudstones and marls are combined in the first category. In general, the combination of colour, presence or absence of pedogenesis, and fossil content of the mudstone/marl suffices to assign encased architectural elements of the other eight categories to one of the major fluvial/deltaic (sub)environments.

Groups, megasequences and cycles

Too often, existing stratigraphic formation or group names have been re-used with considerably modified meaning, for instance the Castissent Formation redefined by Mutti *et al.* (1988) as Castissent (allo)group. Because correlation of sequence boundaries within and between the orogenic basins is controversial, original formation and group names are retained where

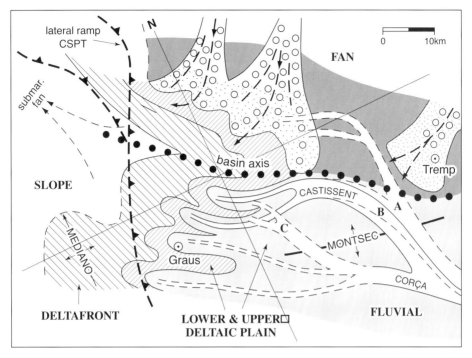

Fig. 8. Palaeodrainage pattern of Castissent Sandstone and related Corça Formation (after Nijman 1989). Castissent units A–C correspond to cycles MM I–III.

possible (Fig. 7). The database allows detailed correlation of facies and sequential patterns crossing formation boundaries from the fluvial into the deltafront domains.

The observed cyclicity is on at least two scales: the larger is given by the presence of eight offlapping megasequences separated by major onlaps. The lower Montanyana Group (LM) is divided into two megasequences, Lower LM and Upper LM, by the mid-lower Montanyana unconformity (MLM-unconformity); the Castissent Formation forms one megasequence (MM: middle Montanyana Group); the upper Montanyana Group (UM) is composed of five megasequences (UM-A to UM-E: upper Montanyana Group A to E).

The megasequences show internal minor on/offlap cyclicity of 40 m scale, comparable to parasequences. In the Castissent Formation and the Campanúe fan delta (Fig. 7), a still higher order of rhythmicity has been observed in the order of less than 20 m thickness (De Boer *et al.* 1991). Sequences and cycles will be further discussed in relation to the fluvial and deltaic architecture in later sections of this paper.

Alluvial fan systems

In the lower Montanyana Group of the eastern part of the basin (Fig. 1), the *Claramunt fan* (new name) is composed of sheets of sandstone and conglomerate encased in floodplain mudstone (Van der Meulen 1986, Puigdefàbregas *et al.* 1991). The fan probably shared its edifice with an early conglomeratic fan in the lower part of the Ager Group, north to northwest of Tremp. The part of the Claramunt fan comprised in the Montanyana Group then merely represents the aftermath of the first major Eocene alluvial fan. Progradation in the LLM-megasequence and lower part of the UM-megasequence is accompanied by northblock-down growth faults (Figs 2b and 19; Nijman 1981; Puigdefàbregas *et al.* 1991). The LLM–ULM transition is marked by pronounced southward progradation of the Claramunt fan over the MLM-unconformity, increase in grainsize and a notable amount of granitic gravel (5 %) (Fig. 3).

The *San Esteban fan* consists of coarse to conglomeratic, arkosic sandstone and related floodplain mudstone. Igneous rock pebbles abound. San Esteban fan deposition began in the upper

Table 1. *Categories of architectural elements*

Sedimentary environments	Categories of architectural elements	Architectural elements*
Interdeltaic mud zone, floodplains, bays, and offshore muddy environments	1. *Mudstones, lime mudstones and marls*	Caliche, lacustrine limestone, floodplain mudstone, inshore transitional oyster-rich mudstone, offshore nummulite marl and slope marl
Alluvial & upper deltaic plain (*subaerial*)	2. *Alluvial sandstones and conglomerates*	Fan lobes, sheetflood deposits, braid sheets, ribbons, meander belts, sheet sandstone (complexes), tide-influenced fluvial channel fills
Lower (inter)deltaic plain (*transitional fluvio-marine and inshore subaqueous*)	3. *Distributary channel fills*	
	4. *Fine-grained depositional lobes and fans*	Fine-grained minor fluvial deltas, mouthbars; crevasse splays and levees
	5. *Coarse-grained (fan) deltas*	E.g. coarse CU-sandstone, with delta foresets
	6. *In/offshore reworked sediments*	Tidal flats, bioturbated sediments, storm deposits
Delta front (*shore and offshore*)	7. *Coastal sediments*	Shoreface deposits, barriers, in-/out-let channels, ebb/flood deltas
Deltaic plain, delta front and slope	8. *Gravity transport deposits*	Mass flows, slump layers, ball-and-pillow beds
In- to offshore marine environments	9. *Biogenic sediments*	Oyster beds, coralgal boundstone, skeletal limestone

* In the logs, architectural elements are further subdivided into or defined by their sedimentary structures such as tabular cross-bed sets, low-angle laminated sets, flaser-linsen cosets etc.

part of the Ager Group in the central sector of the basin (Isábena Valley) where the fan sediment was reworked into a tidal fandelta (San Esteban I = Roda Sandstone; Nijman 1989; Yang & Nio 1989). After a short episode of deficient clastic supply resulting in basin-wide carbonate deposition (Murillo Limestone, Fig. 7; Nijman & Nio 1975) at the base of the Montanyana Group, activity on the San Esteban fan was resumed. The fandelta rapidly prograded into lower Montanyana inshore (San Esteban II) and mid-Montanyana Castissent fluvial/upper deltaic environments (San Esteban III). San Esteban IV tide-modified fandelta lobes have been found in the next megasequence, UM-A. The conglomeratic top has been truncated by the UM-B unconformity (Fig. 10).

In the succeeding megasequences (UM-B to E) fan sedimentation was overtaken by the *Campanúe fan* delta, composed of conglomerate and sandstone derived from Mesozoic and Palaeozoic source rock (Weltje et al. 1996). Most of it intertongues with delta front deposits of the Perarrua Formation (Figs 4, 7 and 10). The base is angular unconformable in the southeast (base UM-B unconformity; see Garrido-Megias & Rios 1972; Mutti et al. 1988) (Fig. 10). From megasequences UM-B to E, the depocentre of the successive Campanúe fan lobes shifts north-westwards; the cumulative thickness of the Campanúe conglomerate amounts to 670 m. The shift tallies with the overall time-transgressive facies pattern in strike section along the northern basin flank, shown by the ensemble of Claramunt, San Esteban and Campanúe fans (Fig. 7). Fan aggradation overlapped major stratigraphic boundaries: the base of the Montanyana Group and the MLM-unconformity by the Claramunt and San Esteban fans, the base and top of the MM-megasequence by the San Esteban fan.

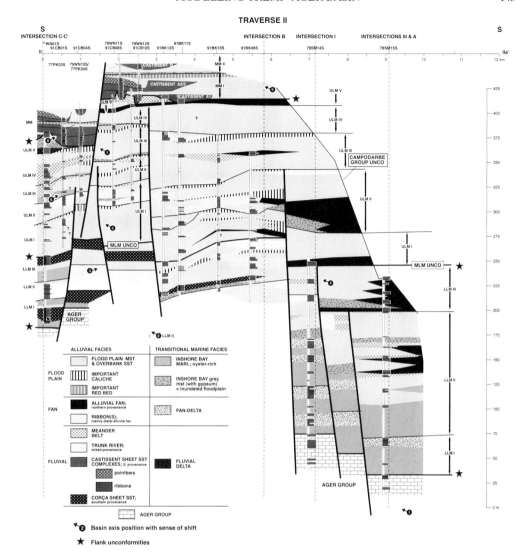

Fig. 9. Traverse II (location in Fig. 1). Dot-arrow symbols indicate position and sense of shift of basin axis in successive cycles. Discussion in text.

Fluvial feeder systems: the Corça and Castissent sheet sandstones

Stacked sheet sandstones of the Corça Formation in the Ager subbasin

The upper part of the Eocene Ager subbasin fill consists of the fluvial sandstones of the Corça* Formation (Mutti et al. 1985; Puigdefàbregas et al. 1991). The Corça Formation contains at least ten stacked though disconnected fluvial sheet complexes[†] (Fig. 11) on top of a thin interval of transitional marine deposits (Ametlla Formation; Dreyer 1994). Palaeocurrents in the lower part of the sequence are due north (Fig. 11: section 2) except for local deflections related to block faulting. Higher in the sequence, strike-parallel northwestward transport directions prevail. The non-carbonate composition of the Corça sandstones (Fig. 12) differs significantly from standard northern alluvial fan supply.

Both transitional and fluvial deposits overlap

---

*Castilian spelling: Corsa.
[†]for fluvial architecture, the nomenclature of Marzo et al. (1988) is used.

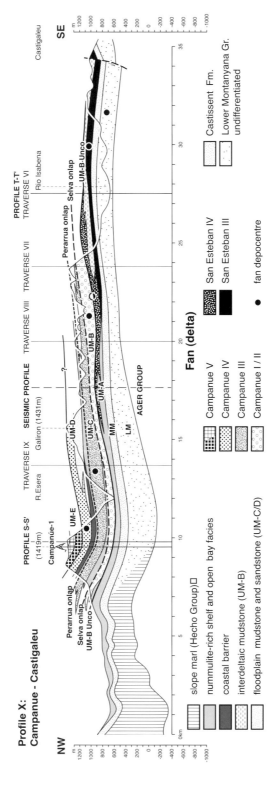

Fig. 10. Longitudinal structural profile X (location in Fig. 1) showing geometry and westward depocentre shift of San Esteban and Campanúe fans in the central and western part of the basin..

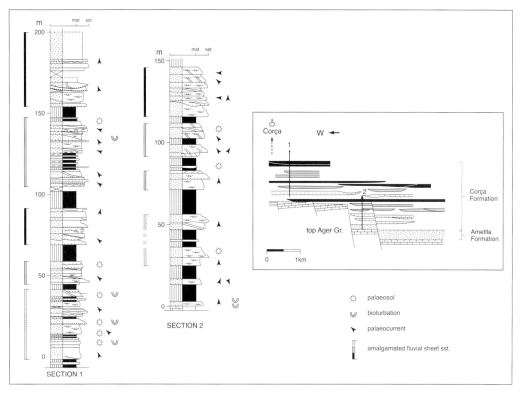

Fig. 11. Stacked fluvial sheet sandstone complexes of the Corça Formation; Ager subbasin. The cross-section is parallel to Montsec thrust and axis of the Ager subbasin. Logs have been taken south of village of Corça. (Data from unpublished MSc reports of De Reuver and Welsink; see also Puigdefàbregas *et al.* 1991).

a north-striking extensional growth fault structure, south of Corça (Fig. 11). A similar relationship between northward sediment input and fault orientation is observed along the northern dipslope of the Montsec thrust wedge near Castelnou de Montsec (see below and Fig. 2b), where white-weathered, distinctively Corça-type (Fig. 12) sheet sandstones are found in the lower Montanyana floodplain of the Tremp subbasin (for details see Puigdefàbregas *et al.* 1991). Setting and palaeocurrent bearings support the inference of a direct connection between the two fluvial systems.

Architecture of the Castissent fluvial sheet sandstone

The Castissent Formation represents a phase of exceptionally strong fluvial progradation within the delta (Figs 5 and 7), that separates the upper from the lower Montanyana Group. A detailed account of its architecture has been given by Nijman & Puigdefàbregas (1978), Marzo *et al.* (1988) and by Puigdefàbregas *et al.* (1991), from which the following summary has been derived. A recent discussion in terms of sequence stratigraphy is found in Emery & Myers (1996).

The major architectural feature of the formation is the superposition of three sheet sandstone complexes (Castissent A-C: MM-I-III), encased in fine-grained, in places vividly red-coloured, floodplain deposits with palaeosols. The floodplain deposits contain layers of dark-grey mudstone characteristic of floodplain inundation and correlatable with brackish-water onlaps (maximum flooding surfaces). The structure of each of these sheet sandstone complexes can be characterized as approximately 25 m thick; 4–6 km wide; and composed of one or more erosively superposed multilateral/multistorey sheet sandstones (amalgamated sheets) with an upward trend from conglomeratic to sandy lenticular-bedded bodies (braided stream deposits) to lateral-accretion-bedded bodies (low-sinuosity coarse-grained meandering stream deposits).

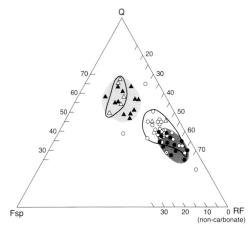

Fig. 12. Non-carbonate composition of Eocene alluvial sandstones of the Tremp–Ager Basin. Standard alluvial fan composition is taken from Escanilla Fm of the Campodarbe Group. (Data from unpublished MSc reports of De Reuver and Welsink).

The sheet sandstone complexes are flanked by ribbon-type channel fills (crevasse channels and braided marginal channels). Downstream (Isábena sector), the complexes expand into a distributive system of less interconnected to disconnected, fine-grained lateral accretion-bedded units.

The sandstone bodies are encased in floodbasin mudstone and siltstone with palaeosol profiles. Minor fine-grained channel fills and coarsening-up siltstone/fine sandstone bundles are the product of crevassing and overbank sedimentation. In the upstream sector of the fluvial system, flood basin deposits have pronounced, red colours, mottled grey and violet by pedogenesis. Vertical cylindrical cracking and calcrete nodule formation are common features of these palaeosols. Basin-wide correlation is achieved by tracing combinations of palaeosol and red bed markers, via grey, brackish water mudstone tongues, which record phases of inundation of the floodbasin due to rising groundwater, to facies indentations in the delta front.

In the Castissent floodplain, alternation of aggradation and non-sedimentation (palaeosols) prevailed. Erosion in the floodplain was restricted to about the scale of the channel depth and generally left the floodplain sequences at the two sides of the river system untouched. Restricted downward and sideward erosion is the main reason for the traceability of onlap phases into the upstream part of the delta and the lower reaches of the alluvial plain. If valley walls ever existed, they must have been defined by the flanks of the thrust sheet syncline, rather than by incision into the substrate. Where the Castissent river passed the Montsec thrustfold from the southeast (Fig. 8), the lower Montanyana Group has been truncated with an angular unconformity (MM-unconformity) and several ULM cycles are missing (Traverse C3/4: Fig. 2b).

Palaeocurrents (Fig. 8), composition (Fig. 12) and fluvial architecture point to a direct connection of the Castissent Sandstone with the Corça Formation. From its entry in the Tremp subbasin the Castissent river first followed a northerly course (MM I) to bend westwards along the northern side of the basin. Direct interference with the San Esteban alluvial fan system prevented development of an interfluvial mud zone (Fig. 8). Subsequently and stepwise, the entire river system shifted to the south and southwest (MM I -> II -> III: see traverses IV, Fig. 3b, and traverse C3/4, Fig. 2b). The uppermost cycle of the megasequence, MM IV (= Castissent-D), is only found in the Isábena sector, and is strongly fan-dominated (San Esteban III).

Farther to the northwest, the Castissent Formation is buried below the younger part of the basin fill. Well log data record the transition from FU-channel fills (borehole Guell: 610–750 m, ENPASA 1966; see profile X, Fig. 10) to CU progradational delta lobes (borehole Graus-1: 540–680 m, ENPASA 1966; see profile Y, Fig. 2a, and profile S, Fig. 4a).

In outcrops along the Esera Valley (right hand side of profile S on Fig. 4a), the Castissent Formation is still very distinctively sandwiched between nummulitid-rich deltafront deposits of the lower and upper Montanyana Group. The facies along this northern basin margin is fan-dominated. Between the Esera Valley and the exposure at the village of Charó in the Ainsa Basin, the entire sandy sequence rapidly fines to fine-grained delta-front sandstones with ample evidence of storm activity between highly bioturbated calcareous sandstone and marls with abundant nummulites. The delta front overlies the lateral ramp structures of the CSPT (Fig. 2). Our correlation with the slope deposits of the Ainsa basin differs from the one shown by Mutti *et al.* (1988) in that we correlate the Atiart unconformity, one of the major unconformities along the base of the Ainsa basin fill, with the base of the ULM megasequence, not with that of the Castissent Formation (see Nijman 1989, fig. 4). Restoration for tilt shows the prodelta slope to have faced northwestwards; it has been eroded not only by delta-front gullies of the Castissent phase of progradation proper, but also, and to considerable depth, by the ones belonging to the upper Montanyana delta slope.

Architecture of the upper deltaic plain

Stratigraphic traverses through the eastern part of the Montanyana Delta

For the following accounts of the deltaic architecture we refer to selected traverses of the database. The alluvial/upper deltaic plain is illustrated by traverses II (Fig. 9), IV/IVa (Fig. 3b) and C (Fig. 2b) in the Ribagorçana sector of the Tremp subbasin, the lower deltaic plain and deltafront (next section) by traverse IX (Fig. 4b) in the Esera sector. The location of the traverses is shown on Fig. 1.

Traverse II (Fig. 9) illustrates the upper deltaic, subaerial floodplain facies of the Lower and Middle Montanyana Group. Traverses IV and IVa (Fig. 3b) are situated along the Ribagorçana Valley, respectively 5 and 9 km downstream with respect to traverse II. Figure 3b represents the most complete cross-section of the Montanyana Group. It records the intertonguing of upper and lower deltaic plain facies. The contact with the underlying Ager Group is extrapolated from the flanks of the basin, accounting for 10° plunge of the thrust sheet syncline. Both traverses show the internal and external architecture of megasequences and cycles and a shift of the basin axis. The relationship between faults and megasequence boundaries in the lower Montanyana Group is illustrated in traverse II (Fig. 9) and in longitudinal traverse C3/4 (Fig. 2b).

Architectural elements

The upper deltaic plain comprises more or less connected channel fills ranging from ribbons to sheet sandstone complexes, floodplain mudstone with overbank siltstone and sandstone, sheet-flood deposits, lacustrine mudstone and limestone, and caliche. Distinction has been made between fan and fluvial systems with northern supply and fluvial systems with southeastern provenance. In one place only, on the south side of cycle UM-A IV in Fig. 3b, fan conglomerate of southern provenance occurs. Most of the meandering channel fills belong to the fluvial feeder system, some seem to be directly related to distal fan facies; a difference to be established with certainty only where channel-fill systems can be sufficiently traced in upstream direction. With few exceptions, ribbon facies is always connected to the fan fringe (examples in Figs 3b and 9). Apart from the amalgamated Castissent and Corça-type sheet sandstones of the fluvial feeder system, a class of multi-lateral/multistorey sheet sandstone with braided and meandering channel-fill components and of mixed provenance occurs intermittently at or near the basin axis. These are interpreted to represent short river tracts at the confluence of the two deltaic components, and are indicated as trunk rivers. They terminate in minor fluvial deltas set in oyster-bearing, brackish water, interdeltaic/-distributary bay marls, which mark the upstream limit of the lower deltaic plain.

The lower Montanyana Group

Megasequences. The two superposed megasequences of the lower Montanyana Group (LLM and ULM) are separated by a discontinuity, the MLM-unconformity, at about 120 m below the base of the Castissent Formation (Traverses II: Fig. 9; IV: Fig. 3b; C3/4: Fig. 2b).

Along the southern basin flank, against the Montsec thrust wedge, the discontinuity becomes a low-angle unconformity and the overlying cycle ULM I contains reworked sandstone, calcrete nodules and caliche of the lower megasequence (LLM). The average palaeocurrent bearing of the lower Montanyana Group in that area is 328° ($n = 205$), generally with a downstream trend to bend parallel to the basin axis.

In the northern basin flank (Fig. 9), the MLM-unconformity is associated with strong southward progradation, and abrupt increase in grainsize and granite pebble content of the Claramunt fanglomerates. The average palaeocurrent bearing of the Claramunt fan is 224° ($n = 76$). Sharp westward to northwestward deflections occur in the drainage pattern where the fan fringes reach the zone of mud accumulation (average palaeocurrent bearing of trunk rivers: 301°; $n = 182$). The formation of the unconformity and the subsequent deposition of sediment have been influenced by an E–W growth-fault pattern, related to differential compaction over a facies change in the underlying Ager Group from competent limestone in the south (Montsec) to thick, incompetent, marine marl in the north (traverse II). The faults acted simultaneously with the afore-mentioned NNW-oriented fault systems at right angles to the strike of the Montsec thrust (Nijman 1989).

Cycles. Within the LM-megasequences, the architectural elements are organized in well recognizable cycles, 19–96 m thick (Fig. 3, see Table 2). In the northern facies belt of the lower Montanyana Group, conglomerate and sandstone sheets and prominent palaeosols determine the cyclicity. Along the southern basin flank (traverse C4: Fig. 3b), fluvial channel fills are concentrated in Corça-type amalgamated sheet sandstones of the lower megasequence (cycles LLM I and III), and in vertical stacks of multi-lateral tabular channel fills with well developed lateral accretion bedding of the upper megasequence (ULM I-IV). Other major sheet sandstones and minor delta complexes correspond to axial trunk river systems which collected coarse clastics from both sides of the basin for distribution into the lower deltaic plain. Mud, on the contrary, was concentrated to extraordinary values of over 90 vol-% in the interfluvial zone of many cycles, presumably by asymmetric overbank deposition at the drainage bends of the feeder systems (good examples in cycles ULM I and II of traverse IV, Fig. 3b; ULM II and III of traverse II, Fig. 9) (Van Hilten 1995).

Cycles show a distinct aggradational sequential pattern (arrows and letter indications in Fig. 3b): (a) sandstone and conglomerate sheets of northern fan provenance and of trunk river systems at the cycle boundary; (b) subsequent vertical aggradation and stacking of fan and fluvial architectural elements in relatively narrow belts along the (sub)basin flanks thereby establishing a central mud zone; (c) deficit of clastic supply and increase of caliche formation; and, finally (d) grey mudstone representing drowning of palaeosols and floodplain inundation up to lacustrine facies; in longitudinal section floodplain inundation facies correlate with brackish water onlaps (traverse C3/4, Fig. 2b: e.g. cycles ULM-II and IV). Trunk river sheet sandstones occur from stage (c) onward, and stages (a) and (d) frequently overlap each other. In terms of sequence stratigraphy, the cycle boundaries should correspond to the changes in the rate of formation of accommodation space visualized by the change from aggradational (in length section often also progradational) behaviour to brackish onlap and retrogradation. The change coincides with a more stabile regime of amalgamation and lateral accretion of trunk river sandstones and of corresponding minor deltas, in the centre of which the cycle boundary should be positioned. For cartographic convenience, the cycle boundaries have been chosen at the base of the brackish water facies and corresponding floodplain inundation facies and minor deltas; the trunk river sandstones are then found in the top of the underlying cycle.

The so-defined *aggradational cycle* is further illustrated and compared with other cycle types in a later section of this paper (see p. 159; cycle architecture), in which also attention will be paid to changes of the position of the basin axis, indicated with dot-arrow symbols in successive cycles of the traverses of Figs 3b and 9.

The middle Montanyana Group

The eastern traverses II (Fig. 9) and IV (Fig. 3b) contain only the MM I and II cycles of the Castissent Formation. The cycle architecture is different from the one in the lower Montanyana Group, in that the trunk rivers have been replaced by much more voluminous amalgamated sheet sandstones of the main fluvial feeder system. Discontinuous aggradation of the floodplain, with palaeosol formation during non-sedimentation, went together with repeated aggradation and degradation of the stream bed accounting for the much larger volume of the Castissent sheet sandstone complexes than of the other channel systems.

With cycle MM I the position of the basin axis has been shifted far northwards (Fig. 8), to reverse towards the southwest during deposition of cycles MM II and III (see Marzo *et al.* 1988).

The upper Montanyana Group

Deposition of the upper Montanyana Group started with a basin-wide marine onlap over the Castissent megasequence. In the eastern sector, the onlap is visible in brackish water deposits in the basal part of cycle UM A-I (traverse C3/4, Fig. 2b; traverse IV/IVa, Fig. 3b) and in grey, inundated floodplain mudstone found as far to the east as the type area of the Castissent Formation (Fig. 2).

Correlation over the Luzas thrustfold (cross-section Y: Fig. 2a) with lower deltaic plain facies of the Isábena and Esera sectors of the basin is based on correspondence of cyclicity and facies pattern. It shows that in traverse IV/IVa the complete UM-A megasequence is exposed together with at least part of the UM-B megasequence, the UM-B unconformity being correlate with the base of the Campanúe Fan in the western part of the subbasin.

Very intense caliche formation at the top of the Montanyana Group in traverse IV/IVa (Fig. 3b: *) records a long period of slow to absent sedimentation at the base of the uppermost Eocene to Oligocene Campodarbe Group. Towards the limbs of the thrust sheet syncline, is appropriate to correlate the lower and middle Montanyana Group. In the upper Montanyana Group, only the stratigraphic sequence of the basin compartments west of the Luzas Fault can be correlated this way successfully. Farther eastwards, cycles of the Capella Formation (Fig. 7) of the Ribagorçana sector of the Tremp basin can only be matched with the UM-A and B megasequences of the much thicker sequence in the west by tracing the UM-B unconformity and distinct marine onlaps in megasequence UM-A. Vertebrate biostratigraphy in the Capella Formation (Cuevas Gozalo 1989) is generally insufficient to support detailed correlation except for the position of the base of the Campodarbe Group (Escanilla Formation); a correlation also confirmed by palaeomagnetic data (Bentham et al. 1992).

Periodic or aperiodic sequentionality and cyclicity?

Data for the correlation scheme of Fig. 13 has been derived from the Ribagorçana sector for megasequences LLM and ULM, from the Isábena sector for MM, UM-A and UM-B, and from the Esera sector for UM-C-E and the Pano Fm. Duration, thickness, and calculated compacted net sedimentation rates of megasequences and cycles are represented in Table 2.

Depocentres of the eight megasequences vary between 148 m (MM) and 404 m (UM-A). The megasequences are aperiodic since they represent periods ranging from 600–700 ka in the lower, 400 ka in the middle, to over 1100 ka in the upper Montanyana Group. No obvious correlation exists between megasequence duration and thickness, and therefore also the net sedimentation rate, which varies between 18.5 and 67.3 cm ka^{-1}, is independent of megasequence duration.

Cycle thickness varies between 12 m (UM-C I) and 100 m (UM-A IV), with an average of 44.5 m. The average cycle duration per megasequence ranges from 100 ka (MM) to 150 ka (LLM), with an overall average of 124 ka. Given the degree of inaccuracy, range and average value might be considered as indicative for periodicity of the cyclicity, i.e. as an approximation of the 100 ka duration of orbitally forced cyclicity (De Boer & Smith 1994). It follows that the considerable variation in thickness between cycles would reflect relatively large fluctuations of sedimentation rate. Part of the thickness variation, however is due to non-sedimentation and/or erosion, often by channel systems.

Cyclicity, basin axis shift and the question of base level control

Figure 13 relates the cross-section of the Montanyana Group with the sea level curve of Haq et al. (1987). To a certain extent, the number of megasequences we distinguish (eight) corresponds with that of third-order cycles (seven) in the sea-level curve, which might indicate a general influence of base-level fluctuations on the sequential architecture. On the other hand, the actual chronostratigraphic resolution always allows correlation between the eustatic curve and the sequence stratigraphy of the Montanyana Group.

Another aspect of the basin architecture is the observed pattern of basin axis shift in cross-section (Fig. 13). In this context, the megasequence boundaries will be discussed in stratigraphic order:

Base LLM-unconformity. The base of the Montanyana Group coincides with southward shift of the (sub)basin axis at the end of the underlying Ager Group. In the lower part of megasequence LLM (Traverse II, Fig. 9), arkosic alluvial fan deltas from northern sources debouched on Corça-type rivers from the south producing a wide amalgamated sandstone sheet with an axis as far southwards (LLM-II in Traverse II, Fig. 9) as the present-day north slope of the Montsec Range, where also the base of the Montanyana Group forms an angular unconformity (Fig. 2b). Before the end of the LLM-megasequence the basin axis shifts about 8 to 9 km northwards (Traverse II, Fig. 9; and Traverse IV, Fig. 3b).

MLM-unconformity. The overlying MLM-unconformity truncates the fluvial system in the south (Figs 2b and 3b), and brings coarsely clastic, partly pebbly granitic material from the north, suggestive of northern basin flank uplift. During deposition of the next megasequence, ULM, a rapid southwards basin axis shift is again followed by an equally well developed northward shift of 11 km (Fig. 3b) just before the formation of the unconformity at the base of the Castissent sheet sandstone. The position of the 52 Ma sequence boundary has been derived from both biostratigraphic and palaeomag data in the Esera Valley.

Although the two megasequences of the lower Montanyana Group can be easily matched with the two third-order sea level cycles between 53.3 and 51.3 Ma (Fig. 13), the rapid lateral shifts in deposystem architecture, combined with flank unconformities, point to structural control rather than to a mere change in sea level.

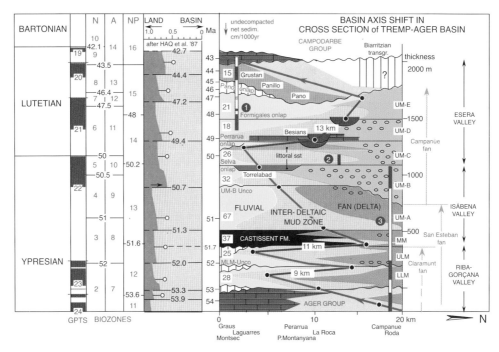

Fig. 13. Composite transverse cross-section of the Montanyana Group in the Tremp subbasin showing relations between magneto/bio/chronostratigraphy, sea-level curve, megasequentionality, rate of sedimentation, and trace of basin axis shift: green trace corresponds to southward shift and fan progradation, red to northward shift in km. Biozonation used for calibrating the sequence stratigraphy from several sources: N, nummulites; A, alveolines; NP, nanoplankton. Magnetostratigraphy after Bentham & Burbank (1996). Megasequences indicated LLM, ULM for lower and upper part of the lower Montanyana Group, MM for the middle Montanyana Group (= Castissent fluvial sheet sandstone), and UM-A to UM-E for the upper Montanyana Group.

Castissent (MM-)unconformity. A correlation of the basal Castissent unconformity with the 51.3 Ma drop in sea level is attractive, but the sheet sandstones are sandwiched between two of the largest marine onlaps in the area. Even between the MM-cycles, onlaps and raised groundwater levels have been recorded far inward into the Montanyana delta. These repeated onlaps indicate a continuously high base level. The Castissent Sandstone is therefore best interpreted as a highstand progradational deposystem (Emery & Myers 1996). The chronostratigraphic data render a match with the 51.7–51.3 Ma sea-level highstand probable. In that case, deposition of the Castissent Formation would span a period of 400 000 years conformable to calculated rates of floodplain aggradation and palaeosol formation by Marzo *et al.* (1988).

Basin-flank erosion, in particular indicated by a sharp increase in Mesozoic and lower Tertiary limestone pebbles in the Castissent A2 sheet (Nijman & Puigdefàbregas 1978), repeated incision of the river system in its own bed, and northward basin axis shift followed by a southward shift, all testify to structural control as a main reason for its formation. The combination of decreasing rate of base-level rise (sea-level highstand) with flank deformation and erosion may account for the exceptionally strong westward progradation and basin-wide emersion of the delta top during Castissent time.

Base UM-A. The boundary between MM and UM megasequences resembles a type II-sequence boundary, flooded rapidly during the initial stage of deposition of the Perrarua Formation. The sea-level curve shows a moderate fall at 51.3 Ma. At this boundary, the trace of basin axis shift shows no reversal as the axis continued to gradually move southward during cycles UM-A and B. No flank unconformities have been formed related to the base of UM-A. During deposition of the UM-A cycle, the rate of net compacted sedimentation, however,

Table 2. *Parameters of megasequences and cycles, also applied in construction of Fig. 13*

Megasequences	Estimated duration (ka)	Thickness (m) used in constructing cross-section of Fig. 13 e.g. (IX) refers to depo-	Average rate of net compacted sedimentation in cross-section (cm ka^{-1})	Cycles	Cycle depocentre (m)	Cycle depocentre asveraged per megasequence (m)[6]	Cycle duration averaged per megasequence (ka)[7]
UM-E[1]	Min.1400	300 (IX-S)	21.4	min. 8	decreasing 40->20 m	?	?
UM-D	1100	204 (IX)	18.5	IVa+b	71	30.4	137
				IIIa+b	39		
				II	70		
				Ic	17		
				Ib	18		
				Ia	28		
UM-C	800	212 (VIII)	26.5	VIa+b	24	33.4	114
				V	38		
				IV	25		
				III	73		
				II	62		
				I	12		
Um-B	500	164 (VI)[2]	32.0	IIa+b+c	93	36.4	100
				Ia+b	89		
UM-A	600	404 (VI)	67.3	IV	100	81.4	120
				IIIb	95		
				IIIa	59		
				II	93		
				I	60		
MM	400	148 (VI)	37.0	IV	56	44.0	100
				III	45		
				II	33		
				I	42		
ULM	700	176 (II)	25.1	V	19	38.0	140
				IV	30		
				III	40		
				II	41		
				I	66		
LLM	600	168 (I)	28.0	III	61	65.3	150
				II	94[4]		
				I	84		
				0	20[5]		

[1] Data partially from Cuevas Gozalo (1989, figs 3–8, 3–12, and 3–17).
[2] Thickness of interdeltaic bay deposit instead of extreme thickness of upstream part of Campanúe Fan (see fig. 13).
[3] Cycles indicated a+b or a+b+c could not be subdivided in the depocentre proper but only elsewhere in the cycle; for calculation of average cycle depocentre (column 7) per megasequence, these subcycles are counted as cycles by dividing the value of cycle depocentre (column 5) by the number of its subcycles; for calculation of column 8 (duration) subcycles are likewise counted as cycles.
[4] Maximum thickness of LLM II is 125 m in hanging wall of growth fault.
[5] Maximum thickness of 100 m attained in slope facies (Hecho Group) in Esera Valley section.
[6] Overall average cycle thickness LLM–UM-D: 44.5 m.
[7] Overall average cycle duration LLM–UM-D: 124 ka.

attained its maximum value (67 cm ka^{-1}), testifying to important aggradation. An alternative explanation of this megasequence boundary will be given in the section on structural control.

UM-B-unconformity. The base of the Campanúe fan coincides with the largest Eocene drop in sea level at 50.7 Ma (arrow in sea-level curve of Fig. 13). A peak supply of conglomerate accompanied this megasequence boundary, which was also coeval with the westward shift from the San Esteban to the Campanúe fan edifice. Flank uplift was highly asymmetric, predominating in the northern flank of the basin. Basin axis shift in a continuously southerly direction slowed down.

The Selva onlap between UM-B and UM-C falls within a third-order cycle. The megasequence boundary does not correspond to any observable change of basin-axis shift.

The Perarrua onlap may coincide with the sea-level rise succeeding the drop in sea level at 49.4 Ma. Limestone deposition concurrent with the onlap would then fit in with the transgressive systems tract, which is also held responsible for the infilling with offshore marl of the deltafront gully at Besians. Nevertheless, the same sequence boundary is clearly related to a change from a fan-dominated to a fluvial-dominated coastline, and to a bend in the pattern of basin axis shift. Again, such a lateral shift of the basin axis of the system cannot easily be explained by the eustatic mechanism and is therefore considered to reflect structural control. Corespondingly, the formation of deltafront gullies can be explained by slope instability during increased tectonic control, and limestone deposition might merely indicate a tectonic disruption of the clastic supply routes.

The Formigales onlap, at the base of megasequence UM-E, has not been studied in detail. It has much in common with the Perarrua onlap, including deltafront gullying, but littoral limestones have not been reported to occur and the onlap does not coincide with a change in architecture and basin axis shift.

The Pano onlap coincided with a southward basin axis shift lasting over a period of 4 Ma (47–43 Ma). Like the shift at the base of the Montanyana Group, it was sufficiently large to put an end to the Montanyana delta system, i.e. to define a group boundary, and to bring the South Pyrenean drainage system out of the Tremp–Ager Basin into a position along the Central South Pyrenean thrust front (Campodarbe Group). This major shift again correlates with a peak in limestone deposition, represented by a fringe of bioconstructed limestones along the CSPT front, which connects the Puy de Cinca Limestone with the Tossa Limestone of the Santa Maria Group in the Catalan basin (Nijman 1989).

Structural control

Structural control of the sedimentation pattern consisted of: (a) westward tilt of the basin, defining the NW facing of the Montanyana delta, and, by enhancing drainage bends in the fan and fluvial feeder systems, considered an important factor to enable the formation of an interfluvial-/deltaic mud zone; (b) fixation of the break-in slope of the deltafront over the active western lateral ramp of the underlying thrust sheet; and (c) complex uplift and subsidence patterns resulting from southward thrust translation.

Modes of delta progradation with tectonic fixation of the break-in slope of delta front. During the Eocene, the shelf break was fixed above the western oblique ramp of the CSP thrust sheet (Nijman & Nio 1975; Nijman 1981). Figure 14 illustrates the influence of the fixation of the break-in-slope of the delta front on the pattern of delta progradation. The lower Montanyana Group was related to the initial stage of delta formation in the Tremp–Ager Basin and therefore had a wide deltafront platform (Fig. 14a), while the shoreline of the upper Montanyana Group rapidly advanced towards the shelfbreak with a corresponding decrease in width of the delta-front platform (Fig. 14b). As a consequence, because of dissipation of energy over the wide delta-front, the lower Montanyana delta was fluvial-dominated with many interdistributary bays and mouthbar-type subdeltas.

The upper Montanyana delta shows the increasing effect of tides and waves in a preponderance of flat-based fan lobe and barrier sandstones over distributary channel sandstone. Progradation occurred by pincer-like closure of interdeltaic bay areas (Fig. 14b) wherever alluvial fan deltas and fluvial deltas met each other in the littoral zone and were further connected by longshore sand transport. Progradation of the northern fans forced the basin axis southwards (Fig. 14c), lateral expansion of the fluvial feeder system had the opposite effect (Fig. 14d). In the latter case, one could argue that a backstep of the alluvial fan fringe would automatically be followed by lateral, i.e. northward expansion of the fluvial system to explain for reversals of basin-axis shift, like the one occurring at the Perarrua onlap. This would only be valid during stages of low rate of aggradation in a shallow accommodation space, because at high aggradation rate the fluvial system would

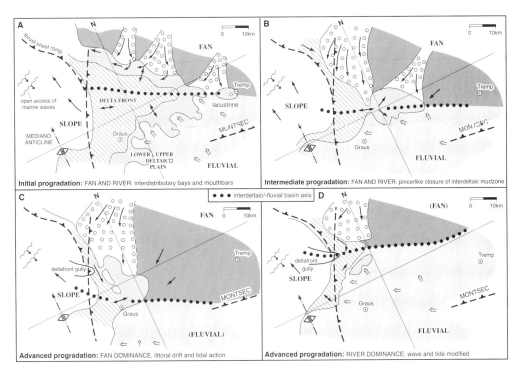

Fig. 14. Patterns of progradation and basin axis shift of Montanyana delta determined by fixation of break-in-slope over western oblique ramp of Central-South Pyrenean thrust sheet. Open arrows refer to fluvial palaeocurrent pattern, black arrows to fan (delta), deltafront and slope transport directions; double-headed arrows indicate tidal transport.

tend to stack vertically without expansion. However, if fan progradation were succeeded by fluvial expansion coeval with flooding and vertical stacking of limestone in the delta front as observed at Perarrua, tectonic forcing seems to be the mechanism.

Modes of megasequence architecture in response to thrust translation. The ECORS traverse (Fig. 1) shows a hinterland-dipping sole thrust connecting the Nogueres antiformal stack in the rear of the CSPT complex with the leading thrusts.

At lithospheric scale, relief and surface gradients were controlled by isostatic uplift in the Axial Zone of the orogen, and foreland basin subsidence due to bending below the thrust sheet front (Fig. 15). The overall result was outward depocentre migration, familiar to many foreland basins. In recent years, the relationship between migration of depocentre, thrust front and forebulge has been widely emphasized (Tankard 1986, Flemings & Jordan 1990, Millán et al. 1995). Reversals of depocentre migration have been explained by relaxation of a viscous–elastic lithosphere and forebulge migration towards the thrust front. This crustal-scale structural control is mainly expressed in the major shifts of basin axis as observed in the succession of the respective piggyback basins.

In the central southern Pyrenees, one has to distinguish between the initial and advanced stage of thrust deformation. The former largely corresponds to inversion of normal faults within the pre-existing shelf without major stratigraphic duplication. This process is linked with thick-skin thrusting in the Axial Zone and the beginning of duplex formation along the rear of the thin-skin thrust system (Puigdefàbregas et al. 1991). Loading and foreland basin subsidence became really important only with the stacking of basement-involved lower thrusts beneath and at the rear of the thin-skin thrust system, and with overriding of the foreland by the upper thrust sheets (Burbank et al. 1992a, p. 1119). For the CSPT the transition between the initial stage and the advanced stage of the thrust system is indicated by the inversion of the Ebro High into a true foreland basin during the late Eocene (Nijman 1989).

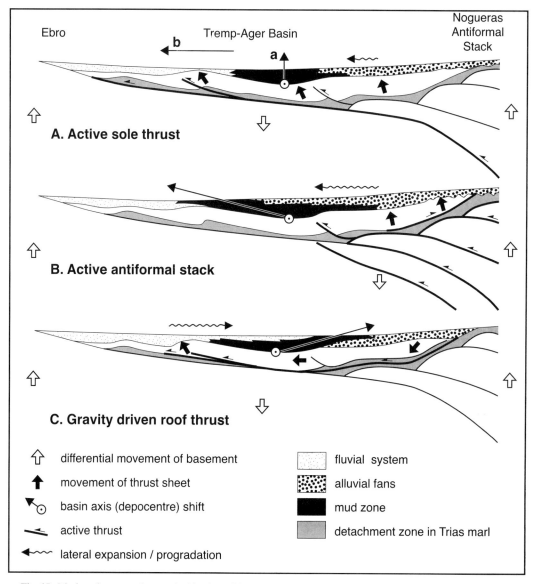

Fig. 15. Modes of structural control of basin architecture and basin axis shift.

At uppermost crustal levels and at the scale of basin architecture represented by the Montanyana database, direct structural control depends on the translation of minor thrust pulses into uplift and subsidence, in particular of the basin flanks. A recent 2D quantitative approach on this subject implying the relationship between fault-bend folding, incremental basin infill, and grainsize distribution is given by Den Bezemer et al. (this volume).

Therefore, in order to explore the possibilities of quantitative modelling with the database of the Tremp–Ager Basin, the surficial thrust effects during the initial stage are considered rather than the complex lithospheric response during the advanced stage of foreland basin development, amply discussed in the literature on basin modelling. Whether thrusting was accommodated on a single element of the thrust complex, or on several elements at a time, may be expected to have determined differences in basin architecture. Although not exhaustively,

Fig. 15 shows schematically the effects of fault motion along particular segments of the thrust system (see also Puigdefàbregas et al. 1991, p.18).

(1) Even if of moderate intensity, thrusting concentrated along the floor thrust (Fig. 15a) will cause upward and outward displacement of the entire thrust wedge, the antiformal stack at the rear and the piggyback basins included. The result resembles a base level lowering, but is counteracted by foreland subsidence probably with a lag effect. Toe structures become more pronounced, which may lead to erosion and unconformity (Zoetemeijer et al. 1993).

Major outward depocentre shifts are related to forward branchline migration (Fig. 6) and so define the major tectono-sedimentary cycles distinguished by Puigdefàbregas & Souquet (1986).

(2) Shortening restricted to the duplex structure below the rear of the thrust sheet, i.e. active growth of the Nogueras antiformal stack (Fig. 15b), generated one-sided uplift of the northern flank of the piggyback basin. This probably went along with backthrusting along the detachment of the upper, Central South Pyrenean thrust sheet (Puigdefàbregas et al. 1991). As a result, erosion is expected to have formed an unconformity at the rear of the thrust sheet supplying enough material to let alluvial fans prograde southwards with a concurrent shift of the basin axis of the piggyback basin. The same mechanism of rear uplift with or without backthrusting could also imply temporary ponding of sediment in the upstream reaches of the supplying streams, allowing for the climatic imprint on the sediment composition which appears discernable on a minor scale of cyclicity than the one discussed in this paper (Weltje et al. 1996). An effect similar to that of thrust activity at the rear of the system would be created by mere isostatic uplift of the axial zone of the orogen.

(3) One-sided uplift at the rear of the thrust system eventually changes the position of the roof thrust at the base of the CSPT into foreland dipping (Fig. 15c). As long as the bending point of the thrust sheet syncline remains fixed by the branchline of the frontal thrust, southward movement of the thrust sheet will be translated in both increasing toe uplift and northward shift of the basin axis relative to its immediate substrate. A comparable reverse shift of basin axis, however, is also computer-predicted to result from out-of-sequence thrusting by Zoetemeijer (1993) and along the back limb of fault-bend folds (Den Bezemer et al. this volume).

Northward basin-axis shift is most explicitly visible in northward expansion of the Castissent fluvial system concurrent with toe uplift and steepening of the gradient of the southern basin flank. Normal southward progradation of the fans will have changed into vertical stacking or, in extreme cases, to retrogradation, because of increased rates of subsidence in the proximal fan areas. In Fig. 15c the CSPT is depicted as a rootless thrust sheet. The foreland dip of its detachment implies the possibility of a gravitational mechanism of emplacement instead of the compressional mechanism acting along a hinterland dipping sole thrust. Ample evidence (e.g. extensional structures in the rear of the thrust sheet: Séguret 1972; recumbent thrusts and slip sheets along and below the western oblique ramp of the thrust sheet: Nijman 1989) suggests that gravitational thrust sheet motion may have occurred at least during some episodes or in segments of the CSPT.

Modes of structural control like the ones discussed determined the megasequence boundaries and architecture to a large extent and in combination. The zig-zag trace of the interdeltaic basin axis in cross-section of the Montanyana deposystem (Figs 13 and 16) correlates with the megasequential order and most of the megasequence boundaries are related to unconformities in one or both of the basin flanks and are marked by onlaps.

Assuming a direct relationship between northward basin axis shift and the rate of thrust sheet motion according to mode C in Fig. 15, the cumulative northward shift of about 33 km (Fig. 13) would correspond to an average translation rate of about 5.5 mm a^{-1} over the 6 Ma of deposition of the Montanyana Group (cf. 2 mm a^{-1} for Swiss Molasse: Sinclair et al. 1991; 2.5–4 mm a^{-1} Montsec thrust front: Burbank et al. 1992; 5 mm a^{-1} Catalan Basin: Vergés et al. 1992). This net effect has been created by three pulses. The first two in the lower Montanyana Group (LLM–MM) caused displacements of respectively 9 and 11 km distance each over an estimated minimum time interval of about two cycles or roughly 200 000 years; the third (UM-C to E) of 13 km distance took 2 Ma. These figures correspond to partial rates of translation respectively of 55, 50 and 5.4 mm a^{-1}, which for the first two pulses is ten times larger than for the third. Such a model might envisage a change from a stick-slip mechanism in the upper Ypresian to slower and more continous thrust translation during the Lutetian. Note, however, that the calculation completely depends on the assumption of gravity-driven thrust motion which perhaps only was operative at the moment of antiformal stacking in the rear and inversion of the Ebro High in the front of the thrust system, late during the Eocene. The

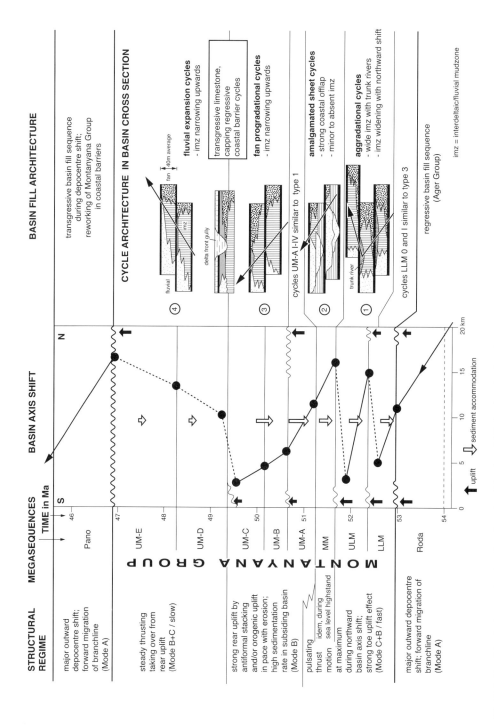

Fig. 16. Relationships between structural regime (see Figs 6 and 15), basin axis shift (see Fig. 13), and megasequence/cycle architecture.

example illustrates the potential of the database to model specific modes of thrust displacement and to compare surficial tectonic relief effects with crustal-scale control.

Combination of sea-level change with structural control created specific architecture; examples are the Castissent Sandstone, the UM-B unconformity, and, possibly, the Perrarua onlap where limestone caps coastal siliciclastic barrier cycles (Fig. 16). The base UM-A megasequence boundary does not correspond to any reversal in basin axis shift, nor to flank unconformity. The associated marine and brackish water onlap did not reach farther inland than the onlaps below and between the underlying Castissent sheet sandstones. Instead of a type II sequence boundary (see previous section), the contact might be explained to have been formed by acceleration of relative sea level rise, probably by tectonic subsidence in response to loading, and diminishing clastic supply. Both factors are the logic consequence of levelling down of the fluvial gradients by erosion and sedimentation during the previous Castissent phase, under conditions of decreasing activity of the thrust system. The subsequent increase in rate of net compacted sedimentation reflects the isostatic uplift in the Axial Zone of the Pyrenees.

Finally, the major southward depocentre migrations at the base and at the top of the Montanyana deposystem due to branchline migration were accompanied by peaks in carbonate sedimentation (Murillo Limestone, Puy de Cinca Limestone, Fig. 7) indicating that the clastic supply systems could not cope with the disturbance caused by tectonically induced depocentre migration.

Cycle architecture. The architecture of the cycles is summarized in Fig. 16. In the lower Montanyana *Group aggradational cycles* prevail characterized by vertical stacking of channel facies after deposition of sheet sandstone/conglomerate, by a wide interfluvial mud zone, and by decelerating aggradation to non-sedimentation, concurrent with flooding, at the top.

The MM megasequence is characterized by repeated aggradation and degradation, and strong progradation (*amalgamated sheet cycles*). The structural regime is interpreted to have been similar to that of the lower Montanyana Group but with an accommodation space restricted by sea-level highstand.

UM-A cycles compare with LM cycles, but show maximum sedimentation rates in response to the preceding tectonic activity.

Finally, the UM-B to E cycles are progradational with definite coarsening-up trends in both feeder systems stacked obliquely to the south during fan dominance (*fan progradation cycles*), obliquely to the north during fluvial expansion and fan retrogradation (*fluvial expansion cycles*). The stacking patterns are determined by the architecture, and hence, structural control of the megasequences.

Often, the cycles appear to mirror the architecture of the megasequence. Steepening of gradients in one or both basin flanks as a consequence of a tectonic pulse would have induced degradation. Whether fans tended to prograde or not in response to upstream erosion largely depended on the accommodation space created by relative base level movement and by the relation of erosion rate to uplift rate (Bentham *et al.* 1992; Emery & Myers 1996). A tectonic pulse, therefore, would have resulted in strong progradation (fan progradation cycles), if at least erosion could keep pace with the flank uplift. If on the other hand erosion could not cope with the hinterland uplift, fans tended to stack vertically or even began to retrograde towards the basin flank during tectonic activity, while coarsening and steepening. This is the observed pattern of the aggradational cycles of the LM and UM-A upper deltaic plain. Return to tectonic quiescence is recorded by sheet conglomerate and sandstone extending far into the basin (e.g. Heward 1978; Heller *et al.* 1988; Flemings & Jordan 1990) when the steepened gradients had been levelled down and accommodation space reduced due to stabilization of the floodplain with respect to base level. Trunk rivers may be expected at this level and palaeosol formation in elevated parts of the floodplain. Though often related to marine and brackish water onlap, floodplain inundation may have been enhanced or even caused by compaction of the interfluvial mud zone.

Conclusion

The construction of a large database for the Montanyana Group allowed recognition and definition of primarily structurally controlled megasequences, of variable thickness and aperiodic, and basin-wide cycles in the order of 40 m thick. The inferred 124 ka periodicity of the cyclicity approximates the 100 ka of orbital forcing (De Boer & Smith 1994). The pattern of stacking of the cycles conforms to the structurally controlled megasequential basin axis shift on which it is superimposed. Coeval sea level fluctuation modified the megasequential architecture accounting for extreme progradation as observed in the Castissent Sandstone. Because the Eocene climate only showed a gradual change from somewhat more humid to seasonal and drier (sub)tropical conditions (Haseldonckx 1973; Plaziat 1981), large-scale climatic fluctuations did not play a prominent role in defining the cyclicity. At the smallest scale of cyclicity, however the influence of orbitally forced climatic variations appears to become perceptible.

The approach exemplified in this paper is considered a step towards 3D numerical tectono-sedimentary modelling of the Tremp–Ager piggyback basin, on a wider scale than pilot studies in the South Pyrenees such as the modelling of the fluvial architecture by Dreyer et al. (1993). It is expected to allow, for instance, a quantitative comparison between behaviour of the sedimentological basin axis as defined in this paper, the depocentre and the structural axis, and the testing of various tectono-sedimentary hypotheses and models involving fault-bend folding, back-thrusting and gravitational thrust sheet emplacement.

For the illustrations a host of data, mainly consisting of sedimentary logs, has been taken from unpublished MSc research reports of Leyden and Utrecht Universities from the period 1971–1996, the titles of which are not given in the list of references. It concerns for Fig. 9: B. Berlie 1991, J. Kloos 1991 and C. van den Bergh 1991; for Figs. 11 and 12: F. de Reuver 1981 and H. Welsink 1979; for Figs 5 and 9: Ph. Kips 1977; for Figs 3 and 9: S. van der Meulen 1978; for Figs 2 and 3: H. Abdul Haziz 1993, M. Felix 1993, H. Frikken 1977, E. R. Kuiper 1993, M. van der Meulen 1993, J. Pronk 1978 and S. van Gaalen 1988; for Fig. 3: A. Visser 1985; for Fig. 4: J. W. Gunster 1971, J. Hommes 1976 and W. Langeraar 1976; and for Table 2: E. Tervoort 1996.

I gratefully acknowledge substantial financial support for the production of the database by Shell Research B.V., and the stimulating contacts with B. Levell, A. van Vliet and A. Speksnijder. Financially, the project was also supported by the European Union as part of the project on Integrated Basin Studies (contract no. Joule 2-CT-92-0110). C. W. M. van Oosterhout of Argo Geologic Consultants, Zeist, the Netherlands, has been the main partner in the construction of the database. The paper benefitted much of the detailed comments of the two reviewers, F. Lafont (ELF-EP, Pau) and P. Joseph (IFP, Paris). The concepts involved in this paper developed during long-term cooperation in the Tremp-Ager Basin with C. Puigdefàbregas (Norsk Hydro University of Barcelona), P. L. de Boer (Utrecht University) and G.-J. Weltje (Technical University, Delft). Thanks are due to T. Senior for correction of the English text. Figures have been produced by the Audio-Visualisation Department of the Institute of Earth Sciences at Utrecht University. This paper is published as publication 970135 of the Netherlands Research School of Sedimentary Geology.

References

ARBUÉS, P., PI, E. & BERÁSTEGUI, X. 1996. Relaciones entre la evolución del Grupo de Aren y el cabalgamiento de Bóixols. *Geogaceta*, **20**, 446–449.

BATES, M. P. 1989. Palaeomagnetic evidence for rotations and deformation in the Nogueras Zone, Central Southern Pyrenees, Spain. *Journal of the Geological Society, London*, **146**, 459–476.

BENTHAM, P. & BURBANK, D. W. 1996. Chronology of Eocene foreland basin evolution along the western margin of the South-Central Pyrenees. *In:* FRIEND, P. F. & DABRIO, C. J. (eds) *Tertiary basins of Spain, the stratigraphic record of crustal kinematics.* World and Regional Geology, **6**. Cambridge University Press, 144–152.

——,—— & PUIGDEFÀBREGAS, C. 1992. Temporal and spatial controls on the alluvial architecture of an axial drainage system: late Eocene Escanilla Formation, southern Pyrenean foreland basin, Spain. *Basin Research*, **4**, 335–352.

BURBANK, D. W., PUIGDEFÀBREGAS, C. & MUÑOZ, J. A. 1992a. The chronology of the Eocene tectonic and stratigraphic development of the eastern Pyrenean foreland basin, northeast Spain. *Geological Society of America Bulletin*, **104**, 1101–1120.

——, VERGÉS, J., MUÑOZ, J. A., & BENTHAM, P. 1992b. Coeval hindward- and foreward-imbricating thrusting in the south-central Pyrenees, Spain: Timing and rates of shortening and deposition. *Geological Society of America Bulletin*, **104**, 3–17.

CUEVAS GOZALO, M. C. 1989. Sedimentary facies and sequential architecture of tide-influenced alluvial deposits. An example from the middle Eocene Capella Formation, South-central Pyrenees, Spain. Geologica Ultraiectana, Utrecht University, **61**.

DE BOER, P. L. & SMITH, D. G. 1994. Orbital forcing and cyclic sequences. *In:* DE BOER P. L. & SMITH, D. G. (eds) *Orbital Forcing and Cyclic Sequences.* Special Publication of the International Association of Sedimentologists, **19**, 11–14.

——, PRAGT, J. S. J. & OOST, A. 1991. Vertically persistent sedimentary facies boundaries along growth anticlines and climate-controlled sedimentation in the thrust-sheet-top South Pyrenean Tremp-Graus Foreland Basin. *Basin Research*, **3**, 63–78.

DEN BEZEMER, T., KOOI, H., PODLACHIKOV, Y. & CLOETINGH, S. 1998. Numerical modelling of growth strata and grain-size distributions associated with fault-bend folding. *This volume.*

DREYER, T. 1994. Architecture of an unconformity-based tidal sandstone unit in the Amettla Formation, Spanish Pyrenees. *Sedimentary Geology*, **94**, 21–48.

——, FÄLT, L. M., HØY, T., KNARUD, R., STEEL, R. & CUEVAS, J.L. 1993. Sedimentary architecture of field analogues for reservoir information (SAFARI): a case study of the fluvial Escanilla Formation, Spanish Pyrenees. *In:* FLINT, S. S. & BRYANT, I. D. (eds) *The geological modelling of hydrocarbon reservoirs and outcrop analogues.* Special Pubications of the International Association of Sedimentologists, **15**, 57–80.

DINARÈS, J., MCCLELLAND, E. & SANTANACH, P. 1992. Contrasting rotations within thrust sheets and kinematics of thrust tectonics as derived from palaeomagnetic data: an example from the Southern Pyrenees. *In:* MCCLAY, K. R. (ed.) *Thrust tectonics.* Chapman & Hall, London, 265–276.

DONSELAAR, M. E. & NIO, S. D. 1982. An Eocene tidal inlet/washover type barrier island complex in the south Pyrenean marginal basin, Spain. *Geologie en Mijnbouw,* **61**, 343–353.

EMERY, D. & MYERS, K. J. 1996. *Sequence Stratigraphy.* Blackwell Science, Oxford.

ENPASA 1966. *Ainsa, Graus y Enclave de Graus, Isobátas H-31B.* Exploración síntesis. Informe **139**.

FLEMINGS, P. B. & JORDAN, T. E. 1990. Stratigraphic modeling of foreland basins: Interpreting thrust deformation and lithosphere rheology. *Geology,* **18**, 430–434.

GARRIDO-MEGIAS, A. & RIOS, 1972. Sintesis geológica del Secundario y Terciario entre los Ríos Cinca y Segre (Pirineo central de la vertiente sur pirinaica, provincias de Huesca y Lérida). *Boletino de Geologia y Miniera,* **83**, 1–47.

HAQ, B. U., HARDENBOL, J. & VAIL, P. R. 1987. Chonology of fluctuating sea levels since the Triassic. *Science,* **235**, 1156–1167.

HARLAND, W. B., ARMSTRONG, R. L., COX, A. V., CRAIG, L. E., SMITH, A. G. & SMITH, D. G. 1990. *A geologic time scale 1989.* Cambridge University Press.

HASELDONCKX, P. 1973. The palynology of some Paleogene deposits between the Río Esera and the Río Segre, southern Pyrenees, Spain. *Leidse Geologische Mededelingen,* **49**, 145–165.

HELLER, P. L., ANGEVINE, C. L., WINSLOW, N. S. & PAOLA, C. 1988. Two-phase stratigraphic model of foreland-basin sequences. *Geology,* **16**, 501–504.

HEWARD, A. P. 1978. Alluvial fan sequence and megasequence models: with examples from the Westphalian D – Stephanian B coalfields, Northern Spain. *In:* MIALL, A. D. (ed.) *Fluvial Sedimentology.* Canadian Association of Petroleum Geologists, Memoirs, **5**, 669–702.

JULIVERT, M. 1978. The area of alpine folded cover in the Iberian Meseta. *In:* LEMOINE, M. (ed.) *Geological Atlas of Alpine Europe and adjoining Alpine areas.* Elsevier, Amsterdam, 93–112.

KAPELLOS, C. & SCHAUB, H. 1973. Zur Korrelation von Biozonierungen mit Grossforaminiferen und Nannoplankton im Paläogen der Pyrenäen. *Eclogae Geologicae Helvetiae,* **66**, 687–373.

KELLER, P. 1992. *Paläomagnetische und strukturgeologische Untersuchungen als Beitrag zur Tektogenese der SE-Pyrenäen.* PhD thesis, ETH Zürich.

MARTINEZ-PEÑA, B., CASAS-SAINZ, A. M. & MILLAN-GARRIDO, H. 1995. Palaeo-stresses associated with thrust sheet emplacement and related folding in the southern central Pyrenees, Huesca, Spain. *Journal of the Geological Society, London,* **152**, 353–364.

MARZO, M., NIJMAN, W. & PUIGDEFÀBREGAS, C. 1988. Architecture of the Castissent fluvial sheet sandstones, Eocene, South Pyrenees. *Sedimentology,* **35**, 719–738.

MILLÁN, H., DEN BEZEMER, T., VERGÉS, J., MARZO, M., MUÑOZ, J. A., CIRÉS, J., ZOETEMEIJER, R., CLOETINGH, S. & PUIGDEFÀBREGAS, C. 1995. Palaeo-elevation and effective elastic thickness evolution at mountain ranges: inferences from flexural modelling in the Eastern Pyrenees and Ebro Basin. *Marine and Petroleum Geology,* **12**, 917–928.

MUÑOZ, J. A. 1992. Evolution of a continental collision belt: ECORS-Pyrenees crustal balanced cross-section. *In:* MCCLAY, K. R. (ed.) *Thrust tectonics.* Chapman & Hall, London, 235–246.

MUTTI, E. & SGAVETTI, M. 1987. Sequence stratigraphy of the upper Cretaceous Aren strata in the Orcau-Aren region, South-Central Pyrenees, Spain: Distinction between eustatically and tectonically controlled depositional sequences. *Annali dell'Università di Ferrara (Nuova Serie), Sezione: Scienze della Terra,* **1**, 1–22.

——, LUTERBACHER, H.-P., FERRER, J. & ROSELL, J. 1972. Schema stratigrafico e lineamenti di facies del Paleogeno marino della zona centrale sud-pirenaica tra Tremp (Catalogna) e Pamplona (Navarra). *Memorie della Società Geologica Italiana,* **11**, 391–416.

——, ROSELL, J., ALLEN, G. P., FONNESU, F. & SGAVETTI, M. 1985. The Eocene Baronia delta-shelf system in the Ager Basin. *In:* MILA, M. D. & ROSELL, J. (eds) *Excursion Guidebook 6th European I.A.S. meeting, Lleida, Spain.* Inst. d'Estudis Ilerdencs, 578–600.

——, SÉGURET, M. & SGAVETTI, M. 1988. *Sedimentation and deformation in the Tertiary sequences of the southern Pyrenees.* Field trip 7, AAPG Mediterranean Basins Conference, Special Publication University of Parma, Italy.

NAGTEGAAL, P. J. C., VAN VLIET, A. & BROUWER, J. 1983. Syntectonic coastal offlap and concurrent turbidite deposition: the Upper Cretaceous Aren Sandstone in the South-Central Pyrenees, Spain. *Sedimentary Geology,* **34**, 185–218.

NIJMAN, W. 1981. Fluvial sedimentology and basin architecture of the Eocene Montañana Group, South Pyrenean Tremp-Graus Basin – *In:* ELLIOTT, T. (ed.) *Field guides to modern and ancient fluvial systems in Britain and Spain. Proceedings of the 2nd. International Conference on Fluvial Sedimentology, Keele UK,* 4.3–4.27.

—— 1989. Thrust sheet rotation? – The South Pyrenean Tertiary basin configuration reconsidered. *Geodinamica Acta,* **3**, 2, 17–42.

—— & NIO, S. D. 1975. *The Eocene Montañana Delta (Tremp-Graus Basin, Prov. Lerida and Huesca, Southern Pyrenees, N.Spain).* 9th International Sedimentological Congress, International Association of Sedimentologists, Nice. Excursion Guidebook, **19**, part B.

—— & PUIGDEFÀBREGAS, C. 1978. Coarse-grained point bar structure in a molasse-type fluvial system, Eocene Castisent Sandstone Formation, South Pyrenean Basin. *In:* MIALL, A. D. (ed.), *Fluvial Sedimentology.* Canadian Society of Petroleum Geologists, Memoirs, **5**, 487–510.

—— & VAN OOSTERHOUT, C. W. M. 1994. *Quantitative*

model study of a thrust sheet-top basin, the Eocene Tremp-Ager Basin, S.Pyrenees, Spain, Phase I (extended version): Database. Department of Geology, Institute of Earth Sciences, Utrecht University, database on diskettes.

PEPER, T. 1993. *Tectonic control on the sedimentary record in foreland basins – inferences from quantitative subsidence analyses and stratigraphic modelling.* PhD thesis, Free University, Amsterdam.

—— & DE BOER, P. L. 1995. Intrabasinal thrust-tectonic versus climate control on rhythmicities in the Eocene South Pyrenean Tremp-Graus foreland basin: inferences from forward modelling. *Tectonophysics,* **249**, 93–107.

PLAZIAT, J. C. 1981. Late Cretaceous to late Eocene palaeogeographic evolution of southwest Europe. *Palaeogeography, Palaeoclimatology, Palaeoecology,* **36**, 263–320.

PUIGDEFÀBREGAS, C. & MASCLE, A. 1998. Introduction: Why forelands? *This volume.*

—— & SOUQUET, P. 1986. Tecto-sedimentary cycles and depositional sequences of the Mesozoic and Tertiary of the Pyrenees. *Tectonophysics,* **129**, 173–203.

——, NIJMAN, W. & MUÑOZ, J. A. 1991. Alluvial deposits of the successive foreland basin stages and their relation to Pyrenean thrust sequences. *In: Fourth International Conference on Fluvial Sedimentology, Sitges, Spain, Guide book Series,* **10** *(2nd ed.),* 80–62 and 113–167.

RIBA, O. 1976. Syntectonic unconformities of the Alto Cardener, Spanish Pyrenees: a genetic interpretation. *Sedimentary Geology,* **15**, 213–239.

ROURE, F., CHOUKROUNE, P., BERASTEGUI, X., MUÑOZ, J. A., VILLIEN, A., MATHERON, P., BAREYT, M., SÉGURET, M., CÁMARA, P. & DERAMOND, J. 1989. ECORS deep seismic data and balanced cross sections: geometric constraints on the evolution of the Pyrenees. *Tectonics,* **8**, 41–50.

SCHAUB, H. 1981. *Nummulites et Assilines de la Téthys paléogène. Taxonomie, phylogenèse et biostratigraphie.* Schweizerische Paläontologische Abhandlungen, **104**. Editions Birkhäuser, Bâle.

SÉGURET, M. 1972. *Etudes tectoniques des nappes et séries décollées de la partie centrale du versant sud des Pyrénées.* Thèse doctorat, ESTELA-Montpellier, Série Géologie Structurale, **2**.

—— 1991. No tectonic control of depositional sequences in front of the Cotiella thrust sheet (Eocene South Pyrenean foreland basin). *Grupo Español del Terciario, Congreso I, Vic, Spain, Comunicaciones,* 320–323.

SINCLAIR, H. D., COACKLEY, B. J., ALLEN, P. A. & WATTS, A. B. 1991. Simulation of foreland basin stratigraphy using a diffusion model of mountain belt uplift and erosion: an example from the central Alps, Switzerland. *Tectonics,* **10**, 599–620.

SIMÓ, A. 1986. Carbonate platform depositional sequences, Upper Cretaceous, South-Central Pyrenees (Spain). *Tectonophysics,* **129**, 205–231.

—— 1989. Upper Cretaceous platform-to-basin depositional sequence development, Tremp Basin, South-Central Pyrenees, Spain. *Society of Economic Paleontologists and Mineralogists, Special Publication,* **44**, 365–378.

SOUQUET, P. & DÉRAMOND, J. 1989. Séquence de chevauchements et séquences de déposition dans un bassin d'avant-fosse. Exemple du sillon crétacé du versant sud des Pyrénées (Espagne). *Comptes Rendus de l'Académie des Sciences, Paris,* **309**, 137–144.

TANKARD, A. J. 1986. On the depositional response to thrusting and lithospheric flexure: examples from the Appalachian and Rocky Mountain basins. *In:* ALLEN, P. A. & HOMEWOOD, P. (eds) *Foreland Basins.* Special Pubications of the International Association of Sedimentologists, **8**, 369–392.

VAN DER MEULEN, S. 1986. Sedimentary stratigraphy of Eocene sheetflood deposits, southern Pyrenees, Spain. *Geological Magazine,* **123**, 167–183.

VAN HILTEN, M. 1995. *3D Facies architecture model of a basin fill increment, Tremp-Ager basin, S. Pyrenees,* MSc Research report, Utrecht University.

VAN HOORN, B. 1970. Sedimentology and palaeogeography of an upper Cretaceous turbidite basin in the South-Central Pyrenees, Spain. *Leidse Geologische Mededelingen,* **45**, 73–154.

VERGÉS, J. & BURBANK, D. W. 1996. Eocene-Oligocene thrusting and basin configuration in the eastern and central Pyrenees (Spain). *In:* FRIEND, P. F. & DABRIO, C. J. (eds) *Tertiary basins of Spain, the stratigraphic record of crustal kinematics.* World and Regional Geology, **6**, Cambridge University Press, 120–133.

—— &, MILLÁN, H., ROCA E., MUÑOZ, J. A., MARZO, M., CIRÉS, J., DEN BEZEMER T., ZOETEMEIJER, R. & CLOETINGH, S. 1995. Eastern Pyrenees and related foreland basins: pre-, syn- and post-collisional crustal-scale cross-sections. *Marine and Petroleum Geology,* **12**, 893–915.

——, MUÑOZ, J. A. & MARTÍNEZ, A. 1992. South Pyrenean fold-and-thrust belt: Role of foreland basin evaporitic levels in thrust geometry: *In:* MCCLAY, K. R. (ed.) *Thrust tectonics.* Chapman & Hall, London, 255–264.

WELTJE, G. J., VAN ANSENWOUDE, S. O. K. J. & DE BOER, P. L. 1996. High-frequency detrital signals in Eocene fan-delta sandstones of mixed parentage (South-Central Pyrenees, Spain): a reconstruction of chemical weathering in transit. *Journal of Sedimentary Research,* **66**, 119–131.

YANG, C. S. & NIO, S. D. 1989. An ebb-tide delta depositional model – a comparison between the modern Eastern Scheldt tidal basin (SW Netherlands) and the Lower Eocene Roda Sandstone in the Southern Pyrenees (Spain). *Sedimentary Geology,* **64**, 175–196.

ZOETEMEIJER, R. 1993. *Tectonic modelling of foreland basins, thin skinned thrusting, syntectonic sedimentation and lithosphere flexure.* PhD thesis, Free University, Amsterdam.

——, CLOETINGH, S., SASSI, W. & ROURE, F. 1993. Modelling of a piggyback-basin: record of tectonic evolution. *Tectonophysics,* **226**, 253–269.

Fluid migration during Eocene thrust emplacement in the south Pyrenean foreland basin (Spain): an integrated structural, mineralogical and geochemical approach

ANNA TRAVÉ[1,2], PIERRE LABAUME[3], FRANCESC CALVET[1], ALBERT SOLER[4], JORDI TRITLLA[4], MARTINE BUATIER[5], JEAN-LUC POTDEVIN[5], MICHEL SÉGURET[2], SUZANNE RAYNAUD[2], LOUIS BRIQUEU[6]

[1]*Departament de Geoquímica, Petrologia i Prospecció Geològica, Facultat de Geologia, Universitat de Barcelona, 08071-Barcelona, Spain (e-mail: trave@antartida.geo.ub.es)*
[2]*Géofluides-Bassins-Eau, CNRS-Université Montpellier II, 34095 Montpellier Cedex 5, France*
[3]*Laboratoire de Géophysique Interne et Tectonophysique, CNRS-Université Joseph Fourier, BP 53X, 38041 Grenoble Cedex 9, France*
[4]*Departament de Cristalografia, Mineralogia i Dipòsits Minerals, Facultat de Geologia, Universitat de Barcelona, 08071-Barcelona, Spain*
[5]*Laboratoire de Sédimentologie et Géodynamique, CNRS-Université Lille I, 59655 Villeneuve d'Ascq, France*
[6]*Géochronologie–Géochimie–Pétrologie, CNRS-Université Montpellier II, 34095 Montpellier Cedex 5, France*

Abstract: In the frontal part of the south Pyrenean Eocene thrust-fault system, syn-kinematic fluid flow during the early compressional deformation of the foreland basin marls is evidenced macroscopically by the abundance of calcite shear veins within the thrust-fault zones and folds.

The geometry and distribution of the veins are indicative of the mechanisms and kinematics of fluid-deformation relationships, and give assessment of the fluid migration paths. The crack–seal mechanism of formation of the shear veins attests to the episodic nature of fault-slip and associated fluid flow in fractures. The distribution of the veins suggests that the main source of fluid was the dewatering of the overpressured, poorly permeable marls from the thrust footwalls, probably related to both (i) vertical compaction due to burial under thrust sheets and (ii) tectonic horizontal shortening. These fluids drained upwards towards the thrust-fault zones, in which they migrated laterally towards the thrust front due to the anisotropy of the fracture permeability in these zones.

The geochemistry of the vein-filling minerals and their comparison with the geochemistry and mineralogy of the host marls are indicative of the fluid types, fluid origins, fluid–sediment interactions, and fluid migration paths. The $\delta^{34}S$ and $^{87}Sr/^{86}Sr$ ratio of the host marl calcite and of the calcite and celestite in the veins away from the thrust-fault zones indicate that the original water trapped interstitially in the marls was Eocene seawater. The elemental composition (Ca, Sr, Mg, Mn, and Fe), $\delta^{18}O$, and $\delta^{13}C$ of the same samples reveal a change of the pore-water composition from marine to formation water during the early burial stage. Fluid-inclusion analyses of the celestite in the veins reveal the presence of a hot, saline ascending fluid restricted to these discontinuities, where it was mixed with the local formation water. These two types of fluids drained towards the thrust-fault zones where they acquired a higher $^{87}Sr/^{86}Sr$ ratio, probably related to local fluid–sediment reactions. Indeed, dickite precipitated during cleavage formation in the most intensely strained part of the fault zones, and its formation was probably mainly controlled by stress. $\delta^{18}O$ depletion in the calcite from the structurally highest/innermost thrust-fault zones suggests also the influence of meteoric water derived from the emerged part of the belt in these structures.

The earlier fluid regime in the Ainsa basin was an intergranular (porous) flow regime (compactional flow) allowing for a pervasive isotopic, and elemental exchange of the marls prior to vein formation. With the onset of compressional deformation, channelized flow along tectonic slip surfaces became dominant.

TRAVÉ, A., LABAUME, P., CALVET, F., SOLER, A., TRITLLA, J., BUATIER, M., POTDEVIN, J.-L., SÉGURET, M., RAYNAUD, S. & BRIQUEU, L. 1998. Fluid migration during Eocene thrust emplacement in the south Pyrenean foreland basin (Spain): an integrated structural, mineralogical and geochemical approach. *In:* MASCLE, A., PUIGDEFÀBREGAS, C., LUTERBACHER, H. P. & FERNÀNDEZ, M. (eds) *Cenozoic Foreland Basins of Western Europe.* Geological Society Special Publications, **134**, 163–188.

Compressional deformation and relief building is thought to be responsible for long-distance migration of large amounts of fluids through thrust belts and their forelands (Oliver 1986; Garven 1995). These large-scale fluid migrations play an important role in the mechanics of seismicity and thrust faulting, diagenetic and metamorphic transformations, and the localization of ore and hydrocarbon accumulations. Hence, understanding interactions between fluid flow and thrust systems has become an important goal in the last years. Recent geophysical investigations and *in-situ* measurements in active accretionary prisms reveal the occurrence of present-day pore-fluid overpressure and circulation in fault zones, and the importance of this circulation in diagenesis and heat transfer (Moore & Vrolijk, 1992; Shipley *et al.* 1994; Moore *et al.* 1995). However, the study of ancient structures remains the main opportunity to unravel complex fluid flow patterns and fluid–rock interactions in thrust systems, through the combination of structural, mineralogical, geochemical, and fluid-inclusion analyses (e.g., Marquer & Burkhard 1992; Guilhaumou *et al.* 1994; McCaig *et al.* 1995; Muchez *et al.* 1995).

In ancient belts, one poorly studied aspect is the fluid-flow pattern at the leading edge of the thrust system, where thrusts propagate through the recently deposited sediments of the foreland basin. Indeed, these very early fluid processes are generally obscured by subsequent deformation and related mineralogical and geochemical transformations when these domains are integrated in internal parts of belts (Marquer & Burkhard 1992), where most studies have been undertaken. However, knowledge of early fluid migrations is important, because deformation-induced compaction of the foreland basin-fill is likely to trigger the expulsion of large amounts of pore fluid, which may be mixed in complex ways with allochthonous fluids derived from deeper settings of the belt or from recycling of meteoric water. Specific aspects to be studied are therefore the respective involvement of fluids from different sources, the structural location of fluid migrations, their timing with respect to burial and thrust propagation, their involvement in deformation mechanisms and kinematics, and the mineralogical and geochemical transformations that may be associated.

In order to define best such early fluid-deformation relationships, we have studied a thrust–fold system in the Eocene south Pyrenean foreland basin-fill, where structures formed during, or soon after, deposition show relatively simple geometries and weak intensity of deformation, and remained shallowly buried and practically unaffected by later processes.

We have combined detailed mapping of the deformation structures with meso/microstructural, petrological, fluid-inclusion, mineralogical and geochemical analyses, to determine the different aspects of the fluid-flow system and its interactions with the thrust system. The geometry of the deformation structures (calcite veins and shear zones in marls), as well as their geometrical relationships with the macrostructures (thrust-faults, footwall synclines), gives information on the deformation mechanisms and kinematics, and a first assessment of the geometry and dynamics of the fluid-flow system through the distribution of the calcite veins. Fluid-inclusion data in veins allow us to determine the nature and temperature of some of the mineralizing fluids; petrological analyses, together with mineralogical analyses, reveal fluid–sediment interactions in certain structural settings. Comparing the geochemical analyses of the vein-filling minerals with those of the host sediment reveals fluid origin, fluid-migration paths and fluid–sediment interactions during this migration.

Geological setting

The Ainsa basin is part of the inner south Pyrenean foreland basin (north Spain; Fig. 1). The Ainsa basin (Fig. 2) is mainly filled by Eocene marls with sandstone and conglomerate intercalations forming turbiditic channel–levee complexes, and contains intercalations of westward-prograding outer-shelf/slope facies in the eastern part (Mutti *et al.* 1988). The substratum of these deposits comprises (from top to bottom) Upper Cretaceous–Palaeocene shelf carbonates, Permo-Triassic marls and sandstones, and a Palaeozoic basement.

The northern part of the Ainsa basin fill is thrusted by the Cotiella Nappe, detached in the Triassic sediments and mainly consisting of the Upper Cretaceous–Palaeocene carbonates (Figs 1 and 2) (Séguret 1972). The nappe moved southwards at least 20 km and was probably rooted in the Palaeozoic basement of the Axial Zone of the belt, although this root cannot be identified at present due to subsequent erosion. At the front of the nappe, a SW-verging imbricated thrust-fold system developed in the basin-fill (Fig. 2). Erosional surfaces and unconformities show that the main movement of the Cotiella Nappe occurred during the early Eocene (Ypresian), and that the frontal thrust-fold system continued to be active during the mid-Eocene (Mutti *et al.* 1988).

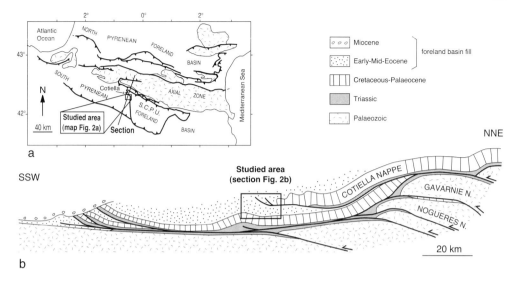

Fig. 1. Structural sketch of the Pyrenean belt (**a**) and cross-section of the south-central Pyrenean foreland basin (**b**), with location of the studied area. SCPU: South-Central Pyrenean Unit. Present-day topography is not shown in the cross-section.

The inner part of the Cotiella Nappe was deformed and uplifted from the mid-Eocene by the emplacement of underlying basement thrust sheets passing southward to décollement in the Triassic materials (the Gavarnie and Nogueres Nappes in Fig. 1) (Séguret 1972). Regional cleavage developed during this emplacement, with the front of this cleavage being located in the northernmost part of the Ainsa basin (Séguret 1972; Holl & Anastasio, 1995).

A broad description of the ancient fluid-flow system at the inner margin of the Ainsa basin, based on petrology, fluid-inclusion data, and stable isotope geochemistry was given by Bradbury & Woodwell (1987) for both the Palaeozoic basement and the Mesozoic–Eocene cover. Fluid flow related to the Gavarnie thrust in, and immediately above, the basement was further investigated by Grant et al. (1990), Banks et al. (1991), McCaig et al. (1995). Fluid flow through the thrusted Cretaceous–Palaeocene limestones was studied by Rye & Bradbury (1988). Ours is the first study dealing in detail with fluid processes in the external parts of the thrust system in the foreland basin-fill. Detail of mineralogical and geochemical analyses are reported in separate papers (Buatier et al. 1997; Travé et al. 1997). In the present paper, we summarize these results and combine them with new detailed structural observations and fluid-inclusion data to draw a general picture of the relationships between structural evolution and fluid processes.

Studied structures

The study, carried out in the thrust–fold system that affects the Ainsa basin-fill at the front of the Cotiella Nappe, focuses on four hectometric-scale outcrops, corresponding to three major thrust-fault zones and one syncline: the Atiart, Samper, and Los Molinos thrusts, and the Arro syncline (AT, ST, MT, and AS, respectively, in Fig. 2).

The Atiart thrust corresponds to the frontal part of the Cotiella Nappe sole thrust. The thrust front is truncated by the major erosional surface corresponding to the basal unconformity of the Castisent group (Mutti et al. 1988), indicating the Ypresian age of the Cotiella Nappe emplacement. Late minor movements of the thrust deformed the erosional surface and the overlying sediments.

The Samper thrust is located in the hanging wall of the Atiart thrust and was formed more recently than the former, during the Ypresian. It affects the lower depositional sequence of the Castisent group, and is truncated by the basal erosional surface of the upper depositional sequence of the Castisent group. Late minor movements deformed the erosional surface and the overlying sediments. Mapping shows that the Samper thrust is an unrooted, 'out-of-the-syncline' thrust related to the late tightening of the major hanging-wall ramp anticlinorium that affects the upper Cretaceous–Palaeocene limestones in the Cotiella Nappe and tilted the overlying Ypresian and Lutetian sediments.

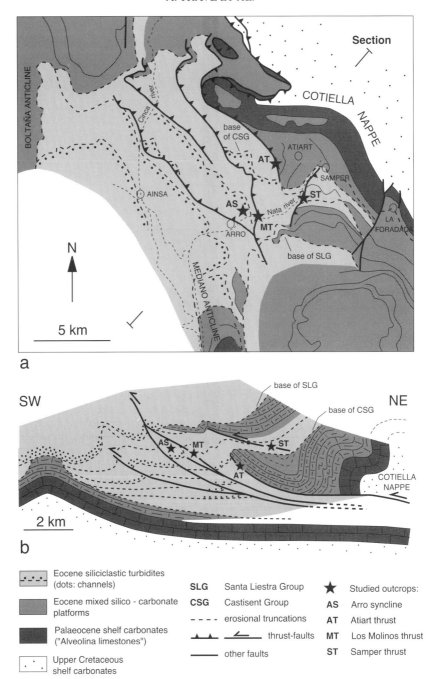

Fig. 2. Geological map (**a**) and cross-section (**b**) of the Ainsa basin, with location of the studied outcrops (stars). The base of the middle Eocene is located in the lower part of the Santa Liestra group (Mutti *et al.* 1988).

The Los Molinos thrust is located in the foreland of the Atiart thrust; the Arro syncline is a kilometre-scale syncline located in the footwall of the Los Molinos thrust. The Los Molinos thrust and Arro syncline affect Ypresian sediments.

On the four studied outcrops, the structures affect Ypresian sediments with mainly marly lithologics. The mesostructures present on these outcrops were formed soon after deposition of the affected sediments, mainly during the Ypresian, and possibly also during part of the Mid-Eocene. Due to the rapid and complex succession of sedimentation, thrusting and erosion events during this period, the burial conditions of these mesostructures during their formation are difficult to constrain between a few hundred metres and 3000 m.

Methods

Detailed structural analysis was carried out in the field on each of the four selected outcrops. The different types of mesostructures demonstrating fluid involvement during deformation (calcite veins and shear zones in host marls) were identified, and their distribution, kinematics and relative chronology were determined. The calcite veins of different generations and the host marls showing varying intensity of deformation were sampled systematically on each outcrop for microstructural, petrographic, mineralogical and geochemical studies. We also sampled undeformed marls away from (a few tens to hundreds of meters) the deformed/veined zones in order to use them as reference material.

Standard and polished thin-sections of approximately 150 samples were examined using petrographic and cathodoluminescence microscopes for petrographic and microstructural study.

Bulk mineralogy of the marls was studied through calcimetry and X-ray diffractometry (XRD). Clay mineralogy was studied through XRD on mineral fractions smaller than 2 μm. Mineral species in the calcite veins and marls, and their microstructural distribution, were also identified through scanning electron microscope (SEM) used in backscattered mode and high resolution transmission electron microscope (HRTEM), both coupled with energy dispersive analyser (EDS). A full description of the clay preparation for the mineralogical study is presented in Buatier et al. (1997).

Porosity of the marls was studied by the mercury injection technique, allowing for the determination of three different characteristics of the connected porous network: total porosity, trapped porosity, and the porosity spectrum. The total, or connected porosity value, is given by the volume of injected mercury at the maximum pressure of 150 MPa. At this pressure, mercury is thought to have reached all the pores (intergranular pores and fractures). The trapped porosity corresponds to the mercury volume which is not expelled from the sample when the pressure is released. This value is comparable to the intergranular pore volume of the rock. The porosity spectrum is the size of the throat diameter reached at each pressure step.

A fluid-inclusion study was undertaken in vein cements to determine the composition and P–T conditions of the vein-forming solutions. Standard and thick sections were used for petrographic examination. About 20 double polished calcite–celestite wafers of 1 cm^2 and exfoliation chips, to avoid deformation, were selected for microthermometric determination. Microthermometric analyses were performed on a Linkam THMS-600 heating–freezing stage calibrated with the triple point of distilled water (0.0 °C) and pure CO_2-bearing fluid inclusions (–56.6 °C). Calibration at high temperatures was made with appropriate chemicals from Merck Corporation. The accuracy in the measurements was about ± 2° C for heating runs.

Isotope composition of the different cements in veins and host marls was determined on powdered microsamples, carefully selected after the petrographic study and picked up from the hand-samples using a diamond-tipped dental drill. Sulphur isotope analyses were performed in vein celestite and are reported in (‰) values relative to the CDT standard. Sr isotope analyses were carried out in vein calcite and celestite, calcite from marls, and bulk marls and are reported as $^{87}Sr/^{86}Sr$ ratio. Carbon and oxygen stable-isotope analyses were carried out in calcite from veins and marls; the results were corrected using standard procedures and are reported in (‰) values relative to the PDB standard. Element (Ca, Mg, Sr, Mn and Fe) composition of calcite from veins and marls was performed using an electron microprobe. Detailed description of the analytical methods and devices used for the isotope and element geochemical analyses are presented in Travé et al. (1997).

Mesostructures and structural evidence for fluid activity

At the outcrop scale, two types of mesostructures are indicative of the deformation and associated fluid activity during thrust-faulting and folding: shear zones in the marls (SZ in Fig. 3a), and tabular calcite shear veins (SV in Fig. 3a). Both types of mesostructures are associated in the thrust-fault zones, however, the shear veins also occur away from these zones, either as isolated small-scale faults or related to flexural slip in fold limbs. Extensional veins are rare and outside the thrust-fault zones there is no cleavage except for spaced fractures with local and irregular development.

The calcite veins are common in the most intensely strained zones (thrust-fault zones and steep limbs of folds), whereas they are less common or absent in other domains. This distribution gives a first, qualitative, assessment of the importance of fluid involvement in the development of tectonic structures.

Following, is a description of (i) the two types of mesostructures of shear deformation and their main microstructural characteristics, and (ii) the distribution of these mesostructures on the studied outcrops.

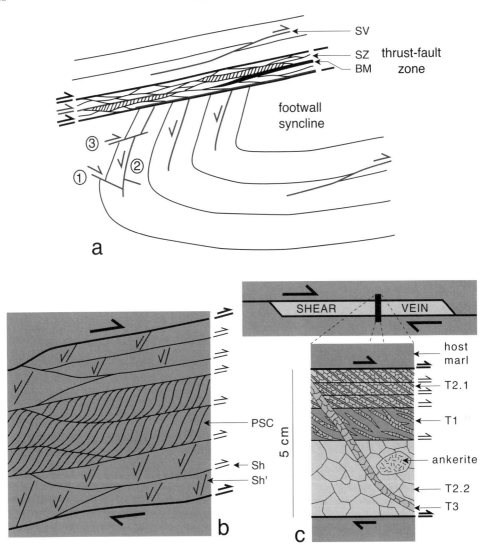

Fig. 3. Synthetic sketch of the structure associations. (**a**) Distribution of the meso-scale structures in the large-scale structures. BM, intensely sheared black marl facies; SV, calcite shear vein; SZ, shear zone in marl; 1, 2, 3, successive generations of shear veins in footwall synclines (see text). Thrust-fault zones range from several meters to tens of meters thick; footwall synclines vary from decametric (Atiart) to kilometric (Arro) size. (**b**) Structural features of the shear zones in marls. PSC, pressure-solution cleavage; Sh, Sh', shear surfaces synthetic and antithetic to the general shearing, respectively. (**c**) Microstructure of the tabular calcite shear veins. T1, T2.1, T2.2, T3, openings of type 1, 2.1, 2.2, 3, respectively, in the calcite shear veins.

Mesostructures of shear deformation

Shear zones in marls. The shear zones in marls correspond to centimetre to metre-thick deformed intervals characterized by the association of shear surfaces and pressure-solution cleavage (Figs 3b and 4a, b). The shear surfaces are shiny and finely striated. They are grouped in two families similar to those already described by Koopman (1983) and Labaume et al. (1990, 1991) in similarly sheared sediments. The main family (Sh, for shear, in Fig. 3b) consists of curved shear surfaces forming an anastomosing network globally parallel to the sheared interval, focus of the main tectonic transport. The other family (Sh' in Fig. 3b) dips at a steeper angle to the previous one, opposite to the sense of general shearing. Displacement on the Sh' surfaces is small (usually millimetric), and corresponds to antithetic extensional sliding with respect to the general shearing.

In the weakly deformed intervals, the lenses bounded by the shear surfaces are made up of macroscopically undeformed sediment, whereas these lenses display a closely spaced, sigmoidal pressure-solution cleavage (PSC in Fig. 3b) in the intervals with stronger deformation. In the latter case, the geometry is analogous to that of S–C tectonites common in metamorphic rocks (Lister & Snoke 1984). The asymmetry of the Sh–Sh' surface network and that of the sigmoidal cleavage indicate the shear sense.

A specific PSC-type shear facies occurs locally in centimetre-thick deformation bands (BM in Figs 3a and 4a), characterized at a macroscopic scale (Fig. 4b) by (i) a very intense sigmoidal pressure-solution cleavage, (ii) a black colour, contrasting with the grey of the neighbouring sheared (GM in Fig. 4a) and undeformed marls, and (iii) numerous calcite veinlets along the cleavage or the shear surfaces. Optical microscope observations show two generations of calcite veinlets (Fig. 4c). The first generation consists of syn-kinematic veinlets corresponding to shear veins along the shear surfaces (mSV in Fig. 4b), and to boudinage veins parallel to the cleavage (i.e., related to cleavage-parallel stretching; BV in Fig. 4b,c). The borders of the latter veins are affected by pressure-solution related to cleavage formation (stylolites with peaks perpendicular to cleavage; Fig. 4c). Second generation veinlets are parallel to the cleavage, and the geometries of the walls show that they result from extension perpendicular to the cleavage (EV in Fig. 4c). In the vein-fill, thin inclusion bands parallel to the walls attest to a multi-stage formation by a crack-seal mechanism (Ramsay 1980). The mode of opening, incompatible with cleavage formation, and the absence of stylolitization of vein walls, indicate that these second generation veinlets are post-kinematic.

Tabular calcite shear veins. These shear veins are centimetre-thick tabular bodies, commonly of metric to decametric extension (SV in Fig. 3a; Fig. 5a). They are bounded by striated shear surfaces and frequently contain internal, mm-spaced striated shear surfaces (Figs 3c and 5b). Vein cement precipitated in rhomb-shaped cavities corresponding to releasing overstep openings along the shear surfaces; the asymmetry of the cavities indicates the sense of shear (Labaume *et al.* 1991; Berty 1992). Each vein results from numerous successive openings, with corresponding incremental crystallization that are chronologically grouped into three types (Fig. 3c).

Type 1 (T1) openings correspond to sub-millimetre-thick irregular veinlets filled by microsparry calcite, and also to diffuse calcite precipitation in the host-marl. T1 openings are characterized by sediment disaggregation showing that vein formation began in poorly lithified sediment.

Type 2 (T2) openings correspond to the main episodes of shear vein formation and developed in two successive stages (Labaume *et al.* 1991; Berty 1992); the first corresponding to a great number of millimetre-sized openings formed successively by a crack-seal mechanism (T2.1 in Figs 3c and 5b), and the second to centimetre-long openings cross-cutting the previous ones (T2.2 in Figs 3c and 5b). T2 openings are filled with subhedral sparry calcite crystals with the C-axis mostly perpendicular to the cavity walls. In some veins, celestite and celestite–barite are associated with calcite; the relationships between the crystals indicate the synchronic origin of both minerals. Locally, the celestite shows partial transformation to strontianite and, to a lesser degree, to pyrite. Calcite in T2 openings also contains, locally, patches of subhedral to anhedral ankerite crystals. The fan geometry of the ankerite crystals in the borders of some patches indicates that ankerite formation results from an advancing front that dissolved the calcite and precipitated the ankerite.

Type 3 (T3) openings, up to 1 mm wide, cross-cut T1 and T2 openings and are filled with equant to subhedral calcite crystals.

Disaggregation of the sediment in T2 and T3 openings is less common than in T1, indicating the progressive induration of the host sediment. However, veins of different generations in the same outcrop show similar chronology of openings. This indicates that sediment induration was restricted to vein vicinity, probably due to cement precipitation in veins, and that the sediments away from the veins remained poorly lithified during the whole deformation sequence.

Distribution of the mesostructures on the studied outcrops

The Atiart thrust. On the studied outcrop (Fig. 6a), the thrust emplaces outer shelf marly limestones and marls over turbiditic marls, with occasional siltstone layers. The thrust forms a flat for the hanging wall and ramps across the reverse limb of a SW-verging syncline in the footwall. In the hanging wall, an upper thrust parallel to bedding occurs about 50 m above the

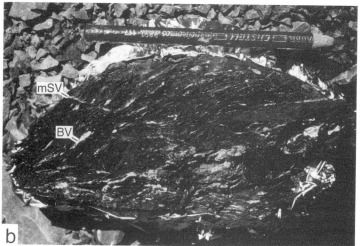

Fig. 5. Tabular calcite shear veins. (**a**) Outcrop view (vein of generation 2 in the steep limb of the Arro syncline; see Fig. 9). (**b**) Micrograph showing type 2 openings (see Fig. 3c). T2.1, millimetric openings formed by a crack-seal mechanism; T2.2, late and larger opening.

Fig. 4. Shear zones in marls. (**a**) Outcrop view (Atiart thrust). BM, intensely sheared black marl facies; GM, sheared grey marl facies; SV, tabular calcite shear vein. (**b**) Detail of the intensely sheared black marl facies. BV, boudinage veinlet; mSV, shear veinlet. (**c**) Micrograph of the intensely sheared black marl facies. BV, boudinage veinlet (with stylolitized borders, i.e., syn-cleavage); EV, extensional veinlet (post-cleavage).

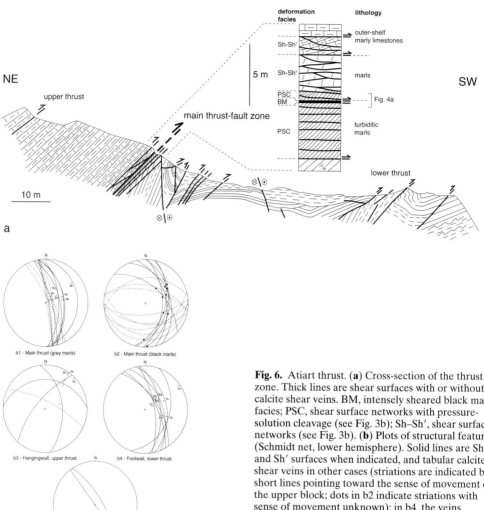

Fig. 6. Atiart thrust. (**a**) Cross-section of the thrust zone. Thick lines are shear surfaces with or without calcite shear veins. BM, intensely sheared black marl facies; PSC, shear surface networks with pressure-solution cleavage (see Fig. 3b); Sh–Sh', shear surface networks (see Fig. 3b). (**b**) Plots of structural features (Schmidt net, lower hemisphere). Solid lines are Sh and Sh' surfaces when indicated, and tabular calcite shear veins in other cases (striations are indicated by short lines pointing toward the sense of movement of the upper block; dots in b2 indicate striations with sense of movement unknown); in b4, the veins marked BS correspond to bedding slip in the footwall syncline, and those marked Th are along the thrust surface. Stippled lines are pressure-solution cleavage surfaces.

main thrust. In the footwall, several minor thrusts are present, the most important associated to a SW-verging syncline in the footwall ('lower thrust' in Fig. 6a).

The main thrust is characterized by a sharply bounded 10m thick fault zone (detail column in Fig. 6a, and Fig. 4a). In the hanging wall, the deformation zone, about 5 m thick, is characterized by Sh–Sh' surfaces in the upper part, and the development of a sigmoidal pressure-solution cleavage between the shear surfaces in the lowermost meter (GM at top of Fig. 4a). The thrust surface is characterized by the presence of a few centimetre thick interval featuring the intensely sheared black marl facies (BM in Fig. 4a). In the footwall, the deformation zone, about 5 m thick, is characterized by numerous Sh–Sh' surfaces cutting the previously tilted beds in sigmoidal lenses, with pressure-solution cleavage parallel to bedding (GM at bottom of Fig. 4a).

Within the main fault zone, tabular calcite shear veins (SV in Fig. 4a) are present at the boundaries of the black marl interval. Outside the latter interval, only isolated millimetric patches of striated calcite on Sh–Sh' shear surfaces and a few subvertical extensional calcite veinlets are present. Tabular calcite shear veins are also very rare in the hanging wall (only two were observed, one being along the upper thrust). By contrast, shear veins are more common in the footwall where they occur along different minor thrusts, as well as parallel to beds in the reverse limb of the lower thrust footwall syncline.

In the main fault zone, the geometry of the Sh–Sh' surfaces and pressure-solution cleavage surfaces in the grey marls (i.e., outside the black-marl facies), as well as the striation direction on the shear surfaces are consistent with a top-to-the-WSW movement (Fig. 6b1). In contrast, the cleavage in the black marl facies dips both northwards and southwards, depending on the different sectors of the outcrop (broken lines in Fig. 6b2). The presently available data do not allow us to know if this geometry corresponds to (i) distinct episodes of thrust movement, (ii) a local, small-scale perturbation of strain during the general westward movement or (iii) a late deformation of an initially eastward-dipping cleavage. Most of the striations on the calcite shear veins bounding the black marl facies have an E–W strike, except two that have a N–S strike (solid lines in Fig. 6b2). Due to their complex internal structure, the sense of displacement and the movement chronology of these veins could not be determined. By analogy with the other kinematic markers in the outcrop, we postulate that the E–W striations correspond to a top-to-the-west displacement, and that the N–S striations are minor late structures, possibly associated to a late strike-slip movement (see below). This late movement may also be responsible for the perturbation of the cleavage orientation in the black marls (cf. the third hypothesis above).

In both the hanging wall and footwall, most structures also indicate a top-to-the-SW thrust movement (Fig. 6b3, 4). The bedding-parallel shear veins in the reverse limb of the lower thrust footwall syncline (BS in Fig. 6b4) are compatible with bedding-slip associated to flexural folding, similar to the case of the Arro syncline (see below). The footwall also contains two steeply dipping NW-striking calcite veins which cut the thrust structures and indicate a late dextral strike-slip movement (Fig. 6b5). This movement is compatible with the dextral strike-slip component of movement which characterises the western boundary of the South-Central Pyrenean Unit at the regional scale (SCPU in Fig. 1 map; Séguret 1972).

Samper thrust. On the studied outcrop (Fig. 7a), the thrust affects turbiditic marls with thin to medium siltstone layers, the latter being more abundant in the upper part of the outcrop. The thrust corresponds to a low-angle ramp for both the hanging wall and the footwall.

The main deformation zone, a few metres thick, is located at the top of the fault zone and features a complex association of Sh–Sh' surfaces networks, pressure-solution cleavage, small-scale folds, and locally the intensely sheared black marl facies. In the lower part of the fault zone, several shear surfaces with flat-ramp geometry climb southwestward across bedding, and are associated to decametric SW-verging folds. Centimetre-thick intervals with pressure-solution cleavage are present locally along these surfaces.

The distribution of the tabular calcite shear veins is very heterogeneous. They are absent in the hanging wall and scarce in the main deformation zone, but common along the shear surfaces in the lower part of the fault zone, as well as in a upper few tens of metres of the footwall, where they disappear downward. In the footwall, most of the veins are parallel or at a low angle to bedding; a few are at a steep angle. The veins in the fault zone, and those sub-parallel to bedding in the footwall indicate a top-to-the-SW movement (Fig. 7b1). The steep-angle veins are late NW-striking dextral strike-slip faults, which are interpreted as having the same regional significance as those found in the Atiart outcrop (DS in Fig. 7b2). A few other late shear veins correspond to minor normal faults cross-cutting the shear surfaces in the lower part of the fault zone (FN in Fig. 7b2).

Los Molinos thrust. On the studied outcrop (Fig. 8a), the thrust affects turbiditic mudstones with intercalations of thin siltstone layers. It cuts the hinge zone of a previously formed, NNW trending symmetric anticline with moderate- to steep-dipping limbs.

The fault zone is 5 to 10 m thick and is sharply bounded by shear surfaces. In the eastern limb of the anticline, the deformation zone is characterized by numerous Sh–Sh' surfaces cutting the previously tilted beds in sigmoidal lenses, with pressure-solution cleavage parallel to bedding in the lenses. Calcite shear veins are frequent, mainly along the Sh surfaces. Both the asymmetry of the sigmoidal structures and the calcite shear veins indicate top-to-the-SW movement (Fig. 8b1). In the hinge zone of the anticline, the fault zone displays metre-scale folds with steep limbs and local axial plane pressure-solution cleavage.

Calcite shear veins, parallel or oblique to bedding, also occur outside the fault zone and become rare a few tens of metres away from it. They are more common in the footwall than in the hanging wall. Most of these veins are compatible with a NE–SW compression (Fig. 8b1),

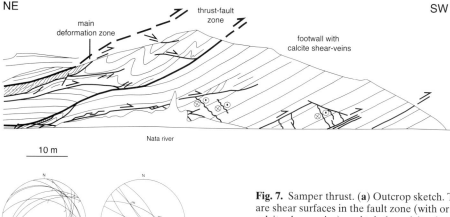

Fig. 7. Samper thrust. (**a**) Outcrop sketch. Thick lines are shear surfaces in the fault zone (with or without calcite shear veins), and tabular calcite shear veins in the footwall. (**b**) Plots of tabular calcite shear veins (Schmidt net, lower hemisphere). Striations are indicated by short lines pointing toward the sense of movement of the upper block. DS, dextral strike-slip faults; FN, normal faults.

except a few veins indicating a (probably late) normal movement (Fig. 8b2).

Arro syncline. On the studied outcrop (Fig. 9a), the syncline affects turbiditic marls. It shows a WSW vergence, with a steeply dipping eastern limb and a shallowly dipping western limb. Three generations of calcite shear veins are differentiated by their cross-cutting relationships (thick lines in Fig. 9).

The first generation is represented in the steep limb by a few veins having at present-days a shallowly-dipping SW-verging normal fault geometry (1 in Fig. 9). They are interpreted as pre-folding SW-verging thrusts rotated by subsequent folding (1r in Fig. 9b2).

The second generation corresponds to several tens of metres long bedding-parallel veins (2 in Fig. 9). In the steep limb, these veins are very common and most of them terminate in the

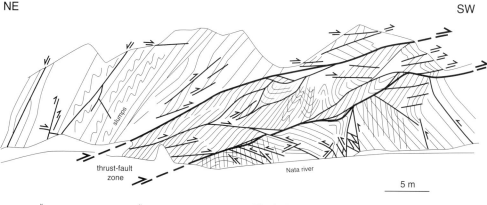

Fig. 8. Los Molinos thrust. (**a**) Outcrop sketch. Thick lines are shear surfaces in the fault zone (with or without calcite shear veins), and tabular calcite shear veins in the footwall and the hanging wall. (**b**) Plots of tabular calcite shear veins (Schmidt net, lower hemisphere). Striations are indicated by short lines pointing toward the sense of movement of the upper block. In b1, veins from inside and outside the fault zone are not differentiated.

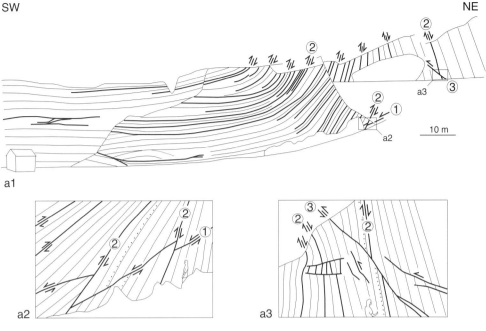

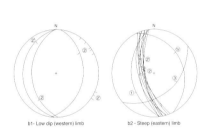

Fig. 9. Arro syncline. (**a**) Outcrop view. (a1) general syncline structure and distribution of the tabular calcite shear veins (thick lines); (a2) and (a3) detail of the intersection of shear veins from different generations (person for scale). 1, 2, 3, shear veins of generation 1 (pre-folding), 2 (syn-folding), and 3 (post-folding), respectively. (**b**) Plots of tabular calcite shear veins (Schmidt net, lower hemisphere). Striations are indicated by short lines pointing toward the sense of movement of the upper block. 1, 2, 3, as in (a); 1r, shear vein of generation 1 restored to its pre-folding geometry; 2': shear veins of generation 2 with top-to-the-SE movement.

hinge zone. Most of the striations are down-dip and correspond to a reverse movement when the beds dip westward (right-way-up beds; 2 in Fig. 9b2), changing to a normal movement when the beds dip eastward (reverse beds). Some of the veins also show striations with a sinistral strike slip component (2' in Fig. 9b2), transition between the two directions being either progressive (curved striations) or abrupt (distinct shear surfaces with different directions of striation). In the low dip limb, the second generation veins are rare and show either a top-to-the-SW movement (2 in Fig. 9b1), a top-to-the-SE movement (2' in Fig. 9b1), or striations corresponding to both movements. The down-dip striations in the steep limb, and those indicating a top-to-the-SW movement in the low dip limb, are compatible with bedding-slip during flexural folding (Berty 1992), whereas the striations with a sinistral strike slip component in the steep limb are compatible with those indicating a top-to-the-SE movement in the low dip limb. No clear chronology could be established between the two movements, and the origin of the top-to-the-SE movement in the regional tectonic frame remains unresolved.

The third generation is represented by a few veins corresponding to post-folding WNW-verging reverse faults in the steep limb (3 in Fig. 9).

Fluid–sediment interactions in thrust-fault zone

Host marl composition and texture

Except the intensely sheared black marl facies, the bulk mineralogy of host marls is homogeneous, both inside and outside the fault zones. The percentage of $CaCO_3$ varies from 13 to 35%

in the turbiditic marls, and from 42 to 50% in the outer-shelf marly limestones. XRD and backscattered SEM images (Fig. 10) show that the framework of the macroscopically undeformed marl is made up of densely packed detritic grains, most of which are smaller than 10–15 μm and made up of calcite, quartz, feldspar (k-feldspar and plagioclases), micas (muscovite and biotite), chlorite, clay minerals, zircon, and rutile. Calcite is common as isolated grains, as well as microcrystalline cement. Neomorphic minerals are: dolomite, quartz, framboidal pyrite and, locally, sphalerite.

Compaction is evidenced by pressure-solution processes and rearrangement of flakey clay minerals surrounding bigger grains. Pressure-solution mainly affected the calcite grains, as shown by the indented grain-to-grain boundaries. Areas with the biggest grains were the least compressible during compaction, and they preserved their intergrain porosity more easily.

Compaction and cementation by calcite, reduced the original total porosity from what was presumably higher than 50% to the very low present-day values varying from 1.32 to 3.67% (the average of 10 samples is 2.63%). Trapped porosity varies from 65 to 99% of the total porosity. Hence, the porous volume essentially consists of intergranular pores, not cracks, with a size of the pore throats from 0.014 to 0.029 μm. Density of these marl samples ranges from 2.56 to 2.64 (the average of 10 samples is 2.60).

Black marl facies in shear zones

Microstructural and mineralogical studies of the black marl facies characterising the most strained part of the Atiart and Samper fault zones, show that this specific shear facies is related to different types of fluid–sediment interactions, specifically mass-transfer during pressure-solution/precipitation processes affecting both calcite and phyllosilicates.

Microstructural studies by optical and electron microscopies reveal that cleavage formation in the black marl facies involves dissolution of calcite. This is shown by (i) the lack or rareness of calcite grains in the black marl, whereas calcite is abundant in other marl facies and (ii) stylolitization of syn-kinematic cleavage-parallel calcite veinlet walls with peaks perpendicular to the walls (BV in Fig. 4c). Synchronously, syn-kinematic calcite precipitated along some shear surfaces (mSV in Fig. 4b) and, mainly, as veinlets in cleavage-parallel, extensional inter-boudin domains (BV in Fig. 4b,c). These observations suggest short distance mass-transfer from high to low normal stress domains (Gratier 1987).

XRD mineralogical analyses and HRTEM show that the main component of the clay fraction in the black marl is dickite, a kaolin polytype, whereas it is absent in the neigbouring marls, sheared or undeformed, where the major components of the clay fraction are illite and

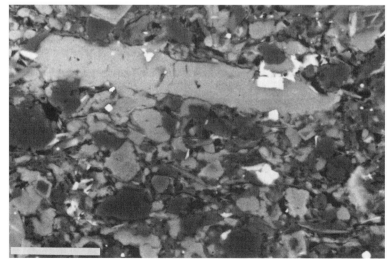

Fig. 10. Back-scattered SEM image showing the texture of macroscopically undeformed marl (Arro syncline). Scale bar is 20 μm.

chlorite (Buatier et al. 1997). SEM and HRTEM images show a close association of dickite with illite, and reveals that dickite is a syn-kynematic mineral crystallized after partial dissolution of illite. The most probable reaction would be:

$$2[KAl_3Si_3O_{10}(OH)_2] + 2H^+ + 3H_2O \longrightarrow 3[Al_2Si_2O_5(OH)_4] + 2K^+.$$

Mass transfer calculations using the Gresens method (Gresens 1967; Potdevin 1993) have been applied using the chemical composition of dickite-bearing and dickite-free marls. These calculations show a loss of K, Na and Mg coupled with a gain of Ca and Sr in the dickite-bearing marls in relation to the dickite-free marls. The loss of K^+ is attributed to the transformation of illite to dickite. K^+ was probably drained out of the system by fluid circulation through the shear zone. The gain in Ca^{2+} is attributed to the abundance of calcite veinlets in this facies, but the mass balance calculation suggests that other minerals were probably involved in the reaction.

Microthermometry of fluid inclusions in shear-vein celestites

The calcite crystals in the shear veins did not lend themselves well to fluid inclusion analysis. Clusters of primary fluid inclusions (3–20 μm) were found in some celestite crystals from Arro syncline veins (T2.2 openings in generation 2 veins), randomly distributed along growth bands. Most of the inclusions have a diameter less than 5 μm, a fact that, together with the turbidity of celestite in these samples, caused optical difficulties and prevented most of the inclusions to be used for microthermometric analyses. Moreover, necking-down phenomena is very important and was observed by the presence of 'tails' or 'clouds' of tiny fluid inclusions around larger ones, or by the variable liquid to vapour ratios, also precluding the use of most of the inclusions for microthermometric measurements. As a result, only a few fluid inclusions, devoid of these problems, were selected for microthermometric analyses.

These inclusions contain two phases at room temperature: an aqueous saline fluid and a non-condensable vapour bubble. Their morphologies are mostly irregular and their degree of filling (liquid/vapour ratio) is approximately between 0.8 and 0.9. Only one inclusion was large enough to perform a complete freezing run. Eutectic temperature was found around –40 °C indicating that this brine contains $CaCl_2$. Hydrohalite melts around –22 °C, hence a $NaCl/CaCl_2$ weight ratio of 0.8 was calculated using the abacus of Oakes et al. (1992). Ice melting temperatures were between –5.5 and –4.3 °C, whereas calculated salinities ranged from 6.8 to 8.5 wt% eq. NaCl. No CO_2 was detected by clathrate melting. Homogenization temperatures by bubble shrinkage ranged between 157 and 183°C (Table 1, Fig. 11a).

The salinities (Table 1, Fig. 11b) found in

Table 1. *Temperatures of homogenization (T_h) and ice melting (T_{fh}), and salinities of the primary fluid inclusions in celestite crystals from the Arro syncline shear veins*

Sample	T_h	T_{fh}	Salinity wt% eq NaCl
Ar-94-24	130.0		
Ar-94-24	158.2		
Ar-94-24	168.6		
Ar-94-24	183.3	–5.4	8.4
Ar-94-24	157.2	–5.0	7.8
Ar-94-24	*87.0*		
Ar-94-24	170.0	–4.9	7.7
Ar-94-24	136.0		
Ar-94-24	158.9	–5.5	8.5
Ar-94-64	181.9	–4.4	7.0
Ar-94-64	176.9	–4.5	7.1
Ar-94-64	179.6	–4.4	7.0
Ar-94-64	181.6	–4.3	6.8
Ar-94-64	*130.8*	*–4.3*	*6.8*
Ar-94-64	*130.0*		
Ar-94-64	179.6	–4.3	6.8
Ar-94-64	175.1	–4.5	7.1

Italicized dates represent partially disintegrated fluid inclusions

these fluid inclusions (about 7 wt% eq. NaCl) indicate that the trapped waters were not pristine seawater (3.5 wt%). Homogenization temperatures, with no pressure correction, were surprisingly high in relation to the temperatures that could be expected according to the geodynamic scenario.

Taking into account the great uncertainty in

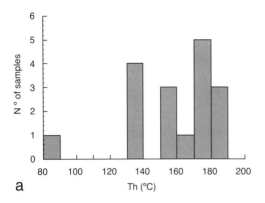

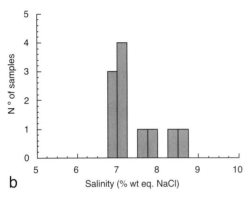

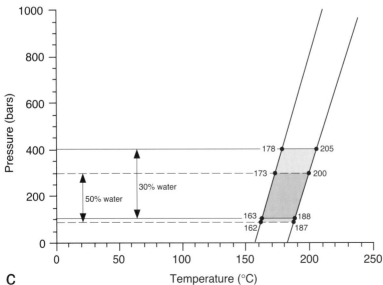

Fig. 11. Microthermometry in primary fluid inclusions in celestite crystals from the Arro syncline shear veins. (**a**) Homogenization temperatures of the fluid inclusions. (**b**) Salinities of the fluid inclusions. (**c**) Corrected temperatures for calculated pressures (see text) and maximum and minimum isochores (Brown & Lamb 1989).

the estimation of depth formation of the studied structures (see above), pressure corrections have been made for depths of 500 and 2000 m. Assuming that during vein formation the fluid pressure was close to, or slightly lower than, the lithostatic pressure (see below), and that the amount of water contained in the marls was between 30 and 50% because the veins were formed in still soft sediment (see above), the calculated sediment bulk densities at vein formation for these two water content values are 2.12 and 1.8 gr cm^{-3}, respectively. The computed lithostatic pressure values are between 88 and 352 bar for 500 and 2000 m, respectively, in marls containing 50% water, and between 104 and 400 bar, in marls containing 30% water. Isochores were calculated using the Brown & Lamb´s (1989) Equation of State for the system H_2O–NaCl–KCl. From the calculated pressures, trapping temperatures can be corrected (Fig. 11c). This correction factor, highly dependent on the percent of water in the marl, is between 2 and 18 °C, a temperature variability that falls within the interval of the measured temperatures of homogenization.

Isotopic geochemistry of veins and host marls

Sulphur isotopes

The celestite samples from T2 openings in all three vein generations in the Arro syncline show $\delta^{34}S$ values varying from 18.3 to 21.7‰ (CDT). These values are consistent with those of the Eocene seawater sulphate (Claypool *et al.* 1980). The dispersion on the $\delta^{34}S$ values (up to 2.4‰) may result from a partial bacteriological reduction of the sulphate to sulfur complexes produced during the late transformation of celestite to pyrite observed in several samples.

Strontium isotopes

In the Arro syncline, the calcite and celestite-barite from T2 openings in all three vein generations, and the calcite fraction of the host marls, show little variation from the $^{87}Sr/^{86}Sr$ ratio (between 0.70774 and 0.70795). This range of values, consistent with the $^{87}Sr/^{86}Sr$ ratio of the Eocene seawater (Katz *et al.* 1972; Burke *et al.* 1982; DePaolo & Ingram 1985), indicates that the source of Sr to precipitate the calcite and celestite–barite in veins and the calcite cement in the host sediment was the same and was controlled by the $^{87}Sr/^{86}Sr$ ratio of the Eocene seawater during the whole evolution of the Arro syncline. Therefore, the vein-forming fluid was probably the interstitial water trapped in the Eocene marine marls derived from the Eocene seawater, or a fluid equilibrated with the Eocene calcitic sediments.

The siliciclastic detrital fraction in the marls has substantially higher $^{87}Sr/^{86}Sr$ ratios (0.71723–0.71894) with respect to the Eocene seawater, giving rise to the more radiogenic $^{87}Sr/^{86}Sr$ ratios in the bulk marl (0.70976–0.70987) than the $^{87}Sr/^{86}Sr$ ratios in the calcite fraction. This more radiogenic $^{87}Sr/^{86}Sr$ ratio from the detrital fraction indicates that this fraction came from the erosion of Palaeozoic rocks.

In the Atiart and Los Molinos fault zones, the $^{87}Sr/^{86}Sr$ ratios in vein calcite (0.70815–0.70917 in T2 openings in shear veins, and 0.70927 in calcite veinlets from the black marl facies) are more radiogenic than the Eocene seawater values. These relatively high values indicate that the vein calcite in fault zones precipitated from a fluid with a different composition than the fluid present outside the fault zones. This may be due to the circulation in the fault zones of a fluid derived from deeper settings, or alternatively, from the enrichment of the fluid in radiogenic Sr related to local water–sediment interactions in the intensely strained sediments (e.g., the illite to dickite transformation). Because the Samper thrust is not a deep-rooted thrust (see above and Fig. 2), the latter interpretation is preferred here.

Oxygen and carbon isotopes

The $\delta^{18}O$ and $\delta^{13}C$ values from the host marls (ranging from –8.2 to –5.8‰, and from –3.3 to –0.7‰ PDB, respectively; Fig. 12) are lower than those of the Eocene marine carbonates which range from –4 to +2‰, and from –0.3 to +2.8‰ PDB, respectively (Shackleton & Kennett 1975; Veizer & Hoefs 1976; Hudson & Anderson 1989). This depletion most probably resulted from burial diagenesis.

At each outcrop, the oxygen isotopic values of the calcite cements in veins are systematically depleted between 1.3 and 2.8‰ PDB in relation to the host marl calcite. This fact indicates that (i) the isotopic composition of the fluid from which the vein calcite precipitated was partially controlled by the isotopic composition of the host sediment, and (ii) the vein calcite probably precipitated from a fluid hotter than the surrounding sediments, probably ascending from deeper zones.

The Atiart and Los Molinos thrusts, which are deep-rooted thrusts, also show a depletion in ^{13}C of the calcite in veins with respect to the host marl calcite. This is interpreted as an input of an external ^{13}C depleted mineralizing fluid channelized through the thrust-fault zones. On the contrary, in the Arro syncline and in the Samper thrust, which have no roots towards deeper areas, the similar δ^{13}C in the calcite from veins and host marls shows that the δ^{13}C of the mineralising fluid was mainly controlled by that of the host sediment.

The isotopic values of calcite from the intensely sheared black marls from the Atiart and Samper thrusts (ranging from –9.0 to –5.8‰ PDB in δ^{18}O, and from –2.6 to –1.5‰ PDB in δ^{13}C, respectively) show no significant differences with respect to calcite from the less deformed marls. However, the calcite veinlets inside these black marls (ranging from –7.3 to –4.4‰ PDB in δ^{18}O, and from –2.5 to –1.6‰ PDB in δ^{13}C) are more enriched in ^{18}O than the calcite veins outside this facies and than the marl calcite. This increase could indicate (i) that the calcite veinlets precipitated from an external fluid circulating along the black marl shear zone and more enriched in ^{18}O than the fluid from which the calcite veins precipitated outside the black marl zone, (ii) that the calcite veinlets precipitated from a fluid similar to that from which the calcite veins precipitated outside the black marl zone but, at a lower temperature, and/or (iii) *in situ* clay–water reactions.

From the three hypothesis, *in situ* clay–water reactions is the only one we have proof of (cf. the dickite formation through partial dissolution of illite). Alteration of rock fragments and feldspars have been proved to produce ^{18}O increase in the fluid (Yeh & Savin 1977; Longstaffe 1993). Although further experimental work should be

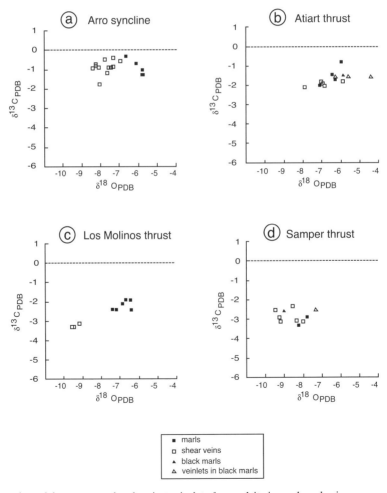

Fig. 12. Cross-plots of the oxygen and carbon isotopic data from calcite in marls and veins.

done to corroborate this hypothesis, in our case, the transformation of illite to dickite may have produced the observed enrichment in ^{18}O of the fluid.

Comparing the values of the four outcrops, both the host marl calcite and the calcite cements in veins show a progressive depletion in both ^{18}O and ^{13}C from the Atiart thrust and Arro syncline to Los Molinos thrust and to Samper thrust. The structural location and burial history of the four outcrops make it difficult to explain the trend of isotopic depletion by differences in burial of the structures. A possible explanation could be a higher meteoric influence in the structures located closer to the emerged part of the belt (i.e., the Samper thrust).

Elemental geochemistry of shear-vein calcite

The Ca, Mg, Sr, Fe and Mn composition of calcite filling all three opening types in shear veins from the four outcrops, as well as a few calcite grains in marls from the Atiart thrust and Arro syncline, were determined through microprobe analyses.

Assuming that precipitation occurred in equilibrium, the elemental composition of the fluid from which calcite precipitated was determined from the elemental geochemistry of the calcite cements, using the distribution coefficient equation of McIntire (1963). The complete elemental geochemical results and the detailed procedure for calculations are presented in a separate paper (Travé et al. 1997). The results show that the elemental composition of the fluid from which all types of calcite precipitated had a Mg/Ca ratio between 0.006 and 1.222, a Sr/Ca ratio between 0.012 and 0.39, a Mg/Sr ratio between 0.089 and 35.9, a Ca/Fe ratio between 79 and 588, and a Mn/Ca ratio between 0.1×10^{-4} and 4×10^{-4}. Most of these values are consistent with formation water and not with unmodified marine waters or oxidizing/reducing meteoric waters (Travé et al. in 1997). The samples from the Atiart thrust and Arro syncline have higher values of Sr/Ca and lower values of Mg/Sr than the samples from the other two outcrops, reflecting original seawater composition to a certain extent.

Discussion of fluid involvement in the geodynamic evolution of the Ainsa basin

The relation between the structures of all scales, mineral precipitations and geochemical signatures allow us to differentiate three stages in the relationships between porosity reduction, water circulation, chemical transformations and mass transfer during synchronous compaction-deformation process (Fig. 13): (i) early compaction by burial, (ii) main tectonic shortening and (iii) after the main tectonic shortening.

Fluids and early compaction by burial

The Ainsa basin is filled with marine turbiditic sediments consisting mainly of marls and thin siltstone layers. During sedimentation the porosity of the marls was, by comparison with recent similar deposits, probably very high (at least 50%). These sediments were progressively buried, and therefore compacted, by subsequent sedimentation of younger sediments.

Petrographic features of the veins and their host marls reveal that during vein formation (i.e., during compressive deformation) the host marls were still poorly cemented and therefore the main episode of marl cementation by calcite occurred after compressive deformation. However, we can use the geochemical signatures of marl calcite (i.e., containing the late cements) to constrain the early diagenetic processes if we can prove that the early geochemical signatures were not changed by later processes.

In the Arro syncline, which is taken as representative of the regional marls not affected geochemically by thrust emplacement, the $\delta^{34}S$ and $^{87}Sr/^{86}Sr$ ratio in marl calcite indicates that the pore-space of the marls was occupied originally by Eocene marine water trapped in the basin-fill.

The $\delta^{18}O$ and $\delta^{13}C$ of the marl calcite from all outcrops reflect an early burial diagenesis, with depletion of both isotope ratios in relation to the Eocene marine carbonates. The initial trapped seawater ($\delta^{18}O = 0‰$ SMOW) would had interacted with the Eocene carbonates with an initial oxygen isotopic composition of $-2‰$ PDB or 28.9‰ SMOW.

The elemental composition of the marl calcite indicates that the interstitially trapped fluid changed from the original seawater composition to formation water composition due to increasing burial conditions.

The highly homogeneous $\delta^{18}O$, $\delta^{13}C$ and elemental composition values of the marl calcite at different distances from the veins indicate that the entire regional marl calcite underwent a pervasive geochemical transformation by equilibration with the diagenetic fluid, and that the final composition of the host marl calcite was attained prior to vein development. For this reason, we interpret the fluid transformation as related to the early burial diagenesis of basin sediments, before the onset of compressive deformation.

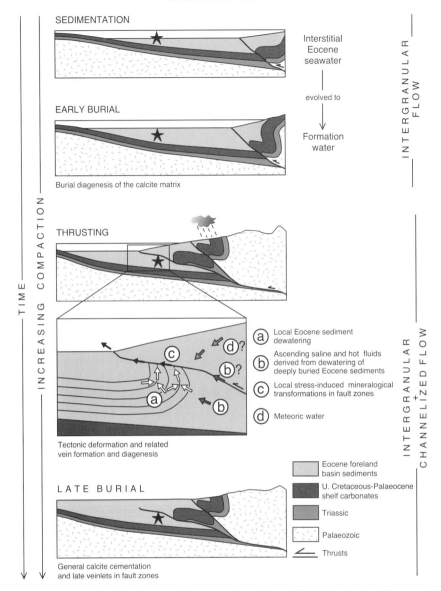

Fig. 13. Synthesis of the relationships between the structural evolution of the Ainsa basin sediments and fluid dynamics.

The pervasive exchange of the isotopical and elemental composition of the whole marl calcite prior to vein formation requires the fluid to have had access to all the grains and therefore, the earlier fluid regime through the basin-fill was an intergranular (porous) flow regime, corresponding to 'compactional flow'.

The $^{87}Sr/^{86}Sr$ ratio of the marl calcite is identical to the $^{87}Sr/^{86}Sr$ ratio of the Eocene seawater. As the $^{87}Sr/^{86}Sr$ ratio of the detritic fraction of the marls is much more radiogenic than that of the calcite fraction, preservation in the calcite of the marine $^{87}Sr/^{86}Sr$ signal indicates the absence of *in-situ* transformations of the detrital components during burial diagenesis.

In conclusion, before the onset of compressive deformation, the sediments underwent only burial compaction and diagenesis associated to dewatering by expulsion of the interstitial fluid (compactional flow).

Fluids and tectonic shortening

Compaction during tectonic shortening. According to the regional tectonostratigraphic pattern, compressive deformation of the Ainsa basin sediments occurred soon after deposition, mostly while sedimentation was still active in the basin. Furthermore, syn-tectonic vein features indicate that deformation occurred before the main episode of marl cementation by calcite. Therefore, compression may have been associated with continuing compaction and water expulsion, involving both (i) vertical compaction related to continuing sedimentation and burial under thrust sheets, and (ii) tectonic horizontal shortening. The importance of the latter process is shown in one place of the Arro syncline by the deformation of vertical burrows contained within marls devoid of apparent cleavage. The systematic oval-shape of burrows (*Diplocraterium*) in the bedding plane corresponds to 20% shortening of the short axis of the ellipse with direction perpendicular to the axial plane of the Arro syncline, which shows that the marls in the syncline underwent at least a 20% volume reduction by pervasive horizontal shortening.

Fluid conditions during shear vein development. Compressive deformation resulted in the onset of channelized flow along interbed planes and newly formed fractures, which acted as discontinuity surfaces for fluid migration. The calcite shear veins are the main macroscopic result of this fluid migration, and their internal geometry and spatial distribution record both the conditions and kinematics of deformation and the geometry of the fluid conduct system.

Microstructural evidence and petrographic features of the different generations of veins from the different structural settings indicate: (i) that the vein formation began in poorly lithified marls, (ii) that progressive induration of the host marl occurred during vein development, (iii) that marl induration was restricted to vein vicinity; marls away from the veins remained poorly lithified during the whole deformation sequence.

Shear vein formation and fluid dynamics. General considerations suggest that fluid pressure was most likely higher than hydrostatic pressure during the compressive deformation of the Ainsa basin-fill. First, theoretical calculations show that excess fluid pressure facilitates thrusting (Hubbert & Rubey 1959), in particular in the case of long-distance displacement along low-angle faults (cf. Cotiella Nappe). Second, the structural context of the studied area is very similar to that of modern accretionary prisms, where structural, geochemical and geophysical evidences indicate frequent fluid overpressures thought to result mainly from rapid tectonic loading of poorly permeable uncompacted sediments (Moore & Vrolijk 1992).

However, the calcite shear vein features suggest that fluid pressure, though at an elevated level, was most often lower than lithostatic pressure during formation of the studied structures. Indeed, the rising of fluid pressure to the level of lithostatic pressure in a thrust context is likely to result in the formation of sub-horizontal extensional veins by hydraulic fracturing (Sibson 1990*a*, *b*; Behrmann 1991), and such veins are absent in the studied area. In contrast, releasing overstep openings such as those of the Ainsa shear veins correspond to domains where local stress-state perturbation results in the lowering of mean stress (Segall & Pollard 1980), allowing local extensional fracturing even under pore pressure lower than lithostatic pressure (Sibson 1986, 1990*b*).

The calcite shear veins formed through an episodic process corresponding to the crack–seal mechanism first described by Ramsay (1980) for extensional veins, and latter recognized in shear veins (Gaviglio 1986; Labaume *et al.* 1991; Jessel 1994). This mechanism implies cyclic rupturing associated with variations of fluid conditions (pressure, permeability, and flow) which, in the case of shear veins, can be described by the following stages: (1) shear rupture, possibly assisted by fluids (maximum pore pressure) inducing releasing overstep openings, (2) drainage of fluids in the opened cavities (maximum fracture permeability), (3) lowering of pore pressure due to drainage that resulted in cessation of shear movement and together with drainage promoted calcite precipitation in the cavities and (4) a rise in pore pressure resulting from sealing by calcite precipitation (reduced permeability), up to a new rupture, etc.

Shear vein distribution, as well as fluid inclusion, mineralogical and chemical analyses, indicate that vein formation allowed long-distance drainage of fluids, from lower to upper structural levels (see below), to take place. The crack–seal mechanism described above implies the episodic nature of this drainage, the shear veins thus having had a fault-valve behaviour (Sibson 1990*a*, *b*) in the fluid drainage system. As the vein openings occurred exclusively at releasing oversteps (cf. the dilational jogs of Sibson 1986, 1990*b*), the invasion of fluids in the openings was mainly triggered by the pore pressure differential between the fast-formed (depressurised) cavities and the surrounding sediments (the suction-pump of Sibson 1986, 1990*b*). Jigsaw breccia of wall-sediments observed in some T2.2 openings attest that the pore pressure differential was sometimes high enough to cause hydraulic implosion during fluid invasion in the cavity (Sibson 1986, 1990b).

Shear vein distribution and fluid drainage pattern. Veins abound inside and over a few tens of metres below the thrust-fault zones, but are rare or absent above them. This heterogeneous distribution indicates that fracture flow during compressive deformation was chiefly concentrated in two structural settings, i.e., the thrust-fault zones and their footwalls. In the latter, the veins occur mainly either as isolated thrust faults (Samper, Atiart and Los Molinos thrusts) and late strike-slip faults (Samper and Atiart thrusts), or as related to bedding slip during flexural folding in the reverse limb of footwall synclines (Atiart lower thrust). On a larger scale, the Arro syncline may be considered a zone of preferential fluid flow in the footwall of Los Molinos thrust, also dominated by bedding slip during flexural folding.

The concentration of veins in thrust-fault footwalls suggests the presence of overpressured fluids in this structural level, due to rapid tectonic loading of the poorly permeable marls during emplacement of the overthrust units. Footwall overpressuring was probably the main triggering mechanism to enhance the expulsion of the interstitial water, producing a gradient of flow from the footwall marls towards the fault zones. The presence of veins in the fault zones, and their absence (or rareness) in the hanging walls, indicates that the fluids arriving in the fault zones flowed preferentially along these zones, which prevented invasion of the hanging walls by the ascending fluids. This drainage pattern shows that fault permeability was higher parallel to than perpendicular to the faults, permeability anisotropy resulting from the effects of both fracture permeability during shear vein formation (see above) and clay particle arrangement in shear zones (Arch & Maltman 1990). Poor development of veins within the hanging-wall material also argues for its compaction and permeability reduction prior to thrusting.

In the case of the footwall synclines, folding kinematics closely controlled drainage of the ascending fluids, as indicated by the preferential occurrence of bedding-parallel calcite shear veins in the steep limb of the synclines. This effect probably resulted from the fact that these knee-shape synclines were similar to kink-bands with fixed hinges (Berty 1992). During the incipient stages of formation of such a kink-band, geometrical constraints impose thickening of the kink-band, an effect likely to pump fluids within the kink-band from surrounding areas (Cosgrove 1993). In the Ainsa case, this process may have controlled drainage of fluids present in thrust footwall domains towards the tilted limbs of the footwall synclines, and facilitated their ascencion towards the overlying thrust-fault zone by flowing along the tilted beds. Beyond a certain amplification, the tilted limb must have thinned, which may have made the pumping of fluids from surrounding areas more difficult, but may have promoted tectonically induced compaction of the tilted limb, hence furnishing fluids for the latest stages of vein formation. However, in the case of the Arro syncline, geometrical analysis of the fold shows that the thinning stage was not reached in the western part of the steep limb where most of the shear veins are concentrated. This observation is consistent with the fact that no geochemical variations have been found in the veins sampled in this area (i.e., the source of fluids remained unchanged during all, or most, of vein formation).

Origin of fluids and fluid-sediment interactions during thrusting. Comparison of the shear vein geochemistry and marl mineralogy from the thrust-fault zones with those of the footwalls reveals both similarities and differences. The similarities indicate that both settings belonged to the same interconnected fluid system; the differences indicate the occurrence of specific fluid-related processes in the fault zones (Fig. 13).

In the thrust-fault footwalls. In the Arro syncline, which is taken as representative of the thrust-fault footwalls, the $\delta^{34}S$ and $^{87}Sr/^{86}Sr$ ratio of the vein calcite are consistent with those of Eocene marine carbonates. $\delta^{18}O$, $\delta^{13}C$ and elemental geochemistry reveal a diagenetic change from the original seawater to formation water compositions during early burial/tectonic compaction. The similarity in elemental composition, $\delta^{34}S$, $^{87}Sr/^{86}Sr$ ratio and $\delta^{13}C$ between the vein and host marl calcite, indicates that the geochemistry of vein calcite was highly controlled by that of the host marls. As the $^{87}Sr/^{86}Sr$ ratio of the detritic fraction of the regional marls is much more radiogenic than that of the calcite fraction, preservation in the calcite from veins of the marine $^{87}Sr/^{86}Sr$ signal indicates that no transformations of the detrital components occurred within this setting during tectonic shortening.

In the Arro syncline as in the three thrust-fault zones, the systematic depletion of $\delta^{18}O$ in calcite from shear veins in relation to their host marls (between 1.3 and 2.8‰ PDB) indicates that the $\delta^{18}O$ composition of the vein calcite was also partially controlled by the isotopic composition of the host marls. The difference in $\delta^{18}O$ can be explained by a higher temperature in the veins than in the host marls, due to the influx in the veins of a hotter ascending fluid (Marshall 1992).

The latter inference is consistent with the high salinities (7 wt% eq. NaCl) and high homogenization temperatures (between 157 and 183 °C) of the fluid inclusions in vein celestite from the Arro syncline, which point to the presence in the veins from this outcrop of a hot, saline solution.

The geochemical characters of the Arro syncline calcite thus indicate that the vein calcite precipitated from the mixing of two different fluids (Fig. 13).

(1) A connate fluid coming from dewatering of the host marls. This fluid is strongly recorded in the vein calcite by the similarity of most of their geochemical markers with those of the host marl. This feature implies that local sediment dewatering was a major mechanism of fluid migration in the thrust footwalls and that a high fluid–rock interaction existed there between the connate fluid and the host marl.

(2) A hot and saline fluid derived from deeper settings which is more subtly recorded in the geochemistry of the vein calcite. It is unlikely that this fluid, whose flux was probably high to maintain a thermic anomaly, was derived from the Palaeozoic or Triassic rocks of the Pyrenean Axial Zone. Indeed, calcite veins which precipitated in the Cretaceous substratum of the Ainsa basin from fluids having this origin show significantly lower Sr content and higher $^{87}Sr/^{86}Sr$ ratio (McCaig et al. 1995) than the Arro syncline veins, where (i) the Sr content is high and (ii) the $^{87}Sr/^{86}Sr$ ratio and the $\delta^{34}S$ of the connate water were not modified. A more probable origin for the deep fluid is the dewatering of the most deeply buried Eocene sediments at the inner edge of the Ainsa basin, where vein calcite shows Sr content and $^{87}Sr/^{86}Sr$ ratio values (McCaig et al. 1995) similar to those found in the Arro syncline calcite. The circulation of this fluid mainly took place within the discontinuity planes formed by the veins, and was expelled upwards without producing any exchange within the marls immediately adjacent to the veins. Fluid mixing may have triggered the precipitation of celestite due to the reverse solubility of this mineral (Sonnenfeld 1984), after the sulphate and the Sr contained in the fluid.

Inside the thrust-fault zones. Elemental geochemistry, and $\delta^{18}O$ and $\delta^{13}C$ isotopes indicate that the shear vein calcite in the thrust-fault zones precipitated from formation water evolved from Eocene marine water, probably expelled from the footwall sediments as well as produced *in situ* by the tectonically induced compaction of the marls in shear zones. Cleavage-forming pressure-solution of calcite and clay (i.e., the illite to dickite transformation) was associated with this local tectonic compaction.

The absence of fluid inclusions hinders us from characterizing the fluid that circulated inside the thrust-fault zones. However, the systematic depletion of $\delta^{18}O$ in the shear vein calcite in the fault zones in relation to their host marls (excluding the calcite veinlets in the dickite-bearing marls) attests that a hot fluid also circulated in these veins. This hot fluid was probably ascending from the local footwalls, but it may also have been partially drained from deeper horizons along the fault zones themselves.

The $^{87}Sr/^{86}Sr$ anomaly in the shear vein calcite in the fault zones probably results from the local clay mineral reactions occurring in the most intensely strained part of the thrust-fault zones, i.e., the illite to dickite transformation. The structural setting where the transformation occurs suggests that it was mainly controlled by the stress conditions, rather than by the temperature or/and composition of the circulating fluid (Buatier et al. 1997).

The increasing depletion of $\delta^{18}O$ in calcite from both the shear veins and the host marls from the Arro syncline to the Atiart, Los Molinos, and Samper thrusts, which is hardly explained by burial differences, is most probably related to the increasing influence of meteoric water in the structurally innermost/highest zones. These waters, derived from the emerged part of the belt, would have been drained laterally by the thrust-fault zones creating barriers hindering their flowing towards the deeper/more external footwalls (the Arro syncline).

Thus, in addition to the fluids produced *in situ* by tectonic compaction of the marls in the shear zones, the thrust-fault zones acted as drains for the fluids derived from both deeper settings (footwall dewatering and hot ascending fluids) and the emerged part of the belt (meteoric fluids). Moreover, specific stress conditions in the most intensely deformed parts of the shear zones were responsible for interactions between the circulating fluid and the deforming marl (dickite formation). Fluid–sediment interactions seem to have been restricted in the thrust-fault zones which acted as channels for fluids, drained towards the more external parts of the belt without affecting all the basin sediments. Ultimately, these fluids were probably discharged at the sea bottom at the thrust front, as is the case in modern accretionary prisms (e.g., Henry et al. 1989).

Fluid behaviour after the main tectonic shortening

Extensional veins in the sheared black marls. In the studied area, notable sets of extensional veins are found only among the very thin veinlets intercalated in the black marl facies, in the most strained parts of the thrust-fault zones (EV in Fig. 4c).

These extensional veinlets are parallel to the cleavage planes and were not affected by pressure-solution related to cleavage formation. Hence, they are considered late features formed by the opening of the pre-existing discontinuities formed by the cleavage planes. Extension direction is parallel to that of the previous shortening responsible for the cleavage formation. This extension may correspond to a stage of compressive stress relaxation, possibly associated with tectonic uplift and relief formation.

The calcite cements in these extensional veinlets are enriched in ^{18}O with respect to the host black marls, and to the shear vein calcite outside the black marl facies. From the different possibilities to explain this $\delta^{18}O$ increase, the most probable is that the fluid was affected by the *in situ* sediment–water interactions (dickite formation). This fluid was probably the same as the fluid with the relatively high radiogenic $^{87}Sr/^{86}Sr$ ratio which circulated along the shear veins in the thrust-fault zones during thrust activity.

Whole sediment cementation. As discussed above, shear vein formation occurred in poorly cemented sediment, and progressive induration during vein formation occurred only in the surrounding marls.

Thus, precipitation of the calcite cement in the whole marl occurred after the main episode of thrusting. The source of carbonate for cement precipitation in the marls was the porewater saturated in carbonate and still remaining in the pore spaces, as well as the detrital and biogenic carbonate components present in the marls (the latter mobilized through a pressure-solution/-precipitation process). As this connate fluid was already equilibrated with the whole marl, no further exchange of the isotopic or elemental composition occurred.

Conclusion

In conclusion, our study shows that fluid evolution in the Ainsa foreland basin involved two main episodes which can be summarized as follows.

During early burial, fluid regime in the Ainsa basin was an intergranular flow regime (compactional flow) which allowed the pervasive isotopic and elemental exchange of the whole marls, prior to vein formation. During this process, the original seawater trapped in the sediment (recorded by $\delta^{34}S$ in vein celestite and $^{87}Sr/^{86}Sr$ ratio in marl and vein calcite and vein celestite) changed into formation water, as indicated by elemental geochemistry, $\delta^{18}O$, and $\delta^{13}C$ in marl and vein calcite.

The localization of deformation in shear zones during thrusting allowed for the onset of a channelized flow, with dewatering of tectonically overpressured underthrust sediments through different types of slip zones, until these fluids reached the main thrust-fault zones which they followed laterally, probably until reaching the seafloor. Channelized flow in tectonically active fractures is recorded by numerous calcite shear veins formed by crack–seal mechanism, which attest the episodic nature of fault-slip and associated fluid migration. During this migration, the local formation water was mixed with a hot, saline fluid derived from deeper underthrusted parts of the basin, and, in some of the thrusts, possibly with meteoric waters. These external waters changed only a few aspects of the geochemical signature of the mineralizing fluid, recorded only by fluid inclusion data and $\delta^{18}O$ depletion. The particular stress conditions in the thrust-fault zones promoted clay sediment–fluid interactions, with dickite precipitation associated with cleavage formation. This reaction is probably responsible for the higher $^{87}Sr/^{86}Sr$ ratio in the vein calcite from the thrust-fault zones, in relation to the same ratio in the footwall marl and vein calcite.

Cementation of the marls by calcite occurred after the main episodes of deformation, within formation water.

This work shows the possibilities offered by a multi-disciplinary approach based on a detailed structural analysis for unravelling fluid-deformation relationships in a thrust system. Specifically, it emphasizes the importance of dewatering of foreland basin sediments along anisotropic fracture permeability associated to early folding and thrust-faulting in mud-dominated systems analogue to some modern accretionary prisms (e.g., the Barbados prism).

This work was realized within the framework of the EBRO Research Network ERBCHRX-CT93-0196: 'Fluids as Agents of Basin Deformation', funded by the Human Capital and Mobility Programme of the European Union. We also acknowledge contribution of DGICYT grant PB94-0868, Grup Consolidat 1996-SGR0086 del Comissionat per Universitats i Recerca de la Generalitat de Catalunya. We thank A. Alvarez, X. Garcia Veigas, X. Llovet and the staff of the Serveis científico-tècnics of Barcelona University for their technical support, C. Recio and J. M. Ugidos at the Laboratorio de Isótopos Estables of Salamanca University for carbon, oxygen and sulphur isotope analyses, and F. Luttikhuizen for the revision of the English version.

Analyses of ICP-AES and X-ray diffractometry were carried out in the Serveis científico-tècnics of Barcelona University. Electron microprobe were performed at the Serveis científico-tècnics of Barcelona University and at Montpellier University. Sr isotopes were analysed in the Laboratoire de Géochimie Isotopique of Montpellier University. Thin-sections were done in Barcelona and Montpellier Universities.

We also wish to thank I. Moretti for helpful reviews of the manuscript.

References

ARCH, J. & MALTMAN, A. 1990. Anisotropic permeability and tortuosity in deformed wet sediments. *Journal of Geophysical Research*, **95**, 9035–9045.

BANKS, D. A., DAVIES, G. R., YARDLEY, B. W. D., MCCAIG, A. M. & GRANT, N. T. 1991. The chemistry of brines from an Alpine thrust system in the Central Pyrenees: An application of fluid inclusion analysis to the study of fluid behaviour in orogenesis. *Geochimica et Cosmochimica Acta*, **55**, 1021–1030.

BEHRMANN, J. H. 1991. Conditions for hydrofracture and the fluid permeability of accretionary wedges. *Earth and Planetary Science Letters*, **107**, 550–558.

BERTY, C. 1992. *Etude des filons de calcite cisaillants dans les nappes superficielles. Applications aux Apennins du Nord et au versant sud des Pyrénées*. Thèse de Doctorat, Université Montpellier II.

BRADBURY, H. J. & WOODWELL, G. R. 1987. Ancient fluid flow within foreland terrains. *In*: GOFF, J. C. & WILLIAMS, B. P. J. (eds) *Fluid Flow in Sedimentary Basins and Aquifers*. Geological Society, London, Special Publications, **34**, 87–102.

BROWN, P. E. & LAMB, W. M. 1989. P-V-T properties of fluids in the system $H_2O \pm CO_2 \pm NaCl$: New graphical presentation and implications for fluid inclusion studies. *Geochimica et Cosmochimica Acta*, **53**, 1209–1221.

BUATIER, M., TRAVÉ, A., LABAUME, P. & POTDEVIN, J. L. 1997. Dickite related to fluid-sediment interaction and deformation in Pyrenean thrust-fault zones. *European Journal of Mineralogy*, **9**, 875–888.

BURKE, W. H., DENISON, R. E., HETHERINGTON, E. A., KOEPINK, R. B., NELSON, H. F. & OTTO, J. B. 1982. Variation of sea-water $^{87}Sr/^{86}Sr$ throughout Phanerozoic time. *Geology*, **10**, 516–519.

CLAYPOOL, G. E., HOLSER, W. T., KAPLAN, I. R., SAKAI, H., & ZAK, I. 1980. The age curves of sulfur and oxygen isotopes in marine sulfate and their mutual interpretation. *Chemical Geology*, **28**, 199–260.

COSGROVE, J. W. 1993. The interplay between fluids, folds and thrusts during deformation of a sedimentary succession. *Journal of Structural Geology*, **15**, 491–500.

DEPAOLO, D. J. & INGRAM, B. L. 1985. High-resolution stratigraphy with strontium isotopes. *Science*, **227**, 938–941.

GARVEN, G. 1995. Continental scale groundwater flow and geologic processes. *Annual Review of Earth and Planetary Sciences*, **23**, 89–117.

GAVIGLIO, P. 1986. Crack-seal mechanism in a limestone: a factor of deformation in strike-slip faulting. *Tectonophysics*, **131**, 247–255.

GRANT, N. T., BANKS, D. A., MCCAIG, A. M. & YARDLEY, B. W. D. 1990. Chemistry, source, and behavior of fluids involved in Alpine thrusting of the Central Pyrenees. *Journal of Geophysical Research*, **95**, 9123–9131.

GRATIER, J. P. 1987. Pressure-deposition creep and associated differentiation in sedimentary rocks. *In*: JONES, M. E. & PRESTON, R. M. F. (eds) *Deformation of Sediments and Sedimentary Rocks*. Geological Society, London, Special Publications, **29**, 25–38.

GRESENS, R. L. 1967. Composition-volume relationships of metasomatism. *Chemical Geology*, **2**, 47–55.

GUILHAUMOU, N., LARROQUE, C., NICOT, E., ROURE, F. & STEPHAN, J. F. 1994. Mineralized veins resulting from fluid flow in decollement zones of the Sicilian prism: evidence from fluid inclusions. *Bulletin de la Société Géologique de France*, **165**, 425–436.

HENRY, P., LALLEMANT, S. J., LE PICHON, X. & LALLEMAND, S. E. 1989. Fluid venting along Japanese trenches: tectonic context and thermal modeling. *Tectonophysics*, **160**, 277–291.

HOLL, J. E. & ANASTASIO, D. J. 1995. Cleavage development within a foreland fold and thrust belt, southern Pyrenees, Spain. *Journal of Structural Geology*, **17**, 357–369.

HUBBERT, M. K. & RUBEY, W. W. 1959. Role of fluid pressure in the mechanics of overthrust faulting. *Geological Society of America Bulletin*, **70**, 115–205.

HUDSON, J. D. & ANDERSON, T. F. 1989. Ocean temperatures and isotopic compositions through time. *Transactions of the Royal Society of Edinburgh*, **80**, 183–192.

JESSEL, M. W. 1994. Bedding parallel veins and their relationship to folding. *Journal of Structural Geology*, **16**, 753–767.

KATZ, A., SASS, E., STARINSKY, A. & HOLLAND, H. D. 1972. Strontium behaviour in the aragonite-calcite transformation: an experimental study at 40-98 °C. *Geochimica et Cosmochimica Acta*, **36**, 481–496.

KOOPMAN, A. 1983. Detachment tectonics in the Central Apennines, Italy. *Geologica Ultraiectina*, **30**, 1-155.

LABAUME, P., BOUSQUET, J. C. & LANZAFAME, G. 1990. Early deformation at a submarine compressive front: the Quaternary Catania foredeep south of Mt. Etna, Sicily, Italy. *Tectonophysics*, **177**, 349–366.

LABAUME, P., BERTY, C. & LAURENT, P. 1991. Syndiagenetic evolution of shear structures in superficial nappes: an example from the Northern Apennines (NW Italy). *Journal of Structural Geology*, **13**, 385–398.

LISTER, G. S. & SNOKE, A. W. 1984. S-C mylonites. *Journal of Structural Geology*, **6**, 617–638.

LONGSTAFFE, F. J. 1993. Meteoric water and sandstone diagenesis in the Western Canada Sedimentary

Basin. In: HORBURY, A. D. & ROBINSON, A. G. (eds) *Diagenesis and Basin Development*. American Association of Petroleum Geologists. Studies in Geology, **36**, 49–68.

MCCAIG, A. M., WAYNE, J. M., MARSHALL, J. D., BANKS, D. & HENDERSON, I. 1995. Isotopic and fluid inclusion studies of fluid movement along the Gavarnie thrust, Central Pyrenees: Reaction fronts in carbonate mylonites. *American Journal of Science*, **295**, 309–343.

MCINTIRE, W. L. 1963. Trace element partition coefficients — a review of theory and applications to geology. *Geochimica et Cosmochimica Acta*, **27**, 1209–1264.

MARQUER, D., & BURKHARD, M. 1992. Fluid circulation, progressive deformation and mass-transfer processes in the upper crust: the example of basement-cover relationships in the External Crystalline Massifs, Switzerland. *Journal of Structural Geology*, **14**, 1047–1057.

MARSHALL, J. D. 1992. Climatic and oceanographic isotopic signals from the carbonate rock record and their preservation. *Geological Magazine*, **129**, 143–160.

MOORE, J. C. & VROLIJK, P. 1992. Fluids in accretionary prisms. *Review of Geophysics*, **30**, 113–135.

MOORE, J. C., & THE ODP LEG 156 SCIENTIFIC PARTY. 1995. Abnormal fluid pressure and fault-zone dilation in the Barbados accretionary prism: evidence from logging while drilling. *Geology*, **23**, 605–608.

MUCHEZ, P., SLOBODNIK, M., VIAENE, W. A. & KEPPENS, E. 1995. Geochemical constraints on the origin and migration of palaeofluids at the northern margin of the Variscan foreland, southern Belgium. *Sedimentary Geology*, **96**, 191–200.

MUTTI, E., SÉGURET, M. & SGAVETTI, M. 1988. *Sedimentation and deformation in the Tertiary sequences of the southern Pyrenees*. American Association of Petroleum Geologists Mediterranean Basins Conference, Field trip No. 7, Nice.

OAKES, C. S., SHEETS, R. W. & BODNAR, R. J. 1992. (NaCl+CaCl$_2$)aq: Phase equilibria and applications. In: *PACROFI IV, Extended Abstracts*. Lake Arrow Head, California, USA, 21–25 May 1992 119–122.

OLIVER, J. 1986. Fluids expelled tectonically from orogenic belts: their role in hydrocarbon migration and other geologic phenomena. *Geology*, **14**, 99–102.

POTDEVIN, J. L. 1993. Gresens 92: A simple Macintosh program of the Gresens method. *Computers & Geosciences*, **19**, 1229–1238.

RAMSAY, J. G. 1980. The crack-seal mechanism of rock deformation. *Nature*, **284**, 135–139.

RYE, D. M. & BRADBURY, H. J. 1988. Fluid flow in the crust: an example from a Pyrenean thrust ramp. *American Journal of Science*, **288**, 197–235.

SEGALL, P. & POLLARD, D. D. 1980. Mechanics of discontinuous faulting. *Journal of Geophysical Reseach*, **85**, 4337–4350.

SÉGURET, M. 1972. *Etude tectonique des nappes et séries décollées de la partie centrale du versant sud des Pyrénées*. Publications de l'Université des Sciences et Techniques du Languedoc (Ustela), 2, Montpellier.

SHACKLETON, N. J. & KENNETT, J. P. 1975. Paleo-temperature history of the Cenozoic and the initiation of Antarctic glaciation: oxygen and carbon isotope analyses in DSDP Sites 277, 279 and 281. *Initial Reports of the Deep Sea Drilling Project*, **29**, 743–755.

SHIPLEY, T. H., MOORE, G., BANGS, N., MOORE, J. C. & STOFFA, P. L. 1994. Seismically inferred dilatancy distribution, northern Barbados Ridge decollement: Implications for fluid migration and fault strength. *Geology*, **22**, 411–414.

SIBSON R. H. 1986. Brecciation processes in fault zones: inferences from earthquake rupturing. *Pure and Applied Geophysics*, **124**, 159–175.

—— 1990a. Conditions for fault-valve behaviour. In: KNIPE, R. J. & RUTTER, E. H. (eds) *Deformation Mechanisms, Rheology and Tectonics*. Geological Society, London, Special Publications, **54**, 15–28.

—— 1990b. Faulting and fluid flow. In: NESBITT, B. E. (ed.) *Fluids in Tectonically Active Regimes of the Continental Crust*. Mineralogical Association of Canada Short Course, Vancouver, May 1990, 93–132.

SONNENFELD, P. 1984. *Brines and evaporites*. Academic Press, Inc. Orlando, Florida, USA.

TRAVÉ, A., LABAUME, P., CALVET, F. & SOLER, A. 1997. Sediment dewatering and pore fluid migration along thrust faults in a foreland basin inferred from isotopic and elemental geochemical analyses (Eocene southern Pyrenees, Spain). *Tectonophysics*, **282**, 375–398.

VEIZER, J. & HOEFS, J. 1976. The nature of O^{18}/O^{16} and C^{13}/C^{12} secular trends in sedimentary carbonate rocks. *Geochimica et Cosmochimica Acta*, **40**, 1387-1395.

YEH H. W. & SAVIN, S. M. 1977. Mechanisms of burial metamorphism of argillaceous sediments: 3. O-isotope evidence. *Geological Society of America Bulletin*, **88**, 1321–1330.

Sequential restoration of the external Alpine Digne thrust system, SE France, constrained by kinematic data and synorogenic sediments

W. HENRY LICKORISH & MARY FORD

Geologisches Institut, ETH-Zentrum, 8092 Zürich, Switzerland

Abstract: The Tertiary foreland basin of the southern Subalpine chains preserves a stratigraphic record of late Alpine deformation, both ahead of the thrust front in the Valensole basin, and in a series of thrust-sheet-top basin remnants. Stratigraphy and growth structures in these basin remnants have been used to identify the location and timing of deformation and hence to constrain the sequential restoration of a cross section through the region.

The thrust belt developed as a single large thrust sheet riding on a weak Triassic evaporite layer. Minor breaching thrusts occur, in particular at Mesozoic normal faults. Kinematic studies of the Digne thrust sheet show a dominantly SW direction of tectonic transport. Where reactivated structures lie at a slightly oblique angle to the transport direction there has been a partitioning of deformation into SW-directed thrusting and a component of dextral strike-slip taken up on minor faults. The total shortening across the fold and thrust belt is 21.5 km, a much lower value than previously estimated but more in line with regional tectonics.

The foreland basin stratigraphy demonstrates that deformation within the Digne sheet occurred in three stages. Firstly, during the late Eocene, Alpine collision in the hinterland caused flexure of the foreland plate to generate a simple broad marine foredeep. Slight detachment above the Triassic evaporites allowed gentle buckling of the floor of the foredeep. Secondly, in the Early to Mid-Oligocene overthrust shear associated with the SW emplacement of the internally derived Embrunais–Ubaye nappes into the foredeep caused substantial deformation in the underlying Digne sheet. A component of flattening strain was associated with overthrust shear beneath the nappes, but ahead of the nappe front gentle, upright folding and thrusting continued. Sedimentation stepped westward, became dominated by continental facies and confined to small thrust sheet top basins (e.g. Barrême). The Mio-Pliocene stage of deformation involved a change from thin-skinned to a thick-skinned out-of-sequence deformation when basement blocks were uplifted on deep-seated structures across the region (e.g. dôme de Chateauredon). Shortening associated with the uplift of the main Argentera basement massif at the internal boundary of the external fold and thrust belt was transferred onto the Triassic detachment from below, and transferred forward to reactivate the frontal edge of the Digne sheet, which was thrust out over the Valensole basin.

Foreland basin systems are elongate areas of flexural subsidence which border collisional orogenic belts and develop principally in response to the orogenic load. The synorogenic sediments can become involved in deformation within the advancing fold and thrust belt to varying degrees. Foreland basin deposits can thus be used to study the progressive deformation of foreland areas during orogenesis. The large scale stratigraphic infill records orogenic processes (e.g. Beaumont 1981; Karner & Watts 1983; Flemings & Jordan 1990; Sinclair & Allen 1992) while more local growth structures record deformational style, chronology and kinematics across the foreland (e.g. Riba 1976; Anadón et al. 1986; DeCelles et al. 1991; Puigdefabrigas et al. 1992; Suppe et al. 1992; Alvarez-Marrón et al. 1993; Burbank & Vergés 1994; Wickham 1995; Vergés et al. 1996; Zapata & Almendinger 1996; Ford et al. 1997). In this paper the stratigraphic record of a series of Tertiary foreland basin remnants are used to constrain the sequential restoration of a profile across the southern Subalpine chains of SE France. This foreland basin succession is particularly useful in section restoration because it records a marine transgression, followed by a deepening phase and later a regression to alluvial and lacustrine conditions. This allows sea-level to be used as a horizontal datum in reverse modelling. The purpose of this work is to clarify the progressive evolution and structural style of this fold and thrust belt and foreland basin, and thus to define the principal controlling parameters.

LICKORISH, W. H. & FORD, M. 1998. Sequential restoration of the external Alpine Digne thrust system, SE France, constrained by kinematic data and synorogenic sediments. *In:* MASCLE, A., PUIGDEFÀBREGAS, C., LUTERBACHER, H. P. & FERNÀNDEZ, M. (eds) *Cenozoic Foreland Basins of Western Europe*. Geological Society Special Publications, **134**, 189–211.

The southern Subalpine fold and thrust belt in SE France (Fig. 1) forms the southern arm of the external arc of the western Alps and comprises principally a single thrust sheet, the Digne thrust sheet, in which Mesozoic and Tertiary strata were carried to the SW on a Triassic evaporitic detachment during late Alpine (Late Eocene to Miocene–Pliocene) convergence (Goguel 1963; Lemoine 1973; Siddans 1979). The area is tectonically overlain by the internally derived

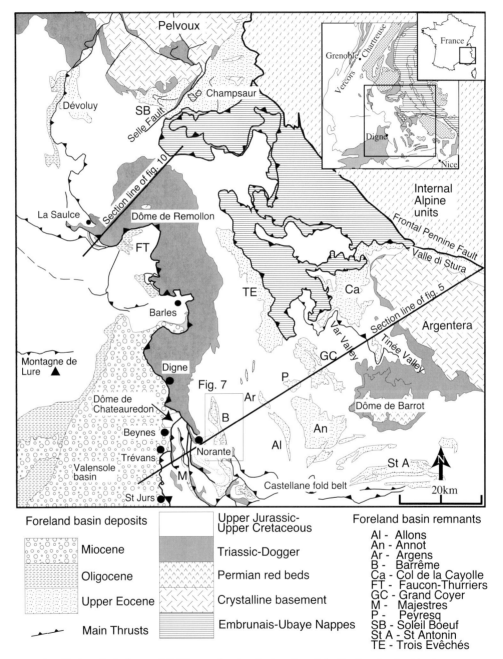

Fig. 1. Simplified geological map of SE France showing the location of the Tertiary foreland basin remnants, and the relation of the Digne thrust sheet to the surrounding Alpine structures. Inset shows the western alpine arc.

Embrunais–Ubaye nappes whose sub-marine emplacement in early Oligocene times onto the foreland basin is recorded by a tectono-sedimentary mélange unit known as the Schistes à Blocs (Kerkhove 1969; Fry 1989). Two crystalline basement massifs, Argentera and Pelvoux (Fig. 1), are now known to have been uplifted principally in late Miocene to Pliocene times from fission-track dating (Mansour et al. 1992; D. Seward pers. comm.). The foreland basin succession is preserved in a series of thrust sheet top remnants, which are best exposed in the southern part of the fold belt. The youngest Mio-Pliocene deposits lie in the Valensole basin in the footwall of the frontal Digne thrust. The frontal Digne thrust records at least 20 km SW displacement, of which at least 7 km is late out-of-sequence movement which carried the thrust system out over the already deformed Valensole basin (Gigot et al. 1974). To the south, the displacement is taken up on small frontal (St Jurs) imbricate faults, the most easterly of which we call the Digne thrust. These frontal imbricates developed above the dôme de Chateauredon (Fig. 1) and were finally emplaced over the Valensole basin. The Digne thrust sheet itself is affected by local mesoscopic fold-thrust structures, the most significant of which occur below the internally derived Embrunais-Ubaye nappes. The Digne system is bound to the south by the E–W-trending, south-verging Castellane arc. North of the dôme de Remollon the Digne thrust has been traced into the complex fault system of the Veynes region (Gidon et al. 1976) and linked, in part, northward along the median Devoluy thrust (3 km displacement; Meckel et al. 1996) into the NE–SW Aspres-les-Corps fault which records a late Alpine dextral strike-slip displacement (Gidon et al. 1970).

Unlike the northern Subalpine chains of the Vercors and Chartreuse regions (Fig. 1, inset) significant facies and thickness changes within the Mesozoic succession here have had an important influence upon structural geometry. Significant deformation occurred across the foreland before the onset of alpine foreland basin subsidence, first during the Eoalpine event (Turonian; Flandrin 1966) and later during the Pyrenean–Provençal (Late Cretaceous–Priabonian) event. Both caused E–W folds, locally intense Siddans 1979; Debrand-Passard et al. 1984). Thus the late Alpine foreland basin and thrust system were superimposed on an already structurally complex crust.

The Embrunais–Ubaye nappes obscure most of the thrust belt in the north but in the south the thrust belt can be observed right back to the Argentera massif (Fig. 1). Our line of section (100 km long) thus crosses this area, cutting through the Valensole basin and a series of thrust sheet top basins (Majestres, Barrême, Argens, Allons, Annot; Fig. 1). It trends N60 °E, parallel to the transport direction, between St Jurs on the thrust front to the west and the Valle di Stura in Italy where it crosses the frontal Pennine fault, the steep boundary between the external and internal zones, thus including the foreland basin deposits in its immediate footwall. The regional geology is shown on the BRGM 1:50 000 map sheets (Moustiers St. Marie, 1978; Digne, 1981; Allos, 1967; Entrevaux, 1980; St Etienne de Tinée, 1970) and the Servizio Geologica Italia 1:100 000 maps (Argentera-Dronero, 1971; Demonte, 1970).

Extensive previous research in this area has focused on Mesozoic passive margin stratigraphy and sedimentology (e.g. Lemoine et al. 1986; Arnaud 1988; de Graciansky et al. 1989), on Alpine structure (e.g. Goguel 1963; Ehtechamzadeh-Afchar & Gidon 1974; Graham 1978; Siddans 1979; Fry 1989) and on foreland basin stratigraphy and sedimentology (Bodelle 1971; Gigot et al. 1974; Elliott et al. 1985; Apps 1987; Pairis 1988; Sinclair 1994). The western part of the cross section has previously been restored by R. Graham using a layer-cake Mesozoic stratigraphy to give a shortening of 65 km (Elliott et al. 1985; Vann et al. 1986; Hayward & Graham 1989). By using a more accurate Mesozoic stratigraphy which accounts for palaeogeographic trends (Fig. 2), and considering the dôme de Chateauradon as a basement uplift, only 21.5 km is now required for the section to balance (see Fig. 5). This lower amount of shortening is consistent with structural observations in other parts of the thrust belt. A three-step history is proposed for the southern Subalpine chains: while gentle folding was on-going throughout early foreland basin subsidence, the main shortening occurred when the Embrunais–Ubaye nappes were emplaced into the basin. The last component of shortening in the Pliocene can be linked to uplift of the Argentera massif.

Mesozoic stratigraphy and palaeogeography

SE France is underlain by Variscan crystalline basement that is exposed in the Pelvoux, Argentera and Maures massifs (Fig. 1; Debrand-Passard et al. 1984). Based on gravity data, basement is predicted to lie at a depth of 3 km along the section line shallowing rapidly to the east of the Annot basin towards the Argentera

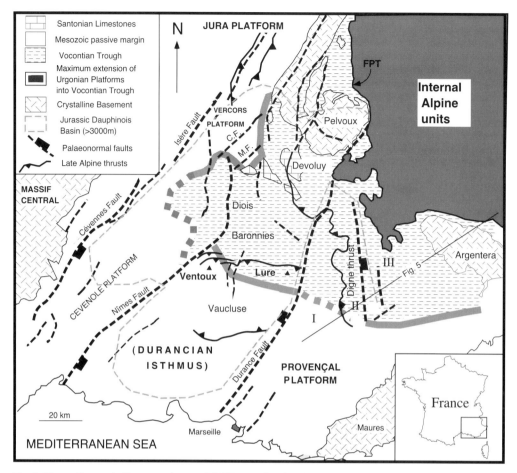

Fig. 2. Non-palinspastically restored map of SE France showing the locations of the principal Mesozoic basin-bounding normal faults and the principal Jurassic and Cretaceous platforms and basins. C.F. Clery fault; M.F. Menée fault; F.P.T. Frontal Pennine thrust I, II and III represent the Provençal-Vocontian facies belts in the Jurassic shown on the section line. Adapted from Jacquin et al. (1991).

massif (Ménard 1979). The crystalline basement is locally overlain by thick Carboniferous and Permian successions (dôme de Barrot) that remained attached to the basement along with the basal Triassic quartzites and red-beds (Elliott et al. 1985). The main Alpine detachment horizon occurs in the overlying Triassic evaporitic succession.

The Mesozoic stratigraphic succession of the western Alps is between 2 and 7 km thick and almost entirely marine comprising largely carbonate sediments deposited across a series of blocks and basins. It has been extensively described, and interpreted as representing the NW Tethyan passive continental margin (e.g. Lemoine et al. 1986; de Graciansky et al. 1987, 1989; Arnaud 1988; Lemoine & de Graciansky 1988; Jacquin et al. 1990). The NE–SW Cevenole fault system, including the Durance, Isère, Nîmes and Cévennes faults largely defined the main depositional areas throughout the Mesozoic (Fig. 2). The largest of these basins, the Jurassic Dauphinois basin, lay between the Cévennes and the Durance faults (Fig. 2; Subalpine basin of Roure et al. 1992; Arnaud 1988). Unconformities of Oxfordian age mark the initiation of Tethyan ocean spreading and thermal subsidence on the passive margin (Arnaud 1988). In the Early Cretaceous the Vocontian basin was the major area of subsidence in SE France (Fig. 2) with thick carbonate platforms (of Urgonian facies) developing to the south (Provençal platform) west and north (Vercors platform; Arnaud 1988; Arnaud-Vanneau &

Arnaud 1990). From the Oxfordian to the Valanginian pelagic sedimentation dominated in the Vocontian trough. The mid-Cretaceous (Albian) was a period of global platform demise, most commonly by drowning, which has been related to the break-up of the southern megacontinent and the initiation of the closure of Tethys (Eberli 1991). Platforms to the north and west of the Vocontian trough were drowned. From the Albian onwards and sedimentation in SE France became more restricted (marnes bleues or Globigerina marl facies), more clastic and strongly influenced by successive deformational phases.

The western half of the main section line crosses three facies belts (Fig. 2), now segmented into thrust sheets; (BRGM 1978, 1981) whose relative positions throughout the Jurassic and Cretaceous were more or less constant. Zone I (Provençal) is characterized by thin lower and middle Jurassic, thickened upper Jurassic limestones, thin (platformal) Cretaceous successions; Zone II (Intermediate) is more or less the same as Zone I; Zone III (Jurassic Dauphinois, Cretaceous Vocontian) is characterized by thick pelagic successions for the lower and middle Jurassic, thin Upper Jurassic (Tithonian) limestones, thick pelagic infill of the Vocontian trough in the lower Cretaceous, thick middle Cretaceous marls and thick Upper Cretaceous limestone succession (Fig. 2).

Kinematics

Fault kinematic data were collected to constrain the direction of thrust motion along the section line. A summary of our data and other available kinematic data is presented in Fig. 3. Unfortunately, due to poor exposure very little data could be collected directly from the main thrusts. However, minor faults and other small-scale criteria can be used to compile a regional overview of the kinematic conditions.

Palaeostress calculations along the St Jurs thrust front made from Cretaceous limestones in the hanging wall (Ritz 1991) indicate a WSW-directed stress along the front (Fig. 3a). Malavieille & Ritz (1989) also record WSW thrust shear in Triassic evaporites along the Digne thrust. Mio-Pliocene sediments in the immediate footwall of the Digne thrust at Cousson, 2 km south of Digne (Fig. 3b1), have abundant slickenside surfaces in overturned strata. These suggest a WSW-directed stress which if occurring prior to the overturning can be unfolded to give a consistent angle with the overlying thrust plane. At Trévans, a large strike-slip fault bounding the northern edge of

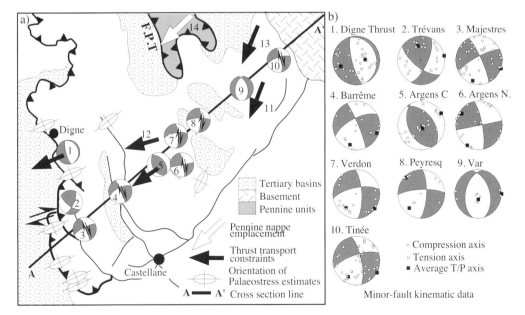

Fig. 3. Summary of kinematic data across the Digne thrust sheet. (**a**) Location of kinematic data showing SW directed thrusting and partitioning of deformation on minor dextral strike-slip faults and WSW thrusting towards the front. 1–11, this study; 12, Platt *et al.* (1989); 13,Graham (1981); 14, Lawson (1987); palaeostress estimates from Ritz (1991). (**b**) Analysis of minor fault slickensides along the line of cross section using Faultkin programme of Allmendinger *et al.* (1989–94).

the St Jurs imbricates (Fig. 3a) has dextral slip indicators implying WSW translation of the St Jurs system to the south of the fault. Minor faults in this fault zone are also consistent with WSW compression along conjugate strike-slip systems (Fig. 3b2). Within the Digne thrust sheet a WSW-directed displacement is also indicated by shear-band and thrust geometries (Platt *et al.* 1989). Extensive shear in the Jurassic and Cretaceous limestones of the Var Valley in the footwall of the Tinée thrust indicate a more SW direction of over-thrust shear. More competent carbonate beds are boudinaged and cut by veins with perpendicular fibre growth showing a NE–SW stretching direction, while the surrounding marls form a strong sub-horizontal bedding-parallel cleavage (Fig. 4). However, a later brittle overprint of slickenside surfaces indicates a more westward direction of stretching strain (Fig. 3b9). Thrust geometries near the Tinée valley (Graham 1981) and in the Nice arc (Malavieille & Ritz 1989) also indicate a SW direction of transport.

Throughout the Digne thrust system the main thrusts dip in a direction between NE and ENE with a range of fold vergence between SW and WSW. This would suggest that tectonic transport towards a SW or WSW direction was dominant; however, the role of strike-slip faulting has also to be considered. Analysis of kinematics on minor fault plane surfaces throughout the Digne thrust sheet indicates a dominance of dextral strike-slip on strike-parallel faults (Fig. 3b). A component of dextral movement has also been suggested for the fault splays immediately below the Digne thrust near Majestres (BRGM 1981; location Fig. 1). This pervasive component of dextral strike slip is hard to quantify, but indicates a more southward component of movement for the eastern part of the sheet. This southward component may in part be related to the south-directed shortening in the Castellane belt to the south. The existence of this strike-slip component may be due to partitioning of a dominantly SW-directed compression on NNW-striking pre-existing Mesozoic faults, (e.g. the Digne thrust, and the faults on the eastern boundaries of the Barrême and Majestres basins; see Fig. 5), into thrusting and dextral strike-slip.

Lying above the Digne thrust sheet, the lower Embrunais–Ubaye nappes comprise internally derived and tectonically interleaved Autapie helminthoid flysch and Subbriançonnais units. These nappes have a very complex internal structure (Kerkhove 1969) and deformational history, and it is not our purpose in this paper to address the problems of the origin and mode of emplacement of these nappes or how much shortening they represent (see Fry 1989 for discussion). However all kinematic data (Schistes à Blocs, footwall ramp orientations, footwall cleavage) show that these nappes were emplaced onto the external zones by SW-directed thrusting (Lawson 1987; Fry 1989; Plotto 1977; J. Bürgisser pers. comm.) and then folded with their external floor before being overthrust by the Parpaillon nappe. Thus the NW component of movement recorded by Merle & Brun (1984) within the thrust stack is unrelated to their emplacement over the external zones.

On the eastern boundary, the external zone stratigraphy is in contact with the Subbriançonnais zones across a steeply dipping fault. A lack of overshear deformation in the footwall of this

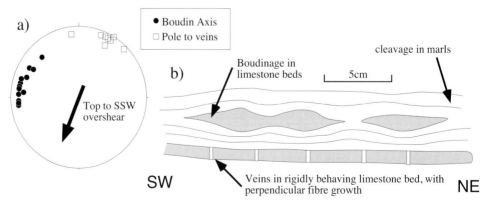

Fig. 4. Strain in the footwall of the Tinée thrust from the Jurassic shale sequence. (**a**) Orientations of boudin axes and stretching veins in more resistant limestone beds. (**b**) field sketch of thecleavage and stretching fabric in marl/limestone beds.

uplift. We believe that this estimate of total shortening does not represent a minimalist solution, but a more reasonable estimate of the actual deformation.

Tertiary stratigraphy and timing of deformation

The chronostratigraphies of the foreland basin remnants which lie along the section line can be integrated (Fig. 6) to show the evolution of the foreland basin of the southern Subalpine chains (adapted from Elliott *et al.* 1985; Pairis 1988; Meckel 1997; Sinclair 1997). The more easterly basin remnants, containing an average stratigraphic thickness of less than 1500 m, contain the typical trinity of the underfilled phase of a peripheral foreland basin, comprising a transgressive limestone facies, deep water hemipelagic marls and deep-water siliciclastic turbidites (Sinclair 1997). We call this the Nummulitic foreland basin. The most easterly remnants of this basin lie on the eastern side of the Argentera massif, in the footwall of the frontal Pennine fault (Fig. 1) but we cannot constrain the original easterly limit of the Nummulitic foreland basin. This phase which occurred principally during the late Eocene, was terminated by the submarine emplacement of the Embrunais–Ubaye nappes in the early Oligocene. The central Barrême basin remnant contains 750 m of strata of late Eocene to late Oligocene age. The Valensole basin to the west preserves 2 km of Early Miocene to Plio-Pleistocene stratigraphy.

The Nummulitic Basin

The base of the Tertiary succession is marked by an unconformity overlain locally by conglomerates derived principally from the underlying Cretaceous limestones, generally believed to be Palaeogene in age. In the study area these are known as the Poudingue d'Argens. They occur sporadically below the Nummulitic limestones across the area. These sediments, which can be up to 500 m thick (e.g. on the eastern limb of the Barrême syncline; Pairis 1971), were deposited by steep-gradient gravel-dominated river systems and appear to fringe palaeohighs (Elliott *et al.* 1985). Locally they record syn-depositional folding (NNW–SSE) as in the Douroulles syncline (see Fig. 8b) or appear to lie in fault-controlled lows (Ravenne *et al.* 1987; Vially 1994). Growth folding in the Poudingue d'Argens is the oldest evidence of deformation with an alpine trend.

In the mid-Eocene transgressive shelf limestones (Nummulitic Limestones, Calcaires Nummulitiques), were diachronously deposited across an uneven erosional surface marking the initiation of loading of the European lithosphere (Apps 1987). These deposits have been dated at Lutetian in the east and Priabonian in the west (Bodelle 1971; Campredon 1977; Pairis 1988). The western limit of the transgression lies somewhere between Barrême and Majestre (Fig. 1; Pairis 1988) and correlates roughly with the location of major facies and thickness changes in the underlying Mesozoic, suggesting that the transgression was impeded from moving further westward by the presence of a palaeogeographic high (the 'bloc crustal de la Durance', Pairis *et al.* 1986). The Nummulitic Limestones are typically 5–50m thick, thinning onto and unconformably overstepping palaeohighs. Locally the limestones record syn-depositional normal faulting principally along 040°–050° trending lineaments of long-lived tectonic activity such as the Var zone (Fig. 1) and the southern margin of the Pelvoux massif (Pairis *et al.* 1986). These deposits also record the development of NW–SE-trending open folds related to thrust activity at depth (Elliott *et al.* 1985).

The Nummulitic Limestones pass rapidly upward into light grey calcareous marls rich in foraminifera known either as the Blue Marls (Marnes Bleues) or the Globigerina Marls (Marnes à Globigérines). Their age is diachronous, being latest mid-Eocene in the east and south and lowest Oligocene in the west (Bodelle 1971; Campredon 1977). Their most westerly occurrence in the study area is in the Barrême basin. In the Annot area, analysis of foraminiferal assemblages indicate that water depth increased upward through the succession reaching a maximum at the base of the Annot Sandstones (Mougin 1978). These marine marlstones chronicle increasing subsidence with time within a starved foreland basin, recording water depths of up to 900 m (Vially 1994; Crampton 1992). The deepest part of the basin is interpreted to be north of the Trois Evêchés area (Fig. 1) where a condensed marl succession was deposited below the carbonate compensation depth (Vially 1994; Ravenne *et al.* 1987). Synsedimentary normal faulting locally affected the Globigerina Marls, for example in the Col de la Cayolle area where over 70 m syn-depositional downthrow is recorded on NW–SE faults (Fig. 1; Elliott *et al.* 1985). Elsewhere the Globigerina Marls show onlapping and thickening geometries recording ongoing gentle fold development below the basin floor (Apps 1987; Elliott *et al.* 1985). Tectonic activity during this time is also recorded by

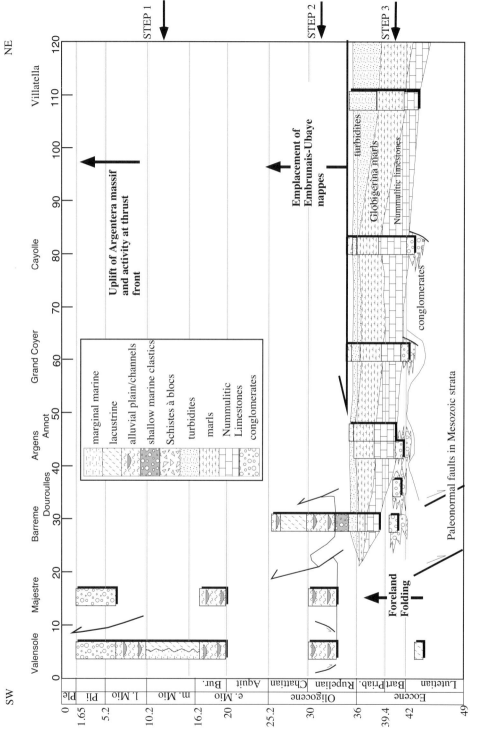

Fig. 6. Chronological diagram of the Tertiary foreland stratigraphy. The Lutetian–Priabonian marine transgression is shown as contemporaneous with conglomeratic deposits associated with growth folding further west. The timing of the main stages of deformation are also indicated.

extensively slumped and slid horizons in the Globigerina Marls.

The top of the Globigerina Marls is marked in many places by a non-erosive disconformity as the increased sedimentation rate of the overlying Annot Sandstones (Grès d'Annot) onlapped and infilled a tectonically induced basin floor topography of up to 250 m (Elliott *et al.* 1985; Sinclair 1994). These siliciclastic turbidites, which can reach 1200 m (Col de la Cayolle, Elliott *et al.* 1985) in thickness, are not well dated due to poor biostratigraphic control, but were mainly deposited from the late Priabonian into the earliest Oligocene. They were deposited by palaeocurrents which flowed to the north to northwest along axes of tectonically controlled sub-basins (Elliott *et al.* 1985; Sinclair 1994). The Annot Sandstones record the onset of clastic sedimentation in the foreland basin due to the emergence of a new source area to the south, usually identified as the Corso-Sardinian massif (Bodelle 1971; Ivaldi 1974). The presence of a second, minor source area in the Argentera massif (Stanley 1965; Jean 1985; Sinclair 1994) is controversial and has been dismissed by Elliott *et al.* (1985) on structural grounds. On the basis of the uplift history of the massif elucidated by fission track analyses (Mansour *et al.* 1992) it must now be concluded that any source in this area was insignificant. Although there is little evidence for local tectonic activity (e.g. active fold growth) within the foreland basin during this time the basin was still subsiding actively, presumably controlled by the orogenic load (Elliott *et al.* 1985). The most proximal facies of the turbidite systems occur at St. Antonin (Fig. 1) with deeper facies appearing northward at Annot, le Grand Coyer and les Trois Evêchés (Fig. 1; Stanley 1975; Elliott *et al.* 1985; Sinclair 1994). Equivalent turbidites of marginal basinal facies are identified in the Barrême basin to the west as the 10 m thick Grès de Ville Formation (Elliott *et al.* 1985; Evans 1987).

Deposition of the Annot Sandstone ceased abruptly in the early Sannoisian with the arrival of the Schistes à Blocs, a tectono-sedimentary mélange comprising a heterolithic assortment of coarse grained conglomeratic and sandstone facies, disrupted to intensely deformed, interbedded with fine micaceous quartz sandstone, siltstones and shales (Kerkhove 1969; Apps 1987). Trace fossils and facies indicate deposition in a deep sea environment (Kerkhove 1969). Material has been derived from the Subbriançonnais, Briançonnais and Helminthoid flysch nappes. Internally the structure of the Schistes à Blocs has been described as grading from olistolithic in character near the base to a structural mélange toward the top (Kerkhove 1969; Apps 1987). This unit underlies the Embrunais–Ubaye nappes everywhere and has been interpreted by Kerkhove (1969) as comprising debris which was shed into the sea floor from the front of these advancing internal nappes.

Thus the Nummulitic foreland basin (Fig. 6) was overidden and 'filled' by internally derived nappes. The timing of emplacement of the Embrunais–Ubaye nappes has been bracketed as ranging from Lutetian to Stampian (upper Rupelian; Fry 1989) and can be further constrained by Sannoisian (lower Rupelian) deposits in the Barrême basin to the west which contain debris derived from these nappes.

Barrême Syncline

In the early Oligocene deposition shifted westward to the Barrême area. The character of the foreland basin changed dramatically from a broad area (> 90 km wide) of relatively homogenous subsidence to a series of much narrower (< 10 km), more localised thrust sheet top basins (Pairis 1988; Debrand-Passard *et al.* 1984) such as Barrême and Dévoluy (Meckel *et al.* 1996). Up to this time both these areas lay on the western edge of the Nummulitic foreland basin. From this time onward the character of sedimentation in these basins was continental.

The Barrême sub-basin (Figs 1 &7) records ongoing deformation from before deposition of the Nummulitic Limestones and throughout basin development (to Chattian; de Graciansky 1972). A detailed sequentially restored cross section through the northern Barrême syncline (Figs 8 & 9) illustrates the early (Nummulitic basin) history of the area. This is the same section line as that presented by Artoni & Meckel (this volume) although our restorations differ. The section line passes through the Clumanc anticline on the eastern limb of the syncline (Fig. 8). During deposition of the Nummulitic Limestone and Globigerina Marls the continuous rotation of the eastern synclinal limb is recorded by growth wedging (Fig. 9e, f). Further south, tectonic activity on the eastern limb of the syncline at this time is recorded by the arrival of submarine glide blocks into the basin. The stratigraphy of the eastern limb is attenuated in comparison to the western limb as suggested by the absence of the Grès de Ville unit (Fig. 9d). In the early Oligocene the eastern limb locked and shortening stepped forward to the Clumanc anticline which we model as a fault propagation fold. This fold was already well developed before deposition of the Upper

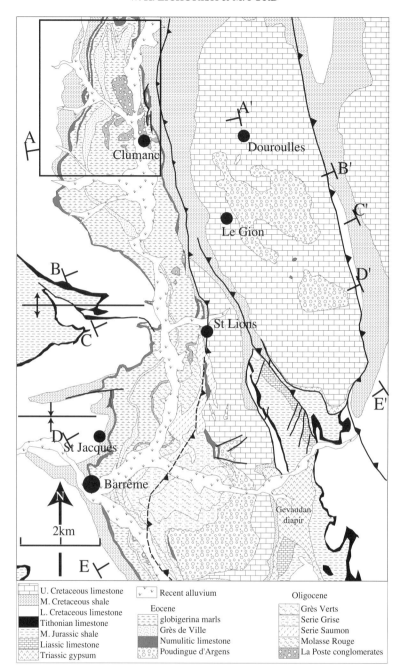

Fig. 7. Geological map of the Barrême syncline showing the cross section lines of Fig. 10. The traces of Cretaceous E–W-trending folds, present on the west side only, are also indicated. (after BRGM 1981). Location of Fig. 8 is shown by the box, and cross sections A to E are in Fig. 10.

Sannoisian La Poste conglomerates during which it developed further to produce the dramatic progressive unconformity across the Barrême syncline (Elliott *et al.* 1985; Figs 8 & 9c, b, a). The stratigraphy on the western limb of the syncline shows no evidence of proximity to a basin margin and therefore we show the basin continuing to the west. The actual western edge

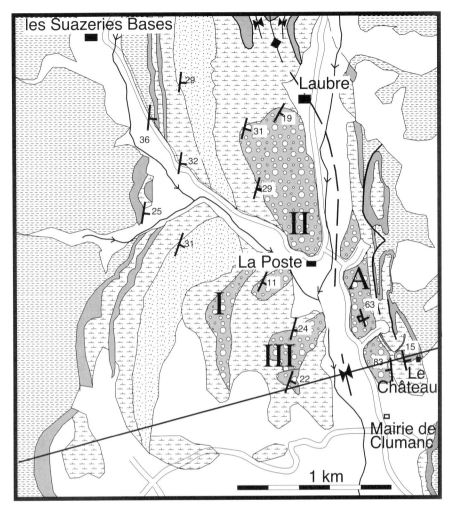

Fig. 8. Detailed map of the northern part of the Barrême syncline showing the trace of the sequentially restored profile of Fig. 9 and the detailed structure of the eastern limb of the syncline. The three La Poste conglomerate units are labelled I, II and III on the western limb of the syncline, on the east the amalgamated conglomerate sequence is labelled A. Key to stratigraphy on Fig. 7.

of the Nummulitic basin was perhaps not more than 10 km to the west. The asymmetry of the Barrême basin and the development of the parasitic Clumanc anticline indicate that compression was migrating westward. A relatively high region certainly lay immediately to the east of the syncline although there is no evidence from the sediments of a major topographic high. The basin must however also have been connected to the larger Nummulitic foreland basin to the east as all the stratigraphic units correlate with Nummulitic basin stratigraphy (Fig 6).

The termination of the marine Nummulitic basin phase of the Barrême syncline is marked by local clastic shoreline deposits (St Lion, and La Poste units) containing ophiolitic fragments, high-pressure minerals and fragments of Helminthoid flysch (Gubler 1958; Chauveau & Lemoine 1961; Bodelle 1971; de Graciansky et al. 1971; Evans & Mange-Rajetzky 1991). These Barrême deposits therefore record the arrival and erosion of the Embrunais–Ubaye nappes to the east. Rupelian–Chattian units, recording the second stage of basin development under continental conditions, are best observed between the villages of St Lions and Barrême (Fig. 7; Molasse Rouge, Série Saumon, Série Grise and Grès Verts). These strata show local growth structures as well as progressive onlap onto the eastern margin of the basin, and a corresponding

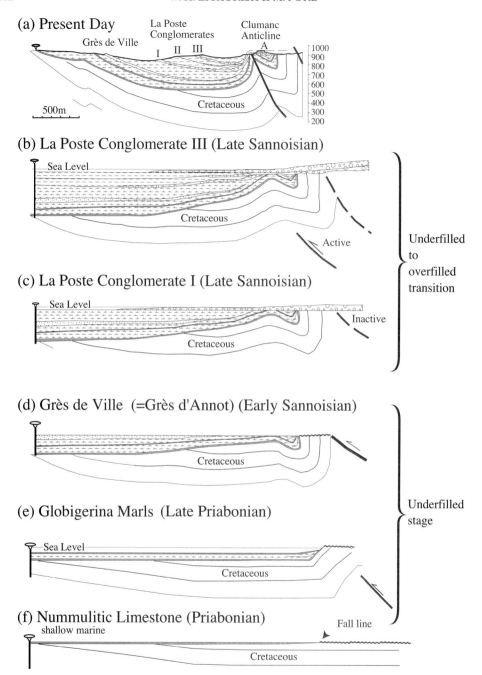

Fig. 9. Sequential restoration of the Clumanc section through the northern Barrême syncline. Deformation in stage (a) corresponds to the regional late basement uplift stage, while stages (b) and (c) are synchronous with the Embrunais–Ubaye nappe emplacement, and stages (e) and (f) are synchronous with the regional foreland folding stage. (**a**) Present-day deformed section showing that the Clumanc anticline tightened further after deposition of the La Poste conglomerates and that the tilting of the western limb of the syncline was also late. (**b**) Late Sannoisian, at the end of the deposition of the La Poste conglomerates. (**c**) Late Sannoisian, during deposition of the first unit of the La Poste conglomerates and concurrent growth of the Clumanc anticline. (**d**) Early Sannoisian, during the deposition of the Grès de Ville turbidites; active growth on the syncline's eastern limb. (**e**) Late Priabonian, basin geometry during deposition of the Globigerina Marls. (**f**) Priabonian, during transgression of the Nummulitic Limestone.

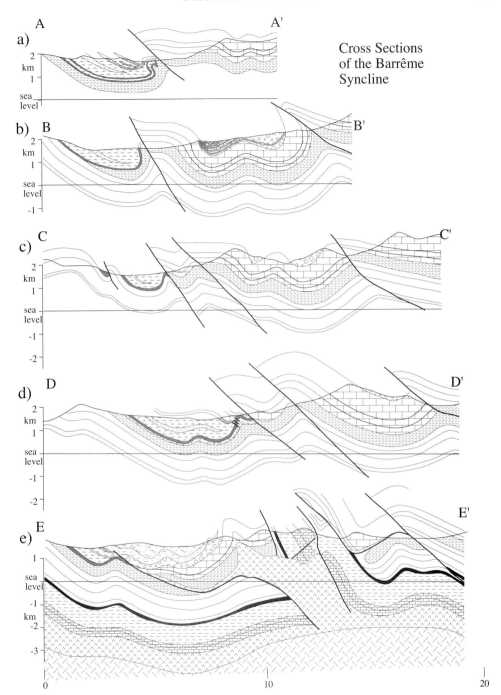

Fig. 10. Structural cross sections through the Barrême syncline (see Fig. 7). (**a**) Clumanc section, showing Oligocene growth on the eastern flank of the basin. Restored sequentially in Fig. 9. (**b**) Douroulles section, showing strong growth of the Douroulles syncline during Palaeogene deposition of the Poudingues d'Argens conglomerates. (**c**) Line of main cross section (Fig. 5). (**d**) St Lions section, showing on-lap of upper Oligocene facies onto the Eocene limestone and the overthrust of Cretaceous limestones onto the eastern margin. (**e**) Barrême section, showing the dramatic eastward shift of the basin depo-centre during the late Oligocene, and the structural complications caused by the intrusion of the Gevaudan evaporite diapir.

eastward migration of the basin depocentre (Fig. 10 d, e) suggesting that at this time the western limb of the syncline began to tilt, forcing the depocentre to shift eastward and become more constricted. We suggest that the underlying Digne thrust became active at this time, translating the basin over a footwall ramp, and thus tilting the western limb.

Valensole–Majestres basins

The Valensole basin lies in the footwall of the frontal thrust and, along with the Tertiary basin remnant at Majestres, contains the youngest foreland sequence of Burdigalian–Pliocene age. Alluvial lower Oligocene (Stampian) conglomerates are present in a small normal-fault bounded basin at Beynes. A depositional hiatus occurred across the foreland during the Aquitanian and sedimentaion resumed only in the Valensole basin in the Burdigalian (Fig. 6). The 2000 m thick Valensole Neogene succession lies directly on Albian limestones and can be divided into two 1000 m thick successions (Debrand-Passard et al. 1984). The lower succession comprises in part fluviatile reddish marls rich in calcareous concretions and with local channelized conglomerates (Burdigalian to Tortonian). From the late Burdigalian to early Tortonian the region was partially covered by a marine transgression depositing sands and marls of a marginal marine facies. During this time the Valensole is believed to have been the deepest and most easterly part of a much larger basin (Debrand-Passard et al. 1984), which extended westward some 60 km.

During the latest Miocene to late Pliocene, activation of the Digne thrust front caused a dramatic change in sedimentation, with dominantly coarse conglomerates (1000 m thick), derived principally from the advancing thrust sheets, filling the Valensole basin.

E–W-trending growth folds at Barles (Fig. 1) involve Upper Miocene to Pliocene strata (Gigot et al. 1974). The Valensole conglomerates are affected by folds (including the Chateauredon dome) of a similar age with fold axes trending NW–SE to NNW–SSE. The Durance fault (Fig. 2) which lies along the western margin of the Valensole basin was reactivated with a west-side-up motion during the late Miocene (Roure et al. 1992) forming a large anticinal uplift which confined the basin to the west. Pliocene sediments are undeformed (Residual basin of Gigot et al. 1974), unconformably overlie the Chateauredon dome and were overthrust by the Digne thrust front at the very end of the Pliocene or early Quaternary (Gigot et al. 1974). The Digne thrust itself was gently folded after emplacement to form the Barles half window and other corrugations in it trace.

Summary of foreland basin evolution

The late Eocene to early Oligocene Nummulitic basin of Haute Provence clearly evolved as part of the flexurally derived foreland basin which developed around the whole alpine arc (Sinclair 1996). As seen elsewhere in the external Alps (e.g. Crampton & Allen 1995), the basin depocentre migrated outward with time. Sinclair (1997) has calculated that the west to southwestward coastal onlap rate was 4.9 to 8.0 mm/a^{-1} in the western Alps. However, unlike the foreland basin of the Central Alps, the character of the foreland basin of the southern Subalpine chains changed dramatically in the early Oligocene. During the Oligocene deposition was confined to relatively small thrust sheet top basins (Barrême and Dévoluy). A depositional hiatus occurred in the Aquitanian across the whole area. From the Burdigalian to the Tortonian the Valensole area was part of a much larger basin which extended to the west. Only in the late Tortonian was basin evolution controlled by the advancing Subalpine thrust front.

Restoration

Using the stratigraphic data on the timing of deformation and foreland basin development, the cross section can be restored sequentially to represent three phases of late alpine deformation (Fig. 5). Each stage is represented by a characteristic style of deformation. Alpine deformation started in the late Eocene after a complex Mesozoic history (Fig. 5e). The section was restored using line-length balancing on the principal limestone horizons, such as the Liassic, Tithonian, Upper Cretaceous and Nummulitic.

Early foredeep phase (Late Eocene–Early Oligocene)

The Alpine foreland basin was initiated by the Lutetian–Priabonian transgression of the Nummulitic Limestone sea from east to as far west as the Digne thrust. Before this transgression NW–SE-trending Alpine structures are recorded by growth in the Poudingue d'Argens conglomerates in the Douroulles syncline. Due to the diachroneity of the transgression it can be envisaged that these conglomerates were deposited in a continental setting synchronously with the development of the marine foreland basin to the east. Throughout the Eocene succession, there is evidence of growth folding and on-lap. This suggests that the foreland basin was divided into a series of sub-basins by gentle folding below the basin floor (Apps 1987). The folds generated in this phase are wide and open with a 5–10 km wavelength (Fig. 5d). No larger structures occur within the Eocene succession, and so most of the main shortening must have occurred later. This deformation propagated as far west as the Digne thrust where, it is suggested, a pre-existing Jurassic normal fault may have cut the Triassic horizon thus blocking further advance of the detachment surface. Although this phase produces a recognisable structural signature it accounts for very little regional shortening, with a total of 1.5 km ($c.$ 1%) regionally across the section. It is suggested that this deformation detached at the Triassic evaporite level.

Nappe emplacement phase (Early–Mid-Oligocene)

The Early Oligocene Schistes à Blocs facies represents the approach and emplacement of the Embrunais–Ubaye nappes toward the SW into the eastern half of the foredeep. Deformation within the Embrunais–Ubaye nappes suggests that there was synchronous deformation in the underlying external cover sequence (Fry 1989). The Tinée thrust and associated structures deform and thus post-date Eocene–early Oligocene deposits. They were later folded by the late Miocene uplift of the Argentera massif thus bracketing their age as Oligocene and probably synchronous with emplacement of the Embrunais–Ubaye nappes. In the footwall of the Embrunais–Ubaye nappes in the Var Valley small scale structures record a flattening and stretching strain (Fig. 4) thinning the strata, while in the Verdon Valley, which lay ahead of the nappe front, small scale deformation above the mid Cretaceous detachment level is in the form of upright folding. These observations indicate that the emplacement of the Embrunais–Ubaye nappes caused synchronous deformation of the external cover.

During this stage of deformation detachment occurred at several levels. The Tinée thrust ramps up from the Triassic evaporite detachment to the mid Cretaceous shales. There is also some additional detachment in both the mid-Jurassic shales and the mid-Cretaceous shales in the footwall of the Tinée thrust, with detachment above the Cretaceous horizon being responsible for the post-Eocene deformation at Argens and in the Verdon valley. Simultaneous with this deformation in the east some deformation was transferred on the Triassic detachment to the folds as far west as the Barrême syncline (Fig. 5c) which records continuous growth throughout the Oligocene.

The emplacement of the Embrunais–Ubaye nappes terminated the Nummulitc basin phase. The nappes 'filled' the basin and the depocentre migrated westwards to Barrême. The Oligocene deposits of the Barrême, Majestres and Valensole basins are largely restricted continental facies and represent a dramatic change in sedimentation pattern from the marine limestone-shale-turbidite sequence of the Nummulitic basin to a succession dominated by fluvial conglomerates and clays.

This phase lasted from the early Sannoisian until at least the Chattian (Fry 1989), and accounts for slightly less than half of the total deformation in the external zones: 9.5 km shortening ($c.$ 9%). However, the emplacement of the Embrunais-Ubaye nappes themselves represents a minimum of about 50 km of SW directed overthrust (and perhaps as much as several hundred) from the internal Alpine zones (Kerkhove 1969).

Basement uplift phase (Late Miocene–Pliocene)

The final phase of deformation in the external zones represents a fundamental change from deformation caused by detachment above the Triassic evaporites, to deeper seated compression causing reactivation of basement faults and uplift of basement blocks and massifs across the region. The late Miocene to Pliocene uplift of the Argentera massif (Mansour et al. 1992) and the dôme de Barrot folded the thrusts and detachments of the Oligocene phase of deformation. At least some of this basement uplift took place on steeply dipping thrusts in Argentera repeating the basement and lower Triassic stratigraphy, and merging into a roof thrust in the mid Triassic evaporites. Contemporaneously with basement uplift, the Digne thrust front was reactivated, shedding conglomerates into the Valensole basin and finally overthrusting the basin in the late Pliocene–Quaternary. Shortening on the basement faults was thus transferred forward on the Triassic detachment to cause deformation at the thrust front (Fig. 5a). Minor reactivation of other basement faults and weaknesses, such as around the dôme de Chateauredon, occurred across the area during the Miocene. Since the dôme de Chateauredon predates most of the deformation in the St Jurs imbricate system (Fig. 5b), basement uplifts in the foreland must have developed out-of-sequence.

This phase accounts for about half of the external zone shortening: 10.5 km shortening (c. 10%). The total shortening in the basement balances with the shortening in the cover sequence above the Triassic detachment even though there are no constraints on individual basement faults.

Regional considerations and conclusions

In this paper a new compilation of stratigraphical and structural data has led to the construction of a three-stage tectonic history for the southern Subalpine chains based on a sequentially restored cross section. The initiation of a very broad (>90 km) foreland basin across SE France in the Late Eocene was marked by a westward migrating marine transgression. Due to the presence of the easy-glide Triassic evaporites, gentle compressional structures formed across the entire width of the basin. This basin continued to develop as a simple underfilled flexural foreland basin (Sinclair 1997) until the Sannoisian. At this time the marine basin was tectonically filled by the emplacement of the Embrunais–Ubaye nappes. Sedimentation stepped westward, became dominated by continental facies and confined to small thrust sheet top basins. The emplacement of the internally derived Embrunais–Ubaye nappes coincided with the first major phase of deformation within the external zones in which all structures can be related to a significant SW-directed overthrust shear. Basement uplifts of varying sizes occurred across the area during the late Miocene and Pliocene. The final Pliocene phase of deformation involved the uplift of the Argentera basement massif with deformation transferred from the basement onto the Triassic detachment. This displacement was transferred to the SW to reactivate the thrust front and cause the Digne thrust and its lateral equivalents at St Jurs, to overthrust the Valensole basin. A two stage evolution of the deformation in the Digne thrust sheet has previously been documented by Ehtechamzadeh-Afchar & Gidon (1974) and Siddans (1979).

All late Alpine deformation of the Digne thrust sheet, including the emplacement of the Embrunais–Ubaye nappes records a SW to WSW directed compression. The total shortening calculated for the external zones below the internally derived nappes (21.5 km) in our restorations is substantially less than that previously estimated (65 km, Elliott et al. 1985).

Minor dextral shear may have been generated by partitioning of SW directed compression on NNW–SSE palaeonormal faults, obliquely reactivated as thrusts. This minor dextral shear moved the thrust sheet southwards causing Alpine tightening of the E–W-trending Upper Cretaceous Pyrenean-Provençal folds in the Castellane belt (Siddans 1979).

The low value of shortening south of Digne correlates well with previously published shortening estimates from the dome de Remollon area to the north (Fig. 1) where approximately 20 km of SW directed shortening has been proposed (Gidon 1975; Gidon et al. 1970). A cross section through the dôme de Remollon (Fig. 11; Siddans 1979) shows that, here the Digne thrust sheet comprises thick Liassic (1200 m) and Dogger of Dauphinois facies describing a dome cored by highly faulted crystalline basement (Mouterde 1961; Gariel 1961; Gidon 1975) overthrust onto a basement high with highly attenuated Mesozoic stratigraphy. The thick Lias and Dogger must derive from the area of deep basement now below the Embrunais nappes to the NE (Fry 1989). In our restoration we interpret the basement block in the core of the dome as having been plucked from the footwall of a palaeonormal fault as the Digne thrust carried the Jurassic basin fill out over the 'Dorsale Dauphinois'. A cleavage-forming penetrative strain increases in intensity toward the NE (Siddans 1979) and is here related to the emplacement of the Embrunais–Ubaye nappes. North of the dôme de Remollon displacement of the Digne thrust abruptly becomes dispersed on a series of splays which carry much thinner Jurassic sequences (Ehtechamzadeh-Afchar & Gidon 1974). Displacement on the main thrust can be traced northward to Dévoluy where a minimum of 3 km westward displacement is recorded (Meckel et al. 1996). Thus the SW–WSW shortening of the Digne thrust system appears to be dispersed and to die out northward.

On the scale of the Alpine orogen it is apposite here to compare the characteristics of the two arms of the external arc. SW directed shortening in the southern Subalpine chains is considerably less than contemporaneous NW directed shortening in the northern Subalpine chains (58 km, Guellec et al. 1990; 100 km, Butler 1984). It also exhibits a different deformation style, being essentially a single thrust sheet rather than an imbricate stack of thrust sheets. SW shortening was largely due to emplacement of the Embrunais–Ubaye nappes onto the foreland basin and to a later phase of basement uplift. In contrast, NW directed thrusting was

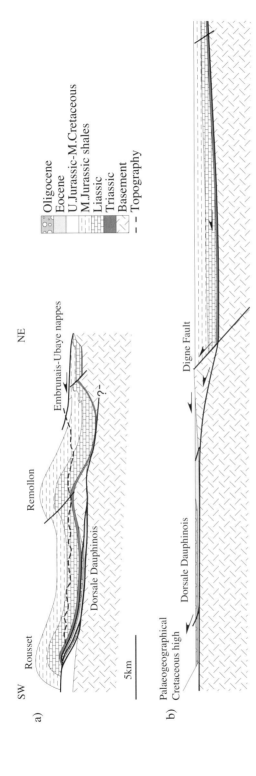

Fig. 11. Cross section through the dôme de Remollon which forms the northern part of the Digne thrust sheet (located on Fig. 1). (a) Present-day deformed state. (b) Restored to end of Cretaceous.

generated principally by the frontal compressional stress system of the orogen. Late out-of-sequence basement uplift also occurred in the northern Subalpine chains (Guellec et al. 1990; D. Seward, pers. comm.). SW shortening of the Digne thrust sheet dies out northward. Thus shortening cannot be continuous around the external arc. These observations have important implications for the evolution of the external Alpine arc. Clearly, deformation was not distributed in a radial fashion around the arc as proposed by Platt et al. (1989). Rather, the arc was generated by two synchronous, but largely independant, orthogonal thrust systems accommodating very different amounts of shortening.

This research was funded by the Swiss Bundesamt für Bildung und Wissenschaft as part of the JOULE II Integrated Basin Studies EU project and by the Swiss National Science Foundation (Project Nr. 2-77-585-92.). We thank E. A. Williams, A. Artoni, L. D. Meckel III, J. Burgisser for their assistence and for discussions on regional tectonics.

References

ALLMENDINGER R. W., MARRETT R. A., & CLADHOUS, T. 1989–94. *FaultKin*, version 3.8a.

ALVAREZ-MARRÓN, J., MCCLAY, K. R., HARAMBOUR, S., ROJAS, L. & SKARMETA, J. 1993. Geometry and evolution of the frontal part of the Magellanes foreland thrust and fold belt (Vicuña area), Tierra del Fuego, Southern Chile. *American Association of Petroleum Geologists Bulletin*, **77**, 1904–1921.

ANADÓN, P., CABRERA, L., COLOMBO, F., MARZO, M. & RIBA, O. 1986. Syntectonic intraformational unconformities in alluvial fan deposits, eastern Ebro Basin margins (NE Spain). *In*: ALLEN, P. A. & HOMEWOOD, P (eds) *Foreland Basins*, Special Publications of the International Association of Sedimentologists, **8**, 259–272.

APPS, G. 1987. *Evolution of the Grés d'Annot basin. SW Alps*. PhD thesis, University of Liverpool.

ARNAUD, H. 1988. Subsidence in certain domains of south-eastern France during the Ligurian Tethys opening and spreading stages. *Bulletin de la Societé géologique de France*, **8**, 725–732.

ARNAUD-VANNEAU, A. H. & ARNAUD, J. 1990. Hauterivian to Lower Aptian carbonate shelf sedimentation and sequence stratigraphy in the Jura and Northern Subalpine Chains (Southeastern France and Swiss Jura). *In*: TUCKER, M. *ET AL*. (eds) *Carbonate Platforms. Facies, Sequences and Evolution*. Special Publication, International Association of Sedimentologists, **9**, 203–233.

ARTONI, A. & MECKEL, C. D. III 1998. History and deformation rates of a thrust-steel-top basin. The Barrême Basin, western Alps, SE France. *This volume*.

BEAUMONT, C. 1981. Foreland basins. *Geophysical Journal of the Royal Astronomical Society*, **65**, 291–329.

BODELLE, J. 1971. *Les Formations nummulitiques de l'arc de Castellane*, Thèse, Université de Nice, France.

BRGM 1967. *Carte géologique de la France au 1:50,000, feuille 919 Allos*. Bureau des Recherches Géologiques et Minières, Orléans.

—— 1970. *Carte géologique de la France au 1:50,000, feuille 920 St. Etienne de Tinée*. Bureau des Recherches Géologiques et Minières, Orléans.

—— 1978. *Carte géologique de la France au 1:50,000, feuille 970 Moustiers Ste. Marie*. Bureau des Recherches Géologiques et Minières, Orléans.

—— 1980. Carte géologique de la France au 1:50,000, feuille 945 Entrevaux. *Bureau des Recherches Géologiques et Minières*, Orléans.

—— 1981. Carte géologique de la France au 1:50,000, feuille 944 Digne. *Bureau des Recherches Géologiques et Minières*, Orléans.

BURBANK, D. W. & VERGÉS, J. 1994. Reconstruction of topography and related depositional systems during active thrusting. *In*: ELLIS, M. A. & MERRITTS, D. J. (eds) *Tectonics and topography Part 3. Journal of Geophysical Research*, **99**, 20281–20297.

BÜRGISSER, J., FORD, M. & MECKEL, L. D. 1995. Kinematics of the western Alpine arc: relations and implications of SW and NW transport directions around the Pelvoux massif. *In: Abstracts, Tectonic Studies Group, 26th Annual Meeting*.

BUTLER, R. W. H. 1984. Balanced cross-sections and their implications for the deep structure of the northwest Alps: Reply. *Journal of Structural Geology*, **6**, 607–612.

CAMPREDON, R. 1977. Les formations Paléogènes des Alpes Maritimes franco-italiennes. *Memoires de la Société Géologique de France*, **9**.

CHAUVEAU, J. C. & LEMOINE, M. 1961. Contribution a l'etude geologique du synclinal tertiare de Barrême (moitie nord). *Bulletin du Service de la Carte Géologique de la France*, **264**, 147–178.

CRAMPTON, S. L. E. 1992. *Inception of the Alpine foreland basin: basal unconformity and Nummulitic limestone*. PhD thesis, University of Oxford.

—— & Allen, P. A. 1995. Recognition of forebulge unconformities associated with early-stage foreland basin development: example from the north Alpine foreland basin: *American Association of Petroleum Geologists Bulletin*, **79**, 1495–1514.

DE GRACIANSKY, P. C. 1972. Le bassin tertaire de Barrême (Alpes de Haute Provence): relations entre déformations et sédimentation; chronologie des plissements. *Comptes Rendus de l'Academie des Sciences, Paris*, **272D**, 2825–2828.

——, BUSNARDO, R., DOUBLET, R. & MARTINOD, J. 1987. Tectogenèse distensive d'âge crétacé inférieur aux confins des Baronnies (chaines subalpines méridionales); liaison avec le rifting atlantique; conséquences sur la tectonique alpine. *Bulletin de la Societé géologique de France*, **8**, 1211–1214.

——, DARDEAU, G., LEMOINE, M. & TRICART, P. 1989. The inverted margin of the French Alps and foreland basin inversion. *In*: COOPER, M. A. & WILLIAMS, G. D. (eds) *Inversion Tectonics*.

Geological Society, London, Special Publications, **44**, 87–104.
——, LEMOINE, M. & SAILOT, P. 1971. Remarques sur la présence de minéraux et de paragénèses du métamorphisme alpin dans les galets des conglomérats Oligocènes du synclinal de Barrême (Alpes-de-Haute-Provence). *Compte Rendus de l'Academie des Sciences, Paris,* **272D**, 3243–3245.
DEBRAND-PASSARD, S., COURBOULEIX, S. & LIENHARDT, M. J. 1984. *Synthèse géologique du sud-est de la France.* Mémoires du Bureau des Récherches Géologique et Minières, **125**, **126**.
DECELLES, P. G., GRAY, M. B., RIDGWAY, K. D., COLE, R. B., PIVNIK, D. A., PEQUERA, N. & SRIVASTAVA, P. 1991. Controls on synorogenic alluvial-fan architecture, Beartooth Conglomerate (Palaeocene), Wyoming and Montana. *Sedimentology,* **38**, 567–590.
EBERLI, G. P. 1991. Growth and demise of isolated carbonate platforms; Bahamian controversies. *In*: MÜLLER, D. W., MCKENZIE, J. A. & WEISSERT, H. (eds), *Controversies in Modern Geology.* Academic Press, 231–248.
EHTECHAMZADEH-AFCHAR, M. & GIDON, M. 1974. Données nouvelles sur la structure de l'extrémité nord de la zone de chevauchements de Digne. *Géologie Alpine,* **276**, 57–69.
ELLIOTT, T., APPS, G., DAVIES, H., EVANS, M., GHIBAUDO, G. & GRAHAM, R. H. 1985. A structural and sedimentological traverse through the Tertiary foreland basin of the external Alps of South-East France. *In*: ALLEN, P., HOMEWOOD, P. & WILLIAMS, G. (eds) *International Symposium on Foreland Basins.* International Association of Sedimentologists, 39–73.
EVANS, M. J. 1987. *Tertiary sedimentology and thrust tectonics in the southwest Alpine foreland basin, Alpes de Haute-Provence, France.* PhD thesis, University of Wales, Swansea.
—— & MANGE-RAJETZKY, M. A. 1991. The provenance of sediments in the Barreme thrust-top bain, Haute-Provence, France. *In*: MORTON, A. C., TODD, S. P. & HAUGHTON, P. D. W. (eds) *Developments in Sedimentary Provenance Studies.* Geological Society, London, Special Publications, **57**, 323–342.
FLANDRIN, J. 1966. Sur l'âge des principaux traits structuraux du Diois et des Baronnies, *Bulletin de la Société géologique de France (7),* **8**, 376–386.
FLEMINGS, P. B. & JORDAN, T. E. 1990. Stratigraphic modeling of foreland basins: interpreting thrust deformation and lithosphere rheology. *Geology,* **18**, 430–434.
FORD, M. & STAHEL, U. 1995. The geometry of a deformed carbonate slope-basin transition: The Ventoux-Lure fault zone, SE France. *Tectonics,* **14**, 1393–1410.
——, WILLIAMS, E. A., ARTONI, A., VERGÉS, J. & HARDY, S. 1997. Progressive evolution of a fault-related fold pair from growth strata geometries, Sant Llorenç de Morunys, SE Pyrenees. *Journal of Structural Geology,* **13**, 413–442.
FRY, N. 1989. Southwestward thrusting and tectonics of the western Alps. *In*: COWARD, M. P., PARK, R. G. & DIETRICH, D. (eds) *Alpine Tectonics,* Geological Society, London, Special Publications, **45**, 83–109.
GARIEL, O. 1961. Le Lias du Dôme du Remollon (Hautes-Alpes). *Mémoires du Bureau de Recherches géologique et minières (France),* **4**, 697–706.
GIDON, M. 1975. Sur l'allochthone du "Dome de Remollon" (Alpes françaises du sud) et ses conséquences. *Comptes Rendus de l'Académie des Sciences, Paris,* **280D**, 2829–2832.
——, ARNAUD, H., PAIRIS, J. L., APRAHAMIAN, J. & USELLE, J. P. 1970. Les déformations tectoniques superposées du Devoluy meridional (Hautes-Alpes). *Géologie Alpine,* **46**, 87–110.
——, PAIRIS, J. L. & APRAHAMIAN, J. 1976. Le lineament d'Aspres-les-Corps et sa signification dans le cadre de l'evolution structurale des Alpes occidentales externes. *Comptes Rendus de l'Académie des Sciences Paris, Serie 2,* **282**, 271–274.
GIGOT, P., GRANDJACQUET, C. & HACCARD, D. 1974. Evolution tectono-sedimentaire de la zone septentrionale du bassin tertiaire de Digne depuis l'Eocene. *Bulletin de la Sociéte Géologique de France,* **16**, 128–139.
GOGUEL, J. 1963. Les problèmes des chaînes subalpines. *In*: DUVAND, DELGA M. (ed.) *Livre à la Memoire du Professeur Paul Fallot,* Vol. II Mémoires, H. S. Sociéte géologique de France, 301–307.
GRAHAM, R. H. 1978. Wrench faults, arcuate fold patterns and deformation in the southern French Alps. *Proceedings of the Geologists' Association,* **89**, 125–42.
—— 1981. Gravity sliding in the Maritime Alps, *In*: COWARD, M. P. & MCCLAY, K. (eds) *Thrust and Nappe Tectonics.* Geological Society, London Special Publications, **9**, 335–352.
GUBLER, Y. 1958. Etude critique des sources de matérial constituant certaines séries détritiques dans le tertiaire des Alpes françaises du Sud: formations détritiques de Barrême, flysch d'Annot. *Eclogae Geologicae Helvetiae,* **51**, 942–977.
GUELLEC, S., MUGNIER, J.-L., TARDY, M. & ROURE, F. 1990. Neogene evolution of the western Alpine foreland in the light of ECORS-data and balanced cross sections, *In*: ROURE, F., HEITZMANN, P, & POLINO, R. (eds) *Deep structure of the Alps.* Mémoires de la Societé géologique de France, **156**, 165–184.
HAYWARD, A. B. & GRAHAM, R. H. 1989. Some geometrical characteristics of inversion. *In*: COOPER, M. A. & WILLIAMS, G. D. (eds) *Inversion Tectonics.* Geological Society, London, Special Publications, **44**, 17–40.
IVALDI, J. P. 1974. Origines du matériel détritique des Séries Grès d'Annot d'après les données de la thermoluminescence. *Géologie Alpine,* **50**, 75–78.
JACQUIN, T., VAIL, P. R., ARNAUD, H., DARDEAU, G., GRACIANSKY, P. C. DE, LEMOINE, M., MAGNIEZ-JANNIN, F., MARCHAND, D. & RAVENNE, C. 1990. Stratigraphic signatures of the tectonic and eustatic effects during the post-rift history of the Tethyan margin in the southern Vercors (France). *American Association of Petroleum Geologists Bulletin,* **74**, 683–684.

JEAN, S. 1985. *Les Grès d'Annot au N-W du Massif de l'Argentera-Mercantour.* Thèse de l'Université Scientifique et Médicale de Grenoble.

KARNER, G. D. & WATTS, A. B. 1983. Gravity anomalies and flexure of the lithosphere at mountain ranges. *Journal of Geophysical Research,* **88**, 10 449–10 477.

KERKHOVE, C. 1969. La "zone du Flysch" dans les nappes de L'Embrunais-Ubaye (Alpes Occidentales). *Géologie Alpine,* **45,** 5–204.

LAWSON, K. 1987. *Thrust geometry and folding in the Alpine structural evolution of Haute Provençe.* Ph.D. thesis, University of Wales, Swansea.

LEMOINE, M. 1973 About gravity gliding tectonics in the western Alps. *In:* DE JONG, K. & SCHOLTEN, R. (eds) *Gravity and Tectonics.* Wiley, New York, 201–216.

—— & DE GRACIANSKY, P. C. 1988. Marge continentale téthysienne dans les Alpes. *Bulletin de la Societé géologique de France,* **8**, 597–600.

——, BAS, T., ARNAUD-VANNEAU, A., ARNAUD, H., DUMONT, T., GIDON, M., BOURBON, M., DE GRACIANSKY, P.C., RUDKIEWICZ, J. L., MEGARD-GALLI, J. & TRICART, P. 1986. The continental margin of Mesozoic Tethys in the western Alps. *Marine and Petroleum Geology,* **3**, 179–199.

MALAVIEILLE, J. & RITZ, J. F. 1989. Mylonitic deformation of evaporites in décollements: examples from the Southern French Alps. *Journal of Structural Geology,* **11**, 583–590.

MANSOUR, M., POUPEAU, G., BOGDANOFF, S., MICHARD, A. & TANE, J. L. 1992. Apatite fission track dating and uplift of the Argentera Massif, western Alps (France, Italy). *In: Geoatelier Alpin.* Géologie Alpine, Série spéciale Resumés de colloques, **1**, 67.

MECKEL, L. D. 1997. *Sedimentological and Structural evolution of the Tertiary Dévoluy Basin, external western Alps, SE France.* PhD thesis, ETH-Zürich.

——, FORD, M. & BERNOULLI, D. 1996. Tectonic and sedimentary evolution of the Dévoluy Basin, a remnant of the Tertiary western Alpine foreland basin. *Géologie de France,* **1996/2**, 3–26.

MÉNARD, G. 1979. *Relations entre structures profondes et structures superficielles dans le sud-est de la France: essai d'utilisation de données géophysiques.* Thèse de 3ème cycle, Université de Grenoble.

MERLE, O. & BRUN, J. P. 1984. The curved translation path of the Parpaillon Nappe (French Alps). *Journal of Structural Geology,* **6**, 711–719.

MOUGIN, F. 1978. *Contribution à l'étude des sédiments tertiaires de la partie orientale du synclinal d'Annot.* Thèse de 3ème cycle, Grenoble, France, 167pp.

MOUTERDE, R. 1961. Variations du Lias supérieur entre Digne, Gap et Castellane. *Mémoires du Bureau des Recherches Géologiques et Minières* (France), **4**, 715–718.

PAIRIS, J.-L. 1971. Tectonique et sedimentation tertiaire sur la marge orientale du bassin de Barrême. *Géologie Alpine,* **47**, 203–214.

—— 1988. *Paléogène marin et structuration des Alpes occidentales Françaises (domaine externe et confins sud-occidentaux du Subbriançonnais.* Thèse de Docteur de Sciences, Université Joseph-Fourier, Grenoble.

——, GIDON, M., FABRE, P. & LAMI, A. 1986. Signification et importance de la structuration nummulitique dans les chaînes subalpine méridionales. *Comptes Rendus de l'Académie des Sciences, Paris,* **303**, 87–92.

PLATT, J. P., BEHRMANN, J. H., CUNNINGHAM, P. C., DEWEY, J. F., HELMAN, M., PARISH, M., SHEPLY, M. G., WALLIS, S. & WESTON, P. J. 1989. Kinematics of the Alpine arc and the motion history of Adria. *Nature,* **337**, 158–161.

PLOTTO, P. 1977. *Structures et déformations des Grès du Champsaur au SE du massif du Pelvoux.* PhD thesis, University of Grenoble.

PUIGDEFABRIGAS, C., MUÑOS, J. A. & VERGÉS, J. 1992. Thrusting and foreland basin evolution in the southern Pyrenees. *In:* MCCLAY, K. R. (ed.) *Thrust Tectonics.* Chapman & Hall, 247–254.

RAVENNE, C., VIALLY, R., RICHÉ, P. & TRÉMOLIÈRES, P. 1987. Sédimentation et tectonique dans le bassin marin Eocène supérieur-Oligocène des Alpes du Sud. *Revue de l'Institut Français du Pétrole,* **42**, 529–553.

RIBA, O. 1976. Syntectonic unconformities of the Alto Cardener, Spanish Pyrenees: a genetic interpretation. *Sedimentary Geology,* **15**, 213–233.

RICOU, L.E. & DE LA MOTTE, F. 1986. Décrochement senestre médio-cretacé entre Provence et Alpes-Maritimes (Alpes occidentales, France). *Revue de Géologie Dynamique et de Géographie Physique,* **27**, 237–245.

RICOU, L. E. & SIDDANS, A. W. B. 1986. Collision tectonics in the Western Alps. *In:* COWARD, M. P. & RIES, A. C.(eds) *Collision tectonics.* Geological Society, London, Special Publications, **19**, 229–244.

RITZ, F. 1991. *Evolution du champ de contraintes dans les Alpes du Sud depuis la fin de l'Oligocène. Implications sismotectoniques.* PhD thesis, Montpellier II.

ROURE, F., BRUN, J.-P., COLLETTA, B. & VAN DEN DRIESSCHE, J. 1992. Geometry and kinematics of extensional structures in the Alpine Foreland Basin of southeastern France. *Journal of Structural Geology,* **14**, 503–520.

SGI 1970. *Carta geologica d'Italia a 1:100,000 Folio 90, Demonte.* Servizio Geologico d'Italia, Rome.

—— 1971. *Carta geologica d'Italia a 1:100,000 Folio 79, Argentera-Dronero,* Servizio Geologico d'Italia, Rome.

SIDDANS, A. W. B. 1979. Arcuate fold and thrust patterns in the subalpine chains of southeast France. *Journal of Structural Geology,* **1,** 117-126.

SINCLAIR, H. D. 1994. The influence of lateral basinal slopes on turbidite sedimentation in the Annot Sandstones of SE France. *Journal of Sedimentary Research,* **A64**, 42–54.

—— 1996. Plan-view curvature of foreland basins and its implications for the palaeostrength of the lithosphere underlying the western Alps. *Basin Research*, **8**, 173–182.

—— 1997. Tectono-stratigraphic model for underfilled peripheral foreland basins: an Alpine perspective. *Bulletin of the Geological Society of America*, **109**, 324–346.

—— & ALLEN, P. A. 1992. Vertical vs. horizontal motions in the Alpine orogenic wedge: Stratigraphic response in the foreland basin. *Basin Research*, **4**, 215–233.

STANLEY, D. J. 1965. Heavy minerals and provenance of sands in flysch of central and southern French Alps. *American Association of Petroleum Geologists Bulletin*, **49**, 22–40.

—— 1975. Submarine canyon and slope sedimentation (Grès d'Annot) in the French maritime Alps. *In*: *9th International Congress of Sedimentology, Abstracts*.

SUPPE, J., CHOU, T. G. & HOOK, S. C. 1992. Rates of folding and faulting determined from growth strata. *In*: MCCLAY, K. R. (ed.) *Thrust Tectonics*. Chapman & Hall, London, 105–121.

VANN, I., GRAHAM, R. H. & HAYWARD, A. B. 1986. The structure of mountain fronts. *Journal of Structural Geology*, **8**, 215–228.

VERGÉS, J., BURBANK, D. W. & MEIGS, A. 1996. Unfolding: an inverse approach to fold kinematics. *Geology*, **24**, 175–178.

VIALLY, R. 1994. The southern French Alps Palaeogene basin: subsidence modelling and geodynamic implications. *In:* MASCLE, A. (ed.) *Hydrocarbon and Petroleum Geology of France*. Special Publication of the European Association of Petroleum Geoscientists, **4**, 367–380.

WICKHAM, J. 1995. Fault displacement-gradient folds and the structure at Lost Hill, California (U.S.A.). *Journal of Structural Geology*, **17**, 1293–1302.

ZAPATA, T. R. & ALMENDINGER, R. W. 1996. Growth stratal records of instantaneous and progressive limb rotation in the Precordillera thrust belt and Bermejo basin, Argentina. *Tectonics*, **15**, 1065–1083.

History and deformation rates of a thrust sheet top basin: the Barrême basin, western Alps, SE France

ANDREA ARTONI & LAWRENCE D. MECKEL, III

Geologisches Institut, ETH-Zentrum, 8092 Zürich, Switerland

Abstract: The geometrical evolution of the 2 km wide Barrême thrust-sheet-top basin has been established over a time interval of 10 Ma. The Barrême basin developed during the late Eocene–Oligocene, during the migration from east to west of the western Alpine orogenic front in SE France. Unlike in other orogenic belts, where foreland basins are often buried and observable only on seismic profiles, the outcrop conditions at Barrême allow the detailed analysis of this basin fragment. The syntectonic sedimentation and growth strata preserved on the east flank of the Barrême basin, on the Clumanc Anticline, are used to restore sequentially a balanced cross-section, thereby constraining the progressive steps of basin development. Numerical ages, obtained by correlating biozones and chronostratigraphy with most recent geochronological calibrations, are assigned to the bases of the mappable growth strata in order to obtain an indication of deformation rates. Shortening rates vary from 0.003 to 0.2 (±0.083) mm a^{-1}, while uplift rates vary from 0 to 0.37 mm a^{-1}. These rates are comparable to those recorded in recently active mountain belts, and represent a significant refinement of previously published rates for the Barrême basin.

The Barrême basin is one of many fragments of the Eocene–Miocene western Alpine foreland basin which outcrop in the hanging wall of the southwest-directed Digne thrust sheet (Elliott *et al.* 1985) (Fig. 1). The Tertiary sediments in the Barrême basin seal a pre-Priabonian palaeotopography, associated with the erosion of large volumes of Mesozoic sediment (Fig. 2). These sediments, upper Eocene to Oligocene in age, expand from east to west as a series of growth strata.

The growth strata on the eastern flank of the Barrême basin record significant syndepositional tectonic activity during the early Oligocene (Evans 1987). Similar growth strata in the Annot basin, 25 km east of Barrême, are pre-Oligocene (Gigot *et al.* 1975; Guillemot *et al.* 1981; de Graciansky *et al.* 1982; Elliott *et al.* 1985; Evans & Mange-Rajetzky 1991) and related to extensional tectonics (Ravenne *et al.* 1987; Vially 1994); whereas, west of the Barrême basin, at the front of the Digne thrust sheet, the oldest growth strata preserved are Mio-Pliocene in age (Elliott *et al.* 1985; Evans 1987), following the Oligocene rifting located in the Barronies–Dios region (Fig. 1) , and on the southern extension of the Bresse graben (Bergerat *et al.* 1990; Guellec *et al.* 1990). Therefore, the east flank of the Barrême basin represents a portion of the incipient Oligocene mountain front in the western Alpine foreland basin. This Oligocene mountain front is growing inside regions with extensional tectonics, similarly to compressive structures within foreland area (Alvarez-Marrón *et al.* 1993; Argnani *et al.* 1993*a*,*b*; Argnani 1996).

Studying the Barrême basin provides an opportunity to observe an outcropping thrust sheet top basin with growth strata. Such strata are buried or are deeply eroded in other mountain fronts such as the central and northern Apennines of Italy and the Neuquin Basin of Argentina (Pieri & Groppi G. 1981; Ricci Lucchi 1986; Beer *et al.* 1990).

Present-day mountain fronts are some of the most active areas on the Earth's surface. The concentration of seismic activity produces differential topographic movements (Bolt 1988), which cause accelerated erosion and landscape modification (Fielding *et al.* 1994; Burbank *et al.* 1996; Keller & Pinter 1996). The processes active in ancient mountain fronts such as Barrême should be comparable to those observed in modern fronts.

In order to constrain this comparison, and in order to understand better the incremental steps of topographic evolution, the geometrical reconstruction of the tectonic evolution, the rates of deformation, and the relationship between deformation and sedimentation experienced by the Barrême basin during the latest Eocene through Oligocene are investigated in this paper. The growth strata are used to reconstruct the differential movements of the preserved Oligocene mountain front and its paleotopography in northern Barrême. Palaeotopographic remnants have already been recognized in the southern part of the Barrême basin (Pairis 1971), but never in the northern portion.

The history of this portion of the Oligocene mountain front in the western Alpine foreland

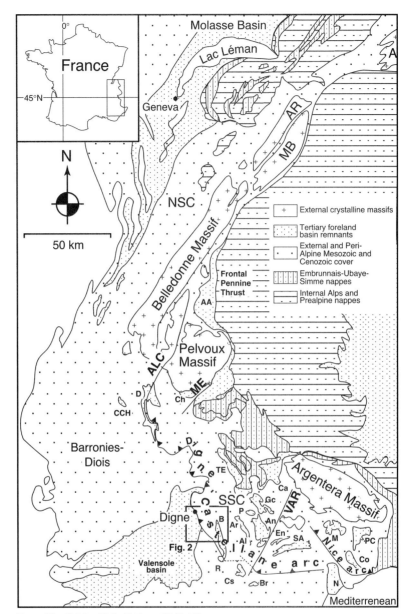

Fig. 1. Location of the Barrême basin in the external Alps of SE France foreland basin. The Barrême basin is part of the Digne Castellane arc. The Digne–Castellane arc and the Nice arc are the two main arcs forming the external Alpine fold and thrust belt of SE France. The box shows the area of Fig. 2.

basin is unraveled by progressively restoring syntectonic unconformities. The methodology of such restorations is described by (DeCelles et al. 1990; Burbank & Verges 1994; DeCelles 1994; Artoni & Casero P. 1997). Numerical ages of the Tertiary stratigraphic units in Barrême are determined by calibrating published biostratigraphic data with the updated Eocene and Oligocene chronologic scales of (Berggren et al. 1992; Odin 1994). From these ages, the velocity of each stage of deformation, as well as the shortening and uplift rates, are calculated.

Regional structures of the western Alpine foreland basin in the vicinity of the Barrême basin

In the external fold and thrust belt of the western Alps in SE France, Eocene–Miocene sediments of the foreland basin follow the arcuate shape of the external Alps (Fig. 1). Two main arcs exist south of the Pelvoux massif: the Digne–Castellane arc and the Nice arc (Fig. 1). The Argens, Allos, Annot, Entrevaux, Barrême, and St Antonin basins trace the shape of the Digne–Castellane arc, and are infilled with coeval Eocene–Oligocene sedimentary deposits.

The Digne–Castellane arc is also parallelled by the folds axes of two deformation phases: W–E-oriented Pyrénéo-Provençale folds parallel the southern W–E oriented portion of the arc (Beaudoin et al. 1975; de Graciansky et al. 1982) and N–S-oriented Alpine folds (Chauveau & Lemoine 1961; Beaudoin et al. 1975; de Graciansky et al. 1982) parallel the N–S-oriented western portion of the arc. The Pyrénéo-Provençale folds are upper Cretaceous–Eocene in age, while the Alpine folds have recognized ages of Early Oligocene–Aquitanian, Mid-Miocene, Late Miocene–Pliocene, and Quaternary (Ritz 1991). The end of the Pyrénéo-Provençale and the beginning of the Alpine deformation is overlapped by the late Eocene–Oligocene extensional event (Ravenne et al. 1987), active to the west and east of the Barrême basin. We will show that an additional Alpine deformation pulse during the Late Eocene–Early Oligocene affected the Barrême basin.

Evidence of both folding phases is preserved around the western Alpine foreland basin. In Barrême, Pyrénéo-Provençale fold axes are clearly observed on the western flank of the basin, but are not present on the eastern flank (Fig. 2a). Instead, here, Alpine fold axes exist (Chauveau & Lemoine 1961; Guillemot et al. 1981) (Fig. 2a). The Barrême syncline is itself also an Alpine structure. No W–E-oriented fold axes occur in the northern portion of the Barrême basin, the subject of this study.

Structural geology and geometry of the northern Barrême basin

The thrust sheet top basin of Barrême lies above the Digne thrust sheet and is aligned with the N–S front of the thrust sheet (Fig. 2). The basin is located west of the Douroulles syncline and the Gèvaudan–Reichard–Blanche–Castellane tectonic units (Fig. 2). The northern splays of the Gèvaudan–Reichard–Blanche–Castellane tectonic units are east of the Clumanc Anticline (Guillemot et al. 1981). The main detachment surface of the Digne thrust sheet is in the Triassic salts, 2 km below the Eocene–Oligocene infill (Fig. 2b). These salts crop out immediately east of Barrême as the Gévaudan salt diapir (Fig. 2a). The 2 km thick Mesozoic stratigraphy below the Barrême syncline does not result from tectonic repetition, as proposed by Hayward & Graham (1989), but rather represents the total thickness of the Mesozoic deposited in the Vocontian domain (de Graciansky et al. 1982).

The northern portion of the Barrême basin is a 2 km wide open syncline (Fig. 3). This syncline has a calculated average orientation of N175°–11°. At the northern termination of the basin, the Nummulitic Limestone shows second-order folding, with axis parallel to the first order syncline axis, related to thrusting in the east limb of the Barrême syncline (Fig. 3). Second-order parasitic folds, about 40 m in amplitude, also affect the more competent units of the Grès de Ville formation and La Poste conglomerate bodies. These parasitic folds formed during the formation of the larger syncline as a result of competence contrasts between the marls and the sandstones (de Lapparent 1938). One of the secondary folds has an average axial orientation of N164°–10°, roughly parallel to the axis of the first-order syncline.

Faults disrupt the eastern flank of the Barrême syncline (Fig. 3). These faults reached the topographic surface after the Oligocene, as deposits of this age are displaced about 25 m on the western limb of the Clumanc Anticline. The Eocene and Oligocene growth strata indicate that the uplift of the Clumanc Anticline was syn-depositional. Moreover, the steep western limb of the Clumanc Anticline and the high cut-off angle (57°) of the thrust cutting the hinge of the Clumanc Anticline are indicative of the formation of this fold as a fault-propagation fold (terminology of Suppe & Medwedeff 1990 and

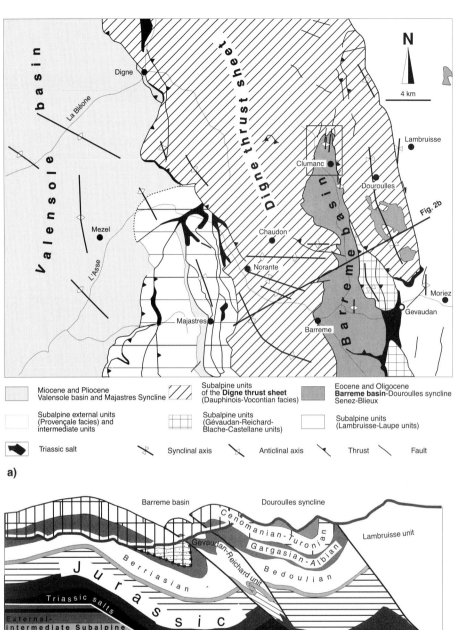

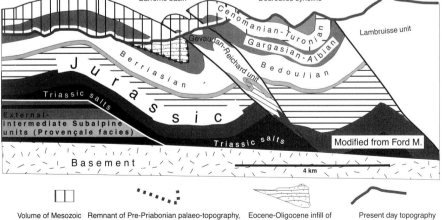

geometrical constraints of Suppe 1985). Additionally, regional cross-sections (Elliott *et al.* 1985; Coward *et al.* 1991) interpret the fold on the east flank of the Barrême basin as a fault-propagation fold. Therefore, we assume that the faults were blind during earlier stages of growth.

Stratigraphy of the Barrême basin

The sedimentary history of the Barrême basin is well studied (Chauveau & Lemoine 1961; de Lapparent 1966; Bodelle 1971; de Graciansky *et al.* 1982; Elliott *et al.* 1985; Evans 1987; Evans & Mange-Rajetzky 1991). Only terminology modifications on the rank of lithostratigraphic units have been introduced here.

The Tertiary stratigraphic succession of the Barrême syncline (Figs 3–5) starts with the Priabonian Nummulitic Limestone formation, a shallow marine transgressive unit (Evans 1987). It is a 6 m thick carbonate bank, uncomformably overlaying Cretaceous limestones and shales (Figs 4 & 5), rich in nummulites on the east side of the basin. The stratigraphic gap at the base of the Nummulitic limestone varies around the basin (Fig. 3) (Evans 1987).

The Blue Marls formation continues the transgressive and deepening trend initiated with the Nummulitic Limestone. Classically, the Blue Marls include a nummulites-rich member, as well as the siliciclastic sediments of the Grès de Ville and La Poste members (de Graciansky *et al.* 1982; Elliott *et al.* 1985; Evans 1987; Evans & Mange-Rajetzky 1991). However, lithological differences and mappability at 1:10 000 scale allows these members to be redefined as separate formations. We therefore define the Blue Marls Formation as the calcareous silty mudstones, with abundant bioturbation and macrofauna, which starts from the top of the Nummulitic Limestone formation for a thickness of about 70 m (Fig. 5), at the base of the Lower Rioux Marl–Siltstones. The Blue Marl Formation includes an anomalous nummulites-rich level, 4–6 m thick, which is here distinguished as the Nummulitic member (Fig. 5). The Nummulitic member of the Blue Marls differs from the Nummulitic Limestone Formation for being more rich in sand-size grains and for being associated with sandstone layers with diagenetic glauconite (Chauveau & Lemoine 1961; Bodelle 1971; Evans 1987). The inferred depositional processes for the first 20 m of Blue Marls is settling of suspended sediments during fair-weather or possibly post-storm conditions, in a shallow, nearshore environment (Evans 1987); whilst the sandy beds are interpeted as being deposited rapidly in association with either storm or seismic events (Evans 1987). The amount of pelagic foraminifera increases upsection, which indicates increasing water depth (Evans 1987).

Above the Blue Marls formation, the Lower Rioux Marl–Siltstone formation (Figs 3–5) is a 78 m thick unit constituted by marls and siltstones with planktic foraminifera (Bodelle 1971), and with layers of thinly-bedded, fine-medium grained, ripple-laminated sandstones. The increase in siliciclastic material is the main lithological difference with the underlying Blue Marls (see e_7–g_1 in de Graciansky *et al.* 1982; Guillemot *et al.* 1981).

The Lower Rioux Marl–Siltstone passes with a transitional contact into the Grès de Ville formation (Fig. 5), which consists of turbiditic sandstones, containing wave-modified current ripples (Evans & Mange-Rajetzky 1991), probably storm generated (Evans 1987). The Grès de Ville is not present on the east side of the basin, which was close to the high formed by extensional faults (Ravenne *et al.* 1987). Above the Grès de Ville, another stratigraphic unit, very

---

Fig. 2. (a) Structural map of the Barrême and Valensole basins (modified by Guillemot *et al.* 1981: BRGM Digne XXIV–41 1:50 000). The Barrême basin is translated on top of the Digne thrust sheet; the Valensole basin on the east side is bounded by the leading edge of the same Digne thrust sheet. The tectonic units correspond to Mesozoic facies distribution (Guillemot *et al.* 1981; de Graciansky *et al.* 1982). The box shows the study area (Fig. 3). The trace of the cross-section of (b) is shown.
(b) Cross-section showing the Mesozoic facies around the Barrême basin (modified from Ford unpublished). The Barreme basin is the thrust-sheet-top basin of the Digne thrust sheet, made of Mesozoic Dauphinois–Vocontian facies and moving on top of the subalpine external–intermediate units of Provençale facies. Along the cross-section (see location in map above) the Digne thrust sheet is divided into three tectonic units: the Barrême basin, Gevaudan–Reichard–Blanche–Castellane units and Douroulles syncline. These tectonic units are controlled by inversion of extensional faults bounding the Gevaudan–Reichard unit. The Cenomanian–Turonian is thick east of the Barrême basin, but is missing under the Nummulitic limestone of the Barrême basin, which lies on top of the Gargasian–Albian. The possible explanations for the missing Mesozoic stratigraphy below the Nummulitic limestone are discussed in the text and shown in Fig. 8. The base of the Nummulitic limestone lies on the remnants of the pre-Priabonian palaeo-topography, which is very smooth at this scale. The palaeo-land surface cut through the Mesozoic units in pre-Priabonian time when both the Gargasian–Albian and the Cenomanian–Turonian were exposed.

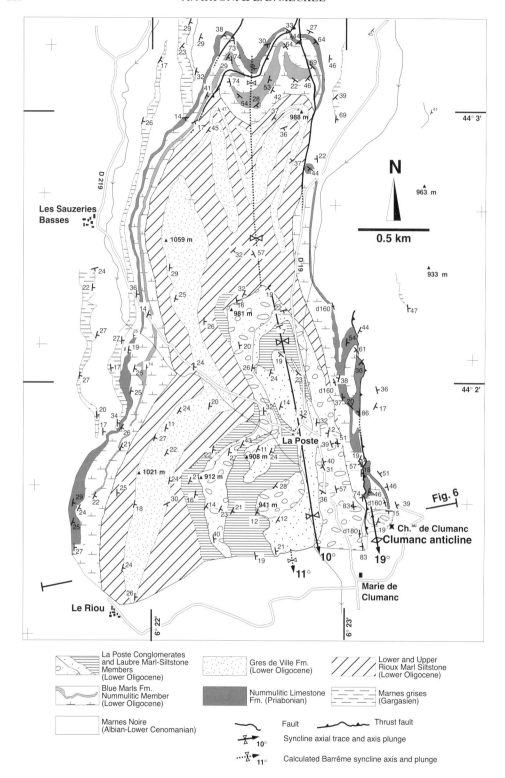

similar to the Lower Rioux Marl–Siltstone is here named Upper Rioux Marl–Siltstone formation (Fig. 5).

The Barrême basin probably reached its maximum depth during the deposition of the Rioux Marl-Siltstone and the Grès de Ville, when the platform was less than 200 m deep (de Graciansky et al. 1982). Afterwards, the stratigraphic succession records a shallowing trend that starts with the La Poste formation and it ends with the deposition of the transitional marine (Grès de Senez and St Lions formations) and continental (Molasse Rouge formation) deposits south of the study area (Guillemot et al. 1981; de Graciansky et al. 1982; Evans 1987).

The La Poste formation, a member of the Blue Marls in the previous works (Bodelle 1971; Chauveau & Lemoine 1961; Guillemot et al. 1981; de Graciansky et al. 1982; Elliott et al. 1985; Evans 1987; Evans & Mange-Rajetzky 1991), is composed of interlayered fine-grained sandstone members and conglomerate members. The fine-grained members, here named the Laubre Marl–Siltstone members (Fig. 5), are more depleted in carbonate content than the underlying fine-grained units (Guillemot et al. 1981; Chauveau & Lemoine 1961; de Graciansky et al. 1982; Evans 1987). The Laubre Marl–Siltstone members are characterized by: (1) clay content increase; (2) thinly bedded, fine- to coarse-grained, normally graded turbidites that occur throughout the unit. The conglomerate members of the La Poste formation (Evans 1987; Evans & Mange-Rajetzky 1991), here named the Lower, Middle, and Upper Conglomerate members, start at their bases with turbiditic-like sandstones and associated pebbly sandstones, and they coarsen upsection into channelized conglomerates and sandstones (Evans 1987). These conglomerate members, 20–40 m thick, have large-scale channel geometries, 750–1200 m wide. They are deposited by high-density gravelly and sandy turbidity currents in submarine channels, and are interpreted as gravelly fan-delta slope deposits (Evans 1987). All the six members of the La Poste Formation (Fig. 5) thin toward the east, against the Clumanc high. The marl and siltstone members are 40–100 m thick in the centre of the basin and thin toward the east, where the conglomerates predominate with three distinguishable bodies between 10 and 20 m thick.

The west to east changes in lithological characters and thinning of the Nummulitic Limestones, Blue Marls, Lower Rioux Marl--Siltstones, Grès de Ville, Upper Rioux Marl–Siltstone and La Poste formation indicate an asymmetry in the sedimentary infill of the Bârreme basin. Such an assymetry and the strong angular unconformity, which separates the La Poste formation from the folded Nummulitic Limestone and Blue Marl formation (Fig. 6), indicates continued deformation of the eastern margin of the basin.

The numerical ages, which can be attributed to these stratigraphic units, will be discussed in the following section in order to date the deformation events.

Attributing numerical ages to the stratigraphic units

The ages of the stratigraphic units are important to define, because they constrain the rates of shortening and folding. To define the numerical ages, we updated published biostratigraphic data (Bodelle 1971, table VIII) with more recent biozonations (Bolli et al. 1985) (Fig. 4). The numerical ages for the updated biozones are taken from Berggren et al. (1992, 1995), integrated with a more recent geological time scale (Odin 1994). The ages of Evans & Mange-Rajetzky (1991) are taken from Harland et al. (1982) time scale, but they are not calibrated to the biostratigraphy of the Barrême basin.

The range chart and the stratigraphic section of Bodelle (1971), compared to the range chart of Bolli et al. (1985) (Fig. 4), allow the definition of the following biozones boundaries (Fig. 4): zone NP24/25; zone P20N1; the zone P16/17 which, together with the Nummulites, gives the Priabonian (Fig. 4). A hiatus of approximately 3.5 Ma exists in the succession because biozones P18–P19 are not present (Fig. 4). The hiatus is possibly due to condensation testified by glauconitic horizon in the Blue Marls below the Nummulitic member. The ages associated with the biozones are from Berggren et al. (1992), and they are: (1) bottom NP 24/25 = 30.5 Ma; (2) top P20N1 = 31.5 Ma; (3) bottom P20N1 = 33 Ma; (4) top P16/17 (top Priabonian) = 36.5 Ma; (5) base Priabonian = 40 Ma. Odin (1994) dates the base of the Rupelian at 33.7 ± 0.5 Ma, which is in agreement with the numerical ages defined by the compared biozones (Fig. 4). A diachronous Rupelian lower boundary, between the Barrême

Fig. 3. Geological map of the northern Barrême basin, mapped at a scale of 1:10 000 from an enlargement of the sheet 3441est-Barrême-Serie Bleue (Institut géographique national, Paris 1988). See Fig. 2 for location. The geology is shown without the alluvial cover. Description of the units in the text.

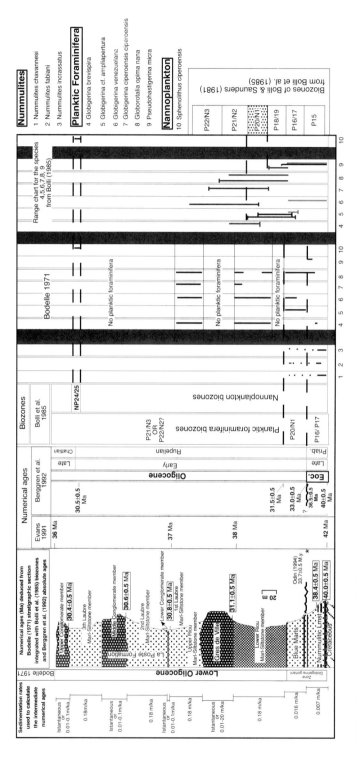

Fig. 4. Numerical ages known from the literature in the stratigraphic column of Les Sauzeries Basse (Bodelle 1971). On the left side, the numerical ages of Evans (1991) following the timescale of Harland *et al.* (1982). On the right side, the biostratigraphic data of Bodelle (1971) are updated to the biozonation of Bolli *et al.* (1985). The biozones are correlated to the timescale of Berggren *et al.* (1992) and Odin (1994). The numerical ages of the stratigraphic contact are calculated assuming a constant rate of sedimentation in the finer-grained units: 0.016–0.007 m ka^{-1} in the Blue Marl; 0.18 m ka^{-1} in the Rioux Marl Silstone and Laubre Marl Silstone. See also Fig. 5 for the numerical ages adopted in this study. The sharp and exact age values are misleading; see text for discussion. However, we give a precision of 100 000 years as a significant time interval for preserved sedimentary stratigraphic events. A specific study on dating the sedimentary infill of the Barrême basin is needed.

basin and the localites dated in Berggren et al. (1992) and Odin (1994), is possible, but it is not the subject of this paper.

To calculate the ages of the stratigraphic boundaries, within the dated surfaces on the Bodelle (1981) stratigraphic section (Fig. 4), we assume a time-averaged, constant sedimentation rate of 0.18 m ka^{-1} during the deposition of the Lower and Upper Rioux Marl–Siltstone formations and the Marl–Siltstone members of La Poste Formation; whilst the Grès de Ville and the conglomeratic members of the La Poste

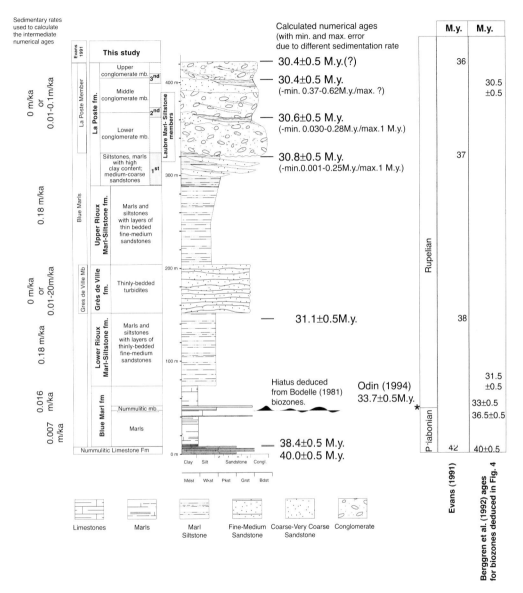

Fig. 5. Composite stratigraphic column of the west Barrême Basin where the maximum thickness of the Tertiary infill is preserved without angular unconformities. The calculated numerical ages of the horizons, which will be restored, are reported on the right side together with the ages of the biozones from Berggren et al. (1992). The ages of Evans (1991) after (Harland et al. (1982), do not agree with the most recent dating of Eocene–Oligocene boundary of Odin (1994). See text for details of stratigraphic units and the calculation of the numerical ages. The decimals in the numerical ages derive from the calculation of the numerical ages between the horizons dated by biozone boundaries (see Fig. 4), but they are not proposed to be so sharp. Absolute ages should be measured by isotope dating and refined biozonations.

formation are assumed to be deposited almost instantaneously. We justify these assumptions as follows: (1) the depositional rate for the Lower and Upper Rioux Marl–Siltstone formations and the Siltstone–Marls members of La Poste formation, calculated by dividing their thickness (180.57 m) by the maximum time they could be deposited (1 Ma), is consistent with documented sedimentation rates (0.01–20m ka^{-1}, (Pickering et al. 1989)) for the continental shelf and borderland basins of the west coast of North America, which are taken as being depositionally comparable to Barrême; (2) the Grès de Ville consists of turbidites, which are known to be event deposits; (3) the La Poste formation consists of fan delta deposits (Evans 1987), which have instantaneous depositional rates of 0.01–0.1m ka^{-1} (Miall 1992). If the Grés de Ville and the conglomeratic members of the La Poste Formation were not deposited instantaneously, as assumed, they represent a time interval, which must be less than 1 Ma, otherwise they overcome the duration defined by the biozones (approximately 1 Ma, Fig. 4). If we apply the sedimentation rates, known in literature (Pickering et al. 1989; Miall 1992), to the coarser units above the Blue Marls (Fig. 4), we define the mimimum and the maximum time during which they could be deposited. In particular, the slowest sedimentation rate of 0.01 m ka^{-1} (Pickering et al. 1989; Miall 1992), gives a total duration of 6.19 Ma, about 5 Ma more than it should be; whereas, the fastest sedimentation rates, 0.1 m ka^{-1} (Miall 1992) and 20 m ka^{-1} (Pickering et al. 1989), give respectively 0.62 Ma and 0.37 Ma. Therefore, the youngest calculated numerical age, 30.4 Ma (Fig. 4), is affected by a minimum error of 0.37 Ma, if the sandstones were deposited at the maximum sedimentation rates of 20 m ka^{-1} (Pickering et al. 1989) and the conglomerates at 0.1m ka^{-1} (Miall 1992); while, the minimum error is 0.62 Ma if the maximum sedimentation rate is 0.1 m ka^{-1} for both conglomerate and sandstone. The minimum errors decrease for older stratigraphic units, closer to the biozone boundary at 31.5 Ma (Figs 4 and 5). The maximum error in numerical ages calculation must not be higher than the time interval defined by the biostratigraphy, 1 Ma. The errors in the calculated numerical ages (Figs 4 and 5) are not given to justify that the employed sedimentation rates are the actual sedimentation rates of the Barrême basin. However, we remark that updating and integrating published data (Bodelle 1971; Bolli et al. 1985; Evans 1987; Pickering et al. 1989; Berggren et al. 1992; Miall 1992) new constraints could be made.

Taking the assumptions into account, the calculated ages are given with the above calculated errors and the average error in geochronological methods (±0.5) (Berggren et al. 1992; Odin 1994) (Fig. 5):

- base Upper Conglomerate member of the La Poste formation = 30.4Ma(-min. 0.37–0.62 Ma/max. 1 Ma) and (±0.5);
- base Middle conglomerate member = 30.6 Ma(-min. 0.030–0.28Ma/max. 1 Ma) and (±0.5);
- base Lower Conglomerate member = 30.8 Ma(-min. 0.001–0.25Ma/max. 1Ma) and (±0.5);
- base Grès de Ville = 31.12 Ma (±0.5);
- top Nummulitic Limestone = 38.4 Ma (±0.5);
- base Nummulitic Limestone = 40 Ma (±0.5).

These contacts correspond to the restoration steps used in Fig. 6. The above numerical ages contain a degree of imprecision because either our assumptions or the absolute dating technique. In order to more accurately date the ages of the units, magnetostratigraphic and isotopic dating data should be collected.

The rates of deformation that will be calculated are affected by the uncertainties in the numerical ages. Although the values for the ages are given to the 100 000 years precision, we repeat that a specific study which integrates absolute dating and sedimentation rates is required to refine the numerical ages.

Methodology of sequential restoration

The progressive formation of the Barrême syncline and the thrust-related growth fold are studied along a sequentially restored ENE–WSW cross-section across the Clumanc Anticline (Fig. 6). Sequential restoration of balanced cross sections has recently been proposed as a way to unravel the evolution of a complexly deformed basin (DeCelles et al. 1990; Burbank & Verges 1994; DeCelles 1994; Artoni & Casero 1997; Ford et al. 1997). Such restorations document the detailed geometrical evolution of the basin by reconstructing the configuration of the basin at the time corresponding to a series of chosen levels.

Restoration requires the cross section to be restored parallel to the tectonic transport direction, usually taken as perpendicular to the major fold axis if fault kinematic data are not available (Dahalstrom 1969; Woodward et al. 1989). However, the fold axes calculated for the different strata in the Barrême basin vary spatially (Fig. 7a–f), and have different angular relationships (Fig. 7g–j). Such differential shifts of the fold axes of stratigraphic units with angular unconformities are well-known (Ramsay & Huber 1987). Therefore, to restore the section, we consider an average fold axis of N175°–11° (Fig. 7n). This fold axis was calculated statistically from the fold axes of the Mesozoic (Fig. 7b), Nummulitic Limestone (Fig. 7d), Grès de Ville (Fig. 7e), and La Poste (Fig. 7f) formations using the program *Stereoplot* (Mancktelow 1995). The average fold axis is located in the middle of the basin where remarkable dip changes occur (Fig. 3). The gentle plunge of the syncline axis and the absence of faults with large strike-slip displacement are supporting a local tectonic transport with no significant out of the section plane movement. However, the regional stress fields, which formed the western alpine arc since the Paleogene, have orientations that could move the Barrême basin both toward south and west.

The restoration is constrained by a 'pin line' that defines the stable points in the cross-section. Dahlstrom (1969) and Woodward et al. (1989) suggest that fold axes are the least deformed parts of deformed belts. Therefore, we locate the pin line in the axis of the Barrême syncline. The migration of the syncline axis during restoration steps 6 and 7 forces the pin lines to be moved (Fig. 6f, g). Shortening and uplift estimates are made relative to the pin lines and the horizontal line coincident with the zero altitude. The absolute movement of the pin line itself can not be measured.

The next stage of sequential restoration is to identify clearly distinguishable and mappable surfaces which mark important evolutionary steps or events in the basin. Stratal boundaries are such surfaces, especially if they correspond to angular unconformities (Artoni & Casero 1997), because they correspond to important changes in depositional processes and also variations in the geometry of the basin. The angular unconformities are accentuated in a growth fold, such as the Clumanc Anticline in Barrême.

The cross-section is then restored by length following the rules of Dahlstrom (1969) and Woodward et al. (1989). The thicknesses of stratigraphic units are maintained using check points (triangles, Fig. 6), such that the restoration preserves a constant volume. No homogeneous strain is considered to have affected the volume, although in Barrême, rare cleavage surfaces oriented N75/47 are found in the core of the Clumanc Anticline.

Variations in vertical thicknesses due to compaction are not included in the restoration, even though they generate errors in uplift calculations. The amount of error introduced by compaction can be sorted out only with an accurate study of the differential compaction in different portions of the basin. Therefore, in this paper, the uplift of different points at the base of the Nummulitic Limestone, a surface which suffered no compaction, are measured (Fig. 8). Four points at the base of the Nummulitic limestone, coincident with the fold hinges, are selected to show particle paths during folding (Fig. 9). These particle paths define the relative amount of shortening and uplift at different positions within the basin.

In this paper, we modify the established technique of restoration back to a flat and pre-established template, which is considered to represent the undeformed stratigraphy (Woodward et al. 1989). In an area of syntectonic sedimentation, the dip of topographic slopes must be taken into account. Thus, flattening a stratigraphic surface that was never horizontal leads to an overrestored section. In the case of sequential restoration, it results in the false representation of the basin at a particular stage and a distortion of the structures below the flattened surface. We therefore modify the sequential restoration technique by introducing a line template that approximates the depositional surfaces (i.e. palaeotopographies).

Because the sediments in the study area are marine, this modification ideally requires a palaeo-bathymetric data set. However, in the Barrême basin, no detailed bathymetry has previously been established for the surfaces which are restored. We therefore adopt a uniform slope of 5° for the dip to which most of the stratigraphic boundaries are restored. Similar values have been documented on the western coast of Oregon, a tectonically active region, but may underestimate local variations, which can reach up to 30° (Pratson & Haxby 1996).

Sequential restoration of the Barrême cross section

The stratigraphic surfaces to which we restore the section (Fig. 6) are described below.

The base of the Nummulitic Limestone, which is coincident with an erosional unconformity (e.g. Baudrimont & Dubois 1977; Ravenne et al. 1987), is chosen as the oldest important step in the evolution of the studied portion of the Barrême basin.

The next relevant surface is considered to be the base of the Grès de Ville Formation, well exposed on the western side of the basin (e.g., at the point labeled 1021 m, Fig. 3). There is no evidence of this unit on the eastern flank of the basin (Fig. 6a). The missing Grès de Ville is due to a high which existed in the Alpine foreland basin at that time (Ravenne et al. 1987). The Grès de Ville must die out at depth on the west limb of the Clumanc Anticline, creating an angular unconformity with the overlaying Lower Conglomerate member of the La Poste formation. This same member overlays an erosional contact with the Blue Marls formation at the hinge of the anticline (Fig. 6a).

The traceable bases of each of the conglomerate members of the La Poste formation are the other stratigraphic surfaces used in the reconstruction of the progressive steps of basin formation. These surfaces were chosen because stratigraphic thinning of the members of the La Poste formation to the east is evidence of a palaeo-topographic gradient dipping toward west. This gradient is considered here to be related to the coeval growth of Clumanc Anticline.

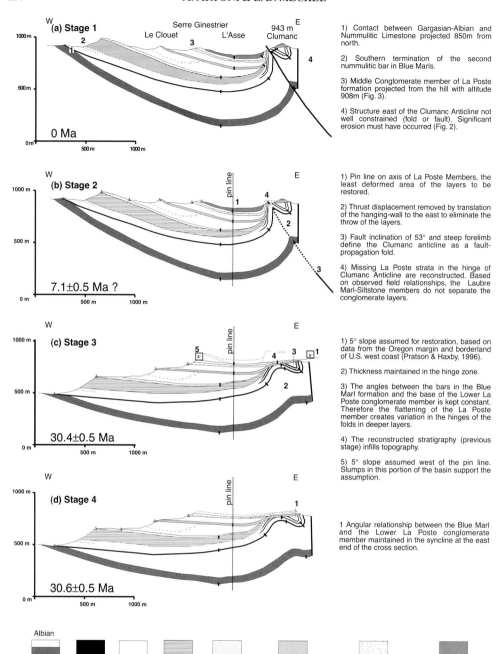

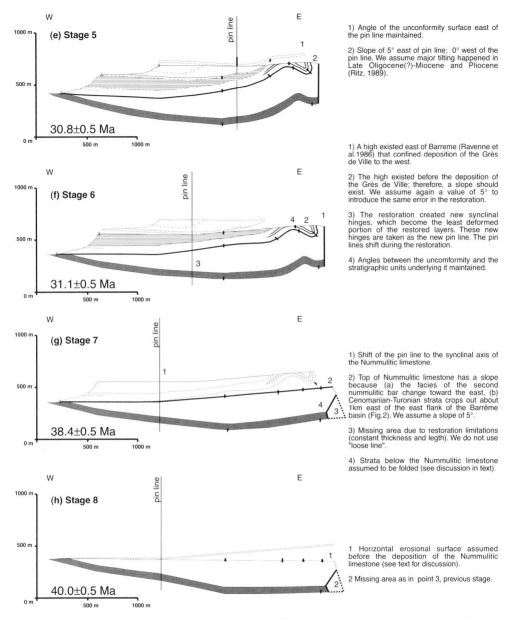

Fig. 6. Sequential restoration of the cross-section across the Clumanc anticline, north Barrême basin. The location of the cross-section is in Fig. 3. In each stage, the dashed lines, on top of the sections, represent the restored stratigraphic unit, not yet deposited at the stage where it is dashed. (**a**) Stage 1, present day; (**b–e**) Stages 2 to 5, tilting on the west limb and thrusting on the east ramp of the Barrême basin; (**f**) Stage 6, Grés de Ville against the east high; (**g**) Stage 7, tilting of the Nummulitic limestones, deposition of the Blue Marls and Lower Riou Marl–Siltstone (**h**) Stage 8, pre-Priabonian erosion.

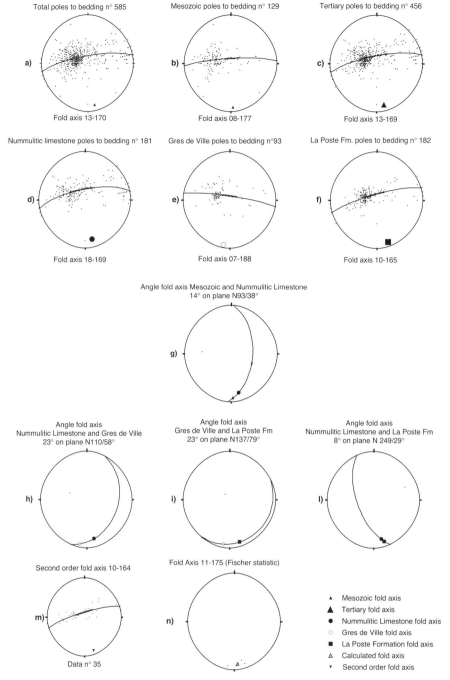

Fig. 7. Fold axis calculated for each stratigraphic unit. After considering all the the dip data (**a**), a subdivision of Mesozoic (pregrowth) (**b**) and Tertiary (growth strata) (**c**) has been made. As the growth strata are all the single stratigraphic unit that infills the basin, the fold axis have been calculated for each Tertiary unit (**d, e, f**). The angles between all the different fold axis calculated is shown in (**g**), (**h**), (**i**), (**l**) and the fold axis of the basin is calculated with Fischer statistic (**n**). The fold axes shift from one stratigraphic units to another because of different number of dip data and because of growth strata unconformably dipping before folding (Ramsay & Huber 1987). All the diagrams and the statistical caculation are made with stereonet 3.0 (Mancktelow 1995). A second-order fold axis is also shown (**m**).

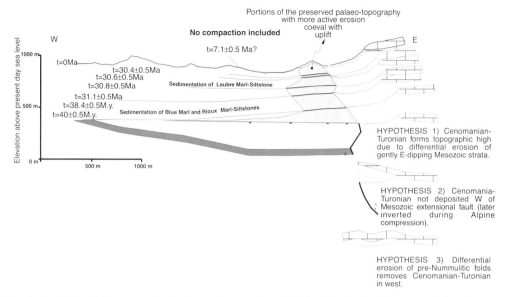

Fig. 8. Evolution of the topography in the northern portion of the Barrême basin. The topographic surfaces correspond to the restored stratigraphic boundary (base Nummulitic limestone, top Nummulitic limestone, base Grés de Ville, base Lower, Middle, Upper conglomerate member of La Poste formation). These surfaces are very smooth because of low resolution in the mapping scale and beacuse of semplification in restoration (see Fig. 6). The surfaces are the result of the infilling of marls and siltstones in the west portion of the basin. The topography east of the studied section is not well constrained. Possible scenarios before the deposition of the Nummulitic Limestone in this unknown portion have to justify the presence of Cenomanian–Turonian, which is missing under the Barrême basin (see Fig. 2).

The last events recorded in the basin are post-depositional: the tilting of the western margin of the Barrême basin and the cutting of the west flank of the Clumanc Anticline by a thrust that was previously blind at depth.

Deformation stage 1: present day (Fig. 6a)

The first step of the restoration is to remove the offset on the thrust cutting the hinge of the Clumanc Anticline (Fig. 6b). The thrust moved following deposition of the Upper Conglomerate member of the La Poste formation (30.4 ± 0.5 Ma). The thrust lies along strike from the faults confining the Gévaudan–Reichard–Blanche–Castellane tectonic unit (Fig. 2), and it is considered a splay of it. Based on the stratigraphical offset, the displacement on the thrust is estimated to be 25 m. The close spatial relationship between this thrust and the Clumanc Anticline makes it reasonable to assume that the thrust was blind during the early stages of fold formation and that the fold developed as a fault-propagation-fold.

Deformation stages 2–5 : tilting of the west limb and thrusting on the east ramp of the Barrême basin (Figs 6b–e)

The second stage of restoration removes the tilt of the western margin of the basin and returns the base of the Upper Conglomerate member to its original topography (Fig. 6c). The bases of the Middle and Lower Conglomerate members are restored to their original geometries in the third (Fig. 6d) and fourth (Fig. 6e) stages.

The age of the tilting of the western limb of Barrême postdates the deposition of the Upper Conglomerate member of the La Poste formation, because there is no evidence of an angular unconformity within the Tertiary succession there (Fig. 6a). The deformation is probably coeval with or older than the Miocene–Pliocene growth strata at the front of the Digne thrust-sheet to the west. Therefore, the tilting probably occurred between the Messinian, or earlier, to the Pliocene, and even in the Quaternary (Ritz 1991). As we cannot specify exactly when the tilt occurred, we consider the first date that it could have been

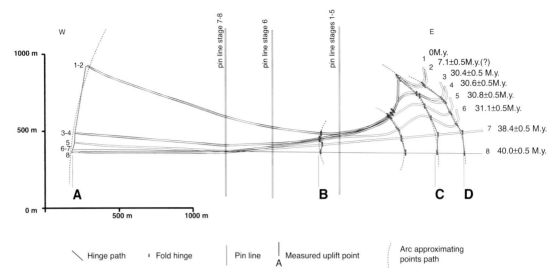

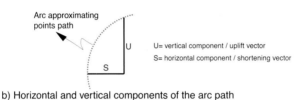

Fig. 9. (a) Progressive folding of the Nummulitic limestone. This stratigraphic unit records all the stages of folding being the oldest. The path of the hinges/points is traced relocating each hinge/point on the restored layer at the same distance from the pin lines. The paths are broken lines which can be approximate by arc of circle. (b) The arcuate paths are decomposed in two components: the shortening component (S) and the uplift component (U). The shortening component of one hinge/point arcuate path is a portion of the total shortening occuring along the cross section. The uplift component of one hinge/point arcuate path represents the amount of uplift in that point of the cross section. It is this uplift component which has been measured to calculate the uplift rates (Fig.10). Instead the shortening component is measured along all the cross section to have the value of the total shortening.

completed as the base of the Messinian (7.1 ± 0.5 Ma, from (Odin 1994; Berggren *et al.* 1995)). The uncertainties in the timing of initiation of thrusting on the east limb and tilting of the west limb implies calculation of rates over a large time interval for the two youngest stages (Fig. 10).

The restored bases of the conglomerate members of the La Poste formation represent the paleo-topography on which the conglomerates were deposited. The surfaces are erosional, and each is an angular unconformity (Figs 6c–e). These angular unconformities are clearly related to the growth of the Clumanc Anticline, where the La Poste Conglomerate Members seal the folded Nummulitic limestone and Blue Marls Formations (see also Chauveau & Lemoine 1961; de Graciansky *et al.* 1982).

Deformation stage 6: the Grés de Ville against the east high (Fig. 6f)

This stage restores the section to the base of the Grès de Ville (Fig. 6f). The Grès de Ville formation, and the Upper Rioux Marl-Siltstone formation that overlays it, are only present west of the Clumanc Anticline (Fig. 7). We assume, therefore, that folding of Clumanc Anticline occured before the deposition of the Grès de Ville Formation (31.1 ± 0.5 Ma), producing a topography on which the Grès de Ville and the

Upper Rioux Marl–Siltstone onlapped. The erosion of the Lower Rioux Marl–Siltstone and the Blue Marl formations in the hinge of the Clumanc Anticline occurred either during the deposition of the Grès de Ville Formation or later during the deposition of the Upper Rioux Marl–Siltstone Formation.

Deformation stage 7: tilting of the Nummulitic limestone, deposition of the Blue marls and lower Rioux marl–siltstone (Fig. 6g)

This stage (Fig. 6g) documents the local tilting of the Nummulitic limestone. This period of tilting is justified for the following reasons. (1) Within the Blue Marls, there is one sandy layer related to either seismic or storm events (Evans 1987) in the west, and two in the east, of the Barrême syncline. This suggests that the eastern side of the basin may have been more proximal to an uplifted source area. (2) Syn-Nummulitic folding occurred east of Clumanc in the Douroulles basin (de Lapparent 1966; Pairis 1971; Evans 1987). (3) Olistoliths in the Nummulitic Limestone in the southern part of the Barrême basin document the uplift of the eastern margin of the basin during this time (Evans 1987).

Deformation stage 8: pre-Priabonian erosion (Fig. 6h)

The final stage shows a flat erosional surface at the base of the Nummulitic Limestone (Fig. 6h). The angle between the Gargasian–Albian strata and this surface demonstrates that folding occurred before erosion. The existence of the pre-Nummulitic Limestone erosional surface is well documented (e.g. Baudrimont & Dubois 1977; Ravenne et al. 1987), but its morphology is poorly defined.

Discussion

Fragments of Palaeogene topography

The topographic surfaces associated with the restored stratigraphic contacts were generated by the progressive deformation of the east flank of the Barrême basin (Elliott et al. 1985). Therefore, the restoration stages show the structure and simplified palaeo-topography of a 2 km long cross-section across the Eocene–Oligocene incipient mountain front of the external Alps of SE France. Additionally, where the Eocene–Oligocene marine sediments were deposited, the topography is a first approximation of different submarine topographies during deposition.

The restoration stages show the evolutionary steps of the topography at eight time steps from the Priabonian to present (Figs 6 & 8). These time steps illustrate a progressive relative uplift of the topography through time, as well as a hypothetical steepening of the mountain front east of Barrême (Fig. 8).

The relative uplift is presented as elevation above present day sea level (Fig. 8). However, these numerical values are a relict of the restoration process, because they are measured relative to the pin lines in the cross sections, the absolute uplift of which cannot be constrained (see above). Thus, the values do not include extra-basinal processes such as the overall uplift of the Digne thrust sheet.

Moreover, because the elevations are calculated only by backstripping the present-day thickness of sediments, and do not include corrections for palaeobathymetry or compaction, they include a second source of absolute error. That is to say, the base of the Nummulitic Limestone, (40 ± 0.5 Ma, Fig. 8) is shown at 400 m above present day sea level (Fig. 8). This surface can be considered to represent Priabonian sea level since the Nummulitic Limestone is the first marine transgressive deposit of the foreland basin. As eustatic sea level is interpreted to have fallen only 100–150 m since the Priabonian (Haq et al. 1988), the differential elevation gives a minimum estimate of error (i.e., not corrected for compaction) of 250–300 m.

Despite the absolute errors in the calculation, and our assumption of 5° slopes, what can be seen is that the topography developed from low relief to higher relief through time (Fig. 8). For instance, whereas the pre-Nummulitic topography is flat, the present-day topography has many peaks and lows. Both surfaces were formed in subaerial continental settings, but the present-day topography is incised into a deformed fold-and-thrust belt, while the pre-Nummulitic topography formed at the western edge, probably on the forebulge, of the external western Alpine foreland basin. The development of the fold and thrust belt during deposition of the various Eocene–Oligocene sediments probably led to the increasing relief preserved in the time steps.

The topography of the mountain front east of Barreme (Fig. 8) is generated based on the fact that there are no marine sediments east of Barrême (Fig. 3). We interpret this absence as being the result of non-deposition, due to the presence of subaerial topography. The subaerial topography is inferred based on the fact that Cenomanian-Turonian strata crop out as a cliff with a higher elevation than the Barrême basin (Fig. 8). When restored back from its present day position, this cliff maintains its higher elevation at all time steps (Fig. 8).

The original presence of the Cenomanian–Turonian cliff may be explained in three ways. (1) The simplest interpretation is that the Cretaceous dipped gently to the west and that the Cenomanian–Turonian therefore suffered more erosion in the east (Fig. 8, hypothesis 1). The Cretaceous strata would have dipped to the east if there were a large wavelength folding, such as is associated with the creation of a forebulge. As the Nummulitic Limestone was probably deposited on the forebulge, this scenario seems likely. (2) A down-to-the-east extensional fault on the eastern side of Barrême could also have caused preferential deposition of the Cenomanian-Turonian to the east (Fig. 8, hypothesis 2). Such extensional faults were active during the Tethyan break-up (Triassic to mid-Jurassic), and were inverted during the Alpine orogeny in the Tertiary (de Graciansky *et al.* 1989; Gidon 1982; Gillchrist *et al.* 1987). (3) The erosion of pre-Priabonian folds could also explain the absence of the Cenomanian–Turonian in Barrême (Fig. 8 hyphothesis 3). One of these folds can be the very open syncline obtained by the restoration (Fig. 6g). The fold axis of this reconstructed syncline should be N–S, perpendicular to the cross section. Such folds occur to the east of Barrême, but not to the west (Fig. 2a). This supports the possibility that the folds were eroded in the west.

The three hypotheses are all geologically reasonable, and there is no strong evidence supporting one to the exclusion of the other two. Even without knowing the exact cause, though, we can confidently state that a higher and steeper topography must have existed on the east side of the Barrême basin. This steeper topography is the source of Upper Cretaceous olistoliths in the Blue Marls close to Barrême village (Elliott *et al.* 1985; Evans 1987).

Generation of the Clumanc Anticline

The geometric evolution of the northern Barrême basin during the Eocene–Oligocene was primarily driven by differential movements associated with the growth of the Clumanc Anticline on the east flank of the basin. These differential movements led initially to non-deposition or erosion near the Clumanc Anticline and deposition and subsidence to the west (Fig. 8). For instance, in the hinge of the anticline, the Lower and Upper Rioux Marls–Siltstone and the Grès de Ville formations are missing (Fig. 6a). In contrast, these formations thicken dramatically to the west (Fig. 6a). Later, during the deposition of the La Poste formation, the sedimentation rate competed with uplift and erosion and the three conglomeratic members of the La Poste Formation are preserved above the hinge of the anticline.

The restoration of the Upper Conglomerate member of La Poste formation to a surface of 5° slope (Fig. 6c) does not change the angles the Blue Marls and the Riou Marl–Siltstone make with the Lower Conglomerate Member. In fact, the Blue Marls Formation is below an erosive unconformity sealed by the base of the Lower Conglomerate Member; no detachment exists on this unconformity surface. Maintaining (1) the angular relationship between the two sandy beds in the Blue Marl and the Lower conglomerate member of La Poste formation and (2) the thickness of the strata, a change in the wavelength of the stratigraphic units below the Upper Conglomerate member is created. The Clumanc Anticline passes from a close and equant fold in stage 2 (Fig. 6b) to an open and broad anticline in stage 3 (Fig. 6c) and an open and wide anticline (Twiss & Moores 1992) at stage 6 (Fig. 6f). The increase in wavelength corresponds to a change in the position of the hinges of the Clumanc Anticline (Figs 6c–g & 9), together with a lengthening and steepening of the fold limb from the older to the younger stages (Fig. 7).

The reconstructed geometries of the Clumanc Anticline (Fig. 6) illustrate some limitations of the restoration technique. For instance, the length of the limb and the position of the hinges at the different stages are constrained by the requirements of constant length and thickness both in the deformed state and in the restored template with respect to the pin lines (Woodward *et al.* 1989). Restoration of a curved surface with respect to the pin line creates new fold hinges on the older surfaces. The new hinges in the Grès de Ville caused by restoring the surfaces of the three members of La Poste conglomerate and keeping the points on the pin line fixed (Fig. 6c, d) reflect this constraint.

The restoration of most of the stratigraphic contacts to a surface inclined 5° is another source of error in the geometries of the basin and folds. Better constraints in palaeobathymetry of the stratigraphic units infilling the Barrême basin could allow such oversimplifications in the sequential restoration to be avoided.

Velocities of deformation in the Barrême basin

Differential movements occuring within the basin can be traced by choosing points on the sequentially restored surfaces (Fig. 9a). The most complete sequence of point positions is recorded at the base of the Nummulitic Limestone.

The paths in the Nummulitic Limestone can be approximated as an arc (Fig. 9a). The arc may be subdivided into horizontal and vertical components, which represent the shortening and uplift vectors, respectively (Fig. 9b). The combination of the horizontal and vertical components produces the folding. This is reflected in the fact that the arcs change their sense of concavity across the axis of the Barrême syncline (right pin line, Fig. 9a). As previously mentioned, the point paths are relative to the position of the pin lines and we have no control to define the absolute movement of the pin lines themself. Therefore, the deformation rates are not representative of the whole western Alpine arc of SE France, except probably for the flanks of the remnant thrust-sheet-top basin distributed around the arc.

The sequential restoration and the numerical ages attributed to the stratigraphic contacts used in the restoration allow the calculation of rates at shorter time intervals and at a smaller scale than has been previously attempted in the Barrême basin. Published shortening rates of 2.2–4.7 mm a^{-1} (Elliott et al. 1985; Evans 1987) were calculated considering three main thrusting events from the Late Eocene to the Lower Oligocene. These events are correlated with the base of the Nummulitic Limestone, a submarine glide block in the Blue Marl that is present south of the studied area, and the base of the La Poste Formation.

Instead, in the time interval from the base of the Nummulitic Limestone to the top of the La Poste Formation, we consider six stages of deformation, which give rates at significantly shorter time intervals. In addition, because we consider the thrust of the west limb of the Clumanc Anticline to be a different event from the tilting of the west limb of the Barrême basin, we infer two stages occurred after the deposition of the Upper Conglomerate member of La Poste formation. The greater number of stages allows detailed quantification of variation of the deformation rates with time. This increased precision allows comparisons with deformation rates for recently active mountain fronts.

Rates of shortening

The shortening rates represent the displacement velocity along the horizontal component of the arcuate point paths (i.e. the shortening vector, Fig. 9b). The calculated shortening rates vary from 0.003 to 0.2 (±0.083) mm a^{-1} on time intervals of 0.2–7.3 Ma (Fig. 10a). Figure 10a plots the shortening rate against the arithmetic mean of the older and younger ages, which gives the time interval. Obviously, the shortening may have been either instantaneous or gradual within this time interval. The calculated shortening rates are strictly related to the sedimentation rate adopted (Pickering et al. 1989; Miall 1992) (Fig. 5). In Fig. 10 for the coarse-grained units, the maximum sedimentation rates are considered (Figs 4 and 5); in fact, the slowest sedimentation rate would have exceed the time represented by the established biozones (Fig. 4). A maximum discrepancy of 0.074 mm a^{-1} shortening rate exists around the average age of 30.4 Ma because of the different sedimentation rates used for ages calculation (Fig. 10).

An increasing shortening rate precedes the deposition of the first siliciclastic deposit, the Grès de Ville formation (Fig. 10a). Maximum shortening rates precede the Lower Conglomerate Member of La Poste Formation (Fig. 10a); whereas a decrease in shortening rate, at 30.8 ± 0.5 Ma (Fig. 10a), corresponds to the deposition of the Middle and Upper Conglomerate Member of La Poste Formation (Fig. 6d). Therefore, the effect of the increasing shortening rates is increased subsidence in the western portion of the basin, coupled with erosion on the eastern side of the basin (e.g., Fig. 8), and does not necessary correspond to coarser grain size of the stratigraphic unit. In fact, both a rapid increase and a rapid decrease in shortening rate correspond to deposition of coarse-grained units (Fig. 10a).

These high-frequency changes in shortening rates during the deposition of the three conglomerate members may also have caused a local deflection of the palaeo-drainage system, seen in a change in source areas of the Grès de Ville and La Poste formations (Evans & Mange-Rajetzky 1991). The Grès de Ville was sourced from the south, along the strike of the Barrême syncline, and infilled the subsiding western portion of the basin (Elliott et al. 1985). In contrast, the clasts of the La Poste formation are sourced from the internal part of the mountain chain, to the east of the basin (Evans & Mange-Rajetzky 1991). The change in shortening rates, and the resulting deflection, may be related to the inception of the Digne–Castellane arc, which seems to follow the distribution of the Oligocene remnant basin of SE France (Fig. 1).

Rate of uplift

The second component of the folding is the vertical component (i.e. the uplift vector, Fig. 9b). The uplift is measured using four points at the base of the Nummulitic Limestone (A, B, C, D in Fig. 9a). Points B, C and D coincide with fold hinges. The points are chosen at the base of the

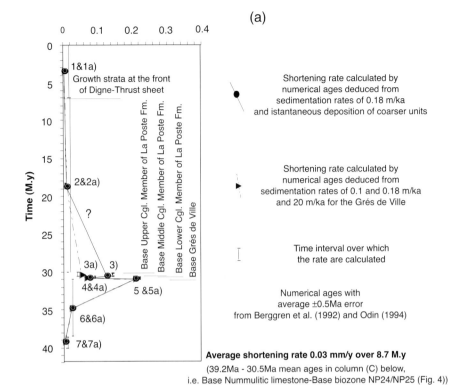

(a)

Average shortening rate 0.03 mm/y over 8.7 M.y
(39.2Ma - 30.5Ma mean ages in column (C) below,
i.e. Base Nummulitic limestone-Base biozone NP24/NP25 (Fig. 4))

	(A) Older age M.a.	(B) Younger age M.a.	(C) Mean age M.a.	(D) Shortening rate mm/y	(E) Shortening rate variations due to numerical age uncertainties mm/y
1)	7.1±0.5	0.0	3.5±0.5	0.002	-0.0002/+0.0001
1a)	7.1±0.5	0.0	3.5±0.5	0.002	-0.002/+0.0001
2)	30.4±0.5	7.1±0.5	18.7±0.5	0.008	-0.003/+0.004
2a)	30.07±0.5	7.1±0.5	18.5±0.5	0.008	-0.003/+0.0004
3)	30.6±0.5	30.4±0.5	30.5±0.5	0.12	-0.10/*
3a)	30.6±0.5	30.07±0.5	30.3±0.5	0.05	-0.03/*
4)	30.8±0.5	30.6±0.5	30.7±0.5	0.07	-0.06/*
4a)	30.8±0.5	30.06±0.5	30.7±0.5	0.06	-0.05/*
5)	31.1±0.5	30.8±0.5	31.0±0.5	0.2	-0.15/*
5a)	31.1±0.5	30.8±0.5	30.9±0.5	0.2	-0.16/*
6)	38.4±0.5	31.1±0.5	34.8±0.5	0.02	-0.002/+0.003
6a)	38.4±0.5	31.1±0.5	34.7±0.5	0.02	-0.002/+0.003
7)	40±0.5	38.4±0.5	39.2±0.5	0.003	-0.001/+0.006
7a)	40±0.5	38.4±0.5	39.2±0.5	0.003	-0.001/+0.006

- value to subtract from the shortening rate, column (D)
+ value to add to the shortening rate, column (D)
* the numerical ages uncertainties are larger than the time interval (Older age (A) - Younger age(B))

HISTORY OF THE BARRÊME THRUST-SHEET-TOP BASIN 233

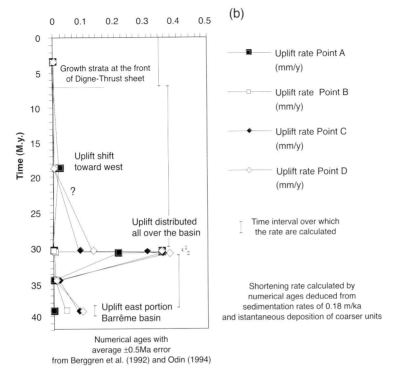

	(A) Older age M.a.	(B) Younger age M.a.	(C) Mean age M.a.	(D) Uplift rate Point A mm/y	(E) Uplift rate Point B mm/y	(F) Uplift rate Point C mm/y	(G) Uplift rate Point D mm/y
1)	7.1±0.5	0.0	3.5±0.5	0.0	0.0	0.001	0.004
2)	30.4±0.5	7.1±0.5	18.7±0.5	0.018	0.0	0.003	0.002
3)	30.6±0.5	30.4±0.5	30.5±0.5	0.0	0.0	0.08	0.125
4)	30.8±0.5	30.6±0.5	30.7±0.5	0.35	0.002	0.3	0.35
5)	31.1±0.5	30.8±0.5	31.0±0.5	0.2	0.002	0.35	0.37
6)	38.4±0.5	31.1±0.5	34.8±0.5	0.0	0.005	0.01	0.006
7)	40±0.5	38.4±0.5	39.2±0.5	0.0	0.035	0.08	0.09

Fig. 10. (a) Shortening rates and (b) uplift rates along the cross-section in the northern part of Barrême basin (Fig. 6). (a) The shortening rates increase together with the increase of siliciclastic input on the trace of the cross-section; with uncertainties in time determination (±0.5 Ma as average error in Berggren *et al.* 1992 and Odin 1994). In the table, the rows 1a–7a) have numerical ages deduced by sedimentation rate 0.1 m ka^{-1}, 0.18 m ka^{-1} and 20 m ka^{-1} starting from the Lower Riou Marl–Siltstone upward (see Fig. 5). (b) The uplift rates are related to four representative points in the Nummulitic limestone along the cross-section (see Fig. 9 to locate them). The uplift rates show the shift toward west of the uplifting during time. The uplift is spread all over the basin during the sedimentation of the siliciclastic deposits of Grés de Ville and La Poste formations.

Nummulitic limestone, where compaction is considered to be negligible. The uplift measured here does not represent the topographic uplift, which must take in account the amount of compaction, erosion, and exhumation (England & Molnar 1990). These factors were not considered in the restoration. Erosion and compaction act against the vertical uplift. Because of the uncertainties in the amount of erosion and non-deposition close to the hinge of Clumanc Anticline, it was not possible to calculate the topographic uplift. Despite the above limitations, the relative differential uplift velocity within different portions of the Barrême basin can be detected.

At the beginning of the Tertiary, the eastern portion of the Barrême basin was uplifting at velocities between 0.001 to 0.37 mm a^{-1} (Fig. 10b). The amount of uplift is 55 m at point B, 125m at point C, and 140 m at point D. Even though the Nummulitic limestone and the Blue Marls are strongly folded between stage 7 and stage 6 (Figs 6e, f), the uplift rate decreases during this time (Fig. 10b), probably because the folding is averaged over a longer time interval (7.3 Ma) than other stages time intervals.

The rate of uplift is approximately constant until stage 3 (Figs 6c & 10b). Eastern Barrême basin had greater relative uplift, as well as uplift rates during this time (Fig. 10b). Point B is an exception; it gives an uplift rate of 0–0.001 mm a^{-1} because it is very close to the pin line. Afterward, the uplift rates in the east decrease and only point A is uplifting. At a mean age of 18.7 ± 0.5 Ma, point A was uplifted 430 m at a rate of 0.018 mm a^{-1} (Fig. 10b). Point A is representative of the differential uplift between eastern and western Barrême basin.

If shortening and uplift rates of recently active mountain fronts are considered, the values obtained are very close despite the fact that the time intervals on the recently active fronts are better constrained than those we can constrain in the Barrême basin. Shortening rates of 0.5–1.0 mm a^{-1} (Lamb & Vella 1987) and 1.4 ± 0.6 mm a^{-1} (Nicol et al. 1994) have been calculated in folds active during the last million years in New Zealand. Similarly, uplift rates of 0.68 ± 0.17 mm a^{-1} on the North Island of New Zealand (Lamb & Vella 1987) and 0–2.16 mm a^{-1} in the Waipara area of New Zealand (Nicol et al. 1994) have been calculated.

Conclusions

Using the technique of sequential restoration, we introduce significant new constraints in the Eocene–Oligocene structural and stratigraphic evolution of the Barrême basin. A total of eight stages of evolution have been established using the Eocene and Lower Oligocene sedimentary infill of the basin. The Clumanc growth anticline, on the east side of the basin, preserves deformation stages which occurred in this portion of the Alpine foreland basin during the Priabonian and Lower Oligocene, when compression reached the east flank of the Barrême basin. The sequential restoration allows the reconstruction of the approximate topography of the Oligocene mountain front at its first stage of formation.

The shoreforedeposits of the Nummulitic Limestone were deposited above a pre-Priabonian subaerial erosion surface. Then, the Nummulitic Limestone was tilted westward and the Blue Marls were deposited on the tilted surface. In subsequent stages, the topographies are controlled by the competition between erosion in the east portion of the Barrême basin and deposition in the west portion. This interplay continued until 30.8 ± 0.5 Ma, when the submarine channels of the La Poste Formation spread over the Clumanc Anticline. Maximum submarine topographies were approximately 200 m at the time of deposition of the Rioux Marl–Siltstone and Grès de Ville (de Graciansky P. C. 1982), but decreased to a poorly defined depth during the deposition of the La Poste Formation (Evans 1987). After approximately 24 Ma, the Barrême basin was transported on top of the Digne thrust sheet and the topography became subaerial again, incising the fold-and-thrust belt of the external Alps foreland basin.

The history of the Barrême basin and its palaeo-topography, coupled with the available and revaluated ages of the stratigraphic units, give the velocities of the geometrical evolution of the basin. The rates of shortening range between 0.003 and 0.2 (±0.083) mm a^{-1}, while the uplift rate varies between 0 and 0.37 mm a^{-1}. The shortening and uplift rates are not constant, but rather change through time, giving a more refined timing of the deformations occurred in the Barrême area.

Increasing shortening rates correspond to increases in clastic sediment input, but not necessary with the grain size, at least along the cross-section studied. A local deflection in the drainage system was probably induced by the submarine morphology during the Oligocene.

Uplift rates change not only in time but also within the Barrême basin. The uplift rates show the shift of the deformation from east to west of the basin since the Priabonian. The shift in the locus of uplift is related to the translation of the Barrême basin on top of the Digne thrust sheet.

The shortening and uplift rates are comparable to shortening and uplift rate of mountain fronts active in the last million years. The restoration method and the integration of stratigraphic data are therefore promising tools for analysing old mountain fronts and understanding the possible evolution of the geometry and topography of the mountain fronts active at present. The similarity of shortening and uplift rates in the Barrême basin to those calculated in recent mountain fronts is the result of a method which give much more importance to all stratigraphic boundaries that can be recognized in the basin, in particular to those which coincide with angular unconformities.

This work was pursued within a project on growth folds funded by the Bundesamt für Bildung und Wissenshaft of Switzerland — BBW Project No. 94.0153, part of the EC – Forshungsprogramm JOULE – Integrated Basins Studies (Phase II), which allowed A.A. to spend more than one year at the Geologisches Institut, ETH-Z, Zurich. It is really time to express my best thanks to the above institutions and organization and to the people and friends met in Zurich. A particular thank to M. Ford for reading an early version of this paper, and to D. Bernoulli and J. P. Burg for supporting and discussing a preparatory pilot study in the Barrême basin. Thanks also to two anonymous reviewers for their useful comments and significant suggestions. This paper could not be realized without the help and support of C. Ohlendorf and his family, D. Rubatto, J. Von Konjienenburg and P. Gawenda who assisted me on many occasions.

References

ALVAREZ-MARRÓN, J., MCCLAY, K.R., HARAMBOUR, S., ROJAS, L. & SKARMETA, J. 1993. Geometry and evolution of the frontal part of the Magellanes foreland thrust and fold belt (Vicuña area), Tierra del Fuego, Southern Chile. *American Association of Petroleum Geologists Bulletin*, **77**, 1904–1921.

ARGNANI A. & FRUGONI, F. 1997. Foreland deformation in the central Adriatic and its bearing on the evolution of the northern Apennines. *Annali di Geofisica*, **40**, 771–780.

——, BONAZZI, C., EVANGELISTI, D., FAVALI, P., FRUGONI, F., GASPERINI, M., LIGI, M., MARANI, M., & MELE G. 1993a. Tettonica dell'Adriatico meridionale. *Memorie della Società Geologica Italiana*, **51**, 227–237.

——, FAVALI, P., FRUGONI, F., GASPERINI, M., LIGI M., MARANI, M., MATTIETTI, G. & MELE, G. 1993b. Foreland deformational pattern in the southern Adriatic sea. *Annali di Geofisica*, **36**, 229–247.

ARTONI, A. & CASERO, P. 1997. Sequential balancing of growth structures, the Late Tertiary Example from the Central Apennines. *Bulletin de la Société Geologique de France*, **1**, 26–41.

BAUDRIMONT A. F. & DUBOIS P. 1977. Un bassin Mésogéen du domain péri-Alpin: le sud -est de la France. *Bulletin du Centre des Recherches Exploration et Production Elf Acquitaine*, **1**, 261–308.

BEAUDOIN, B., CAMPREDON, R., COTILLON, P. & GIGOT, P. 1975. Alpes Méridionales françaises reconstruction du basin de sedimentation. *In*: 9me Congrès International de Sedimentologie, Excursion **7**, Nice 1975, 221.

BEER, J. A., ALMENDINGER, R. W., FIGUEROA, D. E. & JORDAN, T. E. 1990. Seismic stratigraphy of a Neogene Piggyback Basin, rgentina. *AAPG Bulletin*, **74**, 1183–1202.

BERGERAT, F., MUGNIER, J. L., GUELLEC, S., TRUFFERT, C., CAZES, M., DAMOTTE, B. & ROURE, F. 1990. Extensional tectonics and subsidence of the Bresse basin: an interpretation from ECORS data. *In*: ROURE, F., HEITZMANN, P & POLINO, R. (eds) *Deep structure of the Alps*. Société Géologique de France, Memoires, **156**, 146–156.

BERGGREN, A. W., DENNIS, V. K., OBRADOVICH, J. D. & SWISHER, C. C. III 1992. Toward a revised Paleogene geochronology. *In*: PROTHERO, D. R. & BERGGREN, W. A. (eds) *Eocene–Oligocene climatic and biotic evolution.*, Fischer A.G., Princeton Series in Geology and Paleontology, Princeton University Press, Princeton, 29–45.

——, KENT, D. V., AUBRY, M. P. & HANDERBOL, J. 1995. *Geochronology, time scales and global startigraphic correlation*. SEPM, Society of Economic Paleontologists and Mineralogists, Tulsa.

BODELLE, J. 1971. *Les formations nummulitiques de l'arc de Castellane*. PhD Universite de Nice.

BOLLI, H. M., SAUNDERS, J. B. & PERCH-NIELSEN, K. 1985. Plankton stratigraphy. *In*: COOK, A. H., HARLAND, W. B., HUGHES, N. F., PUTNIS, A. & THOMSOM, M. R. A. (eds) *Cambridge Earth Science Series*. Cambridge University Press, Cambridge, 000–000.

BOLT, B. A. 1988. *Earthquakes*. W. H. Freeman, San Francisco.

BURBANK, W. D. & VERGES, J. 1994. Reconstruction of topography and related depositional systems during active thrusting. *Journal of Geophysical Research*, **3**, 1–25.

——, LELAND, J., FIELDING, E., ANDERSON, R. S., BROZOVIC, N., REID, M. R. & DUNCAN, C. 1996. Bedrock incision, rock uplift and threshold hillslopes in the northwestern Himalayas. *Geology*, **379**, 505-510.

CHAUVEAU, J. C. & LEMOINE, M. 1961. Contribution a l'etude geologique du synclinal tertiare de Barreme (moitie nord). *Bulletin du service de la Carte géologique de la France, Comptes-rendus des collaborateurs*, **58**. Libraire polytechnique Ch. Beranger, Paris, 287–318.

COWARD, M. P., GILLCRIST, R. & TRUDGILL, B. 1991. Extensional structures and their tectonic inversion in the Western Alps. *In*: ROBERTS, A. M., YIELDING, G. & FREEMAN, B. (eds) *The Geometry of Normal Faults*. Geological Society, London, Special Publications, **56**, 93–112.

DAHALSTROM, C. D. A. 1969. Balanced cross-section. *Canadian Journal of Earth Sciences*, **6**, 743–757.

DE GRACIANSKY, P. C., DARDEAU, G., LEMOINE, M. & TRICART, P. 1989. The inverted margin of the french alps and foreland basin inversion. *In*: COOPER, M. A. & WILLIAMS, G. D. (eds) *Inversion Tectonics*. Geological Society, London, Special Publications, **44**, 87–107.

——, DUROZOY, G. & GIGOT, P. 1982. *Carte Géologique de la France, Notice explicative de la feuille Digne XXXIV–41 a 1:50.000*. Bureau de Recherches Géologiques et Minières, Orléans.

DE LAPPARENT, A. F. 1938. Etudes géologiques dans les régions provençale et alpines entre le Var et la Durance. *Bulletin du Service Carte Géologique de France*. **40**.

―― 1966. A propos des conglomérates anténummulitiques des Alpes de Provence. *Bulletin de la Société géologique de France*, **7**, VIII, 451–457.

DECELLES, P. G. 1994. Late Cretaceous–Paleocene synorogenic sedimentation and kinematic history of the Sivier thrust belt, northeast Utah and southwest Wyoming. *Geological Society of America Bulletin*, **106**, 32–56.

――, GRAY, M. B., RIDGWAY, K. D., COLE, R. B., SRIVASTAVA, P., PEQUERA, N. & PIVNIK, D. A. 1990. Kinematic history of a foreland uplift from Paleocene synorogenic conglomerate, Beratooth Range, Wyoming and Montana. *Geological Society of America Bulletin*, **103**, 1458–1475.

ELLIOTT, T., APPS, G., DAVIES, H., EVANS, M., GHIBAUDO, G. & GRAHAM, R. H. 1985. A structural and sedimentological traverse through the tertiary foreland basin of the external Alps of south-east France. *Foreland basins Conference, Fribourg – Field trip B*, 39–73.

ENGLAND, P. & MOLNAR, P. 1990. Surface uplift, uplift of rocks, and exhumation of rocks. *Geology*, **18**, 1173–1177.

EVANS, M. J. 1987. *Tertiary sedimentology and thrust tectonics in the southwest alpine foreland basin, Alpes de Haute-Provence, France*. PhD Thesis, University of Wales.

―― & MANGE-RAJETZKY, M. A. 1991. The provenance of sediments in the Barreme thrust-top basin, Haute-Provence, France. *In*: MORTON, A. C., TODD, S. P. & HAUGHTON, P. D. W. (eds) *Developments in Sedimentary Provenance Studies*. Geological Society, London, Special Publications, **57**, 323–342.

FIELDING, E., ISACKS, B., BARAZANGI, M. & DUNCAN, C. 1994. How flat is Tibet. *Geology*, **22**, 163–167.

FORD, M., WILLIAMS, E. A., ARTONI, A., VERGÉS, J. & HARDY, S. 1997. Progressive evolution of a fault-related fold pair from growth strata geometries, Saint Llorenç de Morunys, SE Pyrenees. *Journal of Structural Geology*, **19**, 413–441.

GIDON, M. 1982. La reprise de failles anciennes par une tectonique compressive: sa mise en évidence et son rôle dans les chaîn subalpines des Alpes occidentales. *Géologie Alpine*, **58**, 53–68.

GIGOT, P., GUBLER, Y. & HACCARD, D. 1975. Relations entre sedimentation et tectonique (en compression ou en distension): exemples pris dans des bassins tertiares des Alpes du Sud et de Haute Provence. *IX Congres International de Sedementologie*, Nice 1975, **4**, 157–162.

GILLCHRIST, R., COWARD, M. & MUGNIER, J. L. 1987. Structural inversion and its controls: examples from the Alpine foreland and the French Alps. *Geodinamic Acta*, **1**, 1, 5–34.

GUELLEC, S., MUGNIER, J. L., TARDY, M. & ROURE, F. 1990. Neogene evolution of the western Alpine foreland in the light of ECORS data and balanced cross section. *In*: ROURE, F., HEITZMANN, P & POLINO, R. (eds) *Deep structure of the Alps*. Société Géologique de France, Memoires, **156**, 166–184.

GUILLEMOT, J., COURTILLOT, V., DAGBERT, M., CHERMETTE, J. C., FOUGEIROL, D., DE NAUROIS, G.,

TALLON, J. P., RENARD, A., BOUCHARD, J. P., LESSI, J., GOGUEL, J., LEMOINE, M. & DE GRACIANSKY, P. C. 1981. *Carte Geologique de la France 1/50:000, Sheet Digne XXXIV-41*. Bureau des recherches géologiques et miniéres, Orléans.

HARLAND, W. B., COX, A. V., LLEWELLYN, P. G., SMITH, A. G. & WALTERS, R. 1982. A geologic time scale. Cambridge Earth Science Series, Cambridge University Press, Cambridge.

HAQ, B. U., HANDERBOL, J. & VAIL, P. R. 1988. Mesozoic and Cenozoic chronostratigraphy and cycles of sea-level change. *In*: WILGUS, C. K., HASTINGS, B. S., KENDALL, C. G. C., POSAMENTIER, H. W., ROSS C. A. & VAN WAGONER, J. C. (eds), *Sea-level change: an integrated approach*. SEPM Special Publications, **42**, 71–108.

HAYWARD, A. B. & GRAHAM, R. H. 1989. Some geometrical characteristics of inversion. *In*: COOPER, M. A. & WILLIAMS, G. D. (eds) *Inversion Tectonics*. Geological Society, London, Special Publications, **44**, 17–39.

KELLER, E. A. & PINTER, N. 1996. *Active tectonics, earthquakes, uplift and landscape*. Prentice Hall, New Jersey.

LAMB, S. H. & VELLA, P. 1987. The last million years of deformation in part of the New Zeland plate-boundary zone. *Journal of Structural Geology*, **9**, 7, 877–891.

MANCKTELOW, N. 1995. *Stereoplot 3.0*. ETH-Z, Zurich.

MIALL, A. D. 1992. Alluvial deposits. *In*: WALKER, R. G.& JAMES, N. P., Eds., *Facies Models: response to sea level changes*. Geological Association of Canada, St John's, Newfoundland, Canada, 119–141.

NICOL, A., ALLOWAY, B. & TONKIN, P. 1994. Rates of deformation, uplift, and landscape development associated with active folding in the Waipara area of North Canterbury, New Zeland. *Tectonics*, **13**, 6, 1327–1344.

ODIN, G. S. 1994. Geological time scale. *Comptes Renus de l'Academie des Sciences, Paris*, **318**, Serie II, 59–71.

PAIRIS, J. L. 1971. Tectonique et sedimentation tertiare sur la marge orientale du bassin de Barreme. *Geologie Alpine*, **47**, 203–214.

PICKERING, K. T., HISCOTT, R. N. & HEIN, F. J. 1989. *Deep marine environments. Clastic sedimentation and tectonics*. Unwin Hyman Ltd, London.

PIERI, M. & GROPPI, G. 1981. Subsurface geological structure of the Po Plain. *In*: CNR (eds) *Quaderni della Ricerca Scientifica*, Progetto Finalizzalo Geodinamica, 1–23.

PRATSON, L. F. & HAXBY, W. F. 1996. What is the slope of the U.S. continental slope? *Geology*, **24**, 3–6.

RAMSAY, J. G. & HUBER, M. I. 1987. *Modern structural geology*, **2**. Academic Press Ltd, London.

RAVENNE, C., VIALLY, R., RICHÉ, PH. & TRÉMOLIÈRES, P. 1987. Sédimentation et tectonique dans le bassin marin dans le bassin marin Eocène supérior-Oligocène des alpes du sud. *Revue de l'Institut Français du Pétrole*, **42**, 529–553.

RICCI LUCCHI, F. 1986. The Oligocene to Recent foreland basins of the Northern Apennines. *In*: HOMEWOOD, P., ALLEN, P. A. (eds), *Foreland*

Basins. International Association of Sedimentologists Special Publications, **8**. Blackwell Scientific Publications, 105–139.

RITZ, J. F. 1991. *Evolution du champ de contraintes dans les alpes du sud depuis la fin de l'Oligocene. Implications sismotectoniques*. PhD Thesis, Université Montpelliér II.

SUPPE, J. 1985. *Principles of structural geology*. Prentice-Hall Inc., Englewood Cliffs, New Jersey.

——. & MEDWEDEFF, D. A. 1990. Geometry and kinematics of fault-propagation folding. *Eclogae Geological Helvetiae*, **83**, 409–454.

TWISS, R. J. & MOORES, E. M. 1992. *Structural Geology*. W. H. Freeman and Co., New York.

WOODWARD, N. B., BOYER, S. E. & SUPPE, J. 1989. *Balanced geological cross section: an essential tecnique in geological research and exploration*. American Geophysical Union. (Short Courses in Geology, **6**)

Thin-skinned inversion tectonics at oblique basin margins: example of the western Vercors and Chartreuse Subalpine massifs (SE France)

YANN PHILIPPE[1,2], ERIC DEVILLE[1] & ALAIN MASCLE[1]

[1]*Institut Français du Pétrole, Division Géologie-Géochimie, 1 et 4 Avenue de Bois-Préau, 92852 Rueil-Malmaison Cédex, France*
[2]*Present address: Elf Aquitaine Production, Direction Exploration-Production France, 64018 Paris, France*

Abstract: The western Vercors and Chartreuse Subalpine fold-and-thrust belts form part of the western Alpine foreland that was involved in Neogene compressive tectonics. The deformation front of the two massifs is superimposed on the western margin of the Mesozoic Southeast Basin, which was inverted in the Late Miocene. Obliquity between this NE–SW-directed basin margin and the WNW–ESE Alpine tectonic transport resulted in the development of en-échelon folds and thrusts at the leading edge of the massifs. The gradual change of the tectonic style from the Chartreuse to the southern Vercors belts denotes that the initial configuration of the basin as well as the rheological properties of the basal detachment layer strongly influenced the geometry and the kinematic evolution of the two fold belts. This paper presents the geometry of the inverted western Tethyan palaeomargin through a series of balanced cross-sections and attempts to evaluate the possible boundary conditions that controlled the development of the western Chartreuse and Vercors massifs and the lateral variations of the tectonic style in the study area. Results of an analogue modelling experiment simulating a thin-skinned inversion of an oblique platform–basin transition zone are compared with those of the field study and used to discuss the role of the palaeogeographic inheritance on both the initiation and orientation of thrust faults in décollement tectonics.

The Mesozoic basin of southeastern France corresponds to a complex sedimentary basin, the evolution of which is governed by the opening of the Tethyan and Atlantic (Bay of Biscay) domains. Unlike the other two major onshore Mesozoic basins of France (the Paris and Aquitaine basins), it was intensively affected both by Pyrenean (Late Cretaceous to Eocene) and Late Alpine (Late Miocene) compressive phases, and thus is fully part of the western Alpine foreland. Many structural features correspond to early Jurassic extensional structures related to the Tethyan palaeomargin and/or Oligocene structures related to the Western European rift system, which were inverted during the Neogene Alpine compressive deformation (Butler 1989; Roure *et al.* 1992; Welbon & Butler 1992). The trends and tectonic styles of the Neogene compressive structures are largely induced by the pre-existing geometry of the Southeast Basin, especially near its western and southern former passive margins.

In general, the term 'inversion tectonics' is classically assigned to compressive systems marked by partial or complete reactivations of deep-seated normal faults (*see* Gillcrist *et al.* 1987; de Graciansky *et al.* 1988; Hayward & Graham 1989; Letouzey 1990; McClay & Buchanan 1992). On the other hand, this term can also be ascribed to thin-skinned fold belts where thrust ramps initiate directly above pre-existing discontinuities that form the major boundaries of the inverted sedimentary basin (i.e. normal faults, basement flexures, sedimentary pinch-out), while the detached cover is completely dissociated from its passive basement. The term 'thin-skin inversion tectonics' used here corresponds to the latter definition.

The western Chartreuse and Vercors Subalpine massifs in southeastern France represent typical thin-skinned inverted basins in the Late Miocene, in response to the push of the external crystalline massifs on the east (Fig. 1). Obliquity between the deformation front (NE–SW directed) and the strike of folds and thrusts (N–S to NNE–SSW) within the massifs is a major characteristic of the study area, which can be considered as a large transverse zone developed at the leading edge of a collisional belt (Figs 1 & 2). Subsurface and field data evidence superimposition of the thrust front of the massifs upon the western margin of the Southeast Basin and Chartreuse massif. The latter corresponds to the transition from a platform-domain provided with a thin carbonaceous and marly Mesozoic sequence on the northwest to a deep subsiding

PHILIPPE, Y., DEVILLE, E. & MASCLE, A. 1998. Thin-skinned inversion tectonics at oblique basin margins: example of the western Vercors and Chartreuse Subalpine massifs (SE France). *In:* MASCLE, A., PUIGDEFÀBREGAS, C., LUTERBACHER, H. P. & FERNÀNDEZ, M. (eds) *Cenozoic Foreland Basins of Western Europe.* Geological Society Special Publications, **134**, 239–262.

basin on the southeast, where a thick marly-dominated series accumulated from Early Liassic to Late Cretaceous (Baudrimont & Dubois 1977; Curnelle & Dubois 1986). Such a coincidence between the outer limits of the Alpine orogen and the western boundary of the peri-Tethyan passive margin demonstrates that, while the geometry and kinematics of the Subalpine massifs must be considered in terms of thin-skin tectonics, the rheology of the basal detachment layer and the initial configuration of the inverted basin play a leading role.

This paper deals with geometric, kinematic and dynamic arguments on the western Chartreuse and Vercors tectonics, combining analytical and modelling approaches. Recent seismic-reflection profiles, detailed seriated balanced cross-sections incorporating field, well and older seismic data (Deville *et al.* 1994;

Philippe 1995) imply some constraints on the geometry of the external parts of the two massifs, and help to propose some hypotheses on the boundary conditions that controlled the development of such a transverse zone. A comparison is made between the natural example and a laboratory model in order to investigate the possible mechanisms of thin-skin inversion.

Geological setting and geodynamic evolution of the study area

The Chartreuse and Vercors Subalpine Massifs, classically ascribed to the Dauphino-Helvetic zone of Western Alps, are cut by the Subalpine thrust (the 'Chartreuse orientale' thrust that extends to the northeastern Vercors by the Moucherotte thrust) delimiting the Alpine foreland from the external Alps (Debelmas 1974; Gidon 1981, 1988; Doudoux *et al.* 1982; Butler 1992; Figs 1, 2 & 3). To the west, these massifs are separated from the Tertiary Valence Basin by the Alpine thrust front, whereas they are bounded to the east by the Drac and Isère valleys, respectively, which part them from the so-called 'Liassic Hills' and the Belledonne external crystalline massif (Fig. 3). Moving north, the thrust sheets of the Chartreuse massif west of the Subalpine front extend to the Internal Jura fold-and-thrust belt (Debelmas 1974; Philippe 1994), while inner structures east of the Subalpine thrust connect with those of the Bauges massif (Doudoux *et al.* 1982; Fig. 1). South of the Drôme valley, the Vercors massif is bounded by the Diois and Baronnies massifs where both E–W and N–S structural trends occur (Flandrin 1966; Siddans 1979; Gratier *et al.* 1989).

The present architecture of the Dauphinois Basin (i.e. the area of the Southeast Basin now corresponding to the Subalpine massifs) results from a succession of tectonic events spanning Triassic to Late Miocene (Fig. 4). As a matter of fact, basin development started in the Late Carboniferous, as Stephanian troughs developed along NE–SW wrench-faults related to Late Variscan NNW–SSE compressive events (Arthaud & Matte 1977), and continued in Permian times when the overthickened Hercynian crust started collapsing in response to isostatic readjustments (Ménard & Molnar 1988; Mascle *et al.* 1996). These Late Palaeozoic structures significantly influenced the subsequent segmentation of the Mesozoic Tethyan passive margin as most of the inherited NE–SW strike-slip faults preferentially acted as major normal faults during the Triassic and Jurassic (Baudrimont & Dubois 1977). The Mesozoic and Cenozoic tectonosedimentary evolution of the

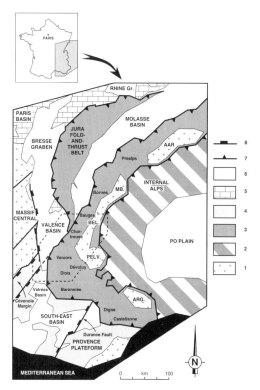

Fig. 1. Schematic structural map of southeastern France and location of the study area.
1, Palaeozoic basement; 2, Internal Alps; 3, External Alps (Dauphino-Helvetic domain) and Alpine foreland (Jura fold-and-thrust belt); 4, Pyreneo-Provençal domain and foreland; 5, stable European craton; 6, Neogene; 7, main Alpine thrust fronts; 8, main Early Mesozoic normal faults. The dotted line indicates the location of Fig. 2.

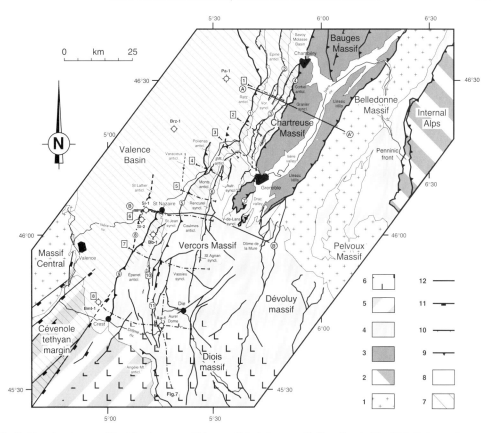

Fig. 2. Structural map of the Chartreuse and Vercors Subalpine massifs (location in Fig. 1). 1, Palaeozoic basement; 2, Alpine metamorphic units; 3, Subalpine zone (*sensu stricto*); 4, detached Mesozoic cover of the Chartreuse and Vercors massifs; 5, autochthonous Mesozoic cover; 6, extension of the Triassic rock salt; 7, approximate extension of the thin 'Jurassian' platform-type and 'Pre-Subalpine' transitional-type Mesozoic series; 8, Cenozoic; 9, major thrusts; 10, minor thrusts and backthrusts; 11, major normal faults; 12, strike-slip and minor faults.

Numbering of thrusts (circled): 1, Ratz thrust; 2, Epine thrust; 3, Voreppe thrust; 4, 'Chartreuse orientale' thrust; 5, Royans thrust; 6, Rencurel thrust; 7, Moucherotte thrust; 8, St Lattier thrust; 9, Barbières thrust; 10, Léoncel thrust; 11, Saillans thrust.

Boreholes: Au-1, Aurel-1; Bb-1, Beauregard–Barret-1, Bmt-1, Montoison-1, Brz-1, Brézins-1, Sl-1, St Lattier-1, Sl-2, St Lattier-2, Pa-1, Paladru-1.

Autr, Autrans; Mt, Monteau; V-de-Lans, Villard-de-Lans; Vor, Voreppe.

Vercors–Chartreuse area can be divided into four main events of unequal duration directly related to the large geodynamic cycles recorded throughout Western Europe.

Liassic to Late Cretaceous rifting and subsidence

Figure 5 shows several reference stratigraphic columns illustrating thickness variations of the Mesozoic series over the study area, on the basis of field outcrops and subsurface data. From the Late Triassic to the end of the Mid-Jurassic, southeastern France underwent a major extension of which the regional minimum stress axis strikes E–W to NW–SE, directly related to the Tethyan rifting preceding the opening of the Ligurian ocean from the Callovian (Lemoine 1985; Lemoine *et al.* 1986; Coward & Dietrich 1989). Inherited NE–SW Late Variscan wrench faults acted as major normal faults, notably the Cévenoles and Durance fault zones that flank the NE–SW subsiding axis of the basin (Baudrimont & Dubois 1977; Curnelle & Dubois 1986; Roure *et al.* 1992; Figs. 1 & 2). During the Late Jurassic and Early Cretaceous, the Chartreuse

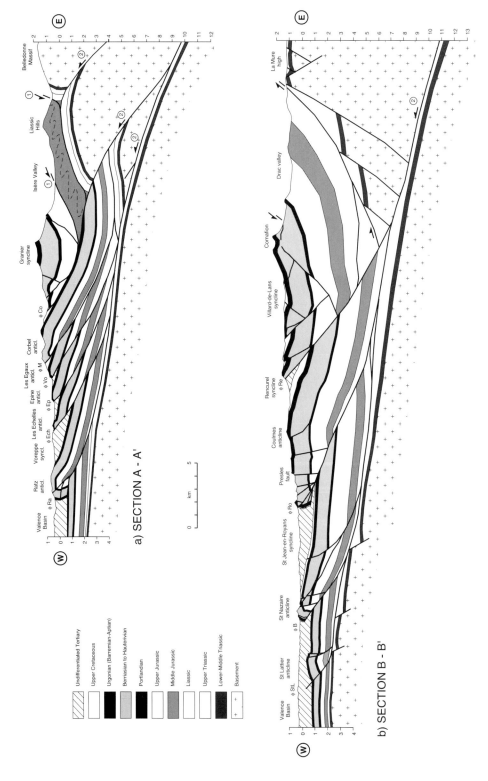

Fig. 3. Regional balanced cross-sections through the Chartreuse and Vercors massifs (Deville *et al.* 1994; modified). Location in Fig. 2. Major thrust faults: φB, Barbières thrust; φCo, Chartreuse orientale thrust; φEch, Les Echelles thrust; φEp, Epine thrust; φM, Median thrust; φRa, Ratz thrust; φRe, Rencurel thrust; φRo, Royans thrust; φSl, St Lattier thrust; fφVo, Voreppe thrust.

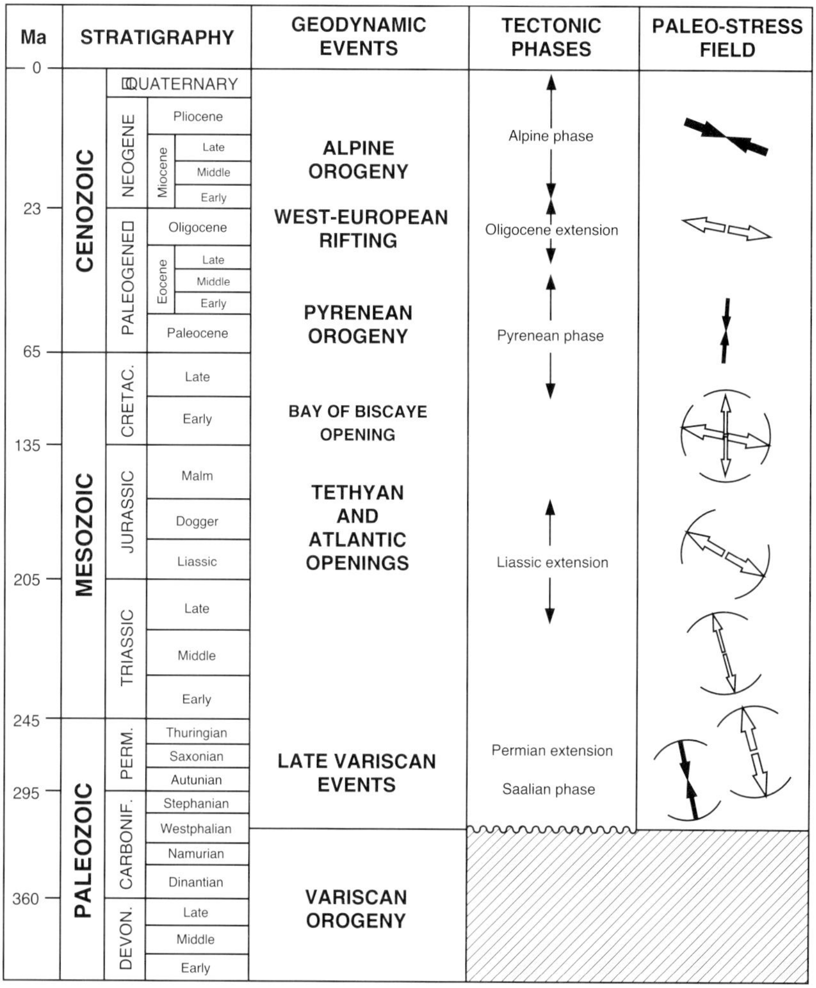

Fig. 4. Successive tectonic phases recorded in the Chartreuse and Vercors massifs. Large arrows correspond to significant events.

and Vercors regions underwent a regional subsidence associated with the opening of the Ligurian ocean, so that the Late Jurassic–Early Cretaceous sequence thickens to the ESE without local evidence of active normal faults. In contrast, the Diois–Baronnies–Dévoluy province (corresponding to the so-called 'Vocontian trough') was affected by a significant extensional phase spanning Oxfordian to Aptian attested by block tilting and activation of E–W normal faults (Dardeau et al. 1988; de Graciansky & Lemoine 1988). At larger scale, such a N–S-trending extension (Fig. 4) is believed to be a consequence of a rifting event well recorded in the Aquitaine Basin (Boillot et al. 1984), related to the opening of the Bay of Biscay (de Graciansky & Lemoine 1988).

Late Cretaceous to Early Eocene N–S compression ('Pyrenean phase')

From the Barremian, a N–S compressive tectonic regime initiated S and SE of the Vercors massif ('Devoluy phase'), then reached a climax from Campanian to Mid-Eocene (Trémolières pers. comm. 1994). This N–S shortening associated with the late stages of the Pyrenean collision between Iberia and Europe, completed the development of the so-called 'Pyreneo-Provençal' belt consisting of E–W folds and thrusts that propagate northward up to the southernmost part of the Vercors massif prior to the Lutetian (Flandrin 1966; Siddans 1979; Riché & Trémolières 1987). This compressive phase is vaguely recorded in the Chartreuse and

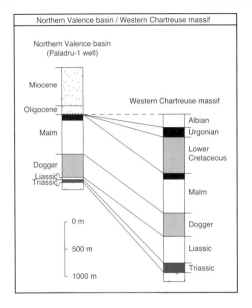
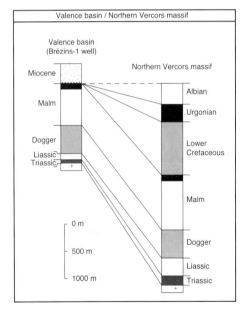
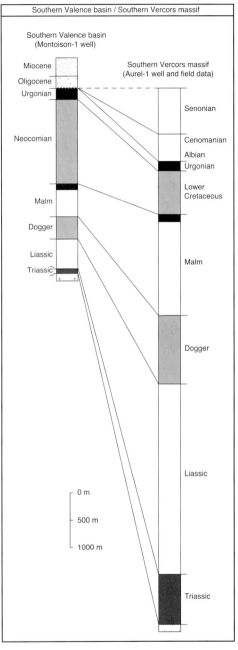

Fig. 5. Seriated WNW–ESE stratigraphic correlations characterizing the Jurassian and Dauphinois Mesozoic domains of southeastern France, from north to south. Thicknesses are based on subsurface (boreholes, seismic-reflection profiles) and field data.

Vercors thrust belts and has no influence upon the present structure of the massifs, excepted in the southern edge of the Vercors massif (Aurel Dome, see Fig. 8). Northwest of the Vercors massif, the pre-Miocene NNE–SSW sinistral strike-slip faults accompanied by NE–SW en-échelon folds observed (e.g. the Presles fault, Arnaud 1973, 1981; see Fig. 7, cross-sections 5 and 6) are probably induced by the N–S compressive Pyrenean phase (Arpin 1988). In the Vercors massif, this event is generally represented by N–S-directed tension gashes and stylolitic peaks. Rare conical E–W folds in the vicinity of NNE–SSW sinistral strike-slip faults and large E–W arches are attributed to this phase, on the basis of relative chronology arguments (Philippe 1995).

Late Eocene to Early Miocene E–W extension ('Oligocene rifting')

This event led to the development of the Valence Basin that broadly corresponds to an asymmetric half-graben tilted westward, its evolution controlled by a N–S-trending normal fault located along its western edge (Mascle *et al.* 1992, 1994). In the Valreas Basin located south of the Valence Basin, N–S- to NNE–SSW-trending normal faults related to this extension are sealed by Late Burdigalian and Helvetian deposits (Riché & Trémolières 1987). East of the Valence Basin, restricted N–S roll-over grabens formed as eastward gravity sliding occurred above Triassic layers (e.g. the St Jean-en-Royans syncline, Figs 2 & 7, section 6). Field, seismic and well data clearly establish that these en-échelon half-grabens developed from Late Eocene to Early Miocene. In the western Chartreuse and Vercors massifs the Oligocene extension is not significantly recorded with the exception of few small NNE–SSW-directed normal faults, (Butler 1987). Microtectonic analysis indicates an horizontal minimum stress axis (σ_3) ranging from N095° to 100°E.

Late Miocene to Recent E–W compression ('Alpine phase')

This corresponds to the main orogenic phase recorded in the Alpine foreland, which resulted in the development of the Jura fold-and-thrust belt and the Chartreuse and Vercors massifs. Folding and thrusting of these massifs initiated in the Late Miocene, in response to the WNW–ESE horizontal push of the Belledonne massif (Ménard 1979, 1988; Beach 1981; Doudoux *et al.* 1982: Ménard & Thouvenot 1987), while gravity sliding may account for the latest stages in the development of the Vercors massif (Gamond 1994). On the basis of brittle microtectonic analysis, palaeo-stress calculations, by using the method of direct inversion (Angelier 1984), indicate that maximum horizontal stress axes (σ_1) related to the Alpine orogeny are relatively uniform over the study area, with a mean trend ranging from N105° to 115°E (Fig. 6). Folds and thrusts are believed to develop normal to the stress axis as no significant deflections of the σ_1 axis are observed over the study area. This allows the restoration of the cross-sections as they subparallel the regional tectonic transport.

Most of the folds are related either to forethrusts or backthrusts cross-cut by subsidiary dextral NE–SW and sinistral NW–SE wrench faults. The frontal thrust of the western Chartreuse and Vercors massifs is split up into three en-échelon NNE–SSW-directed thrusts (Fig. 2): (1) the Barbières thrust to which the Épenet – St Nazaire anticline is related; (2) the Royans thrust to which the Coulmes and Poliénas anticlines are related; (3) the Ratz thrust to which the Monteau and Ratz anticlines are related. In the centre of the two massifs, the main thrust faults illustrated on the seriated sections are: the Léoncel thrust, the Saillans backthrust, the Rencurel thrust that extends in the Chartreuse massif by the Voreppe thrust, the Epine thrust and the Subalpine frontal thrust ('Chartreuse orientale' thrust and Moucherotte thrust). West of the Vercors frontal thrust, a set of en-échelon N–S anticlines emerge through the Cenozoic fill of the Valence Basin: from SW to NE, these are the St Lattier, Varacieux and Poliénas anticlines (see BRGM 1975; Gamond & Odonne 1984), the significance of which is discussed in the following chapters.

Several stratigraphic intervals propitious to décollement tectonics have been encountered within the Mesozoic sequence either in the field (Butler 1987; Mugnier *et al.* 1987; Arpin 1988) and on seismic-reflection profiles. Their involvement is commensurate with their respective spatial distribution over the study area. We can distinguish from bottom to top: (1) Triassic rocksalts (Early Keuper), present in the southernmost Vercors massif (Curnelle & Dubois 1986) and north of the western Chartreuse massif (i.e. beneath the Jura fold-and-thrust belt and Molasse Basin; Lienhardt 1984); (2) Liassic marls and shales (Toarcian–Aalenian) in the northern Vercors and Chartreuse massifs, where they replace the classic Keuper basal detachment layer as Triassic evaporites and rocksalt are lacking; (3) Bathonian to Early Oxfordian black shales, especially in the southern half of the Vercors massif where their thickness is maximum; and (4) Berriasian marls.

Description of the serial cross-sections

Several balanced cross-sections were constructed from north to south (Figs 2 & 7), on the

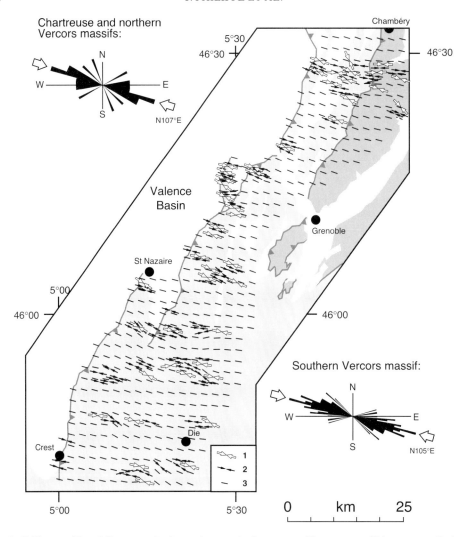

Fig. 6. Grid-map of Late Miocene main shortening axes in the western Chartreuse and Vercors massifs. 1, σ1 axis calculated from fault slip data (obtained from Angelier's inverse method INVD; Angelier 1984); 2, stylolitic peak measured; 3, σ1 axes interpolated from stylolitic peaks.

basis of field observations, but are partly calibrated by industrial seismic-reflection profiles and rare well data. Hence four seismic profiles that subparallel sections 1, 3, 6 and 8, respectively, provided useful information about the structural style at depth and served to constrain the basement topography interpolated between each of the profiles. A fifth N–S-directed seismic profile passing through the Diois massif at the southern edge of the study area also supports our interpretation, especially in the vicinity of the Aurel dome (Fig. 7, section 8 and Fig. 8).

Each cross-section was balanced by using the LOCACE® software (Moretti & Larrère 1989), taking the top of Urgonian (Barremian–Aptian) limestones as reference level, except for section 8 where restoration was made at the end of the Jurassic. Accordingly, the amount of Late Eocene–Early Miocene extension is not quantified because only section 6 is retrodeformable with acceptable accuracy at the end of the Aquitanian. All this means that the amount of the Late Miocene shortening indicated with each section is slightly underestimated.

Section 1

This profile is located north of the study area where the Jura orocline branches off from the External Alps. The thickness and composition of the deformed sedimentary sequence is similar to those found in the Jura Mountains (Debelmas 1974; Philippe 1994). The autochthonous basement dips gently eastward under the allochthonous Chartreuse massif. The Mesozoic and Cenozoic fill of the northern Valence Basin has been imbricated into a consistently west verging stack of narrow thrust sheets (e.g. Ratz, Les Echelles, Epine and Les Egaux ramp-related anticlines), which are detached from the autochthon at an Upper Triassic or more likely at a Lower Jurassic (Aalenian?) level. Triassic series are extremely reduced and devoid of evaporites and rocksalt. This strong basal detachment layer is responsible for the development of a typical high-tapered fold belt achieved by imbricate foreland-verging thrusts, in good agreement with analytic models of fold-and-thrust belts and accretionary wedges (Davis *et al.* 1983; Davis & Engelder 1985; see below). The individual thrusts ramp up through the entire Mesozoic sequence and do not employ subsidiary potential detachment levels provided by Callovian–Oxfordian and Berriasian marls.

Section 2

The profile cross-cuts the southern part of the western Chartreuse massif, and exhibits two major imbricate ramp anticlines (the Ratz and Grande Sure anticlines) separated by the Voreppe syncline. The Echelles and Epine anticlines have disappeared with regard to the previous section, but their southern termination is likely masked by the Grande Sure anticline. Both Mesozoic sequence and tectonic style are comparable to profile 1; the entire amount of shortening is accommodated by foreland-verging thrust faults rooted in (Triassic –) Liassic shales.

Section 3

This section parallels the Isère valley which separates the Vercors massif from the Chartreuse massif. The Mesozoic cover is still characterized by Jurassian affinities and a reduced thickness. The Monteau anticline represents the southern continuation of the Ratz anticline observed on sections 1 and 2. It corresponds to a typical fault-propagation fold related to the Ratz thrust that ramps up to the west from Lower Liassic layers. This fault enables the duplication of the cover thanks to an upper footwall-flat hosted at the base of the Neocomian marls. Minor upper thrusts develop in the footwall of the Ratz thrust from subsidiary décollement at the level of Oxfordian and Lower Cretaceous strata. The westernmost thrust, observed on a seismic profile, vanishes northward and connects to the south with the Vercors thrust front (Fig. 2). The easternmost thrust equates for the southern prolongation of the Voreppe thrust west of the Chartreuse massif and the Veurey-Voroize syncline is the lateral equivalent of the Voreppe syncline observed on section 2.

Section 4

The Mesozoic series is quite similar to the northern profiles, but gradually thickens eastward. One can recognize the two major thrusts encountered on the previous section, i.e. the Rencurel thrust that prolongs the Voreppe thrust and the southern continuation of the Ratz thrust. The Monts anticline represents the lateral equivalent of the Monteau anticline of section 3. The westernmost structure corresponds to the Polienas anticline which forms part of a set of N–S en-échelon folds located at the eastern edge of the Valence Basin (Polienas, Varacieux and St Lattier anticlines; Fig. 1). According to Gamond & Odonne (1984) these folds are yielded by a basement dextral wrench-fault coinciding with the Isère river ('faille de l'Isère'). Based on seismic data (see sections 3 and 6), there is no evidence of such a deep-seated fault. We therefore consider these structures as oblique folds related to thrust faults which ramp up from the Triassic/Liassic regional décollement level. One possible interpretation of the Polienas anticline as illustrated on section 3 is in a triangle zone developed thanks to a lower forethrust involving Jurassic strata (the 'Royans thrust') relayed by a superficial backthrust originating in Berriasian marls.

Section 5

The thickness of the Mesozoic cover is significantly greater than in the northern sections. East of the profile, superficial forethrusts rooted in the Berriasian marls ramp up through the Neocomian sequence belonging to the hanging wall of the Rencurel thrust. To the west, the Coulmes anticline represents a wide ramp anticline (related to the Royans thrust), that corresponds to the merging of the Polienas and Monteau anticlines, as the Ratz thrust connects with the Royans thrust between sections 3 and 4 (Fig. 2). On top of the Coulmes anticline, an apparent

NNE–SSW-directed normal fault separates Urgonian from Hauterivian limestones. It corresponds to the Prelses fault that acted as a sinistral wrench-fault prior to the Miocene (Arnaud 1973, 1981; Arpin 1988).

Section 6

This is the most constrained profile of all thanks to a recent seismic section (Deville *et al.* 1994) that clearly displays both the gradual eastward thickening of the Mesozoic cover and the influence of the Oligocene extensive structures on the Late Miocene Alpine deformation. The thickness of the Triassic–Aptian sequence is about 1700 m west of the profile and reaches 5200 m east. We can distinguish four major thrust sheets and the Tertiary St Jean-en-Royans syncline that is one of the most interesting features of the section. To the east one can identify the Chalimont anticline related to the Rencurel thrust, the westward displacement of which decreases with respect to northern profiles, and the Coulmes anticline related to the Royans thrust, previously encountered on section 5. The Urgonian limestones are cut by steep faults, especially by a set of apparent normal faults, NNE–SSW-directed, including the Presles fault, which is an inherited 'Pyrenean' sinistral strike-slip fault (Arnaud 1973, 1981) probably reactivated as a normal fault in the Eocene–Oligocene.

The St Jean-en-Royans syncline is characterized by an asymmetric Tertiary fill as the Late Eocene to Early Miocene strata thin toward the east of the structure. It corresponds to a folded half-graben bounded on the west by a NNE–SSW normal fault facing east, that outcrops in a quarry at St Nazaire-en-Royans. The fault plane exhibits dip-slip slickensides and separates Eocene red sandstones in the hanging wall from Hauterivian limestones in the footwall. From the fan shape of the Tertiary infill of the graben, we infer that this fault flattens at depth and connects with the basal detachment layer. The same occurs at the level of the St Lattier anticline: the SL-1 well indicates a restricted Eocene–Oligocene asymmetric basin related to a listric normal fault facing east. Unlike the fault observed at St Nazaire-en-Royans, this fault was partially inverted during the Late Miocene Alpine phase, completing the development of the St Lattier anticline: the SL-2 well (also located on top of the St Lattier anticline) encountered the same fault at depth where it corresponds to a normal fault. This characterizes the partial reactivation of an inherited normal fault, as the late reverse offset which permitted the development of the anticline does not completely cancel the normal downthrow at the lower part of the fault.

The section precisely illustrates the transition from platform-type Jurassian Mesozoic series to thick infill of the Dauphinois basin to the

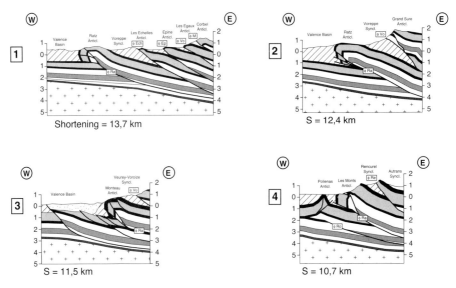

Fig. 7. Serial balanced cross-section through the western Chartreuse and Vercors massifs. Location in Fig. 2. Major thrust faults: φB, Barbières thrust; φCo, Chartreuse orientale thrust; φEch, Les Echelles thrust; φEp, Epine thrust; φL, Leoncel thrust; φM, Median thrust; φRa, Ratz thrust; φRe, Rencurel thrust; φRo, Royans thrust; φSl, St Lattier thrust; φVo, Voreppe thrust.

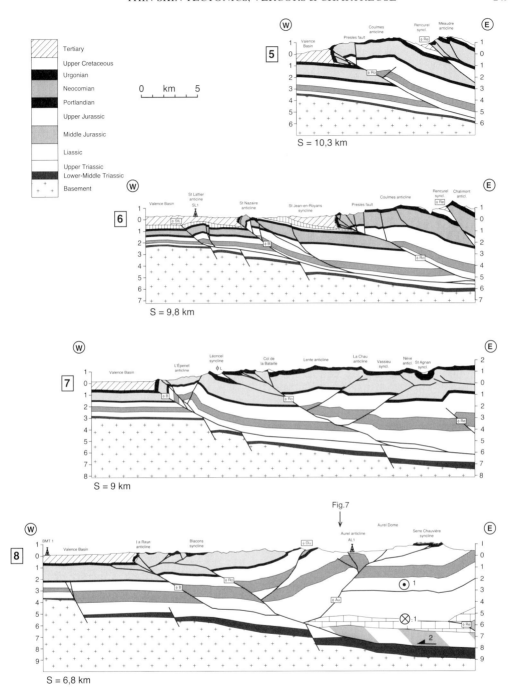

southeast. Based on the seismic profile parallel to the section, the northwest margin of the Southeast Basin appears as a gradual deepening of the basement accommodated by minor normal faults facing east, rather than controlled by any major abrupt fault, in good agreement with sections 1 and 3 that are also constrained by seismic data.

Section 7

Like the previous section, the Mesozoic sequence thickens dramatically from the Jurassian domain below the Valence Basin (2500 m thick) to the Dauphinois Basin in the east (7500 m thick). The Rencurel thrust practically disappears as the St Agnan syncline, representing the southern continuation of the Rencurel syncline, corresponds to an upright box-fold unaffected by a reverse fault along its eastern flank. The westward displacement along the Royans thrust also decreases with respect to previous sections and it emerges less clearly in the field. South of the St Jean-en-Royans syncline, which does not appear on this profile, the Royans thrust splits into several minor faults, the westernmost corresponding to the Leoncel thrust. The wide anticline in the hanging wall of the Royans thrust (i.e. the Coulmes anticline) is cut by east-verging backthrusts. These backthrusts are evidenced near the surface by folds involving the Urgonian limestones, like the Col de Chau and Nève anticlines. The decrease of the westward displacement along the Royans thrust is accommodated by the development of backthrusts in its hanging wall. In contrast to previous sections, this creates a certain symmetry of the thrust belt. At the 'Col de la Bataille' located in the centre of the section, one can note the presence of a passively transported normal fault facing east, separating Urgonian limestones in the footwall from Late Cretaceous layers in the hanging wall (see also Butler 1989). This fault is probably similar to the Presles fault observed further north, but in the present case the pre-Miocene left-lateral offset along the fault seems to be less. The westernmost structure of this section is represented by the Épenet anticline, a direct continuation of the St Nazaire anticline on profile 6. It corresponds to a wide fault-propagation box-fold superimposed upon a basement faulted zone. On this profile, the transition from Jurassian platform domain to Dauphinois Basin is sharp, and this faulted zone represents the northeastern prolongation below the Tertiary infill of the Valence Basin of the NE–SW-directed Cévenole margin, which

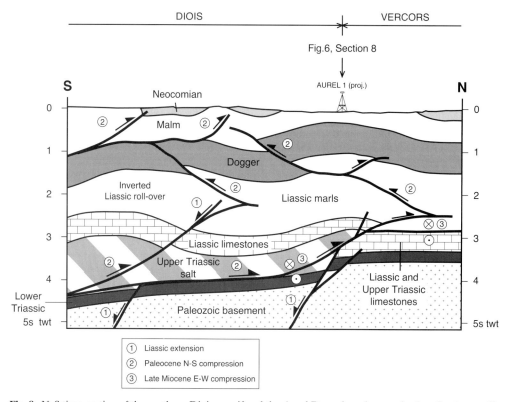

Fig. 8. N–S time-section of the northern Diois massif and the Aurel Dome based on a seismic-reflection profile (location in Fig. 2). Note the southward deepening of the basement–cover interface and the development of Triassic massive salt layers south of the Aurel Dome.

delimits the Southeast Basin from the Massif Central (Figs 1 & 2). On this section, the Vercors thrust front initiates above inherited Jurassic basement faults, deepening the basal detachment layer eastward and causing a rapid change in the thickness of the sedimentary cover.

Section 8

This section passes through the Aurel Dome and parallels the Drôme valley up to the Montoison-1 well to the west. It is calibrated by two seismic profiles, one coinciding with the section and one N–S-directed located in the vicinity of the Aurel-1 well, both of which provided structural information about the geometry at depth of the Aurel Dome. Again, the section reveals a significant eastward thickening of the Mesozoic cover, from 3400 m in the Montoison well-1 to over 9000 m in the east. There is no evidence of west-verging thrusts in the field: backthrusts are predominant, the hanging walls of which correspond to the Aurel and Saillans anticlines. According to our interpretation of the seismic section, the Saillans fault is not presently rooted in the basement as before suggested on the basis on gravimetric and aeromagnetic data (Flandrin & Weber 1966), and the top of the basement deepens eastward without any relationships to thrust faults in the cover. This confirms the complete decoupling between the Mesozoic cover and its pre-Triassic basement in the Vercors massif as already advocated by Butler (1987).

West of the section, the frontal thrust is not superimposed upon the southwestward continuation of the Isère fault observed on section 7. The margin of the Dauphinois Basin is located farther west as a major fault corresponding to the northeastward prolongation of the Cévennes fault (Fig. 2) controls the eastward thickening of the Triassic–Jurassic sequence. This fault passes through the southern Valence Basin east of the Montoison-1 well. East of the thrust front which does not emerge, one can observe short-wavelength backthrust folds related to superficial décollement levels hosted in the Neocomian marly layers.

The wide Aurel Dome denotes a relative elevation of the Mesozoic cover as the base Dogger, for instance, culminates at –2000 m (absolute depth) under the Serre–Chauvière syncline, whereas the depth of the same stratigraphic horizon is about –4000 m east of the previous section. This vertical uplift is interpreted, on the basis of a N–S-trending seismic profile running through the Aurel-1 well (Fig. 8), as an interference pattern resulting from the N–S Palaeocene compression and E–W Late Miocene phase. The deformation front of the Pyrenean fold-belt in southeastern France is superimposed onto the Drôme valley (Fig. 2), south of which early E–W-directed folds and thrust occur in the Diois and Baronnies massifs. Some of these Pyrenean structures are accentuated by the Late Alpine compression, as Miocene layers that unconformably overlay Cretaceous or Eocene strata crop out as E–W-trending subvertical beds. The leading edge of the Pyrenean orogen in the study area correlates with the northward disappearance of Triassic rock salts (Fig. 2), which is superimposed on a basement flexure inclined southward, as the absolute depth of the Palaeozoic basement exceeds 10 km below the erosion level in the Diois and Baronnies massifs (Philippe 1995). In these massifs, detachment of the Mesozoic cover above its basement initiated in the Palaeocene in relation to the N–S Pyrenean compression, since the main E–W-directed thrust faults developed at this time are rooted at the level of Keuper rock salt.

Conclusions on the cross-sections

According to seismic and well data, the geometry of the 'Cévenole margin' varies laterally from SW to NE: to the south, it corresponds to a NE–SW-directed normal fault zone, separating the Jurassian platform domain from the subsiding Mesozoic Southeast Basin, whereas to the north of the Vercors Plateau and in the Chartreuse massif, transition from Jurassian sequence and Subalpine series is accommodated by a smooth flexure zone. The so-called 'Isère fault' rapidly vanishes NE and was not reactivated as right-lateral strike-slip fault during the Late Miocene 'Alpine' phase. En-échelon folds located east of the Valence Basin (St Lattier, Varacieux and Polienas anticlines) are considered to be fault-related anticlines, which developed near the oblique thrust front of the Vercors massif. As such, the Épenet–St Nazaire anticline shows particular inner structures (Fig. 9) that confirm that en-échelon folds are not necessarily generated by a basement dextral wrench-fault, but can develop in case of complete detachment of the sedimentary cover: as depicted on profile 7, this anticline corresponds to a fault-propagation fold superimposed upon the oblique western margin of the Southeast Basin, but Neocomian marly layers permit decoupling between the stiff Urgonian limestones (Barremian/Aptian) and the underlying sequence. Accordingly, pre-Urgonian beds outcropping in the core of the Épenet anticline are deformed as second-order en-échelon folds

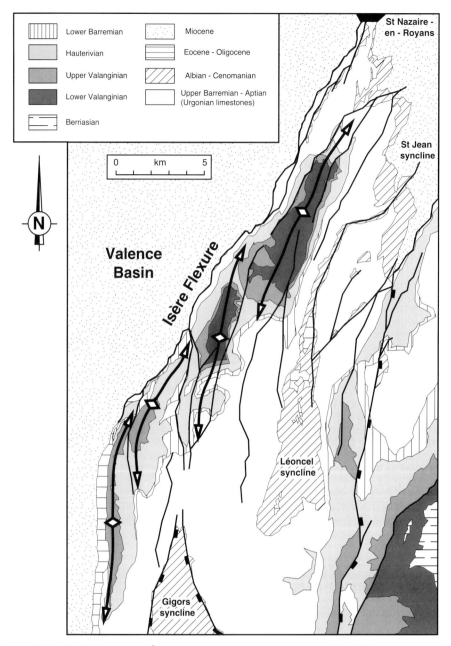

Fig. 9. Detailed geological map of the Épenet–St Nazaire fault-propagation anticline (after 1:50,000 geological maps edited by the Bureau de Recherches Géologiques et Minières). Note the superimposition of a wide box-fold parallel to the basin margin, implying the Urgonian limestones, upon second-order en-échelon folds that affect Lower Cretaceous layers in its core.

oblique to the inverted basin margin, while Urgonian and post-Urgonian strata are bent as a large box-fold parallel to the latter.

As revealed by field and well data, some of these folds (i.e. the St Lattier and St Nazaire anticlines) correspond to pre-existing Eocene–Oligocene roll-over structures that were partially or completely reactivated as thrust folds during the Late Miocene compression. In the western Vercors massif (at least), detachment of the cover above its pre-Triassic basement was initiated before the onset of the Alpine event, for two reasons: (1) south of the Drôme river, E–W-trending folds and thrusts developed largely during the Late Cretaceous–Palaeocene compressive phase, above a sole thrust hosted in the Late Triassic rocksalt; (2) in the centre and north of the massif, the E–W extension, spanning Late Eocene to Early Miocene, enabled gravity sliding of the Mesozoic series eastward, using Triassic and/or Liassic marly layers. This led to the development of asymmetrical grabens related to listric normal faults involving the 'allochthonous' cover. Such half-grabens (St Lattier and St Jean-en-Royans synclines) are arranged 'en-échelon', directly above the Southeast Basin margin, as a result of the obliquity between the Oligocene extension and the western boundary of the basin (see analogue modelling experiments of oblique gravity extension by Gaullier et al. 1993). The thin-skin inversion at the western edge of the Southeast Basin have been controlled in a large part by the 3D geometry of the previous extensional faults.

The study area is characterized by a notable changeover of the tectonic style along-strike. Northern sections (Fig. 7, sections 1, 2 and 3) exhibit short-wavelength structures consisting of a series of thrust sheets verging west bounded by thrust faults characterized by large displacements. Southern sections (Fig. 7, sections 7 and 8) display a more symmetrical thrust belt as the main structures evoke large pop-up structures limited by antithetic forethrusts and backthrusts (apart from superficial folds and thrusts related to intra-Cretaceous or intra-Oxfordian décollement levels). Significant shortening is accommodated by eastward back-thrusting which hardly accounts for the deformation at the southern part of the Vercors massif. This lateral changeover of tectonic style correlates with two essential facts: (1) an along-strike variation of the thickness of the detached cover and (2) a longitudinal decrease of the shear-strength of the Liassic shales acting as basal detachment layer, as they thicken from north to south, in addition to the development of Triassic salt in the southernmost Vercors (see next section).

Numerical and analogue considerations

The serial cross-sections demonstrate that the tectonic styles of the western Chartreuse and Vercors massifs vary along-strike from north to south. As mentioned above, two main parameters together differentiate the northern and southern parts of the study area: (1) the nature of the basal detachment layer that evolves from reduced Liassic shales in the north to Triassic rocksalt and thick Liassic shales in the south; and (2) the thickness of the allochthonous Mesozoic cover. The critical taper model of Davis et al. (1983) stipulates that internal deformation results in the development of a triangular cross-section accretionary fold-and-thrust belt that continues to thicken until a critical taper is attained. Depending on the rheology of the detachment layer, the basal shear strength τb influences both the plunge of the $\sigma 1$ axis (Mandl & Shippman 1981; Fig. 10a) and the critical taper ($\alpha + \beta$) of the belt (Davis & Engelder 1985). If the basal shear strength is high (no evaporites), the $\sigma 1$ axis plunges steeply forward. The dip of the backthrusts is so high so that they rarely develop, while slightly inclined forethrusts accommodate most of the shortening. Deformation cannot propagate far from the rear of the wedge, resulting in a high-tapered and asymmetric fold belt as only foreland-verging thrusts form (Fig. 10b, case 1). By contrast, if the shear-strength of the detachment layer is low, (e.g. presence of rock salt and/or increase of the thickness of the sheared basal layer, (Chapple 1978), the $\sigma 1$ axis remains nearly horizontal so that the fore- and backthrusts have comparable dips (δf and δb, respectively, in Fig. 10a). Deformation propagates far from the rear of the wedge, of which the taper remains low, as symmetric compressive structures form (i.e. pop-up structures, Fig. 10b, case 2). The thickness of the allochthonous sedimentary pile represents the second major parameter controlling thrust propagation within a fold-and-thrust belt: brittle analogue experiments demonstrate that for a given basal shear strength, roughly symmetrical pop-ups develop in a perfect piggy-back sequence, and the spacing of two successive thrusts is governed by the thickness of the detached sedimentary cover (Colletta et al. 1990; Fig. 10c). As such, even if the amount of shortening is constant, the greater the initial thickness of the allochthonous cover is, the further from the rear of the wedge the deformation propagates. According to these results, both the north to south variation of the tectonic style and the offset of the thrust front of the western Chartreuse and Vercors massifs are believed to be induced directly by the combined increase of the thickness of the Mesozoic cover and the longitudinal decrease of the basal shear strength, as the detachment layer becomes progressively more efficient in the south of the study area (Fig. 10b).

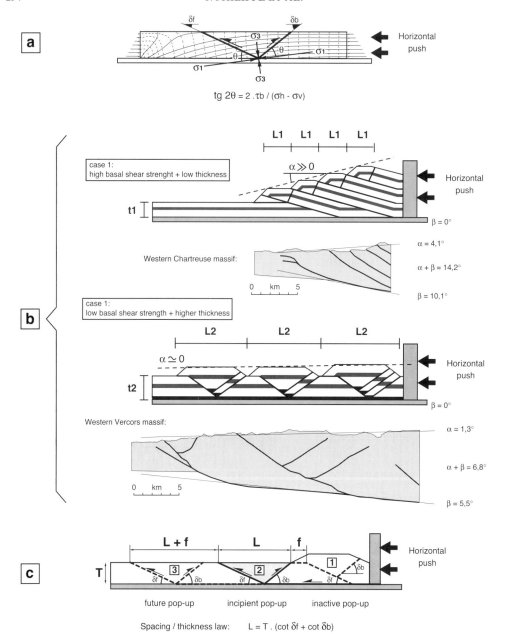

Fig. 10. (a) Stress trajectories within a rectangular body horizontally shortened by a vertical mobile back-stop and location of potential conjugated forethrust and backthrust (Mandl & Shippman 1981; modified). (b) Scheme illustrating the sequential development of pop-up structures bounded by conjugated thrust and backthrusts in a homogeneous model, the spacing of which is related to the thickness of the initially rectangular body (Colletta et al. 1990; modified). (c) Application of the Coulomb wedge model to the western Chartreuse and Vercors massifs, based on numerical models (Davis et al. 1983; Davis & Engelder 1985) and analogue models (Colletta et al. 1990; Calassou et al. 1993; Liu Huiqi et al. 1992; Philippe 1994, 1995). In these cases, the slope at the base of the wedges β equals to zero.

Experimental approach

In order to validate our interpretation of the mechanism of thin-skin inversion in the Vercors and Chartreuse massifs (notably that en-échelon folds and thrusts were able to develop along the transverse margin of the basin without the participation of basement wrench fault), a multi-layered analogue model was constructed, in which limestones and sandstones are represented by sand and glass powder and detachment layer by glass microbeads. It was designed to study thrust propagation in a simple brittle model characterized by longitudinal thickening of the detached cover.

Experimental set-up and boundary conditions

The bottom of the model is made of polystyrene and roughly mimics the basement topography beneath the western Chartreuse and Vercors massifs (Fig. 11). It displays an elevated part representing the Jurassian domain on the left and a deeper part simulating the thick Mesozoic fill of the southern Vercors area on the right. The higher compartment is underlain by a brittle cover 9 mm thick, consisting of two layers of sand and glass-powder. The other compartment is made of three layers of sand and glass powder reaching a total thickness of 24 mm, with a thin basal layer of glass microbeads to simulate the weak Triassic and Liassic layers where the sole thrust is hosted. These two domains are separated by a flexure zone inclined to the right that becomes progressively steeper toward the lower edge of the box, corresponding to the western margin of the Southeast Basin. The model was uniformly deformed at high strain rates (10 cm h^{-1}) by displacing a vertical back-stop from right to left, to simulate the horizontal push of the external Alpine massifs. During the deformation of this model, transverse cross-sections were acquired at several stages using an X-ray scanner. On completion of the experiment, serial cross-sections 3 mm apart were gathered, allowing a complete 3D analysis (for further explanations about the X-ray tomography technique, see Colletta *et al.* 1991).

Main results of the experiment

Thanks to the three series of developing cross-sections (Fig. 12), we first observe that frontal ramps propagate according to a forward mode and the thrust spacing is related to the thickness of the deformed sand-cake: in the thicker part of the model (section C–C'), the deformation front is located at a greater distance from the mobile back-stop than in the reduced compartment (section A–A').

The surface views (Fig. 13) show that an oblique ramp (thrust no. 2) develops directly above the flexure zone and first emerges in the

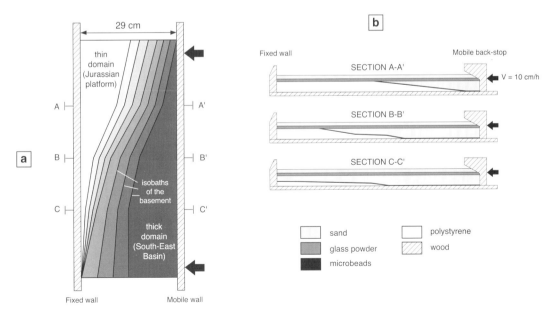

Fig. 11. (**a**) Surface view of the bottom of the model. (**b**) Three seriated cross-sections of the model showing along-strike thickening of the sedimentary cover from the upper edge to the lower edge of the model.

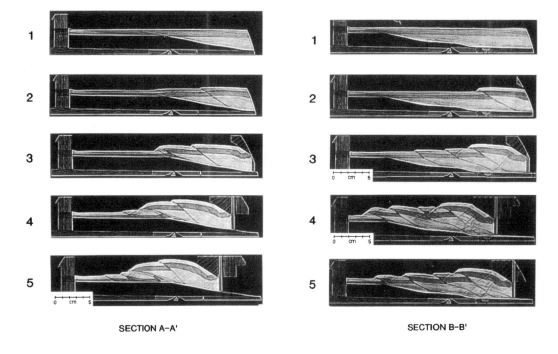

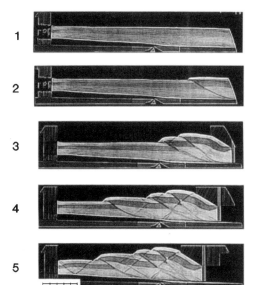

Fig. 12. Serial evolutive cross-sections of the model during the experiment. Location in Figs 11 and 13. Deformation is time-independent.

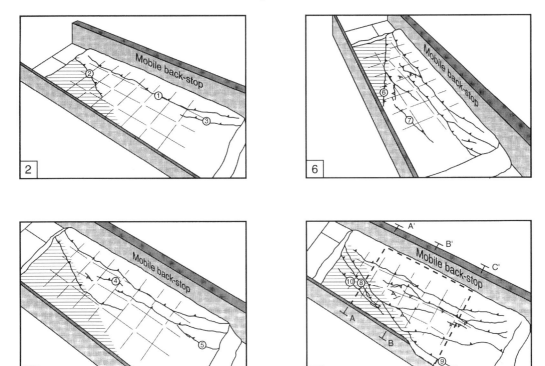

Fig. 13. Surface views of the model at four steps of the experiment among nine stages of deformation (drawn after photographs). Numbers of thrusts and backthrusts are related to their order of appearance. The cross-hatched area corresponds to the thin compartment simulating the Jurassian platform-domain. Discontinuous lines are passive markers forming squares of 5 cm × 5 cm. The dotted area in the centre of the model at the final stage (no. 9) indicates the zone covered by a 3D acquisition (see Figs 14 and 15).

vicinity of the upper border of the box, where the sedimentary wedge is very close to the backstop. Further on, this oblique ramp no. 2 propagates along-strike toward the lower edge of the model. In stage 4, the oblique ramp no. 2 transits to a frontal ramp that vanishes laterally, while several frontal ramps develop in the deep domain at the back of the transverse zone (ramps nos 3 and 4). In stage 6, a new oblique ramp forms (thrust no. 6) from a branch-point with the pre-existing oblique ramp no. 2. Accordingly, it propagates far from the mobile backstop in comparison with a new frontal ramp that develops in the inner compartment (ramp no. 7). The latter emerges at a distance from the frontal ramp no. 3 governed by the spacing/thickness law (Fig. 10b), and propagates along-strike to connect with the oblique ramp no. 6 at stage 9. In the final stage of the experiment (stage 9), the outermost frontal thrust of the thick compartment (thrust no. 9) forms in the lower side of the model and trends perpendicular to the shortening axis. Two en-échelon thrusts appear within the reduced domain in front of the transverse zone (thrusts nos 8 and 10).

The following observations can be made from the evolutive cross-sections and surface views: (1) a major oblique ramp develops directly above the transition zone between the two compartments of the model; (2) this fault propagates at each step at a greater distance from the mobile back-stop than frontal ramps that develop later within the thick domain, and some of them branch off the transverse ramp; (3) small oblique thrust sheets form within the external compartment in front of the major transverse ramp, as the tectonically thickened inner domain plays the role of an oblique rigid back-stop (for detailed explanations, see Philippe 1995).

Seriated cross-sections in the model in the final stage (Fig. 14) show the longitudinal evolution of the thrust sheets conditioned by the

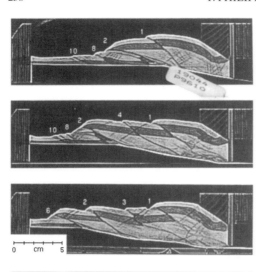

Conclusions on the experiment

The boundary conditions of the experiment obviously fail to account for all the parameters that influenced the deformation of the western Chartreuse and Vercors massifs. We incorporated neither intermediate décollement levels within the sedimentary series (simulating Oxfordian and Neocomian marly layers), nor preexisting extensive structures (representing the St Jean-en-Royans and St Lattier Oligocene half-grabens). Moreover, no viscous layer was placed at the bottom of the model near the lower lateral edge of the box, such as silicone putty, to simulate ductile Triassic rock salt. As shown on the balanced cross-sections, all these parameters have closely conditioned the location of the main thrust faults, the tectonic style and the evolution of the two massifs. Despite this, the model displays some major analogies with the natural example: (1) frontal thrust sheets are wider in the thick domain and laterally evolve in narrow thrust sheets in the transverse sedimentary wedge; (2) the amount of displacement along the transverse ramp is higher in the vicinity of the upper lateral edge of the box (Figs 11 & 12, cross-section A–A') and progressively decreases toward the lower part of the model (Fig. 12, cross-section C–C') where the same shortening is accommodated by a wider thrust belt; (3) the outermost thrust sheet is located at the edge of the thick compartment and laterally vanishes above the oblique flexure zone, agreeing well with the northern termination of the Épenet–St Nazaire anticline; (4) thrust faults developed in the thin compartment of the model (faults nos 8 and 10 in Fig. 13) strike obliquely to the transverse ramp. These small thrust sheets are quite similar to the N–S-directed en-échelon folds west of the Vercors front in the Valence Basin (St Lattier, Varacieux and Polienas anticlines). We consider these folds as ramp-related anticlines branched to the regional sole thrust. The results of the analogue model fully support this interpretation.

Conclusions and discussion

The leading edge of the Chartreuse and Vercors fold-and-thrust belts represents a transverse zone characterized by a progressive westward offset of the thrust front from north to south arising from the palaeogeographic configuration of the western part of the Southeast Basin prior to its Neogene inversion. The location of this transverse segment results from the oblique western margin of the Dauphinois Basin which corresponds either to a gradual transition zone

Fig. 14. Seriated cross-sections between sections A–A' and C–C' at the final stage showing spatial evolution along-strike of the internal structures. Numbers of thrusts and backthrusts are identical to those in Fig. 12.

thickness of the detached cover. In particular, the pop-up structure located at the leading edge of the model near the lower border of the model, which is related to thrust no. 9 and its conjugated backthrust, laterally terminates in the vicinity of the oblique thrust developed at the limit of the inner domain. This periclinal closure of a fault-bend fold above the transverse leading thrust displays some analogies with the northern part of the Épenet–St Nazaire anticline which vanishes northward, like an en-échelon fold. A comparison between a horizontal section in the centre of the model and a map view of the middle part of the western Vercors and Chartreuse transverse zone (Fig. 15) reveals geometrical similarities. In particular, each of the ramps nos 8 and 10 located in front of the transverse zone corresponds to small en-échelon thrust faults which combine to form a single ramp near the surface.

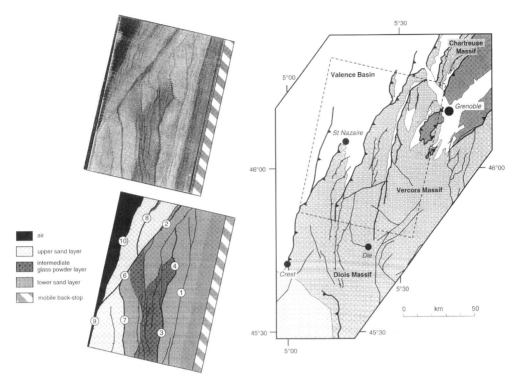

Fig. 15. Comparison between an horizontal section in the centre of the model (location in Fig. 13) and a map view of the middle part of the western Vercors and Chartreuse transverse zone. Numbers of thrusts and backthrusts of the model are identical to those in Figs 13 and 14.

(Chartreuse and northern Vercors), or to an abrupt basement normal fault separating two domains of contrasted sedimentary infill (southern Vercors). The along-strike thickening of the allochthonous Mesozoic cover and the decrease of the basal shear-strength (related to the thickening of the Liassic shales and the appearance of the Triassic evaporites) cause an increase in the wavelength of the thrust sheets and lead to the development of a more symmetrical and wider thrust belt, from north to south.

Serial balanced cross-sections demonstrate that the Mesozoic cover is detached above the stable Palaeozoic basement thanks to a regional sole thrust hosted in the Triassic evaporites and/or Liassic marls. The amounts of shortening apparently decrease from north (13–14 km) to south (7 km). This apparent along-strike variation of the westward displacement is possibly linked to smaller amounts of shortening at a crustal-scale, i.e. beneath the External crystalline massifs to the east. This argues in favor of a counterclockwise rotation of the Pelvoux–Belledonne–Mt Blanc–Aiguilles Rouges backstop about a pole located south, in good agreement with the model proposed by Vialon *et al.* (1989) and Vialon (1990) stipulating that significant amounts of rotation account in the western Alpine tectonics. However, it remains very difficult to quantify precisely the shortening amounts of the Subalpine massifs, since the relationships between the allochthonous crystalline massifs and their detached sedimentary cover still await clarification. Below the eastern Chartreuse massif, field data indicate that both the allochthonous Mesozoic cover and the external segment of the Belledonne massif are significantly displaced above large thrust faults verging west (Fig. 3a). In contrast, the western edge of the Belledonne massif east of the Vercors Plateau represents a crustal wedge above which the Mesozoic cover is backthrust eastward (Butler 1987; Arpin *et al.* 1988; Gamond 1994; Fig. 3b). Drastic variations of the Early Jurassic sequence from the Drac valley to the Mure Dome (Butler 1987; Bas 1988) provide evidence of active Liassic blocktilting in the eastern Vercors: the Belledonne massif can be interpreted as a Liassic palaeohorst short-cut by a deep thrust fault verging west (Butler 1987; Fig. 3b). Displacement of the western Belledonne massif is compensated by backthrusting of the thick Mesozoic cover of the eastern Vercors Plateau to the east, creating a triangle zone geometry. Unfortunately, the quantification of the displacement along this backthrust remains quite imprecise, as it depends on the tectonic style adopted for the construction of the section.

The thickness of the Mesozoic sequence in the southern part of the study area (exceeding 12 km in the Diois and Baronnies massifs) helps to rank the Southeast Basin among the deepest Mesozoic basins in the world. Due to the complexity of its present architecture which results from a polyphase evolution since the Late Palaeozoic, no detailed consideration has yet been given to the mechanical behaviour of the crust during the Jurassic (WNW–ESE directed) and Cretaceous (N–S directed) stretching events. Moreover, the western Alpine foreland is characterized by a considerable northward thinning of the Mesozoic cover from the Southeast Basin to the Molasse Basin east of the Jura fold-and-thrust belt: such a variation of the sedimentary/granitic–granulitic ratio of the European crust could have played an important role in the evolution of the Alpine foreland during the Oligocene and Miocene (see Karner & Watts 1983): as deep thrusting prevailed in the inner Alps, a broad foredeep established above a thick elastic crust as in northern Switzerland where the Mesozoic cover is thin. In contrast, no wide foreland trough was able to develop where the lower–middle crust is topped by a very thick Mesozoic sedimentary fill as in southeastern France. This certainly accounts, at least in part, for the southwestward disappearance of the Molasse Basin north of the Vercors and Chartreuse massifs (Figs 1 & 2).

This study was part of a thesis granted by and carried out in the Department of Geology and Geochemistry of the Institut Français du Pétrole. Y.P. wishes to thank B. Colletta and P. Balé for their support and constructive remarks at each stage of the work. The authors are indebted to the reviewers A. Artoni and J.L. Mugnier for providing helpful comments and suggestions.

References

ANGELIER, J. 1984. Tectonic analysis of fault slip data sets. *Journal of Geophysical Research*, **89**, (B7), 5835–5848.

ARNAUD, H. 1973. Mise en évidence d'un important décalage anté-Miocène de sens sénestre, le long de la faille de Presles (Vercors occidental). *Comptes Rendus de l'Académie des Sciences, Paris*, **276**, 2245–2248.

—— 1981. De la plate-forme urgonienne au bassin vocontien – Le Barrémo-Bédoulien des Alpes occidentales entre Isère et Buech (Vercors méridional, Diois oriental et Dévoluy). *Géologie Alpine, Mémoires*, **12**, University of Grenoble, France.

ARPIN, R. 1988. *Déformations et déplacements des massifs subalpins de Vercors et Chartreuse*. Thèse de Doctorat, Université de Grenoble.

——, GRATIER, J. P. & THOUVENOT, F. 1988. Chevauchements en Vercors-Chartreuse déduits de l'équilibrage des données géologiques et géophysiques. *Comptes Rendus de l'Académie des Sciences, Paris*, **307**, II, 1779–1786.

ARTHAUD, F. & MATTE, P. 1977. Late Paleozoic strike-slip faulting in southern Europe and northern Africa: result of a right-lateral shear zone between the Appalachians and the Urals. *Geological Society of America Bulletin*, **88**, 1305–1320.

BAS, T. 1988. Rifting liasique dans la marge passive téthysienne: le haut-fond de la Mure et le bassin du Beaumont (Alpes occidentales). *Bulletin de la Société géologique de France*, **8**, IV, 717–723.

BAUDRIMONT, A. F. & DUBOIS, P. 1977. Un bassin mésogéen du domaine péri-alpin: le Sud-Est de la France. *Bulletin des Centres de Recherches de l'Exploration – Production, Elf-Aquitaine*, **1**, 261–308.

BEACH, A. 1981. Thrust tectonics and crustal shortening in the external French Alps based on a seismic cross-section. *Tectonophysics*, **79**, T1–T6.

BOILLOT, G., MONTADERT, L., LEMOINE, M. & BIJU-DUVAL, B. 1984. *Les marges continentales actuelles et fossiles autour de la France*. Editions Masson, Paris.

BRGM 1975. *Carte géologique de la France au 1/50,000, feuille Romans-sur-Isère*. Bureau de Recherches Géologiques et Minières, Orléans.

BUTLER, R. W. H. 1987. Thrust evolution within previously rifted regions: an example from the Vercors, French Subalpine Chains. *Memorie della Societa Geologica Italiana*, **38**, 5–18.

—— 1989. The influence of pre-existing basin structure on thrust system evolution in the Western Alps. *In*: COOPER, M. A. & WILLIAMS, G. D. (eds) *Inversion Tectonics*. Geological Society, London, Special Publications, **44**, 105–122.

—— 1992. Structural evolution of the western Chartreuse fold and thrust system, NW French Subalpine chains. *In*: MCCLAY, K. R. (ed.) *Thrust tectonics*. Chapman & Hall, London, 287–298.

CALASSOU, S., LARROQUE, C. & MALAVIEILLE, J. 1993. Transfer zones of deformation in thrust wedges: an experimental study. *Tectonophysics*, **221**, 325–344.

CHAPPLE, W. 1978. Mechanics of thin skinned fold and thrust belts. *Geological Society of America Bulletin*, **89**, 1189–1198.

COLLETTA, B., BALLARD, J. F., BALÉ, P., BÉNARD, F., LETOUZEY, J. & PINEDO, R. 1990. Thrust propagation and non-cylindrical structures in small-scale models: kinematics and 3D analysis by X-ray tomography. *In*: MCCLAY, K. (org) *Thrust Tectonics Conference*, London, 1990. Abstracts with program, 64.

——, LETOUZEY, J., PINEDO, R., BALLARD, J. F. & BALÉ, P. 1991. Computered X-ray tomography analysis of sandbox models: examples of thin-skinned thrust systems. *Geology*, **19**, 1063–1067.

COWARD, M. P. & DIETRICH, D. 1989. Alpine tectonics: an overview. *In*: COWARD, M. P., DIETRICH, D. & PARK, R. G. (eds) *Alpine Tectonics*. Geological Society, London, Special Publications, **45**, 1–29.

CURNELLE, R. & DUBOIS, P. 1986. Evolution mésozoïque des grands bassins sédimentaires français: bassins de Paris, d'Aquitaine et du Sud-Est. *Bulletin de la Société géologique de France*, **8**, II, 529–546.

DARDEAU, G., ATROPS, F., FORTWENGLER D., GRACIANSKY DE, P. C. & MARCHAND, D. 1988. Jeu de blocs et tectonique distensive au Callovien et à l'Oxfordien dans le bassin du Sud-Est de la France. *Bulletin de la Société géologique de France*, **8**, IV, 771–777.

DAVIS, D., SUPPE, J. & DAHLEN, A. 1983. Mechanics of fold-and-thrust belts and accretionary wedges. *Journal of Geophysical Research*, **88**, B2, 1153–1172.

DAVIS, D. M. & ENGELDER, T. 1985. The role of salt in fold-and-thrust belts. *Tectonophysics*, **119**, 67–88.

DE GRACIANSKY, P. C. & LEMOINE, M. 1988. Early Cretaceous extensional tectonics in the southwestern French Alps: a consequence of North-Atlantic rifting during Tethyan spreading. *Bulletin de la Société géologique de France*, **8**, IV, 5, 733–737.

——, DARDEAU, G., LEMOINE, M. & TRICART, P. 1988. De la distension à la compression: l'inversion structurale dans les Alpes. *Bulletin de la Société géologique de France*, **8**, IV, 5, 779–785.

DEBELMAS, J. 1974. Le passage du Jura aux Chaînes subalpines. *In*: DEBELMAS, J. (ed.) *Géologie de la France, vol. 2: Les chaînes plissées du cycle alpin et leur avant-pays*, Doin, Paris, 461–464.

DEVILLE, E., MASCLE, A., LAMIRAUX, C. & LE BRAS, A. 1944. Tectonic styles, reevaluation of petroleum plays of southeastern France. *Oil and Gas Journal*, **31**, 53–58.

DOUDOUX, B., MERCIER DE LEPINAY, B. & TARDY, M. 1982. Une interprétation nouvelle de la structure des massifs subalpins savoyards (Alpes occidentales): nappes de charriage oligocènes et déformations superposées. *Comptes Rendus de l'Académie des Sciences, Paris*, **295**, 63–68.

FLANDRIN, J. 1966. Sur l'âge des principaux traits structuraux du Diois et des Baronnies. *Bulletin de la Société géologique de France*, **7**, VIII, 376–386.

—— & WEBER, C. 1966. Données géophysiques sur la structure profonde du Diois et des Baronnies. *Bulletin de la Société géologique de France*, **7**, VIII, 387–392.

GAMOND, J. F. 1994. Normal faulting and tectonic inversion driven by gravity in a thrusting regime. *Journal of Structural Geology*, **16**, 1–9.

—— & ODONNE, P. 1984. Critères d'identification des plis induits par un décrochement profond: modélisation analogique et données de terrain. *Bulletin de la Société géologique de France*, **7**, XXVI, 115–128.

GAULLIER, V., BRUN, J .P., GUÉRIN, G., LECANU, H. & COBBOLD, P. 1993. Raft tectonics: the effects of residual topography below a salt decollement. *Tectonophysics*, **228**, 363–381.

GIDON, M. 1981. La structure de l'extrémité méridionale du massif de la Chartreuse aux abords de Grenoble et son prolongement en Vercors. *Géologie Alpine*, **57**, 93–107.

—— 1988. L'anatomie des zones de chevauchement du massif de la Chartreuse (Chaînes subalpines septentrionale, Isère, France). *Géologie Alpine*, **64**, 27–48.

GILLCRIST, R., COWARD, M. P. & MUGNIER, J. L. 1987. Structural inversion and its controls: examples from the Alpine foreland and the French Alps. *Geodinamica Acta*, **1**, 5–34.

GRATIER, J. P., MÉNARD, G. & ARPIN, R. 1989. Strain-displacement compatibility and restoration of the Chaînes Subalpines of the western Alps. *In*: COWARD, M. P., DIETRICH, D. & PARK, R. G. (eds) *Alpine Tectonics*. Geological Society, London, Special Publications, **45**, 65–81.

HAYWARD, A. B. & GRAHAM, R. H. 1989. Some geometrical characteristics of inversion. *In*: COOPER, M. A. & WILLIAMS G. D. (eds) *Inversion Tectonics*. Geological Society, London, Special Publications, **44**, 17–39.

KARNER, G. D. & WATTS, A. B. 1983. Gravity anomalies and flexure of the lithosphere. *Journal of Geophysical Research*, **88**, B12, 449–477.

LEMOINE, M. 1985. Structuration jurassique des Alpes occidentales et palinspastique de la Téthys ligure. *Bulletin de la Société géologique de France*, **8**, 1, 126–137.

——, BAS, T., ARNAUD-VANNEAU, A., ARNAUD, H., DUMONT, T., GIDON, M., BOURBON, M., GRACIANSKY DE, P. C., RUDKIEWICZ, J. L., MÉGARD-GALLI, J. & TRICART, P. 1986. The continental margin of the mesozoic Tethys in the Western Alps. *Marine and Petroleum Geology*, **3**, 179–199.

LETOUZEY, J. 1990. Fault reactivation, inversion and fold-thrust belt. *In*: LETOUZEY, J. (ed.) *Petroleum and tectonics in mobile belts*. Editions Technip, Paris, 101–128.

LIENHARDT, M. J. 1984. Trias - Puisssance et faciès de la partie supérieure. *In*: DEBRAND-PASSARD, S., COURBOULEIX, S. & LIENHARDT, M. J. (eds) *Synthèse géologique du Sud-Est de la France*, vol. 2. Mémoires du BRGM, **126**, planche T2.

LIU HUIQI, MCCLAY, K. R. & POWELL, D. 1992. Physical models of thrust wedges. *In*: MCCLAY, K. R. (ed.) *Thrust tectonics*. Chapman & Hall, London, 71–81.

MANDL, G. & SHIPPMAN, G. K. 1981. Mechanical model of thrust sheet gliding and imbrication. *In*: MCCLAY, K. R. & PRICE, N. J. (eds) *Thrust and nappe tectonics*. Geological Society, London, Special Publications, **9**, 79–98.

MASCLE, A., BERTRAND, G. & LAMIRAUX, C. 1994. Exploration for and production of oil and gas in France. A review of the habitat, present activity and expected develoment. *In*: MASCLE, A. (ed.) *Hydrocarbon and petroleum geology of France*. European Association of Petroleum Geologists, Special Publications, **4**, 3–28.

——, JIMENEZ, C., DUVAL, P., BIJU-DUVAL, B., TRÉ-MOLIÈRES, P., ARNAUD, H. & CARRIO, E. 1992. *Field trip to: the Western Alps and their foreland in France*. EAEG–EAPG Annual meeting, Paris.

——, VIALLY, R., DEVILLE, E., BIJU-DUVAL, B. & ROY, J. P. 1996. The petroleum evaluation of a tectonically complex area: the western margin of the Southeast Basin (France). *Marine and Petroleum Geology*, **13**, 941–961.

MCCLAY, K. R. & BUCHANAN, P. G. 1992. Thrust faults in inverted extensional basins. *In*: MCCLAY, K. R. (ed.) *Thrust tectonics*. Chapman & Hall, London, 93–104.

MÉNARD, G. 1979. *Relations entre structures profondes et structures superficielles dans le Sud-Est de la France. Essai d'utilisation de données géophysiques*. Thèse 3° cycle, Université de Grenoble.

—— 1988. *Structure et cinématique d'une chaîne de collision: les Alpes occidentales et centrales*. Thèse de Doctorat d'Etat, Université de Grenoble.

—— & MOLNAR, P. 1988. Collapse of a Hercynian Tibetan Plateau into a late Paleozoic European basin and range province. *Nature*, **334**, 235–237.

—— & THOUVENOT, F. 1987. Coupes équilibrées crustales: méthodologie et applications aux Alpes externes: un modèle cinématique. *Geodinamica Acta*, **1**, 35–45.

MORETTI, I. & LARRÈRE, M. 1989. LOCACE: computer-aided construction of balanced geological cross sections. *Geobyte*, Oct. **89**, 16–24.

MUGNIER, J. L., ARPIN, R. & THOUVENOT, F. 1987. Coupes équilibrées à travers le massif subalpin de la Chartreuse. *Geodinamica Acta*, **1**, 125–137.

PHILIPPE, Y. 1994. Transfer zone in the southern Jura thrust belt (eastern France): geometry, development and comparison with analogue modelling experiments. *In*: MASCLE, A. (ed.) *Hydrocarbon and petroleum geology of France*. European Association of Petroleum Geologists, Special Publications, **4**, 327–346.

—— 1995. *Rampes latérales et zones de transfert dans les chaînes plissées: géométrie, conditions de formation et pièges structuraux associés*. Thèse de Doctorat, Université de Savoie.

RICHÉ, P. & TRÉMOLIÈRES P. 1987. *Tectonique synsédimentaire sur la bordure orientale du bassin tertiaire de Valréas*. IFP-ENSPM internal report, 34 943.

ROURE, F., BRUN, J. P., COLLETTA, B. & VAN DEN DRIESSCHE, J. 1992. Geometry and kinematics of extensional structures in the Alpine Foreland Basin of southeastern France. *Journal of Structural Geology*, **14**, 503–519.

SIDDANS, A. W. 1979. Arcuate fold and thrust patterns in the Subalpine Chains of Southeast France. *Journal of Structural Geology*, **1**, 117–126

VIALON, P. 1990. Deep Alpine structures and geodynamic evolution: an introduction an outline of a new interpretation. *In*: ROURE, F., HEITZMANN, P. & POLINO, R. (eds) *Deep structure of the Alps*. Mémoires de la Société géologique de France, Paris, **156**; Mémoires de Société géologique de Suisse, Zürich, **1**; Vol. Speciale della Società Geologia Italiana, Roma, **1**, 7–14.

——, ROCHETTE, P. & MÉNARD, G. 1989. Indentation and rotation in the western Alpine arc. *In*: COWARD, M. P., DIETRICH, D. & PARK, R. G. (eds) *Alpine Tectonics*. Geological Society, London, Special Publications, **45**, 329–338.

WELBON, A. I. & BUTLER, R. W. H. 1992. Structural styles in thrust belts developed through rift basins: a view from the western Alps. *In*: LARSEN, R. M., BREKKE, B. T. & TALLERAAS, E. (eds) *Structural and tectonic modelling and its application to petroleum geology*. Norwegian Petroleum Society (NPF), Special Publications, **1**, 469–479.

Horizontal shortening control of Middle Miocene marine siliciclastic accumulation (Upper Marine Molasse) in the southern termination of the Savoy Molasse Basin (northwestern Alps/southern Jura)

CHRISTIAN BECK[1], ERIC DEVILLE[2], ERIC BLANC[1], YANN PHILIPPE[3], MARC TARDY[1]

[1]*Laboratoire de Géodynamique des Chaînes Alpines, UPRES A 5025 C.N.R.S., Université de Savoie, Campus Savoie-Technolac, F-73 376 Le Bourget du Lac Cedex, France*
[2]*Institut Français du Pétrole, 1-4, Av. de Bois-Préau, F-92 506 Rueil-Malmaison, France*
[3]*D.E.P.F., Elf-Aquitaine (Production), Boussens, F-31360 Saint-Martory, France*

Abstract: Between the southern Jura and the Subalpine Chains (Bornes, Bauges, and Chartreuse Massifs), the Burdigalian–Langhian–Serravalian tidal siliciclastic 'Upper Marine Molasse' (UMM) outcrops in N–S synclines between N–S-trending ramp anticlines separating southward-narrowing synclines. A surface survey and an interpretation of transverse seismic-reflection profiles across the latter confirm that they represent progressively individualized and separated sedimentation areas. Combined seismic configuration and surface geometry show westward migration of depo-axes, and coeval progressive steepening close to each ramp, upon blind thrusts. Several thick earthquake-disturbed layers offer further evidence of a syntectonic sedimentation. A regime of flexure–subsidence prevailing during the Chattian–Aquitanian (the so-called Lower Freshwater Molasse, LFM) was overprinted by thin-skin shortening. The latter, combined with Burdigalian–Langhian higher eustatic sea-level, favoured the accumulation and the preservation of abundant tidal bedforms and sequences, which are generally poorly preserved in stable settings.

The northwestern portion of the Alpine Orogen, results from the collision (still in progress) between the European and the Apulian marginal lithospheres, which were previously separated by the Tethyan ocean. These two domains are presently represented within the Outer and Inner Zones of the Alps (Fig. 1), with the remnants of Tethyan ocean belonging to the Inner Zones. The Outer Alps are represented, in the French–Swiss part, by the so-called Subalpine chains (Fig. 1); the latter, together with the Molasse Basin and the Jura Mountains, represent the foreland fold-and-thrust belt discussed in the present article. The crustal scale structure was expressed by the ECORS-CROP deep seismic profile and its interpretation (Guellec *et al.* 1990 a, b; Tardy *et al.* 1990).

From the External Crystalline Massifs to the front of the Jura, the westernmost and latest (Oligocene–Miocene) horizontal shortening (several tens of kilometres) was accommodated by a basal décollement of the Mesozoic–Cenozoic sedimentary pile (Laubscher 1992), associated thrusts and ramp anticlines, transcurrent faults and lateral ramps (Burkhard 1990; Deville *et al.* 1994; Laubscher 1974, 1981, 1992; Mugnier & Ménard 1986; Philippe 1994; Wildi *et al.* 1991; Signer & Gorin 1995; and many others). The aforementioned foreland is widely developed in the Swiss part of the western Alps (approximately north of Geneva, Fig. 1), whereas it narrows southward.

A major component of the foreland is a very thick Oligocene and Miocene, mainly siliciclastic, accumulation, either in continental lacustrine or in marine epicontinental environments known as the Molasse Basin. Several kilometers thick and about one hundred kilometers wide in the northern part (Bern area), it thins progressively and narrows southward, where the southern termination of the Molasse Basin (known locally as the Savoy Basin) is more deformed (Lamiraux 1977; Gidon 1990; Blanc 1991; Allen & Bass 1993; etc.). In this region, two major lithostratigraphic components crop out, the Late Oligocene–Early Miocene Lower Freshwater Molasse (LFM) and the Mid–Late Miocene Upper Marine Molasse UMM). The latter is considered as approximately coeval with compressive tectonics in the study area (Guellec *et al.* 1990*a*).

The area framed in Fig. 1 was chosen to investigate the relationships between the molassic sedimentation and the foreland tectonics (see

BECK, C., DEVILLE, E., BLANC, E., PHILIPPE, Y. & TARDY, M. 1998. Horizontal shortening control of Middle Miocene marine siliciclastic accumulation (Upper Marine Molasse) in the southern termination of the Savoy Molasse Basin (northwestern Alps/southern Jura). *In:* MASCLE, A., PUIGDEFÀBREGAS, C., LUTERBACHER, H. P. & FERNÀNDEZ, M. (eds) *Cenozoic Foreland Basins of Western Europe.* Geological Society Special Publications, **134**, 263–278.

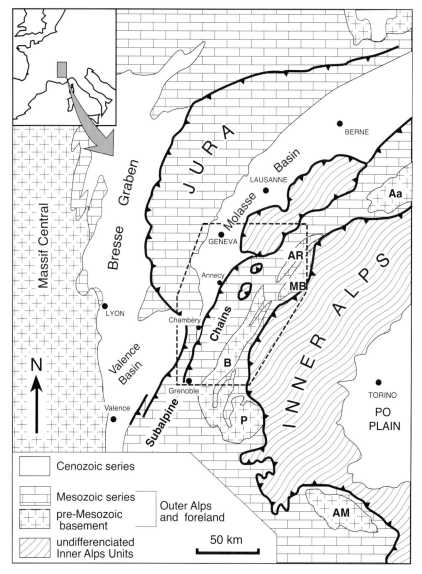

Fig. 1. Geological and structural framework of the northwestern Alps and their foreland. Aa, Aar Massif; AR, Aguilles Rouges Massif; MB, Mont-Blanc Massif; B, Belledonne Massif; P, Pelvoux Massi; AM, Argentera–Mercantour Massif.

also Guellec *et al.* 1990*a*; Blanc 1991; Allen & Bass 1993; Deville *et al.* 1994). A structural, stratigraphical and sedimentological analysis has been carried out to determine how a shallow-water siliciclastic sedimentation accommodated the compressive deformation of underlying upper crust, and what structural information can be inferred from the internal architecture of the sediments. Both approaches were aimed to ascertain the timing of deformation, and test previous estimation of the horizontal shortening velocity (Deville *et al.* 1994). Two sets of data are presented: (1) present-day geometrical setting of the different molassic units and (2) internal configuration of these different units, based on a combination of surface geological surveying and oil-industry seismic-reflection data. Sedimentological characteristics of the Upper Marine Molasse, and its relationship with the underlying Lower Freshwater Molasse, were investigated by one of us (Blanc 1991) for the area framed on Fig. 2.

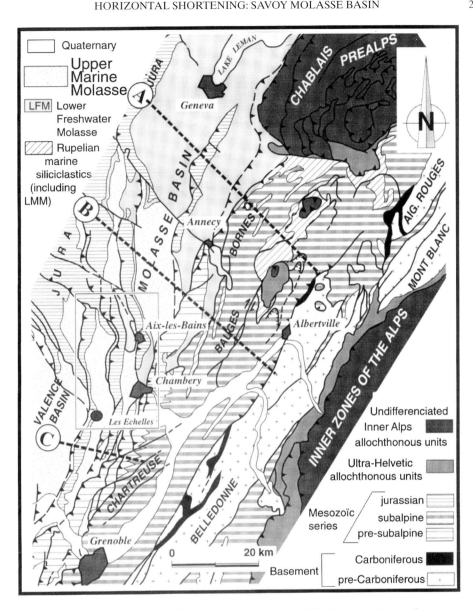

Fig. 2. The southern Jura-Subalpine Massifs junction (with location of Fig. 4 cross sections, and area presented in Figs 5 and 7).

Detailed work on the sedimentology and basin analysis of the UMM in the whole region has been published by Allen & Bass (1993). Thus, types of sedimentary bodies and corresponding depositionnal environments will be directly mentioned according to these previous works and will not be discussed here. Conversely, their respective positions within structural sets will be presented as arguments for their syn-tectonic deposition.

Structure and Tertiary stratigraphy of the southern Jura–western Alps junction

The structural map (Fig. 2), and the stratigraphic chart (Fig. 3) were synthesized from data published in Deville *et al.* (1995).

The Savoy area of the Molasse Basin is longitudinally divided into two portions (Fig. 2).

(1) North of Annecy, between the innermost Jura anticline (to the west) and the Subalpine

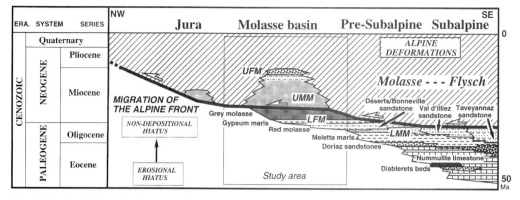

Fig. 3. Synthetic Cenozoic chronostratigraphic chart across southern Jura/northwestern Alps borders. The area discussed concerns the framed median part (Molasse Basin).

frontal thrust (to the east), the Molasse Basin shows a relatively simple structure with only a major thrust fault and associated ramp anticline. The molassic infill is almost exclusively represented by the Chattian-Aquitanian Lower Freshwater Molasse. East of Geneva, the Molasse Basin is directly bounded by allochtonous units belonging to the inner zones of the Alps (Prealpine Chablais klippes).

(2) Between Annecy and the northern Chartreuse Massif, the molassic infill is preserved in N–S-elongated outcrop belts (synclines) separated by several westward-thrusted ramp anticlines; locally (Part III-B-1), eastward-verging thrusts occur, due to NW–SE transfer faults. Here, between Annecy and Les Echelles, the Upper Marine Molasse is the major outcropping component.

Although the Mesozoic series have not been subdivided in Fig. 2, it must be pointed out that, on the eastern (or innermost) boundary of the Jura, the Late Cretaceous and Early Cretaceous disappear progressively westward. Several authors (Allen *et al.* 1991; Sinclair *et al.* 1991; Crampton & Allen 1995) have associated this fact with forebulge erosion related to tectonic loading and flexure of the Subalpine marginal crust. In the inner Jura (Fig. 3), this event led to an erosional hiatus of Cretaceous sediments and could explain the scarcity of Oligocene–Miocene deposits. Progressive westward migration of LFM sedimentation (framed part of Fig. 3) and pinch-out, are syndepositionnal and related to westward thrusting in the Subalpine realm (Late Oligocene); the UMM deposition covered part of the Inner Jura. At the difference, in the latter, the lack of Cretaceous deposits is considered as erosional.

The area discussed below corresponds to the central part of the stratigraphic chart (Fig. 3) made along a NW–SE (transverse) profile (Deville *et al.* 1995). The Lower Freshwater Molasse (LFM) or 'Obere Meeresmolasses' (OMM) of germanic authors and the Upper Marine Molasse (UMM) or 'Untere Süsswassermolasses' (USM; see Berger 1992) are the main components. In the following, we will use the terms LFM and UMM. The geometry and structural settings of these Molasse divisions, and especially of the UMM, is now presented.

Biostratigraphic and other data of previous studies (Rigassi 1957; Lamiraux 1977; Burbank *et al.* 1992; Berger 1992; Allen & Bass 1993) are used below to constrain the aged of the LFM and UMM. Detailed sampling done by one of us (Blanc 1991) did not provide more accurate biostratigraphic data with respect to Lamiraux's work (1977). In the Chambéry area, the UMM is Burdigalian and Langhian in age, whereas further westward (north of Les Echelles), the Serravalian is also represented. West of Annecy, the LFM is Late Chattian and Aquitanian in age (Burbank *et al.* 1992).

Present-day geometry of Upper Marine Molasses units in the southern termination of the Savoy Basin

Structural setting

Three cross-sections are presented (Fig. 4): between Annecy and Geneva (section A), between Aix-les-Bains and Annecy (section B), and south of Les Echelles (section C). They were prepared and balanced using surface surveys, well logs, and various unpublished seismic-reflection profiles (Philippe 1994); for this procedure (see also Deville *et al.* 1994), the UMM was treated as a unique volume.

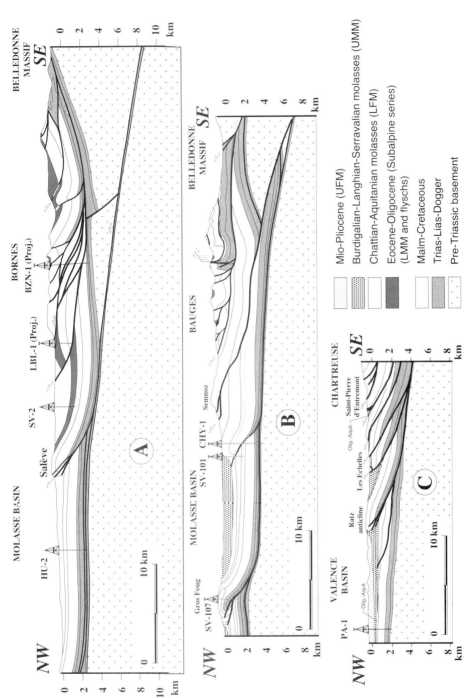

Fig. 4. Transverse sections across the southern Jura/Outer Alps borders (location on Fig. 3). These sections are based on surface data, seismic-reflection data and oil exploration wells.

The basal décollement of the whole system (prolongating intra-basement thrusting below the subalpine realm) is located within Keuper evaporitic and argillaceous layers). Out-of-sequence thrust planes cut the whole Mesozoic pile and induced fault-propagation folds; secondary décollements (not represented) locally played a minor role: between Early and Mid-Jurassic, and between Late Jurasssic and Neocomian (Philippe 1994). Concerning the Tertiary series, the frontal thrust of the Subalpine realm (between SV-2 and LBL-1 wells, on Sections A; east of CHY-1 well on Section B) cutted the LFM (Chattian–Aquitanian). For the Molasse Basin (Gros Foug and Salève, and other similar structures; sections A and B), the tectonic structures and sediments architecture envisaged hereafter is slightly different from previous works: ramps cut a major part of the LFM but do not reach the UMM basal layers; they end as blind thrusts parallel to bedding.

Several differences can be identified from north to south.

(1) WNW–ESE horizontal shortening increases especially from section B toward section C, by mean of several ramp anticlines and, in the southern end, of a thrust pile.

(2) The Salève anticline (section A), which can be considered as the easternmost (or innermost) Jura structure approaches the Subalpine frontal thrust (section B), and disappears between sections B and C.

(3) Upper Eocene–Oligocene marine sediments are well developed in the Bornes massif (section A) and poorly represented in the Bauges massif (section B); they are unknown in the Chartreuse massif (section C); the continental part (LFM, Chattian-Aquitanian) of the molassic infill is well developed between Geneva and Aix-les-Bains; the marine part (UMM) overlies the latter west of Annecy and is present on the western flank of the Chartreuse massif.

(4) Between sections B and C, the LFM with its clear westward pinch out (Homewood et al. 1986), disappears as a continuous formation; local Oligocene freshwater marls and conglomerates are found west of the LFM pinline (see below).

(5) The Savoy Molasse Basin sensu stricto disappears between section B and C; the Valence basin, as well as the Bresse basin (see also Fig. 2), belong to the undeformed foreland.

(6) In the three sections, the LFM (Chattian–Aquitanian) appears completely affected by thrust-faults, while the UMM (Burdigalian–Langhian–Serravalian) is often only folded and not directly cut by thrust faults. The ramps may not reach the surface and end as 'blind' subhorizontal thrusts (see Fig. 4, Section B). We shall deal further with this particular situation.

Relations with the pre-Burdigalian substratum

In the framed area of Fig. 2, the relationship between the UFM and its substratum was surveyed in detail. A map of the pre-Burdigalian surface is presented in Fig. 5. This reconstruction (see also Allen & Bass 1993) shows a rather complicated situation possibly implying pre-Burdigalian deformations. The latter could be the consequence of the general Chattian–Aquitanian flexure of the foreland, but very local thick (25 m) conglomeratic discharges (up to 1 m blocks), within lacustrine marls, involving local provenance imply active faulting. Locally, these deposits rest directly on topmost Jurassic limestones (western edge of surveyed area, Fig. 5), by mean of a steep palaeosurface; they mainly rework neocomian limestones. The horizontal distribution of the conglomeratic discharges, together with the direction of their fluxes (Fig. 5), led us to propose a palaeotectonic scheme.

Considering the general westward pinching out of the Oligocene–Early Miocene LFM (Homewood et al. 1986), the lacustrine marls and intercalated conglomerates west of Chambéry (Fig. 5) appear to lie outside (west) of

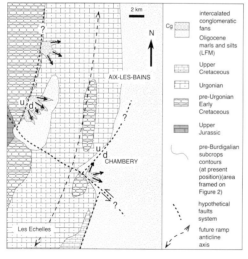

Fig. 5. Restoration of the pre-Burdigalian transgression surface in the southern termination of the Jura. Modified from Blanc (1991) (area located on Fig. 2). This subcrop map is based on detailed field surveys. Burdigalian basal calcareous shelly sandstones rest on different formations from Uppermost Jurassic to Oligocene.

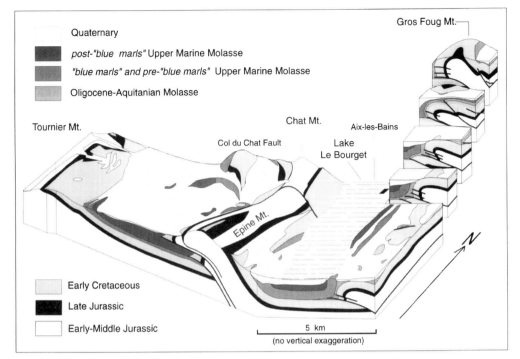

Fig. 6. Three-dimensional view of the southernmost Jurassian structures and associated molassic units. Modified from Blanc (1991) The Upper Marine Molasse (Burdigalian–Langhian–Serravalian) appears in two different situations: either folded but not affected by ramp thrusts (Gros Foug anticline), or cut by subhorizontal prolongations of ramp thrusts (Chat Mount anticline).

the flexure domain. This peculiar situation was observed only at the southern termination of the Jura, close to the northeastern corner of the Valence basin (Fig. 2). Hence, this pre-Burdigalian deformation could more likely be related to the Oligocene–Aquitanian extensional tectonics of the Valence and Bresse Basins belonging to the west European rift system (Bergerat et al. 1990). The Les Echelles–Chambéry area (Figs 2 and 5) could represent a zone of interference between reduced Alpine foreland flexure and the general Oligocene extensional regime (with normal and strike slip faulting) affecting the eastern edge of the Massif Central.

Thus, focusing on the UMM (Burdigalian–Langhian–Serravalian), we can identify the main units on both sides of the Gros Foug anticline (section C, Fig. 4, framed area in Fig. 2) and of the Chat-Epine anticline. These structures are shown in three dimensions in Fig. 6.

Internal configuration of main Upper Marine Molasse units

Two sets of data and analysis (presented below) were compiled.

(1) Lithostratigraphy and sedimentology of all available outcrops between Aix-les-Bains and Les Echelles (framed area in Fig. 2, and Fig. 3) across a Jura anticline.

(2) interpretation of seismic reflection data between Aix-les-Bains and Annecy. The first set of data, although laterally discontuous, illustrates the effect of ramp-anticline building on distribution and thicknesses of sedimentary bodies; the second provides a continuous information across the main growth syncline.

Lithostratigraphical profile across a Jura ramp-anticline and adjacent synclines: field data

General statements. In terms of chronology, the UMM has been divided into two major stratigraphic units (Fig. 6). They are separated by a thin prominent horizon of laminated silty clays known as the 'blue marls' (Lamiraux 1977) or 'Montaugier lithosome' (Allen & Bass 1993), in which Late Burdigalian-Langhian calcareous nanoplankton have been reported.

A set of logs corresponding to field sections

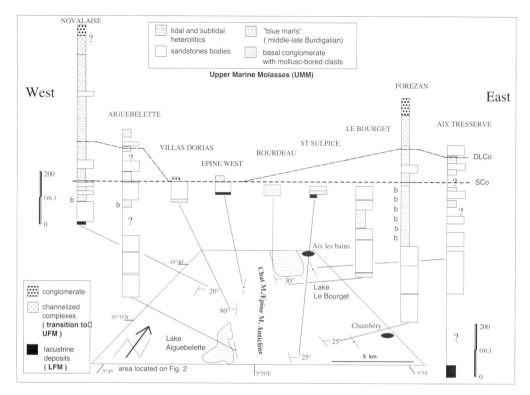

Fig. 7. Stratigraphy of the Upper Marine Molasses through a transverse profile of the southernmost Jurassian structures. (area located in Fig. 2)
b, bioturbations; SCo, litho- and biostratigraphical correlation; DLCo, correlation based on an earthquake-disturbed layer.

are drawn in Fig. 7; considering the N–S trend and cylindricity of the structures, these logs are projected on a W–E profile across the Chat–Epine anticline and neighbouring synclines (same area as for Fig. 5, location on Fig. 2). For each section (Blanc 1991), lithology, sedimentary features (related to depositional processes and bioturbations), and types of stratification, allow the reconstruction of the lateral and vertical evolution of the sedimentary environments. They range from supratidal to subtidal, the 'blue marls' representing a deeper deposit; evidence for palaeo-tidal regime is abundant, especially in the Nan Forezan section (see also Allen & Bass 1993; Assémat 1991). Different sequences bearing tidal cyclicities were found in this section, characterized and analyzed using the method of Tessier et al. (1989). This tidal environment, observed in different parts of the Alpine marine Mid–Upper Miocene (Homewood & Allen 1985; Tessier & Gigot 1989) was recently modeled by Martel et al. (1994); they proposed a combination of two tidal waves (a northward one from the Tethys marine-oceanic realm, a southwestward one from the Paratethys realm), with 2–3 m amplitudes.

Correlations. The logged sections are simplified in Fig. 7; they were selected when their lower contact with the LFM or older formation was observed (see below); for several sections, the transition to Upper Frewater Molasse (UFM) was also observed. A distinction is drawn between the main sandstone bodies (tidal bars, subtidal hydraulic dunes, see Berné et al. 1989) and the intervals consisting of different types of alternating sand, silt, and clayey silt (the 'heterolitics' of Allen & Bass 1993). Due to sedimentary discontinuities and rapid lateral variations (either longitudinal, N–S, or tranverse, W–E) especially for sandstones bodies, lithostratigraphic (geometric) correlations could not be established among the whole set of logs. Chronostratigraphic correlations may rely only on poor biostratigraphic data. Thus, we choose (see also Blanc 1991) the 'blue marls' horizon both for its depositional environments (clearly deeper than the other UMM lithosomes, and

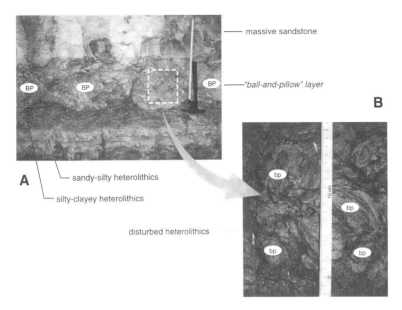

Fig. 8. Earthquake-disturbed layers in Burdigalian deposits. (Nant Forezan section; see Fig. 7). BP refers to decimetric (first order) ball-and-pillow structures; bp refers to centimetric (second order) ball-and-pillow structures. Within the latter, the original stratification (microprogradation within sand lenses or ripples?) is preserved.

widely regionally developed; see also Allen & Bass 1993) and for its constrained age. This correlation appear as SCo on Fig. 7.

Except for the western Novalaise log, placing the 'blue marls' (SCo) as a horizontal reference shows a thickness decrease in the anticlinal position. This is not caused by subsequent erosion as in the Aiguebelette and Forezan sections the transition to the braided alluvial complexes of the UFM base was observed. We interpret this depositional geometry as directly related to progressive ramp-associated folding and relative subsidence on both sides, during the Burdigalian–Langhian–(Serravalian *pro parte?*). The maximum thickness measured for the UMM in the survey area is close to 800 m and the overall global eustatic rise amplitude estimated for the Burdigalian–Langhian period (Haq *et al.* 1988) is about 100 m. Thus, three processes could account together for the UMM infilling, which relative roles will be further discussed: general subsidence, folding, and sea-level changes.

The westernmost surveyed section (Novalaise) could indicate a minor pre-'blue marls' subsidence and a major post-'blue marls' subsidence; we interpret this fact as due to possible significant E–W diachronism of horizontal shortening.

Occurrence of earthquake-disturbed layers.
During the field survey, different disturbed layers were observed, consisting of sandy 'clasts' incorporated within silty clay or clayey silt. They are laterally continuous at scale of the checked outcrops (few tens of metres in Nan Forezan stream). Some of them are almost isopach layers (up to 1 m thick) made of a roughly isotropic mixture (sandy pebbles with sizes ranging from 1 cm to 20 cm); other disturbed layers exhibit the 'ball-and-pillow' structure. We choose to illustrate this case (Fig. 8); it was found in the Nan Forezan stream. The reworking of sandy material within silty-clayey material appears at two scales: (1) tens of centimetre-scale large 'pillows' regularly spaced (BP on Fig. 8), and (2) inside the latters, few centimetre-scale large sandy 'clasts' with preserved millimetric stratification (bp on Fig. 8) sometimes with cross-bedding. We interpret these second-order structures as remnants of ripples with internal micro-progradation; the whole initial level could be tidalites with dominant sand at the top and clayey silt dominant in the lower part. These 'ball-and-pillow' structures can be interpreted as earthquake-originated (see Montenat 1980). We discard here the possibility for these disturbances, with respect to their shallow water environment, to have been induced by storms or tidal waves (Tessier & Terwindt 1994).

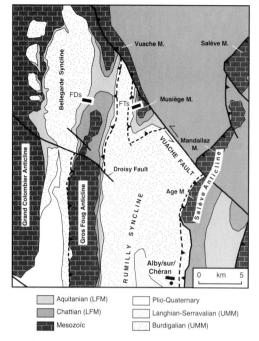

Fig. 9. Structural sketch map of the Gros Foug-Salève area. FDs, Findreuse river section; FTs, Fornant river section.

turbance was found on both sides of the Chat–Epine anticline, with same characteristics and same lithostratigraphic position above the 'blue marls'; considered as a unique layer, it was used for correlation (DLCo on Fig. 7) (Blanc 1991). Using this layer for correlation appears compatible with the previous correlation as it also underline thickness increasing on both side of the Chat–Epine mount. Furthermore, with respect to lateral E–W variation, the relative deepening in synclinal position (Aiguebelette and Forezan–Aix–Tresserve sites) is accentuated.

Following this interpretation for observed UMM post-depositionnal disturbances, we can use the latters as an additional argument for a significant seismo-tectonic activity during sedimentation.

Internal configuration of the Upper Marine Molasses in the Albanais–Rumilly syncline

Seismic-reflection data: Several seismic-reflection profiles by the Elf-Aquitaine (Production) Company (see also Deville *et al.* 1994, 1995; Philippe 1994) cross the Savoy Molasse Basin, in the region of the Gros Foug anticline and the Albanais-Rumilly syncline (Fig. 9). These structures are interrupted by the important Vuache strike slip Fault (northeastern corner of Fig. 9) (Blondel *et al.* 1988).

The Gros Foug anticline (see also Fig. 6) is cut by the Droisy Fault which transfers a relative westward-verging thrust fault (south) to an eastward back-thrust. Close to the Vuache Fault, the Musiège anticlinal structure is symmetrical with minor eastward and westward reverse faults.

As already suggested by different authors in other sedimentary basins (see Dreesen *et al.* 1989), earthquake-induced disturbances, with sufficient lateral extension and continuity, can be considered as isochrones, and used for stratigraphic correlation. One of the UMM strong dis-

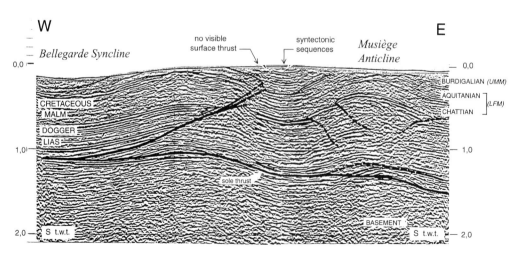

Fig. 10. E–W seismic profile between the Bellegarde syncline and the Musièges anticline.

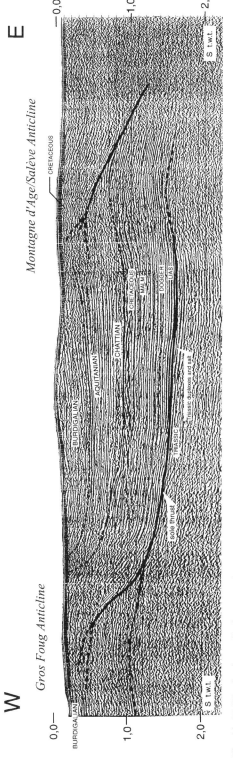

Fig. 11. E–W seismic profile between the Chautagne and Annecy area.

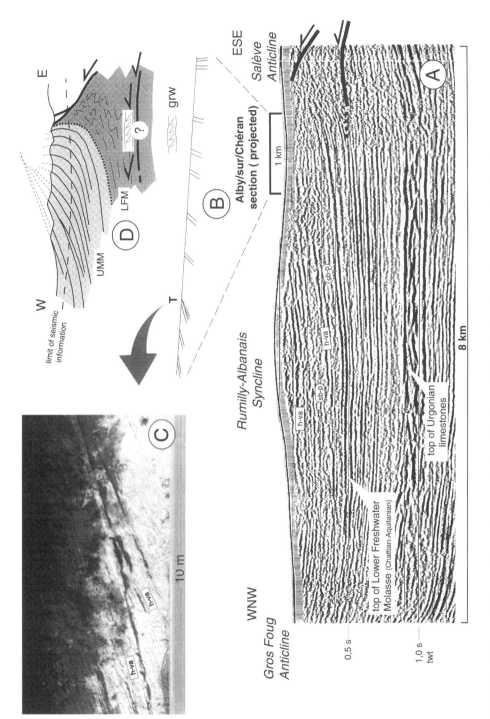

Fig. 12. Seismic-reflection profile across the southern termination of the Savoy Molasse Basin and complementary surface data. h-va, heterolithics with vertical aggradation; sb-p, sandstone bodies with progradation; grw, dissolved gypsum-rich water circulation. The schematic section (Part D) is built combining the seismic line and the Alby-sur-Chéran surface section. A similar geometry can be drawn for the western flank of the Gros Foug anticline along which fibrous gypsum veins were observed within the upper part of the LFM.

These structures appears on the seismic profile shown on Fig. 10, an E–W section across the Gros Foug and Musiège anticlines, north of the Droisy Fault. The Rumilly syncline (also called the Albanais syncline) and its bounding anticlines appear on the seismic profile of Fig. 11, an E–W section crossing the southern termination of the Age Mt (with Cretaceous outcrop).

On the profile shown in Fig. 12, we selected the central portion to study the Molasse structure (Fig. 12, Part A). Although acquisition and processing were not dedicated to detailed seismic stratigraphy, differences between internal configurations of the LFM and the UMM appear clearly.

The first is represented by a lower part (Chattian) with poor continuity of low-amplitude reflections, and a upper part (Aquitanian, sandstone-rich) with good continuity and strong amplitude reflections; the whole shows an eastward-divergent configuration.

The second is made of rather strong and continuous reflectors (or sets of two parallel reflectors) separating sets of low amplitude poorly organized reflectors. Within the latter, west-dipping oblique reflectors and associated truncations can be observed in places. By comparison with surface observations, we associate the two configurations of the UMM respectively with heterolithics succession with dominant vertical aggradation (h-va, Fig. 12, Part A) and large sandbodies with progradational structure (sb-p, Fig. 12, Part A).

The configuration of the whole UMM shows a tendency to westward migration of depo-axis; this general tendency is accommodated, at a more detailed scale, by westward thinning of sand bodies and by troncations within heterolithics. This architecture was observed in several other UMM units within the survey area (west of the Gros Foug anticline).

Surface data: the Alby-sur-Chéran section. South of the seismic profile shown in Fig. 12, the western approach to Alby-sur-Chéran exposes a 800 m cross-section within the UMM. This section is close to the southern termination of the Salève anticline (see location on Fig. 9); using bedding directions and following the structural axis of the Rumilly–Albanais syncline, we projected this surface section at the corresponding place on the seismic profile located a few kilometres to the north (Fig. 12; Parts A and B).

Part B shows the change in true dip (section orthogonal to bedding direction) and a clear internal unconformity (T, Fig. 12, Parts B and D). Lithology and bedding are identical above and below this surface: parallel stratification, few decimetres to few centimetre-thick alternations of slightly rippled sandstones and clayey silts or flaser bedding. Thus, this unconformity occurred within the deposition of a heterolitic (with vertical aggradation) episode. Interpreting this geometry demands tilting of the lower part and erosion. Due to lack of precise time-control, we propose two distinct explanations: (1) this syn-depositionnal movement was fast with respect to the duration of a single depositionnal sequence (third or fourth order if using sequence stratigraphy concepts); (2) the tilting was more progressive and a sequence boundary may account for the erosion. In both cases, we relate the geometry to folding close to the active ramp; choosing the second hypothesis leads to combine this deformation with eustatic fluctuations. Although only one unconformity could be observed on the Alby-sur-Chéran cross-section, the eastern flank of the Rumilly syncline (Fig. 12, Part D) probably included several similar structures. It could be compared to well-documented progressive unconformities described in the southern side of the Pyrenées (Riba 1976; Specht 1989; Poblet & Hardy 1995; Vergés *et al.* 1996; etc.).

Décollement horizons below the Upper Marine Molasse and fibrous gypsum occurrences.

The above-proposed interpretation of the Rumilly syncline (and similar structures in the Savoy Molasse) has a corrolary: the existence of disharmony and blind thrusting between the top of the Urgonian limestones and the base of the UMM (Fig. 12, Part A and D). Structural interpretation of seismic data (Fig. 11) led to subdivision of the upper part of the ramp into two fault planes; we consider the lower one, getting horizontal and parallel to bedding as a minor décollement. It should be located within the LFM just below the base of the UMM with progressive westward disappearance. We relate this problem to the occurrence of fibrous gypsum precisely at this level for the following.

Close to the northern termination of the Gros Foug anticline (site FDs in Fig. 9), the Findreuse river reference section (Burbank *et al.* 1992; Berger 1992) shows abundant gypsum in the upper part of the LFM, consisting of anastomosed, slightly oblique to bedding plane, several cm-thick, veins of fibrous gypsum, with high angle inclination of fibres with respect to the vein surfaces. In the Fornant section (site FTs in Fig. 9), a few similar gypsum veins were observed below a minor back-thrust plane. In both sections, the veins were observed within thick (up to 25 m for the Findreuse section) silty marls and clays (top of LFM) below thick sandstone layers (base of UMM). Recently, one of us (C.B.) found fibrous gypsum in 100 m long cores taken (for geotechnical purpose) in the LFM, close to the western front of the Salève Mt (Fig. 9). This gypsum occurs, either as millimetric bundles on low inclination fault planes and parallel to slickensides, or as thicker veins (1 cm) with high-angle inclination of fibres with respect to the vein surfaces; in the second case, the veins have the same dip as the fault surfaces. Thus, we consider these differences occurrences of gypsum are secondary and have a tectonic and hydraulic origin. In the different cross-sections (Figs 4 and 12, Part D) these gypsum occurrences probably correspond to low-angle blind thrust zone. Pending confirmation by a chemical signature (see Moore et al. 1989), we interpret these veins as the remnants of overpressured fluid flow along incipient or active thrust planes. As the latter represent zones of separation within the upper part of the LFM (Aquitanian) or close to the base of the UMM (Burdigalian), they may be considered as minor décollements acting along short distances and allowing, above them, the progressive folding of the upper part of the molassic series, while the lower part is tectonically thickened with disharmonic style. If so, this gypsum has two possible origins: (1) dissolution of evaporites deposited within the LFM (Reggiani 1989) and short distance displacement; (2) dissolution of deeper evaporites belonging to the Mesozoic and longer displacement along main décollements and thrusts.

Discussion and conclusions

In the narrow southern part of the Upper Marine Molasse domain, siliciclastic sedimentation developed on a complex, tectonically active, upper crust (Fig. 13). During the Burdigalian–Langhian, elongated furrows (corresponding to incipient synclines) accumulated thick subtidal to supratidal deposits. These deeper areas were probably connected above transverse (transfer) structures. The general subsidence, continuation of Chatian–Aquitanian flexure of the Alpine foreland and a few eustatic rises allowed episodes of deeper homogeneous sedimentation in the whole domain (basal transgression and 'blue marls'). In the Langhian and Serravalian, this morphology was accentuated and led to separate depositional areas whose western and eastern margins were affected by the development of ramp anticlines. Thus, the internal geometry of the molassic units represents a direct record of this deformation, reflected by depo-axis migration and internal unconformities; this reinforces the estimations of shortening velocities made by Deville et al. (1994). A few strongly disturbed layers are attributed to major seismic activity, further suggesting syntectonic deposition.

On the western (or eastern in case of backthrust) flank of each ramp, thick clayey–marly intervals in the LFM show disharmonic behaviour below the mainly sandy UMM. In general, the ramps do not reach and cut the UMM (as shown by accurate surface mapping), and are prolonged over a short distance (1–2 km) by subhorizontal blind thrusts. The latter acted as minor décollement horizons, possibly with overpressured fluid containing dissolved gypsum.

With regard to the Burdigalian palaeogeography, the Savoy basin, in front of the Jura mountains, appears as a strait between the wider and deeper Swiss part of the UMM domain and the southern French Alps portion (Tessier & Gigot 1989; Allen & Bass 1993). The occurrence of field criteria for a tidal regime with axial sediment transport, the conclusions of tidal-circulation modelling by Martel et al. (1994), and the

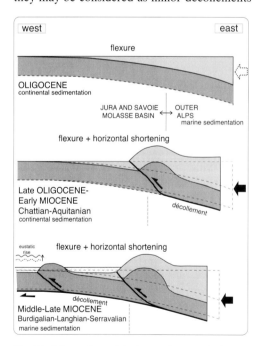

Fig. 13. Schematic model of tectonic control of Miocene sedimentation in the northwestern Alps foreland. During the Burdigalian (third stage), the horizontal NW–WE to WNW–ESE shortening begins to affect the Mesozoic–Cenozoic sedimentary pile west of the Subalpine frontal thrust; this progressively separates areas with relative subsidence within Upper Marine Molasse deposition domain.

size of the whole perialpine molassic basin, could identify the English Channel and North Sea shallow siliciclastic sedimentary domain as a possible present-day example for comparison (Allen & Bass 1993). In the latter, however, 'definitive' accumulation of sand bodies or tidal heterolitics is very localized, and the different deposits and bedforms of the coastal domain are progressively eradicated with its landward migration. Here, the particular geodynamic conditions of the molassic sedimentation (general subsidence and, above all, faster localized subsidence between elongated positive reliefs) produced the rapid and exceptional preservation of many different sequences and tidally cyclic bedforms. Conversely, this Burdigalian–Langhian–Serravalian sedimentation, with rapid imprint of minor eustatic fluctuations and of substratum deformation, progressively recorded and sealed the horizontal shortening.

Surface surveys were conducted on grants from the Elf-Aquitaine (Production) Company and the Institut Français du Pétrole; we also acknowledge the Elf-Aquitaine (Production) Company for allowing the use and publication of seismic data. We are grateful to J. Vergés and to an anonymous colleague for review and improvement of our first manuscript. Great thanks to B. Doudoux for precious help during part of the field-survey and to B. Tessier for tidal cyclicities analysis.

References

ALLEN, P. A. & BASS, J. 1993. Sedimentology of the Upper Marine Molasse of the Rhône-Alpes region, eastern France: implications for basin evolution. *Eclogae Geologicae Helvetiae*, **86**, 121–172.

——, CRAMPTON S. L. & SINCLAIR, H. D. 1991. The inception and early evolution of the North Alpine foreland basin, Switzerland. *Basin Research*, **3**, 143–163.

ASSEMAT, S. 1991. *Recherche et analyse des tidalites dans la molasse marine miocène du domaine subalpin dans la région de Frangy-Rumilly-Aix-les-Bains.* Mémoirs D.E.A. Géodynamique, Université de Savoie.

BERGER, J.-P. 1992. Correlative chart of the European Oligocene and Miocene: application to the Swiss Molasse Basin. *Eclogae Geologicae Helvetiae*, **85**, 573–609.

BERGERAT, F., MUGNIER, J.-L., GUELLEC, S., TRUFFERT, C., CAZES, M., DAMOTTE, B. & ROURE, F. 1990. Extensional tectonics and subsidence of the Bresse basin: an interpretation from ECORS data. *In*: ROURE, F., HEITZMANN, P. & POLINO, R. (eds) *Deep structure of the Alps*, Mémoires de la Société Géologique de France **156**, Mémoires de la Société Géologique de Suisse **1**, Vol. 2 Speciale *Società Geologica Italiana* **1**, 165–184.

BERNE, S., ALLEN, G., AUFFRET, J.-P., CHAMLEY, H., DURAND, J. & WEBER, O. 1989. Essai de synthèse sur les dunes hydrauliques géantes tidales actuelles. *Bulletin de Société de la Géologique de France*, **6**, 1145–1160.

BLANC, E. 1991. *Evolution sédimentaire syntectonique au front d'une chaîne de collision en environnement littoral. Le cas du Miocène (molassse marine) de la région de Chambéry, Savoie, France.* Mémoires D.E.A. Géodynamique, Université de Savoie.

BLONDEL, T., CHAROLLAIS, J., SAMBETH, U. & PAVONI, N. 1988. La Faille du Vuache (Jura méridional): un exemple de faille à caractère polyphasé. *Bulletin de la Société Vaudois des Sciences Naturelles*, **79**, 65–91.

BURBANK, D. W., ENGESSER, B., MATTER, A. & WEIDMANN, M. 1992. Magnetostratigraphic chronology, mammalian faunas, and stratigraphic evolution of the Lower Freshwater Molasse, Haute-Savoie, France. *Eclogae Geologicae Helvetiae*, **85**, 399–431.

BURKHARD, M. 1990. Aspects of the large-scale Miocene deformation in the most external part of the Swiss Alps (Subalpine Molasse and Jura fold belt). *Eclogae Geologicae Helvetiae*, **83**, 559–593.

CRAMPTON, S. & ALLEN, P. A. 1995. Recognition of forebuldge unconformities associated with early stage foreland basin development: example from the North Alpine foreland basin. *American Association of Petroleum Geologists Bulletin*, **79**, 1495–1514.

DEVILLE, E., BLANC, E., TARDY, M., BECK, C., COUSIN, M. & MENARD, G. 1994. Thrust propagation and syntectonic sedimentation in the Savoy Tertiary Molasse Basin (Alpine Foreland). *In*: MASCLE, A. (ed.) *Hydrocarbon and Petroleum Geology of France*. European Association of Petroleum Geologists Special Publications, **4**, 269–280.

DEVILLE, E., PHILIPPE, Y., BECK, C. & TARDY, M. 1995. Field-Trip Guide in the Western Alps: thrust dynamics of the outer alps. *American Association of Petroleum Geologists Conference*, Nice, September 14–16 1995.

DRESSEN, R., PAPROTH, E. & THOREZ, J. 1989. Events documented in Famennian sediments (Ardenne-Rhenish massif, Late Devonian, NW Europe). *In*: MACMILLAN, N. J., EMBRY, A. F. & GLASS, D. J. (eds) T*he Devonian of the World, Vol. 3 Sedimentation*, 295–308.

GIDON, M. 1990. Les décrochements et leur place dans la structuration du massif de la Chartreuse (Alpes occidentales françaises). *Géologie Alpine*, **66**, 39–56.

GUELLEC, S., MUGNIER, J.-L., TARDY, M., ROURE, F. 1990a. Neogene evolution of the western Alpine foreland in the light of ECORS data and balanced cross-section. *In*: ROURE, F., HEITZMANN, P., POLINO R. (eds) '*Deep structure of the Alps*, Mémoires de la Société Géologique de France **156**, Mémoires de la Société Géologique de Suisse **1**, Speciale *Società Geologica Italiana* **1**, 165–184.

——, LAJAT, D., MASCLE, A., ROURE, F. & TARDY, M. 1990b. Deep seismic profiling and petroleum potential in the western Alps: constraints with

ECORS data, balanced cross-sections and hydrocarbon modiling. In: PINET, B. & BOIS, C. (eds) *The Potential of Deep Seismic Profiling for Hydrocarbon Exploration*, Editions Technip, Paris, 425–438.

HAQ, B. U., HARDENBOL, J. & VAIL, P. R. 1988. Mesozoic and Cenozoic chronostratigraphy and cycles of sea-level change. In: WILGUS, C. K, POSAMENTIER, H., ROSS, C. A. & KENDALL, C. G. (eds), *Sea-level change: an integrated approach*, Special Publications of SEPM, **42**, 71–101.

HOMEWOOD, P & ALLEN, P. 1985. Wave-, tide-, and current-controlled sandbodies of Miocene Molasse, western Switzerland. *American Association of Petroleum Geologists Bulletin*, **65**, 2534–2545.

——, —— & WILLIAMS, G. D. 1986. Dynamics of the Molasse Basin of western Switzerland. *International Association of Sedimentinertologists Special Publications*, **8**, 19–217.

LAMIRAUX, C. 1977. *Géologie du Miocène des chaînons jurassiens méridionaux et du Bas-Dauphiné nord-oriental entre Chambéry et La Tour du Pin*. PhD Thesis, Université de Savoie.

LAUBSCHER, H. P. 1974. Basement uplift and décollement in the Molasse Basin. *Eclogae Geologicae Helvetiae*, **67**, 531–537.

—— 1981. The 3D propagation of décollement in the Jura. In: MCLAY, K. R. & PRICE, N. J. (eds), '*Thrust and nappes tectonics*'. Geological Society of London Special Publications, **9**, 311–318.

—— 1992. Jura kinetics and the Molasse Basin. *Eclogae Geologicae Helvetiae*, **85**, 653–675.

MARTEL, A. T., ALLEN, P. A. & SLINGERLAND, R. 1994. Use of tidal-circulation modelling in paleogeographical studies: an example from the Tertiary of the Alpine perimeter. *Geology*, **22**, 925–928.

MONTENAT, C. 1980. Relations entre déformations synsédimentaires et paléoséismicité dans le Messinien de San Miguel de Salinas (Cordillères bétiques orientales, Espagne). *Bulletin de la Société Géologique de France*, **XXII**, 3, 501–509.

MOORE, J. C., MASCLE, A., TAYLOR, E. & ODP LEG 110 SHIPBOARD SCIENTIFIC PARTY. 1989. Tectonics and hydrogeology of the northern Barbados Ridge: results from Ocean Drilling Program Leg 110. *Geological Society of American Bulletin*, **100**, 1578–1593.

MUGNIER, J.-L. & MENARD, G. 1986. Le développement du bassin molassique suisse et l'évolution des Alpes externes: un modèle cinématique. *Bulletin du Centre Recherches Exploration Production Elf-Aquitaine*, **10**, 167–180.

PHILIPPE, Y. 1994. Transfer zones in the southern Jura belt (eastern France): geometry, development and comparison with analogue modelling experiments. In: MASCLE, A. (ed.) *Hydrocarbon and Petroleum Geology of France*. European Association of Petroleum Geologists, Special Publications, **4**, 269–280.

POBLET, J. & HARDY, S. 1995. Reverse modelling of detachment folds: applications to the Pico de Aguila anticline in the South Central Pyrenees (Spain). *Journal of Structural Geology*, **17**, 1707–1724.

REGGIANI, L. 1989. Faciès lacustres et dynamique sédimentaire dans la molasse d'eau douce inférieure Oligocène (USM) de Savoie. *Eclogae Geologicae Helvetiae*, **82**, 325–350.

RIBA, O. 1976. Tectogenèse et sédimentation: deux modèles de discordances syntectoniques pyrénéennes. Bulletin BRGM, s.2, **1**, 383–401.

RIGASSI, D. 1957. Le Tertiaire de la région genevoise et savoisienne. *Bull. Ver. Schw. Petr.-Geol. u. Ing.*, **24**, 19–34.

SIGNER, C. & GORIN, G. E. 1995. New geological observations between the Jura and the Alps in the Geneva area, as derived from reflection seismic data. *Eclogae geologicae Helvetiae*, **88**, 235–265.

SINCLAIR, H. D., COAKLEY, B. J., ALLEN, P. A. & WATTS, A. B. 1991. Simulation of foreland basin stratigraphy using a diffusion model of mountain belt uplift and erosion: an example from the central Alps, Switzerland. *Tectonics*, **10**, 599–620.

SPECHT, M. 1989. *Tectonique de chevauchement le long du Profil ECORS-Pyrénées: un modèle d'évolution de prisme d'accrétion continental*. PhD Thesis, Université de Bretagne Occidental.

TARDY, M., DEVILLE, E., FUDRAL, S., GUELLEC, S., MENARD, G. & VIALON, P. 1990. Interprétation structurale des données du profil de sismique réflexion profonde ECORS-CROP Alpes entre le front Pennique et la ligne du Canavese (Alpes Occidentales). In: ROURE, F., HEITZMANN, P & POLINO, R. (eds) *Deep structure of the Alps*, Deep structure of the Alps, Mémoires de la Société Géologique de France **156**, Mémoires de la Société Géologique de Suisse **1**, Speciale Società Geologica Italiana **1**, 217–226.

TESSIER, B. & GIGOT, P. 1989. A vertical record of different tidal cyclicities: an example from the Miocene Marine Molasse of Digne (Haute-Provence, France). *Sedimentology*, **36**, 767–776.

——& TERWINDT, J. 1994. Un exemple de déformaion synsédimentaire en milieu intertidal: l'effet du mascaret. *Compte Rendue Academie de Sciences Paris*, **319**, 217–223.

——, MONTFORT, Y., GIGOT, P. & LARSONNEUR, C. 1989. Enregistrement des cycles tidaux en accrétion verticale, adaptation d'un outil de traitement mathématique. Exemples en Baie du Mont Saint-Michel et dans la molasse marine miocène du bassin de Digne. *Bulletin de la Société Géologique de France*, 8, **5**, 1029–1041.

VERGES, J., BURBANK, D. W. & MEIGS, A. 1996. Unfolding: an inverse approach to fold kinematics. *Geology*, **24**, 175–178.

WILDI, W., BLONDEL, T., CHAROLLAIS, J., JAQUET, J.-M. & WERNLI, R. 1991. Tectonique en rampe latérale à la terminaison de la Haute-Chaîne du Jura. *Ecologae Geologicae Helvetiae*, **84**, 265–277.

Evolution of the western Swiss Molasse basin: structural relations with the Alps and the Jura belt

MARTIN BURKHARD & ANNA SOMMARUGA

Institut de Géologie, rue E. Argand 11, CH-2000 Neuchâtel, Suisse

Abstract: The tectonic evolution of the NW alpine front and foreland basin is reviewed in the light of new structural and chrono-stratigraphical data. Seismic-reflection profiles from the Jura fold thrust belt and Molasse basin, surface-geology and thrust-system considerations lead to a complete cross-section of the NW Alpine front including the Helvetic domain. Restoration of this section places individual Cenozoic formations in their approximate palaeogeographic position. 'Geohistory' plots are constructed for five profiles along a SE–NW transect. Thrust front, onlap and forebulge advanced at high rates of 10–20 km/Ma^{-1} at the onset of foreland basin formation in the late Eocene/ early Oligocene (40–30 Ma). In these early stages, the foreland basin is an underfilled flexural trough with about 100 km width, less than 600 m water depth at the deepest point and less than 200 m of total accumulated sediments. From 30 to 22 Ma, thrust front and 'pinch-out' migrate at a decreased rate of about 5 km/Ma^{-1} northwestward. The basin width remains constant at around 100 km; an increased total subsidence (*c.* 2.7 km) is compensated by sedimentation. At around 22 Ma, the thrust front seems to come to a halt southeast of Lausanne, whereas a strong subsidence trend prevails. After the Serravallian (*c.* 12 Ma) the Alpine thrust front jumps by about 100 km northwestward from a position southeast of Lausanne to the external Jura leading to thrust related uplift, deformation and concomitant erosion of the entire basin fill. No new flexural foreland basin in response to the modified thrust- and load-geometry has yet been developed. The present-day Molasse basin is only a small remnant of a much larger foreland basin in a very advanced stage of its evolution.

The North Alpine Molasse (Fig. 1) has long been recognized as the 'detritus washed down from the rising Alps' and deposited in a peripheral depression related to Tertiary Orogeny (Heim 1921). In contrast with the nearby Extensional Grabens of the Oligocene Rhine–Bresse system (Fig. 1), flexural bending of the European, down-going plate under the thrust load advancing from the south is proposed as the principal (Molasse) foreland basin forming mechanism (Price 1973; Dickinson 1974; Turcotte & Schubert 1982) . This led to the characteristic downwarp flexure responsible for the wedge-shaped geometry of the basin (in cross-section) and an upward flexure, the forebulge, located in the foreland. Sedimentology, stratigraphy, subsidence profiles and gravity anomaly studies corroborate this interpretation (Lemcke 1974; Karner & Watts 1983; Naef *et al.* 1985; Homewood *et al.* 1989). The North Alpine Molasse basin fits perfectly with the idealized foredeep as proposed by Bally (1989) showing all the characteristic units and evolutionary steps (Fig. 2). (1) A *crystalline basement* is dipping gently from the foreland (Fig. 1, Vosges, Black Forest, Massif central) toward the orogen. (2) Localized grabens filled with Carboniferous and Permian clastics, could be regarded as a *syn-rift series*. The majority of Alpine authors, however, consider these grabens as the last expression of the Variscan rather than the onset of the Alpine orogenic cycle (Trümpy 1980). (3) Above a major 'post-Variscan' erosional unconformity, Middle Triassic to Upper Cretaceous shallow-marine carbonates represent a *passive margin series* with a total thickness varying from 1 to 2.5 km (Wildi *et al.* 1991; Loup 1992) deposited at the southern margin of the European plate, including the future Jura, Molasse, 'Autochthonous'- and Helvetic domains. (4) Tertiary clastic wedges are deposited above a pronounced *foreland unconformity* (Fig. 3) which can be traced from the external Jura southward into the most internal Helvetic domain (Boyer & Elliott 1982; Herb 1988; Homewood *et al.* 1989). This unconformity marks the beginning of the flexural response to Alpine loading of the European plate. (5) The lateral continuity between Helvetic and Molasse parts of the foreland basin has been obscured and partly destroyed by incorporation into the advancing tectonic prism. Tectonic activity started in the internal parts and progressively advanced toward the northwest including Penninic, Helvetic and Subalpine Molasse thrust systems (Pfiffner 1986; Burkhard 1988b). With the advent of Jura folding starting in the Late Serravallian, the entire western Molasse basin is incorporated into the Alpine thrust system and

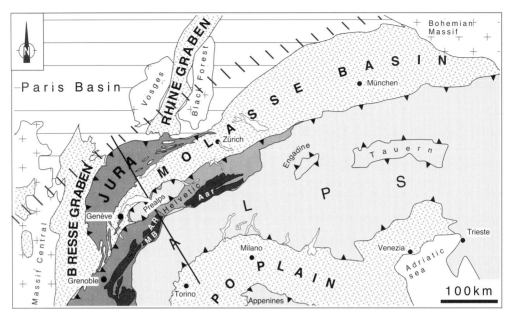

Fig. 1. Location of the major Cenozoic sedimentary basins: Rhine–Bresse Grabens, Molasse basin and Po Plain with respect to the Alpine orogen. Late Alpine culminations and External Crystalline Massifs are highlighted: AR, Aiguilles Rouges massif; MB, Mont Blanc massif. The cross-hatched band in the foreland indicates the supposed present-day position of the lithospheric forebulge.

displaced by up to 30 km to the northwest (Laubscher 1965, 1992; Mugnier & Ménard 1986; Guellec et al. 1990). The present-day western Swiss Molasse basin is riding in a piggy back fashion above a décollement horizon in Triassic evaporite series (Deville et al. 1994). Thrust involvement has put an end to the subsidence history of the Molasse basin – no sediments younger than uppermost Serravallian have been found. This indicates that the northwestern Alpine foredeep has been bypassed and cannibalized since the early Tortonian. Accordingly, the present-day Molasse basin is but a small remnant of a much larger foreland basin in a very advanced stage of its evolution. An important step in unraveling the foreland basin's history is the restoration of the initial position of the various tectonic units involved (Boyer & Elliott 1982; Homewood et al. 1986; Pfiffner 1986; Burkhard 1988b, 1990).

In this paper, we examine the present-day geometry of the northwestern front of the Alps in a regional cross section from the undeformed foreland through the Jura, Molasse basin, External crystalline massifs and Helvetic nappes (Fig. 4). The evolution of this portion of the Alps since the mid-Eocene is discussed on the basis of a restored section where the

Fig. 2. Schematic stratigraphic/time section summarizing the most important Mesozoic and Cenozoic lithostratigraphic formations involved in Tectonics of the NW Alpine foreland including the Jura, Molasse basin, External Crystalline Massif and Helvetic domains. The horizontal axis in km, measured from a pin in the foreland, is a restoration of the different palaeogeographic domains (compare Fig. 6). Vertical axis is time according to Palmer (1983). Major hiatiuses are distinguished with different shades of grey, corresponding to (a) post-Variscan peneplenation, (b) non-deposition or erosion during Mesozoic times including the so-called 'Alemanic land' and other important lacuna in the Cretaceous series on the craton side, (c) a pronounced foreland unconformity separating Mesozoic platform carbonate series from Tertiary foreland formations and (d) Alpine collision and erosion. Reference stratigraphic columns have been compiled for Les Verrières (Le Locle), Yverdon, Lausanne (Savigny), Mt Pèlerin, Morcles, Wildhorn. For details of the Tertiary series, see Fig. 7.

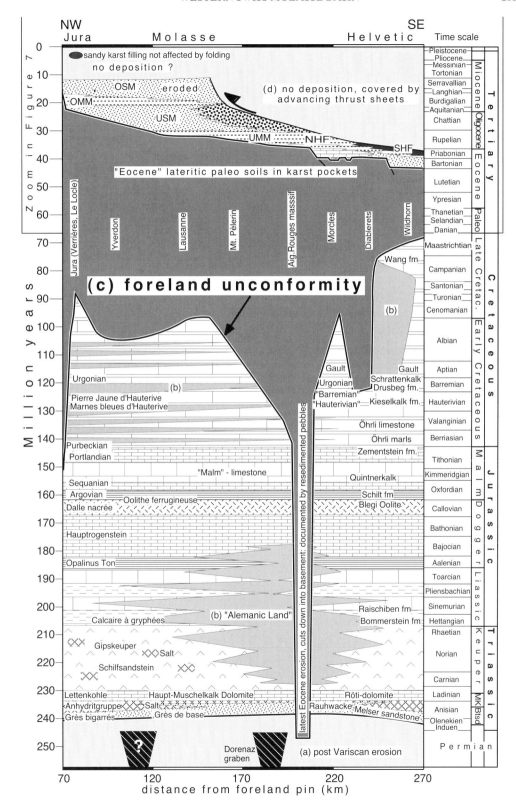

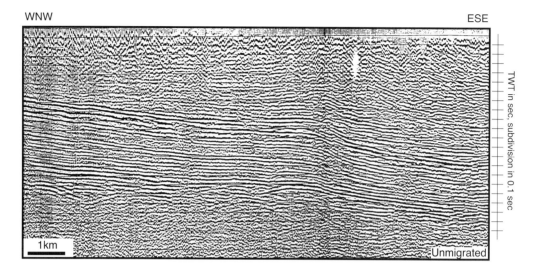

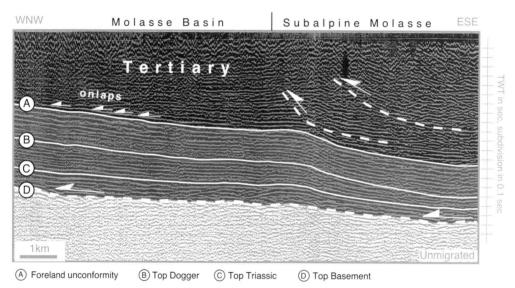

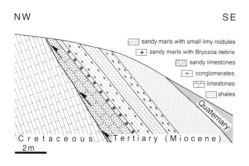

Fig. 3. (a) Seismic expression of foreland unconformity below the Molasse basin of western Switzerland (section published by Gorin *et al.* 1993). For location of this section, see Fig. 5. (b) Field example of foreland unconformity in the folded Haute Chaîne Jura. At Les Verrières, Upper Marine Molasse (OMM) rests unconformably above lower Cretaceous limestones (modified from Martin *et al.* 1986).

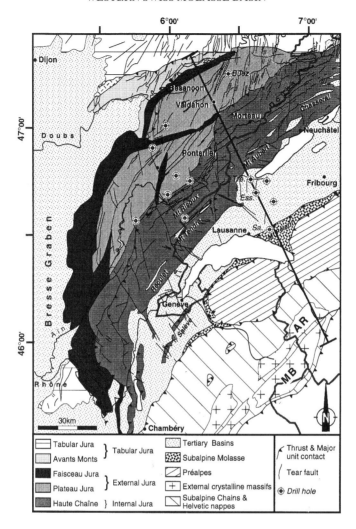

Fig. 4. Tectonic map of Western Switzerland, with trace of cross-section presented in Fig. 6. Drill holes referenced in Fig. 6 are labeled: Tre, Treycovagnes; Ess, Essertines; Sa, Savigny.

progressive onlaps and the nature of increasingly younger foredeep series are put in relation with the advance of individual thrust fronts. Detailed structural insights have become possible during the last decade with the release of petroleum industry seismic data from the Molasse basin (Ziegler 1990) and the Jura fold thrust belt (Gorin et al. 1993; Sommaruga 1995, 1997). Major progress has been made in dating of the Tertiary Molasse series, mostly by detailed palaeontological work (Berger 1992, 1995) and palaeomagnetic investigations (Burbank et al. 1992; Schlunegger et al. 1996). An increasing accuracy in both spatial and temporal resolution of the (restored) sediment bodies provides the necessary input data for a better understanding of the basin-forming mechanisms, notably the progressive flexing of the European lithosphere under the load of the advancing alpine thrust sheets. We hope that our compilation will help modellers to present refined models of the foreland basin evolution (Stockmal et al. 1986; Sinclair et al. 1991; Beaumont et al. 1994; Crampton & Allen 1995).

Deformation affecting the Molasse basin and the Jura fold and thrust belt

Cenozoic and Mesozoic series of the western Swiss Molasse basin and the Jura fold belt have been severely deformed over a main décollement zone located within Triassic evaporites. This deformation started after the Serravallian and apparently put a halt to the subsidence history of the Molasse basin. Surface observations and especially seismic subsurface data from the central Jura and the Molasse basin display two different types of folds (Fig. 5): evaporite-related and thrust-related. The first type has been observed in the Plateau Jura at the very front, to the NE of the Haute Chaîne Jura folds, as well as beneath the Molasse basin. Thrust-related folds are predominant in the Haute Chaîne Jura (see Figs 4 and 5). These two types of folds are respectively interpreted as embryonic and evolved stages of the late Miocene cover deformation (Sommaruga 1997).

Folds are thought to evolve from low-amplitude, apparently symmetric, buckle folds in response to layer-parallel compression. The very weak evaporite rocks of the Triassic series infilled the space generated at the base of the sedimentary cover by flow mechanism. The core of the folds present a thickening within the very weak basal zone, whereas the strong layers buckled with no change of thickness (concentric folding). With progressive deformation, fault ramps nucleate in the hinge area at the base of the strong layers in order to accommodate further shortening. Fault ramps then propagate upward within the stiff layers or bend their trajectory within overlying weak or incompetent layers. With further transport, the cover may be doubled, as observed in the present-day Haute Chaîne Jura folds (Fig. 5b). This tentative explanation of the evolution of thrust-related folds is based on the observation of various evolutionary stages across the western Jura fold thrust belt. Fold style, however, varies not only with time but is also critically dependent on the relative thickness of competent and incompetent layers above the main décollement horizon. Along strike in the Jura, a tendency from asymmetric fault propagation folds to more symmetrical fold shapes (lift-off-folds, detachment folds) can be observed from west to east. This evolution seems to be related to an north-eastward decrease in the thickness of the stiff Malm layer and a concomitant increase of both number and relative proportion of incompetent Layers (Muschelkalk, Keuper, Aalenian).

It is generally admitted that fold and thrust belts evolve by progressive deformation from the hinterland to the foreland (Bally et al. 1966; Dahlen et al. 1984). The folds beneath the Molasse basin are located in a more internal position of the Alpine foreland belt and could therefore be expected to be more evolved than the Jura folds. Molasse folds, however, represent typically early stage buckle folds and their Triassic cores seem to be filled with well-organized evaporite duplexes (Fig. 5c). The most obvious reason for this apparent contradiction seems to be the load of the Tertiary sediments, which prevented the Molasse basin buckle folds to further evolve into thrust-related folds. With progressive deformation and concommitant erosion of the Tertiary wedge sediments, Molasse basin anticlines may have the possibility to evolve into Jura folds. The present day taper-geometry of the Alpine front would probably favour a back-stepping thrust sequence with the most frontal thrust emerging close to the Molasse basin–Haute Chaîne Jura border (cf. Mugnier & Ménard 1986).

Jura–Alps cross-section of the NW Alpine front

A regional scale cross section of the present day NW Alpine front of western Switzerland is presented in Fig. 6 (see Figs 1 and 4 for location). Only the northwestern half of this section is discussed below. However, we have roughly filled in the space occupied by Penninic and Austroalpine units. For details within this part of the section, the reader is referred to Escher et al. (1993).

With regard to previous interpretations (Aubert et al. 1980; Boyer & Elliott 1982; Mugnier & Vialon 1986) this section takes into account new seismic and drill hole data. The Jura part is based on geological maps (Muhlethaler 1930; BRGM 1963, 1965, 1968, 1969, 1972; Jordi 1994, 1995; Rigassi & Jaccard 1995), completed with information from seismic-reflection lines (Sommaruga 1997). The external Plateau Jura is composed of flat areas, separated from each other by the so called 'Faisceaux', narrow areas affected by strong deformation. The Plateau Jura presents broad anticlines cored with Triassic evaporite stacks (compare Fig. 5). The Haute Chaîne Jura is characterized by thrust-related folds (Fig. 5) with kilometric throw to the NW and occasional back-thrusts to the SE. Seismic-reflection lines and drill holes revealed the presence of a very important cushion of Triassic evaporites below the central Jura with considerable lateral thickness variations from 0.5 to 1.4 km. The thickness of this ductile décollement series is minimal below synclines and below some of the 'Haute Chaîne'

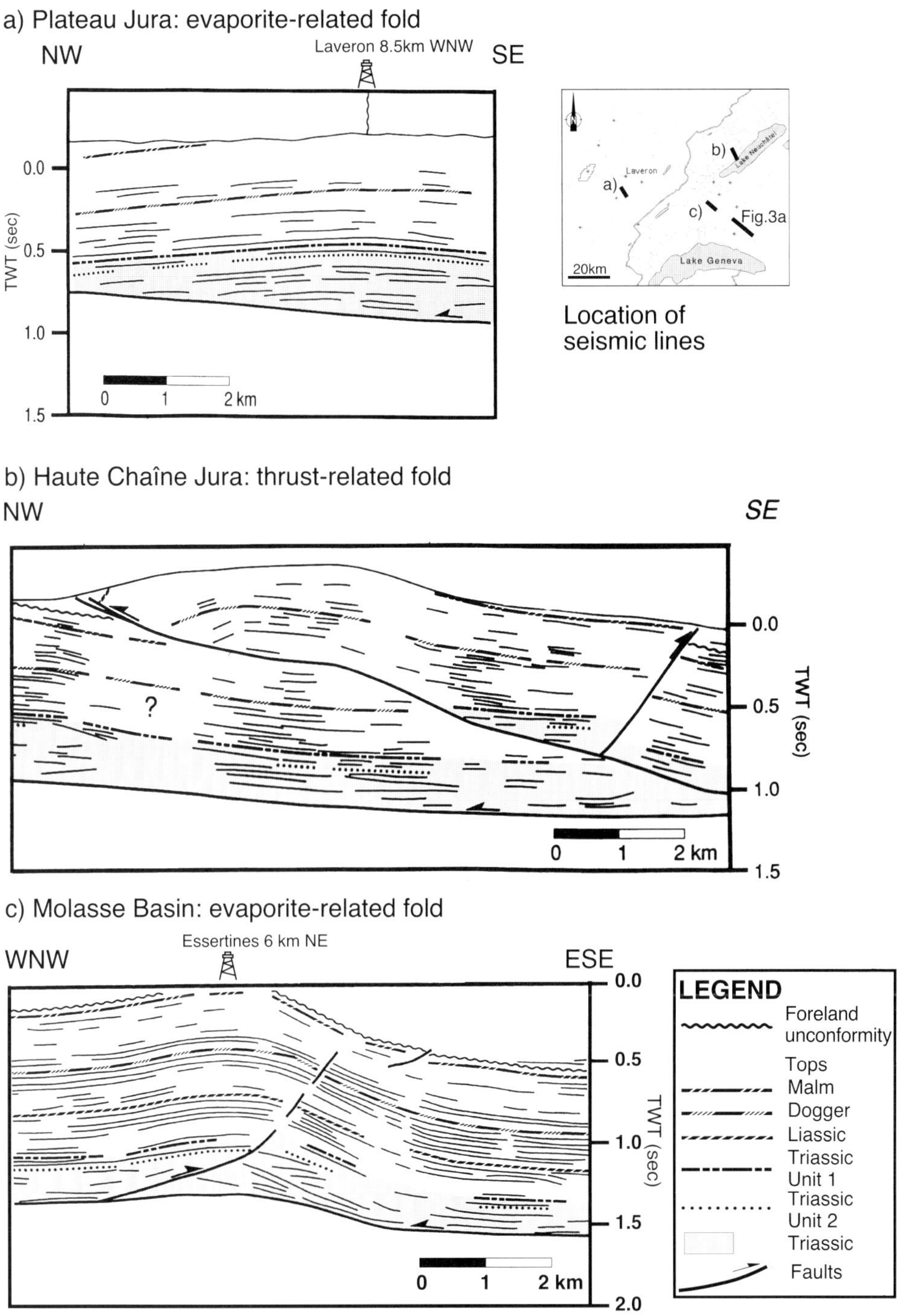

Fig. 5. Line drawings of examples of typical fold structures in the Jura and the Molasse basin. Horizontal scale is in km, vertical scale is in seconds two way travel time. Triassic series are highlighted in grey. Plateau Jura and Molasse basin examples represent low-amplitude evaporite-related 'buckle' folds. The Haute Chaîne Jura example presents a high amplitude thrust-related fold (modified from Sommaruga 1995).

Jura thrust-related anticlines whereas anticlines both in the external 'Plateau' Jura and the Molasse basin are cored by important accumulations of Triassic 'evaporite pillows' (Lienhardt 1984). These thickness variations are important in terms of balancing cross-sections; they have not been considered in previous balancing work in the Jura (Laubscher 1965; Mugnier & Vialon 1986; Philippe 1995).

The central portion of the Molasse foreland basin is well constrained by a series of drill holes (Essertines, Treycovagnes, Savigny, Fig. 4) and a grid of seismic strike and dip lines (Weidmann 1988; Gorin et al. 1993; Jordi 1993, 1994, 1995; Signer & Gorin 1995). It is characterized by gentle folds with several kilometre wavelengths and amplitudes of less than 500 m. These 'buckle' folds are compensated by thickness variations within Triassic evaporite series. The top of the basement, including crystalline and sedimentary rocks older than Triassic, forms an essentially smooth surface gently dipping to the SSE to SE. The northern limit of the present-day Molasse basin corresponds to an erosional limit along the most internal high-amplitude folds of the Jura belt. Cenozoic Molasse sediments are found much further to the north, preserved as small remnants within synclines of the 'Haute Chaîne' Jura.

The frontal parts of the Subalpine Molasse, southeast of Lausanne are well constrained from surface observations and a few seismic lines (Weidmann 1992; Gorin et al. 1993; Sommaruga 1995, 1997; Mosar et al. 1996). A rather surprising feature in this part of the section is the discovery of a very broad anticline within the Mesozoic layers below the Mt Pèlerin conglomerate fan. This anticline has been interpreted as being cored by an inverted Permo-Carboniferous halfgraben (Gorin et al. 1993); alternatively, it could be cored by another huge pillow of Triassic evaporites or, correspond to a basement ramp. Unfortunately, only the northern half of this structure appears on seismic lines, which do not continue further south. The size of the structure as it appears on the seismic lines is certainly exaggerated due to a velocity pull up below the Pèlerin conglomerat fan. Accordingly, the least-constrained part of our cross section lies between the front of the Prealps and the northern flank of the Aiguilles Rouges massif. Despite the results of the deep seismic-reflection campaign of the Swiss NFP20 program (Mosar et al. 1996; Pfiffner et al. 1997) there remains quite some ambiguity in the interpretation of individual reflectors and even first order features such as the top basement are ill constrained between Mt Pèlerin and St Maurice. South of St Maurice, the cross-section relies largely on surface observations made within the Helvetic nappes. However, information is projected over some 10 to 20 km southwestward above actual topography (Langenberg et al. 1987; Burkhard 1988a, b; Dietrich & Casey 1989; Escher et al. 1993).

An important additional constraint for interpretations of this cross-section is provided by data regarding the depth of the Moho discontinuity (Baumann 1994). The Moho trend drawn in Fig. 6 is based mostly on refraction seismic data (projected horizontally over <60 km), interpolated regionally to satisfy the Bouguer anomaly using a Vening-Meinesz isostatic compensation model (Baumann 1994). 'Moho depth' and 'top crystalline basement' together define a crustal thickness that is of the order of 26 km in the northern part of the section. A modest reduction to about 24 km in the extreme northwest might be interpreted as an effect of Oligocene thinning in relation with the Rhine–Bresse Graben opening; however, accuracy of the determined Moho depth is of the order of ±3 km at best (Baumann 1994). Southeast of Lausanne, depending on the interpretation of the above-mentioned anticline, crustal thickness increases considerably to reach more than 35 km below the external crystalline massifs. This is a clear indication for late Alpine 'thick-skinned' involvement of European crystalline basement leading to the external crystalline massifs, which are interpreted as huge fault-related folds (cf. Rodgers 1995).

Restored section

A restored section is presented in Fig. 6. In order to visualize better stratigraphic thickness changes and thrust system organization, the restored section has been vertically exaggerated. The foreland unconformity has been chosen as an arbitrary horizontal reference line. Our restoration yields a gross total shortening of 26 km for the Jura fold thrust belt between Besançon and Essertines. This estimate is in line with previous determinations by Laubscher (1961), Mugnier & Vialon (1986) and Philippe (1995). The main difference with these is a revised stratigraphic thickness column, with a considerably increased, but laterally variable thickness of Triassic evaporite series. These variations, as well as the structural style of folding and thrusting within the Jura are documented by a grid of seismic lines (Sommaruga 1997). Shortening within the gently folded Mesozoic series of the Molasse basin between Essertines and Lausanne is minimal, 'smoothing' these broad folds yields less than 2 km. This estimate might be slightly increased if an important internal deformation by pressure solution is assumed (Engelder *et al.* 1981). Horizontal stylolites are frequently observed in Malm limestones cropping out in the entire Jura (Plessmann 1972; Tschanz 1990). Horizontal shortening of up to a few percent is also seen within coarse sandstones of the Molasse basin (Hindle & Burkhard 1995), this tectonic 'horizontal compaction' could in part account for anomalously high seismic velocities in Molasse series (Kaelin *et al.* 1992). Horizontal shortening deformation is, however, certainly less than 5% and it could account at most for another 3 km of horizontal shortening, if integrated over a horizontal distance of 60 km.

The various slivers of Subalpine Molasse southeast of the Savigny drill hole require a minimum of some 10 km of horizontal shortening, which is less constrained than shortening estimates in the Jura, however. According to seismic-reflection profiles, this deformation is restricted to the Tertiary series and must therefore cut down into the Mesozoic cover further south. The southward continuation of thrusts responsible for the Subalpine Molasse slivers is not observable and can only be inferred. The most important uncertainties in the restored section are the distance between Mt Pèlerin and St Maurice (Fig. 6). An unknown amount of shortening within Mesozoic and Tertiary series may be hidden below the Pennine Prealps Klippen (Mosar *et al.* 1996).

In summary, 35 km is a certain minimum estimate of cover shortening observed between the external Jura and the crest of the 'Autochthonous' cover series of the Aiguilles Rouges massif; given all the uncertainties, mostly within the subalpine Molasse, this shortening could be as large as 50 km.

Crustal-scale thrust organization below the external crystalline massifs is still a matter of debate which has been fueled in the last few years by the availablility of deep seismic-reflection profiles of ECORS and NFP20 (Mugnier *et al.* 1990; Tardy *et al.* 1990; Pfiffner *et al.* 1997). However, interpretations of the external crystalline massifs still vary from stacks of thin, upper crustal 'flakes' detached at a few km depth within the upper crust (Boyer & Elliott 1982; Laubscher 1992) to 'whole-crustal' thrust ramps (Tardy *et al.* 1990). Despite these uncertainties, an estimate of the crustal shortening involved with the formation of the external crystalline massifs can be obtained from drill-hole and seismic data regarding the basement top and a knowledge of the Moho geometry in the NW Alpine foreland. Assuming an original basement thickness of 26 km throughout and taking the present-day Moho-geometry to define the 'regional top crust restored', an 'excess area' of at least 570 km^2 related with crustal shortening is determined for the External Crystalline Massifs (Aiguilles Rouges and Mt Blanc, Fig. 6). Although the 'top crystalline basement' is ill defined below the Prealps, the corresponding uncertainty in the 'excess area' is small and the total surface might be some 60 km^2 larger (if more important basement culminations were present, instead of the assumed volume of 'hidden' proximal USM, UMM and NHF series, see Table 1). Based on this 'excess area', a gross horizontal crustal shortening can be evaluated: shortening = excess area/depth to detachment. Assuming a mid-crustal detachment, above the well-layered lower crust as seen on most deep seismic-reflection profiles crossing the Alps (Guellec *et al.* 1990; Mugnier *et al.* 1990; Pfiffner *et al.* 1990), i.e. at a depth of 10–15 km, bulk horizontal shortening estimates of 40–60 km are obtained. These values include shortening associated with the thrust emplacement of the lowermost Helvetic Morcles nappe. The geometric solution proposed in Fig. 6 is therefore compatible with gross horizontal shortenings measured along individual major thrust systems (Jura, Subalpine Molasse, basal Helvetic), and thrust system organization (Boyer & Elliott 1982) as well as timing (Burkhard 1990). Notably, the onset of Jura folding and thrusting in the Late Serravallian correlates well with cooling, which is interpreted as uplift-related unroofing of the External Crystalline massifs. This cooling is well documented by apatite and some scarce zircon fission-track ages (Soom 1990). Apatite fission-track ages show a strong altitude dependence between 6 Ma (at 3000 m) and 3 Ma (at 500 m). If interpreted as the time elapsed since cooling below a 120°C temperature, these ages indicate that the presently exposed external crystalline massifs must have started their cooling from more than 300°C considerably earlier than 6 Ma, most probably before 15 Ma. This age is compatible with the onset of thrusting deformation within and below the external crystalline massifs and uplift/unroofing of the massifs was certainly active during thrusting of the Jura.

The restoration of the Helvetic nappes above and behind the Aiguilles Rouges massif is facilitated by excellent outcrop conditions. Multiphase and partly strong internal deformations complicate the picture, however, and prohibit the application of simple line length balancing of the stiff Malm, Öhrlikalk and Urgonian layers. Volume restoration of the Morcles, Diablerets and Wildhorn nappes yields some 20–30, 15 and 20–25 km respectively. In summary, the Helvetic domain south of the Aiguille Rouges Massif had an initial horizontal width of 55–70 km (Huggenberger 1985; Burkhard 1988b, fig. 3). The corresponding basement is only partly cropping out in form of the Mt Blanc massif, representing, at least in parts, the basement core to the thick skinned Morcles nappe (Escher et al. 1993). Higher Helvetic nappes, such as the Diablerets and Wildhorn are entirely thin skinned with a major detachment within 'Aalenian' black shales. The southward continuation of this basal Higher Helvetic thrust systems as well as the internal basement units (Mt Chetif/Gotthard) representing the homeland of these nappes, are still deeply buried below Penninic nappes. The present-day cross-section (Fig. 6) shows, however, that there is very little space to account for the more internal european basement units. Two alternative solutions to this problem can be proposed: (a) important parts of the European continental (lower) crust may have been subducted into the mantle or (b) the apparently missing space in this section is due to out of section (eastward) movements of the 'missing' internal crystalline basement units.

Foredeep sediments in space and time

The evolution of the Tertiary foreland basin can be reconstructed in detail by the analysis of successive lithostratigraphic units and their lateral correlation (Fig. 7). In the Helvetic domain, the classical Tertiary 'trilogy' consists of sandy Nummulite limestones, black Globigerina marls and South Helvetic Flysch series (SHF) (Herb 1988). This ensemble is cut at the top by thrusts associated with so-called 'wild-flysch' sedimentary and/or tectonic mélanges (Bayer 1982). This 'trilogy' is slightly diachronous across the Helvetic domain, starting in the upper Eocene in the southeast and ending in the lowermost Oligocene in the northwest (Herb 1988; Lateltin 1988). Some remarkable steps in the 'onlap curve' (Figs 2 & 7) and corresponding steps in the footwall ('foreland unconformity') are associated with syn-sedimentary normal faults such as the well-documented Hohgant–Rawil fault (Herb 1988). We propose similar faults as northern limit of the Morlces domain and their presence can also be inferred in the Jura, where documentation is not as thorough as in the Wildhorn nappe. In both areas, breccias and immature conglomerates document the presence of some relief and strong erosion. In the case of the Morcles nappe, conglomerates include crystalline pebbles clearly stemming from the northwestern Aiguilles Rouges massif (Badoux 1971; Mayoraz 1995). We interpret this denudation as erosion of tilted blocks, in an extensional environment (the forebulge region), related to the flexuring of the European crust.

Across most of the Helvetic domain, flysch series show a very modest development with generally less than 200 m thickness. Only the Northernmost Helvetic Flyschs (NHF) show some increase in thickness with the Taveyannaz and Val d'Illiez formations, deposited on the future Morcles nappe. Further to the north, these deposits evolve into the Molasse series which are found in front of the Alpine chain and extend northward into the Jura fold and thrust belt. In particular, the North Helvetic Val d'Illiez sandstone is regarded as the lateral equivalent of the so called Lower Marine Molasse (UMM from the German terminology). This is followed by littoral and fluvial systems of the Lower Freshwater Molasse (USM), indicating the transition from an underfilled to an overfilled stage of the foreland basin (Allen et al. 1986; Beaumont et al. 1994). Distal parts of the Molasse basin remained at very low altitude above sea level during deposition of the USM, since shallow marine conditions were established again during the following OMM stage. The latter stage is most likely due to the Aquitanian / Burdigalian sea-level highstands (Keller 1989). The Upper Freshwater Molasse (OSM) is very scarce in western Switzerland and only some distal remnants are found within synclines of the Jura fold thrust belt. In more proximal positions, OSM sediments have been eroded. Their existence can be inferred, however, from comparisons with central and eastern Switzerland and estimations of overburden removed from the presently exposed OMM and USM in western Switzerland (Lemcke 1974; Schegg et al. 1997).

Ages, palaeogeographic position and thicknesses of the various sediment bodies have been compiled into a synoptic view of the foreland basin's evolution through time from 50 Ma to 10 Ma (Figs 7 & 8). Data have been compiled from Steffen (1981), Herb (1988) and Menkveld-Gfeller (1994) for the Wildhorn and Diablerets nappes; from Lateltin (1988) for the Diablerets and Morcles-nappes and Aiguilles Rouges cover; from Weidmann (1988, 1992) for the Subalpine and Plateau Molasse near Lausanne; from Jordi (1995) for the Molasse basin near Yverdon and from Aubert (1975) and Martin et al. (1991) for the Molasse remnants in the folded Jura. Absolute ages are according to a revised version of the correlative chart compiled by Berger (1992, 1995), who takes the most recent

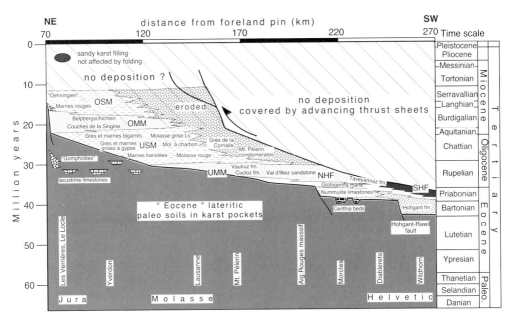

Fig. 7. Chronostratigraphy of Tertiary formations placed on a restored NW-SE section. This figure is an enlarged detail of Fig. 2. Compilation of data from (Herb 1988; Lateltin 1988; Weidmann 1988, 1992; Martin *et al.* 1991; Berger 1992, 1995; Jordi 1995; Schlunegger 1996). Legend : *OSM*, Upper Freshwater Molasse; *OMM*, Upper Marine Molasse; *USM*, Lower Freshwater Molasse; *UMM*, Lower Marine Molasse; *NHF*, North Helvetic Flysch; *SHF*, South Helvetic Flysch; *fm*, formation; *Mol.*, Molasse; LS, Lausanne.

palaeomagnetic data into account (Schlunegger *et al.* 1996).

Geohistory plots or 'total subsidence' curves for the different domains have been constructed (Fig. 8a) from these data with the aim of determining the evolution of different reference points of the foreland through space and time (Fig. 8b). Parameters considered for the construction of these curves are: (1) age of the oldest foreland sediments found above the foreland unconformity, (2) ages and (3) thicknesses of lithostratigraphic units and (4) estimated water depth for marine or 'elevation a.s.l.' for fluvial and lacustrine series. No corrections for sediment compaction nor eustatic sea-level changes have been applied.

Water depths are well constrained for marine shallow water or littoral series such as Nummulite or Lithothamnia limestones of the Helvetic 'trilogy' and OMM beach sandstones. The depth of deposition of Flysch series on the one hand and the palaeo-elevation of fluvial Molasse series on the other hand are rather ill constrained. Palaeoecology of foraminifera (Steffen 1981; Lateltin 1988) indicate that Helvetic flyschs were deposited in relatively shallow troughs of less than 600 m water depth – an average depth of 400 m has been used in the construction of the 'total subsidence' curves of Fig. 8. The fact that the Aquitanian/Burdigalian sea-level highstands led to a marine transgression in large parts of the Molasse basin is taken as an indication for a limited elevation of less than 100 m above sea level during the previous USM stage for distal Molasse sections (Yverdon, Lausanne) (cf. Berger 1996). The OSM stage of the Molasse basin evolution is even less constrained in Western Switzerland, since most of the relevant sedimentary series have been removed by post-Serravalian erosion. Curves have been drawn nevertheless, based on vitrinite reflectance data which indicate that up to 2.6 km of overburden has been removed from the Molasse basin in the Lausanne and Essertines areas (Schegg *et al.* 1997). Similarly, subsidence curves are extended toward the SE below the advancing thrust front. Tectonic burial of the Helvetic nappes is quite well constrained by their lowest-grade metamorphism and some radiometric data providing age constraints for the 'peak of metamorphism' in the Late Oligocene (Burkhard 1988b; Kirschner *et al.* 1995).

Total subsidence curves are arranged in Fig. 8b according to their palaeogeographic position

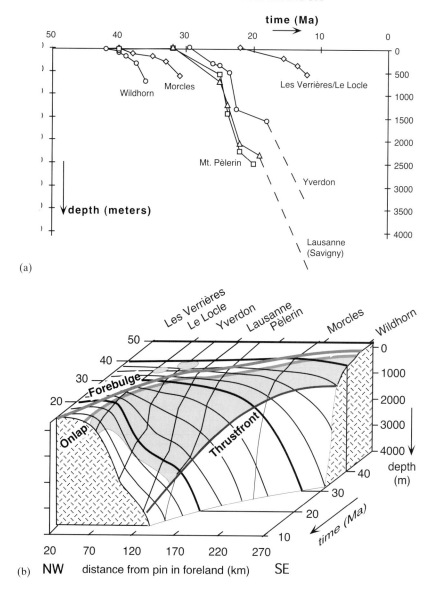

Fig. 8. (a) Geohistory plots or 'total subsidence' curves for a selection of seven sites within the NW Alpine foreland basin. Vertical axis is burial depth of the foreland unconformity in metres with respect to an assumed constant sea level = 0 m. Burial depth calculated from sediment thicknesses and estimated waterdepth or elevation but without corrections for eustasy or compaction (see text for details). Maximum burial at Lausanne (Savigny drill hole) and Yverdon is constrained by vitrinite reflectance data (Schegg *et al.* 1996). (**b**) Block diagram constructed to visualize the evolution of the shape and size of the foreland basin through time. Horizontal axis to the right is distance from a foreland pin in km (compare Fig. 6). Vertical axis is burial depth in meters (same as a). Oblique axis from rear to front is time from 50 to 10 Ma. 'Skewed' Burial curves of a) are arranged in their restored position and thus allow for the construction of the geometry of the foreland unconformity through time : indicated with thin (every 2.5 Ma) and thick (every 10 Ma) lines in the 'distance v. time space'. The area of this 3D geohistory surface documented by the preserved sedimentary record is shaded in grey. The inferred position of thrust front, onlap and forebulge are indicated with thick grey lines.

along a restored NW–SE section. In addition to ordinary geohistory plots (Fig. 8a) for various points in the Molasse basin, this 3D arrangement yields palaeoprofiles of the 'foreland unconformity'. These profiles show the evolution of the shape and size of the foreland basin through space (migrating from southeast to northwest) and time (from 50 to 10 Ma). The grey shaded area shows the portion of this evolution which is truly documented by the sedimentary record. Additional information can be integrated as follows. The top surface of this space–time diagram is constrained to the left by the pre-Late Eocene erosion leading to the foreland unconformity (Figs 2 & 7). In the absence of any detailed information about this period, which may have lasted since the latest Cretaceous, the foreland top is simply assumed to be in a close to horizontal position near, but (mostly) above sea-level. This first interpretation is corroborated by subsidence analyses of the Mesozoic series, which indicate clearly decreseasing rates of subsidence to the end of Cretaceous in all but the southernmost Helvetic domain (Loup 1992; Wildi & Huggenberger 1993). Interestingly, the youngest sediments found below the 'foreland unconformity' are of Turonian, locally even Maastrichtian age both in the Jura and in some isolated areas of the Helvetic domain (Fig. 2). The following hiatus could be due to the first-order regression cycle starting in the Latest Cretaceous when large parts of central Europe fell dry. Alternatively, the hiatus contained in the foreland unconformity could also be due to a short erosional event related to the passage of the forebulge in the Latest Eocene to Oligocene (Crampton & Allen 1995). The presence of relatively 'old', i.e. early Oligocene, lacustrine series in the Yverdon area (schematically represented by patches in Figs 7 and 8b) is an argument against this interpretation. Due to their thinness, these sediments do not appear on the burial curve for the Yverdon area (Fig. 8a). However, their old age and northern position speaks in favour of a deposition on the northwestern side (or on top of) the Oligocene forebulge (compare Fig. 8b). If subsequent erosion in the forebulge region was strong and responsible for the foreland unconformity, this erosion should also have removed these sediments. The precise position of the forebulge in space and time is not directly constrained by any sedimentary data however. We indicate it as just preceding the onset of foreland sedimentation. It is noteworthy to mention that even the present-day position of a forebulge in the NW Alpine foreland is not easily determined. In the profile under consideration, the forebulge could lie somewhere near our 0 km 'pin', a saddle region between the Molasse basin to the SE, the Paris basin to the NW, the Vosges culmination to the NE and the Massif Central to the SW (cross-hatch pattern in Fig. 1). The map scale trend of the forebulge to the NE is also ill defined. Interference with the Oligocene Rhine/Bresse-Graben rift system and associated shoulder uplifts (Massif Central, Vosges, Black Forest) obscures (and/or amplifies?) the flexural forebulge of the Alpine system (Karner & Watts 1983; Laubscher 1992).

Quantitative aspects of Foreland basin evolution

'Thrust advance' and 'pinch-out' migration rates

The chronostratigraphic chart (Fig. 7) and geohistory plots (Fig. 8) give quantitative summary overviews of the evolution of the Alpine foredeep in time and space and permit calculation of rates at which the foreland basin evolved (Allen & Allen 1990). In particular 'Orogenic wedge tip' or 'thrust-advance' and 'pinch-out-migration' rates are determined directly from the sedimentary record. The accuracy of these rates depends critically on at least the three following factors : (a) biostratigraphic control, (b) biostratigraphy / chronostratigraphy correlations and (c) geometric restoration. The overall trends shown in Fig. 7 are rather well documented and seemingly 'smooth and reasonable'. However, inaccuracies and uncertainties do remain. As an example, consider the case of the North Helvetic Taveyannaz Flysch formation. Based on nannofossils (stages NP20 and NP21), this formation is proposed as deposited at the Eocene–Oligocene boundary (Latetlin 1988). In terms of absolute age, this corresponds to an age span from 36 to 32.5 Ma according to different authors (Berger 1992, 1995). Some hornblendes from andesitic pebbles found within the Taveyannaz formation (of the Chaînes subalpines) have been K/Ar and Ar/Ar dated at 32 Ma (Fischer & Villa 1990). The slight discrepancy of 0.5 to 4 Ma may seem small on a chronostratigraphic chart (Fig. 7). However, when the thrust front advance rate for some 50 km width of the Helvetic domain is calculated, this makes for a considerable variation from about 10 km Ma^{-1} (young Taveyannaz fm) to almost 20 km Ma^{-1} old Taveyannaz fm). Similarly, 'pinch-out rates' should not be calculated from Fig. 7 over too short horizontal distances. As mentioned above, the restoration is particularly uncertain in the area between Lausanne and Morcles. Accordingly, the marked slowing of the thrust advance, which seems to come to a relative halt at around 20 Ma southeast of Lausanne, is only circumstantially constrained. Despite these restrictions, Figs 7 and 8 show some important tendencies which can be summarized as follows.

At around 40 Ma, the foreland basin is a narrow, shallow, underfilled trough less than 50 km wide, maybe some 400 m deep and with less than 600 m of total subsidence.

From 40 to 30 Ma, the thrust front and the 'pinch-out' advance at considerable rates of about 10 and 20 km Ma^{-1} respectively, leading to an increasing width (100 km) of the still underfilled basin with less than 1000 m total subsidence.

From 30 to 22 Ma both thrust front and 'pinch-out' migrate at the same decreased rate of around 5 km Ma^{-1} northwestward. Consequently, the basin width remains constant at around 100 km; the increased total subsidence (2.7 km) is totally compensated by sedimentation and the basin will be over-filled for the rest of its evolution.

From 22 to 12 Ma, the thrust front seems to be in a fixed position southeast of Lausanne. The 'pinch-out' migration to the NW seems to be markedly slowed down too. The basin remains either at a constant width (c. 100 km) or increases slightly (to less than 140 km). However, a strong subsidence trend continues (compare Fig. 8).

After the Serravallian at c. 12 Ma, the Alpine thrust front jumps more than 100 km northwestward from a position southeast of Lausanne to the external Jura. This thrust involvement puts a halt in the Molasse basin evolution. Subsidence is replaced by thrust-related uplift, deformation and concomitant erosion of the basin fill.

No new flexural foreland basin in response to the new thrust and load geometry has yet been developed.

Flexing of the European lithosphere

Width, depth and shape of the foreland basin are largely dependent on the rheology of the flexed foreland lithosphere. Assuming an elastic behaviour of a thin lithosphere three parameters at least determine the profile shape: elastic thickness (T_e), Young's modulus (E) and load (Turcotte & Schubert 1982). Based on these assumptions, the evolution of the basin's shape through time as reconstructed in Fig. 8b can be used to estimate (palaeo-) elastic parameters of the involved foreland lithosphere. We tried to model individual palaeoprofiles by 'best-fitting' calculated flexural profiles according to Turcotte & Schubert (1982). Model calculations have been done by freely varying T_e and E of the flexed lithosphere, influencing on the profile shape as well as thrust load, responsible for the maximum depth of the foreland basin. Modelled curves are compared with observed ones using the solver function in an EXCEL spreadsheet (available from the first author upon request – please send a formatted diskette).

In agreement with previous studies considering the basin shape (Karner & Watts 1983; Lyon-Caen & Molnar 1989; Allen & Allen 1990; Sinclair et al. 1991; Gutscher 1995; Sinclair 1996) we find rather small elastic thicknesses of 5–15 km and low Young's modulii between 4 and 6 × 10^{10} Pa. Based on the widening (from 40 to 20 Ma) of the foreland basin with time we expected to find an increasing 'Rigidity' with thrust advance to the northwest. Such a trend could be expected if the southernmost European margin (Ultrahelvetic and Southhelvetic domain) is assumed to be tectonically thinned during the Mesozoic, and if this lithosphere remained thin and weak till the onset of alpine collision. In contrast to Sinclair's conclusion (1996, fig. 8) our model calculations failed to show any significant trend of increasing T_e with time. This is mainly due to the modest depth of the early foreland basin, which, in the Late Eocene/Early Oligocene, is only a couple of hundred metres deep at the deepest point documented by sedimentary series (Fig. 8a). Any calculations based on these early profiles are fraught with large uncertainties, due to the very weakly pronounced flexural shape. The later deepening of the basin (from 30 to 12 Ma) at an apparently constant width is best explained with an increased thrust load and does not require a lateral change in rheology.

Elastic parameters determined from the geometry of the foreland flexure as described above, are ill constrained, however. This can be illustrated using the present-day geometry of the top crystalline basement in the alpine foreland (Fig. 6). In Fig. 9, actual data for the top basement are compared with two extreme models of elastic flexure. Model A with 10 km elastic thickness is opposed to model B with a 25 km elastic thickness. The main difference between the two models is the wavelength, i.e. the distance between the foredeep and the forebulge. This distance increases with increasing rigidity of the lithosphere. In reality, the forebulge region is affected by extension and strike slip faults in relation with the Rhine–Bresse–Graben transfer zone (Bergerat 1987). Accordingly, the top of the basement in this region deviates from an ideal smooth elastic flexure and the plate may have to be considered as being 'broken'.

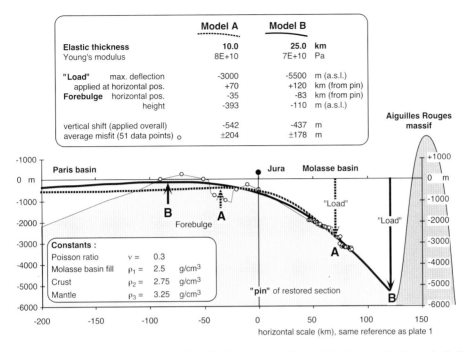

Fig. 9. Comparison of the present-day Molasse basin geometry with predicted geometries from elastic flexural models. South of the pin line, basement top data (dots) are determined from seismic reflection data projected over ≤20 km laterally onto the profile. Basement top north of pin line is from Debrand Passard et al. (1984, pl. G3) and Rangin (1992, pl. 2a). Two different elastic flexural profiles have been calculated according to Turcotte & Schubert (1982): parameters are indicated in boxes. Best-fit curves were calculated by minimizing the squared difference between calculated and observed (dots) basement tops.

Furthermore, the observed descent of the top basement to the NW (dipping into the Paris basin) is compensated by an northwestward increasing thickness of Mesozoic sediments, resulting in an essentially horizontal configuration for the foreland unconformity. For this reason, we did not include any data from this area in our best-fitting procedure. In any case, calculated data points can be shifted laterally and vertically such as to satisfy either Model A or B quite accurately. The main problem in the comparison of the modeled, with the observed data, lies indeed in the localisation of the forebulge and the 'point load' applied at the right end of the model. Neither position is easily determined in the present-day Alpine foreland and accordingly it is even more difficult to infer for older stages of the Molasse basin evolution. Figure 1 shows a probable location of the present-day forebulge in the NW Alpine foreland (Karner & Watts 1983; Allen & Allen 1990; Laubscher 1992), but Fig. 9 illustrates the difficulty in locating this position on any profile across the NW Alps. Geologically, it might seem reasonable to take the position of the external crystalline massif culminations as a line-load applied to the southeastern edge of the 'broken' european lithosphere. The distance between this line-load and the forebulge is on the order of 200 km (Figs 1 & 9), considerably more than the width of even the restored foreland basin fill. Taken together, these considerations are clearly in favour of Model B (Fig. 9) and would indicate a relatively large elastic thickness for the European crust on the order of 25 km – in agreement with calculations based on the Bouguer gravity anomaly (Macario et al. 1995). The same considerations apply to the restored basin shapes (Fig. 8b) and any attempts at determination of paleo elastic parameters. The 'true' distance between line load and forebulge was certainly at all times larger than the rather small documented width of the foreland basin. Calculations of elastic parameters based on sedimentary data will therefore tend to yield smaller values than thoses considering crustal or lithospheric architecture (Bouguer anomaly, Moho geometry, top crystalline basement).

Similar comments apply to Sinclair's (1996) estimate of the Early Oligocene plate rheology and the apparent increase of T_e with time from Mid-Miocene to Recent could be an artefact. The small T_e values of 5–15 km obtained at 17 Ma (Sinclair 1996, figs 7 & 8) are based on the assumption of a coherent, European plate and its large scale curvature around the arc of the western Alps. The most important restriction to this assumption is the uncertain continuity between the westernmost Molasse basin and the Digne–Valensole basin to the southwest of the Alpine arc, both riding on the same flexed european foreland lithosphere. This lithosphere seems to be severely broken in front of the northwestern subalpine chains where synsedimentary strike-slip faults are described, i.e. the Salève region (Wildi et al. 1991; Deville et al. 1994). Such faults thus could explain smaller curvature than predicted for an intact elastically flexed sheet.

Conclusions

The restoration of a complete section across the northwestern frontal Alpine thrust system from the external Jura to the Penninic front in the Valais, places Neogene foreland sediments in their supposed original palaeogeographic position (Fig. 6) and permits to draw qantitative conclusions regarding the evolution of this foreland basin. Restoration of the Jura thrust system reveals 26 (± 3) km of cover shortening, due to thrusting and folding above an important cushion of Triassic evaporite series with variable thickness, in response to deformation. This thrusting is younger than Serravallian as documented by some rare erosional remnants of Upper Freshwater Molasse preserved in Jura synclines. The Subalpine Molasse thrust system accounts for some 10 (+10, –1) km of shortening within the most internal Molasse series. Both Jura and Subalpine Molasse thrusts root within the frontal, non-exposed Aiguilles–Rouges Massif. Thrust ramps with more than 36 km (up to 50 km) of combined horizontal shortening are responsible for the formation of the External Crystalline Massif culmination, with an excess area of 570 (+60, –10) km². Assuming a mid-crustal detachment level, this excess area is compatible with some 50 km total northwestward transport of the Morcles nappe with its crystalline core, the Mt Blanc massif. Additional restoration of the overlying higher Helvetic nappes, leads to an initial width of some 250–300 km for the entire north Alpine realm from the external Jura to the southern edge of the Helvetic domain.

Foreland sedimentation starts in the late Eocene in the most internal (Helvetic) domain with a shallow (<600 m deep) and narrow (<100 km wide) underfilled trough, which receives only a very modest amount of South Helvetic Flysch before being covered by the rapidly advancing Penninic (including Ultrahelvetic) thrust front.

In the latest Eocene (from c. 40 to 30 Ma) thrust front and 'pinch out' advance with 10 to 20 km Ma^{-1} northwestward to reach the Morcles/Aiguilles Rouges Massif realm by Early Oligocene. During the same period, the basin width remains constant at about 100 km and depth and sedimentation increase only slightly.

In the early Oligocene, thrust front and 'pinch-out' migrate at a decreased rate of c. 5 km Ma^{-1} northwestward to come to a halt somewhere southeast of Lausanne by about 22 Ma. The basin width remains constant at some 100 km whereas an increased total subsidence (c. 2.7 km) is compensated by sedimentation indicating increasing thrust load and topography in the Hinterland.

The time span from 20 to 12 Ma is badly constrained in the western Swiss Molasse basin, since most of the corresponding sediments have subsequently been removed by erosion. Some distal erosional remnants in Jura synclines as well as the degree of burial below younger Molasse series testify to a strong ongoing subsidence trend with probably more than 2.5 km of Upper Freshwater Molasse deposited in the Lausanne area. The maximum width of the basin seems to increase, but was probably never larger than 150 km. Thrusting may have taken place within the Subalpine Molasse zone in this time interval, this activity is not properly documented, however.

Jura folding and thrusting started after the Serravallian (c. 12 Ma), indicating a major 'jump' of the Alpine thrust front by about 100 km northwestward from a position southeast of Lausanne to the external Jura. This 'jump' is explained by the presence of an important accumulation of Triassic evaporite series in the Alpine foreland. The latest, most external Alpine thrust system uses this weak basal décollement horizon to propagate at a very shallow angle NW ward. This event leads to thrust related uplift, deformation and concomitant erosion of the entire basin fill and bypassing of the Molasse basin since the Serravallian. No new flexural foreland basin in response to the modified thrust and load geometry has been developed yet. The present-day Molasse basin is only a small remnant of a much larger foreland basin in a very advanced stage of its evolution.

The restored geometry of the foreland basin and sediment thicknesses together with the most recently available chronostratigraphic data have been used to construct 'subsidence' curves for various points in the basin along a south–north transect. The arrangement of these curves in a three-dimensional synoptic diagram (Fig. 8) summarizes the most important data about the geometric evolution of the North Alpine foreland basin in western Switzerland.

The present-day geometry of the flexural shape of the Foreland basin has been used to estimate the elastic parameters of the European lithosphere. Elastic thickness and flexural rigidity are ill constrained and depend largely on the assumed but unknown positions of the applied load and forebulge respectively. Our favoured model has an elastic thickness of about 25 km. The narrow width of the remaining Molasse basin in western Switzlerland is due to thrust involvement both at the external (Jura) and internal (Alps) edges of the basin and should not be mistaken as an indication for a strongly flexed, thin elastic plate.

The authors thank the following persons, institutions and companies for kindly giving access to seismic data: British Petroleum, Shell International (SIPM), Swisspetrol, Forces Motrices Neuchâteloises and the Musée de Géologie at Lausanne in the person of the Director A. Baud; the PNR 20 project (Swiss National Project). The authors are indebted to A.W. Bally, A. Jordi, P. Lehner, J. Mosar, G. Schönborn, R. Schoop and M. Weidmann for many stimulating discussions. Special thanks go to M. Baumann, J.-P. Berger, R. Schegg and F. Schlunegger who made unpublished manuscripts and data available. M. Weidmann, E. Deville and Y. Philippe are thanked for constructive reviews. Financial support by Swiss National Science Foundation (Grants No 21-37'366.93) and by Neuchâtel University are gratefully acknowledged.

References

ALLEN, P. A. & ALLEN, J. R. 1990. *Basin Analysis, Principles and Applications.* Blackwell Scientific Publications, Oxford.
——, HOMEWOOD, P. & WILLIAMS, G. D. 1986. Foreland basins: an introduction. *In:* ALLEN, P. A. & HOMEWOOD, P. (eds) *Foreland basins* Special Publication of the International Association of Sedimentologists, **8**, 3–12.
AUBERT, D., AYRTON, S., BEARTH, P., BURRI, M., CARON, C., ESCHER, A., SCHAER, J.-P. & WEIDMANN, M. 1980. Excursion No. II: Geotraverse of Western Switzerland. *In:* KOMMISSION, S. G. (ed.) *Geology of Switzerland – a guide-book. Part B: Geological excursions.* Wepf, Basel, New York, 155–181.
BADOUX, H. 1971. *Notice explicative, feuille 1305 Dt. de Morcles.* Commission Géologique Suisse.
BALLY, A. W. 1989. Phanerozoic basins of North America. *In:* BALLY, A. W. & PALMER, A. R. (eds) *The geology of North America–An overview.* Geological Society of America, Boulder, Colorado, 397–446.
——, GORDY, P. L. & STEWART, G. A. 1966. Structure, seismic data and orogenic evolution of southern Canadian Rocky Mountains. *Bulletin of Canadian Petroleum Geology,* **14**, 337–381.
BAUMANN, M. 1994. *Three-dimensional modeling of the crust-mantle boundary in the alpine region.* PhD thesis, ETH, Zürich.
BAYER, A. 1982. *Untersuchungen im Habkern-Mélange ('Wildflysch') zwischen Aare und Rhein.* PhD thesis Nr. 6950, ETH Zürich.
BEAUMONT, C., FULLSACK, P. & HAMILTON, J. 1994. Styles of crustal deformation in compressional orogens caused by subduction of the underlying lithosphere. *Tectonophysics,* **232**, 119–132.
BERGER, J.-P. 1992. Correlative chart of the European Oligocene and Miocene: Application to the Swiss Molasse Basin. *Eclogae Geologicae Helvetiae,* **85**, 573–609.
—— 1995. updated version of Berger (1992), unpublished.
—— 1996. Cartes paléogéographiques-palinspastiques du bassin molassique suisse (Oligocène inférieur – Miocène moyen). *Neues Jahrbuch für Geologie und Paläontologie, Abhandlungen,* **202**, 1–44.
BERGERAT, F. 1987. Paléo-champs de contrainte tertiaires dans la plate-forme européenne au front de l'orogène alpin. *Bulletin de la Société géologique de France,* **8**, 611–620.
BOYER, S. E. & ELLIOTT, D. 1982. Thrust systems. *American Association of Petroleum Geologists Bulletin,* **66**, 1196–1230.
BRGM. 1963. *Ornans.* Carte géologique détaillée de la France 1:50'000, feuille No530. Service de la Carte géologique de la France.
—— 1965. *Vercel.* Carte géologique détaillée de la France 1:50'000, feuille No503. Service de la Carte géologique de la France.
—— 1968. *Morteau.* Carte géologique détaillée de la France 1:50'000, feuille No531. Service de la Carte géologique de la France.
—— 1969. *Pontarlier.* Carte géologique détaillée de la France 1:50'000, feuille No557. Service de la Carte géologique de la France.
—— 1972. *Baume les Dames.* Carte géologique détaillée de la France 1:50'000, feuille No 473. Service de la Carte géologique de la France.
BURBANK, D. W., ENGESSER, B., MATTER, A. & WEIDMANN, M. 1992. Magnetostratigraphic chronology, mammalian faunas, and stratigraphic evolution of the Lower Freshwater Molasse, Haute-Savoie, France. *Eclogae Geologicae Helvetiae,* **85**, 399–431.
BURKHARD, M. 1988a. *Horizontalschnitt des Helvetikums der Westschweiz (Rawildepression). Beiträge zur Landes Hydrologie- und Geologie,* **4**.
—— 1988b. L'Helvétique de la bordure occidentale du massif de l'Aar (évolution tectonique et métamorphique). *Eclogae Geologicae Helvetiae,* **81**, 63–114.
—— 1990. Aspects of the large scale Miocene deformation in the most external part of the Swiss Alps

(Subalpine Molasse to Jura fold belt). *Eclogae Geologicae Helvetiae*, **83**, 559–583.

CRAMPTON, S. L. & ALLEN, P. A. 1995. Recognition of Forebulge Unconformities Associated with Early Stage Foreland Basin Development: Example from the North Alpine Foreland Basin. *American Association of Petroleum Geologists Bulletin*, **79**, 1495–1514.

DAHLEN, F. A., SUPPE, J. & DAVIS, D. M. 1984. Mechanics of fold-and-thrust belts and accretionary wedges (continued): Cohesive Coulomb theory. *Journal of Geophysical Research*, **88**, 1153–1172.

DEBRAND-PASSARD, S., COURBOULEIX, S. & LIENHARDT, M.-J. 1984. Synthèse géologique du Sud-Est de la France. Stratigraphie et Paléogéographie, BRGM Mémoires, **125**, Orléans.

DEVILLE, E., BLANC, E., TARDY, M., BECK, C., COUSIN, M. & MÉNARD, G. 1994. Thrust propagation and syntectonic sedimentation in the Savoy Tertiary Molasse basin (Alpine Foreland). *In:* MASCLE, A. (ed.) *Hydrocarbon and Petroleum Geology of France*. Special Publications of the European Association of Petroleum Geoscientists, **4**. Springer-Verlag, 269–280.

DICKINSON, W. R. 1974. Plate tectonics and sedimentation. *In:* DICKINSON, W. R. (ed.) *Tectonics and Sedimentation*. Special Publications of the SEPM, **22**, 1–27.

DIETRICH, D. & CASEY, M. 1989. A new tectonic model for the Helvetic nappes. *In:* COWARD, M. P., DIETRICH, D. & PARK, R. G. (eds) *Alpine Tectonics*. Geological Society (London) Special Publications, **45**, 47–63.

ENGELDER, T., GEISER, P. A. & ALVAREZ, W. 1981. Penrose conference report: role of pressure solution and dissolution in Geology. *Geology*, **9**, 44–45.

ESCHER, A., MASSON, H. & STECK, A. 1993. Nappe geometry in the Western Swiss Alps. *Journal of Structural Geology* **15**, 501–509.

FISCHER, H. & VILLA, I. M. 1990. Erste K/Ar und $^{40}Ar/^{39}Ar$-Hornblende-Mineralalter des Taveyannazsandsteins. *Schweizerische Mineralogisch Petrographische Mitteilungen*, **70**, 73–75.

GORIN, G. E., SIGNER, C. & AMBERGER, G. 1993. Structural configuration of the western Swiss Molasse Basin as defined by reflection seismic data. *Eclogae Geologicae Helvetiae*, **86**, 693–716.

GUELLEC, S., MUGNIER, J. L., TARDY, M. & ROURE, F. 1990. Neogene evolution of the western Alpine foreland in the light of ECORS data and balanced cross sections. *In:* ROURE, F., HEITZMANN, P. & POLINO, R. (eds) *Deep structure of the Alps*. Mémoires de la Société géologique Suisse, **1**, 165–184.

GUTSCHER, M. A. 1995. Crustal structure and dynamics of the Rhine Graben and the Alpine foreland. *Geophysical Journal International*, **122**, 617–636.

HEIM, A. 1921. *Geologie der Schweiz. Band I Molasseland und Juragebirge*. Tauchniz, Leipzig.

HERB, R. 1988. Eocaene Paläogeographie und Paläotektonik des Helvetikums. *Eclogae Geologicae Helvetiae* **81**, 611–657.

HINDLE, D. & BURKHARD, M. 1995. Internal deformation in the Molasse of western Switzerland. *Terra Abstracts* **EUG - VI**, 269.

HOMEWOOD, P., ALLEN, P. A. & WILLIAMS, G. D. 1986. Dynamics of the Molasse Basin of western Switzerland. *In:* ALLEN, P. A & HOMEWOOD P. (eds) Special Publications of the international Association of Sedimentologists **8**, 199–217.

——, RIGASSI, D. & WEIDMANN, M. 1989. Le bassin molassique Suisse. In: *Dynamique et méthodes d'étude des bassins sédimentaires* Association française des sédimentologistes eds) Technip, Paris, 299–314.

HUGGENBERGER, P. 1985. *Faltenmodelle und Verformungsverteilung in Deckenstrukturen am Beispiel der Morcles Decke (Helvetikum der Westschweiz)*. PhD thesis, ETH Zürich.

JORDI, H. A. 1993. Tectonique du bassin molassique et de son substratum jurassique-crétacé dans la région Orbe-Yverdon-Grandson. *Bulletin de la Société Vaudoise des Sciences naturelles* **82**, 279–299.

—— 1994. Yverdon-les-Bains. Atlas géologique Suisse, feuille No. 94. Service hydrologique et géologique national.

—— 1995. Yverdon-les-Bains. Atlas géologique Suisse, feuille No. 94. Notice explicative. Service hydrologique et géologique national.

KAELIN, B., RYBACH, L. & KEMPTER, E. H. K. 1992. Rates of deposition, Uplift and Erosion in the Swiss Molasse Basin, estimated from Sonic and Density Logs. *Bulletin der Vereinigung Schweizerischen Petroleum-Geologen und -Ingenieure*, **58**, 9–22.

KARNER, G. D. & WATTS, A. B. 1983. Gravity anomalies and flexure of the lithosphere at mountains ranges. *Journal of Geophysical Research*, **88**, 10'449–10'447.

KELLER, B. 1989. *Fazies und Stratigraphie der oberen Meeresmolasse (unteres Miozän) zwischen Napf und Bodensee*. PhD thesis, Bern.

KIRSCHNER, D. L., SHARP, Z. D. & MASSON, H. 1995. Oxygen isotope thermometry of quartz-calcite veins: Unravelling the thermal-tectonic history of the subgreenschist facies Morcles nappe (Swiss Alps). *Geological Society of America Bulletin*, **107**, 1145–1156.

LANGENBERG, W., CHARLESWORTH, H. & LA RIVIERE, A. 1987. Computer-constructed cross-section of the Morcles nappe. *Eclogae Geologicae Helveticae*, **80**, 655–667.

LATELTIN, O. 1988. *Les dépôts turbiditiques oligocènes d'avant-pays entre Annecy (Haute-Savoie) et le Sanetsch (Suisse)*. PhD thesis, Univ. Fribourg.

LAUBSCHER, H. P. 1961. Die Fernschubhypothese der Jurafaltung. *Eclogae Geologicae Helvetiae*, **54**, 221–280.

—— 1965. Ein kinematisches Modell der Jurafaltung. *Eclogae Geologicae Helvetiae*, **58**, 232–318.

—— 1992. Jura kinematics and the Molasse basin. *Eclogae geologicae Helvetiae*, **85**, 653–676.

LEMCKE, K. 1974. Vertikalbewegungen des vormesozoischen Sockels im nπrdlichen Alpenvorland von Perm bis zur Gegenwart. *Eclogae Geologicae Helvetiae* **71**, 121–133.

LIENHARDT, M. J. 1984. Trias: puissance et faciès de la partie supérieure. In: DEBRAND-PASSARD, S., COURBOULEIX, S. & LIENHARDT, M.-J. (eds) Synthèse géologique du Sud-Est de la France. Stratigraphie et Paléogéographie, Mémoires **125**, Orléans, 00–00.

LOUP, B. 1992. Evolution de la partie septentrionale du domaine helvétique en Suisse occidentale au Trias et au Lias: contrôle par subsidence thermique et variations du niveau marin. PhD thesis, Genève.

—— 1992. Mesozoic subsidence and stretching models of the lithosphere in Switzerland (Jura, Swiss Plateau and Helvetic realm). *Eclogae Geologicae Helvetiae*, **85**, 541–572.

LYON-CAEN, H. & MOLNAR, P. 1989. Constraints on the deep structure and dynamic processes beneath the Alps and adjacent regions from an analysis of gravity anomalies. *Geophysical Journal International*, **99**, 19–32.

MACARIO, A., MALINVERNO, A. & HAXBY, W. F. 1995. On the robustness of elastic thickness estimates obtained using the coherence method. *Journal of Geophysical Research*, **100**, 15 163–15 172.

MARTIN, J., CHAUVE, P. & SEQUEIROS, F. 1986. Le contexte polyphasé du faisceau salinois. *Annales scientifiques de l'Université de Besançon Géologie*, **4**, 43–47.

——, PHARISAT, A. & RANGHEARD, Y. 1991. Le synclinal des Verrières (Haute-Chaîne jurasienne): nouvelle interprétation structurale. *Annales scientifiques de l'Université de Besançon Géologie*, **4**, 99–112.

MAYORAZ, R. 1995. Les brèches tertiaires du flanc inverse de la nappe de Morcles et des unités parautochthones (Bas Valais, Suisse). *Eclogae Geologicae Helvetiae*, **88**, 321–345.

MENKVELD-GFELLER, U. 1994. Die Wildstrubel-, Hohgant- und die Sanetschformation: Drei neue lithostratigraphische Einheiten des Eozäns der helvetischen Decken. *Eclogae geologicae Helvetiae*, **87**, 789–809.

MOSAR, J., STAMPFLI, G. M. & GIROD, F. 1996. Western Préalpes Médianes Romandes: timing and structure. A review. *Eclogae Geologicae Helvetiae*, **89**.

MUGNIER, J. L., GUELLEC, S., MÉNARD, G., ROURE, F., TARDY, M. & VIALON, P. 1990. A crustal scale balanced cross-section through the external Alps deduced from the ECORS profile. In: ROURE, F., HEITZMANN, P. & POLINO, R. (eds) *Deep structure of the Alps*. Mémoires de la Société géologique Suisse, 203–216.

—— & MÉNARD, G. 1986. Le développement du bassin molassique suisse et l'évolution des Alpes externes: un modèle cinématique. *Bulletin des Centres de Recherche et Exploration-Production d'Elf-Aquitaine*, **10**, 167–180.

—— & VIALON, P. 1986. Deformation and displacement of the Jura cover on its basement. *Journal of Structural Geology*, **8**, 373–387.

MUHLETHALER, C. 1930. *Les Verrières – La Chaux*. Atlas géologique Suisse, feuille No. 2. Commission Géologique Suisse.

NAEF, H., DIEBOLD, P. & SCHLANKE, S. 1985. *Sedimentation und Tektonik im Tertiär der Nordschweiz*. Nagra, Baden.

PALMER, A. R. 1983. *Geologic Time Scale*. Decade of the North American Geology. The Geologic Society of America.

PFIFFNER, O. A. 1986. Evolution of the north Alpine foreland basin in the Central Alps. In: ALLEN, P. A. & HOMEWOOD, P. (eds) Special Publications of the International Association of Sedimentologists **8**, 219–228.

——, FREI, W., VALASEK, P., STÄUBLE, M., LEVATO, L., DUBOIS, L., SCHMID, S. M. & SMITHON, S. B. 1990. Crustal shortening in the alpine orogen: results from deep seismic reflection profiling in the eastern swiss Alps line NFP 20-east. *Tectonics*, **9**, 1327–1355.

——, LEHNER, P., HEITZMANN, P., MUELLER, S. & STECK, A. 1997. *Deep Structure of the Swiss Alps, results of NRP 20*. Birkhäuser, Basel.

PHILIPPE, Y. 1995. *Rampes latérales et zones de transfert dans les chaînes plissées: géométrie, conditions de formation et pièges structuraux associés*. PhD thesis, Chambéry (Savoie, France).

PLESSMANN, W. 1972. Horizontal-Stylolithen im französisch-schweizerischen Tafel-und faltenjura und ihre Einpassung in den regionalen Rahmen. *Geologische Rundschau*, **61**, 332–347.

PRICE, R. A. 1973. Large-scale gravitational flow of supracrustal rocks, southern Canadian Rockies. In: SCHOLTEN, K. A. D. J. A. R. (ed.) *Gravity and tectonics*. Wiley and Sons, New York, 491–502.

RIGASSI, D. & JACCARD, M. 1995. *Ste-Croix*. Atlas géologique Suisse, feuille No. 95. Service hydrologique et géologique national.

RANGIN, F. P. 1993. *Sismotectonique de la France métropolitaine*. Mémoire de la Société Géologique de la France **164**.

RODGERS, J. 1995. Lines of basement uplifts within the external parts of orogenic belts. *American Journal of Science*, **295**, 455–487.

SCHEGG, R., LEU, W., CORNFORD, C. & ALLEN, P. A. 1997. New coalification profiles in the Swiss Molasse Basin: Implications for the thermal and geodynamic evolution of the Alpine Foreland. *Eclogae Geologicae Helvetiae*, 79–96.

SCHLUNEGGER, F., BURBANK, D. W., MATTER, A., ENGESSER, B. & MÖDDEN, C. 1996. Magnetostratigraphic calibration of the Oligocene to Middle Miocene (30–15 Ma) mammal biozones and depositional sequences of the Swiss Molasse Basin. *Eclogae Geologicae Helvetiae*, **89**, 753–788.

SIGNER, C. & GORIN, G. E. 1995. New geological observations between the Jura and the Alps in the Geneva area, as derived from reflection seismic data. *Eclogae Geologicae Helvetiae* **88**, 235–265.

SINCLAIR, H. D. 1996. Plan-view curvature of foreland basins and its implications for the paleostrength of the lithosphere underlying the western Alps. *Basin Research*, **8**, 173–182.

——, COAKLEY, B. C., ALLEN, P. A. & WATTS, A. B. 1991. Simulation of foreland basin stratigraphy using a diffusion model of mountain belt uplift and erosion: an example from the central Alps, Switzerland. *Tectonics*, **10**, 599–620.

SOMMARUGA, A. 1995. Tectonics of the Central Jura and the Molasse Basin. New insights from the interpretation fo seismic reflection data. *Bulletin Société Neuchâteloise des Sciences Naturelles*, **118**, 95–108.

—— 1997. *Geology of the central Jura and the Molasse Basin: new insight into an evaporite-based foreland fold and thrust belt*. Mémoire de la Société des Sciences naturelles de Neuchâtel, **12**.

SOOM, M. A. 1990. *Abkühlungs- und Hebungsgeschichte der Externmassive und der penninischen Decken beidseits der Simplon-Rhonelinie seit dem Oligozän: Spaltspurdatierungen an Apatit/Zirkon und K-Ar-Datierungen an Biotit/Muskowit*. PhD thesis, Bern.

STEFFEN, P. 1981. *Zur Stratigraphie und Paläontologie des helvetischen Eozäns in der Wildhorndecke des Berner Oberlands*. PhD thesis, Bern.

STOCKMAL, G. S., BEAUMONT, C. & BOUTILIER, R. 1986. Geodynamic models of convergent tectonics: the transition from rifted margin to overthrust belt and consequences for foreland basin development. *American Association of Petroleum Geologists Bulletin*, **70**, 181–190.

TARDY, M., DEVILLE, E., FUDRAL, S., GUELLEC, S., MÉNARD, G., THOUVENOT, F. & VIALON, P. 1990. Interprétation structurale des données du profil de sismique réflexion profonde ECORS-CROP Alpes entre le front Pennique et la ligne Canavese (Alpes occidentales). *In:* ROURE, F., HEITZMANN, P. & POLINO, R. (eds) *Deep structure of the Alps*. Mémoires de la Société géologique Suisse, 217–226.

TRÜMPY, R. 1980. *Geology of Switzerland - a guidebook. Part A: An outline of the geology of Switzerland. Part B: Geological excursions*. Wepf, Basel, New York.

TSCHANZ, X. 1990. Analyse de la déformation du Jura central entre Neuchâtel (Suisse) et Besançon (France). *Eclogae Geologicae Helvetiae*, **88**, 543–558.

TURCOTTE, E. & SCHUBERT, G. 1982. *Geodynamics - applications of continuum physics to geological problems*. John Wiley, New York.

WEIDMANN, M. 1988. *Lausanne*. Atlas géologique Suisse, feuille No. 85. Service hydrologique et géologique national.

—— 1992. *Châtel-St-Denis*. Atlas géologique Suisse, feuille No. 92. Service hydrologique et géologique national.

WILDI, W. & HUGGENBERGER, P. 1993. Reconstitution de la plate-forme européenne anté-orogénique de la Bresse aux Chaînes subalpines: éléments de cinématique alpine (France et Suisse occidentale). *Eclogae Geologicae Helvetiae*, **86**, 47–64.

——, BLONDEL, T., CHAROLLAIS, J., JAQUET, J. & WERNLI, R. 1991. Tectonique en rampe latérale à la terminaison occidentale de la Haute Chaîne du Jura. *Eclogae Geologicae Helvetiae*, **84**, 265–277.

ZIEGLER, P. A. 1990. Molasse Basin, Conference proceedings. *Eclogae Geologicae Helvetiae*, **85**, 511–797.

Eustatic versus tectonic controls on Alpine foreland basin fill: sequence stratigraphy and subsidence analysis in the SE German Molasse

J. ZWEIGEL, T. AIGNER & H. LUTERBACHER

Institut für Geologie und Paläontologie, Universität Tübingen, Sigwartstrasse 10, D-72076 Tübingen, Germany

Abstract: A sequence-stratigraphic and subsidence analysis in the SE German Molasse basin was carried out based on well and seismic data. The sequence-stratigraphic study revealed that the basin fill is composed of five second-order sequences. Sequences 3 and 4 can be subdivided into several third-order sequences. Seismic facies maps illustrate the temporal and spatial evolution of the study area. Comparison of an accommodation curve of the Molasse basin with the eustatic sea-level curve of Haq *et al.* (*Science*, 1987, **235**, 1156–1167) showed, that the sequence stratigraphic subdivision was mainly controlled by eustatic sea-level fluctuations.

Subsidence analysis revealed a major flexural event in the SE German Molasse basin from late Eocene to early Miocene (late Egerian), followed by a possible visco-elastic relaxation of the lithosphere responsible for the formation of an angular unconformity between the two classical transgressive–regressive megacycles of the Molasse stratigraphy (early Miocene, late Egerian–Eggenburgian boundary). A phase of relatively uniform, decreasing subsidence followed until the end of Molasse sedimentation, when an isostatic rebound resulted in the uplift of the basin. The subsidence analysis also revealed, that even though eustacy controlled sequence boundary formations, the interplay between subsidence and sedimentation rates, both governed by tectonics, controlled the transgressive-regressive megacycles. However, the coupling of the two factors depends on the tectonic process.

The study area is situated in the SE German Molasse basin (Fig. 1) which is part of the northern Alpine foreland basin, a classical asymmetric foreland basin (Homewood *et al.* 1986). Its deepest part lies in the south in front of the Alpine nappes. It is filled with predominantly clastic sediments of Tertiary age, mainly the debris of the rising Alps, reaching thicknesses up to 5000 m in front of the orogen.

The Tertiary Molasse basin fill is traditionally divided into two transgressive–regressive cycles separated by an unconformity (Eisbacher 1974). This classical Molasse stratigraphy (Lower Marine Molasse (LMM) – Lower Freshwater Molasse (LFM), Upper Marine Molasse (UMM) – Upper Freshwater Molasse (UFM)) can be followed from western Switzerland into Bavaria. However, east of Munich marine or even deep-marine conditions prevailed during the Lower Freshwater Molasse until the Mid-Miocene (Fuchs 1976; Betz & Wendt 1983; Fig. 2). The study area covers the transitional zone between these two realms, an area well suited for sequence stratigraphic studies. Reflection seismic lines and well data, generously supplied by German oil companies, provided the basis for the study (Fig. 3).

In this paper we try to refine and extend previous sequence stratigraphic interpretations (Bachmann & Müller 1991; Jin 1995; Jin *et al.* 1995) to get a better impression of the temporal and spatial development in the study area. Moreover, a subsidence analysis based on sequence-stratigraphic results intended to quantify the influence of subsidence rates and sedimentation rates on the stratigraphic development of the basin and to discriminate tectonic from eustatic signals in basin dynamics.

Sequence stratigraphy

Lithostratigraphy

Well-log interpretation incorporating published data (Füchtbauer 1964; Müller & Blaschke 1971; Wagner 1980; Buchholz 1986; Kronmüller & Kronmüller 1987; Lemcke 1988; Bachmann & Müller 1991) allowed the distinction of several informal litho-units and their interpretation in terms of palaeo-environments. The litho-units formed the basis for the well correlations (Fig. 4). These units and their stratigraphic position are shown on Table 1.

A north-south transect (Fig. 4a), situated in the shelf area, documents a southward increase in thickness towards the orogen typical for a

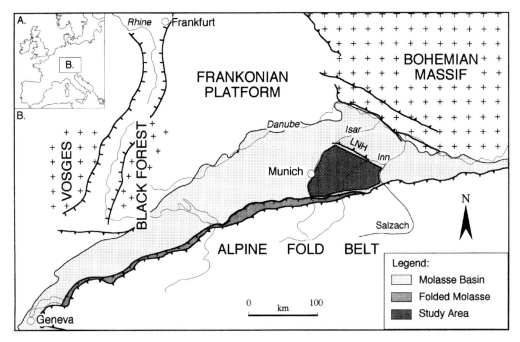

Fig. 1. Map of the German Molasse basin with the study area indicated; LNH, Landshut-Neuöttinger Hoch.

foreland basin. The basal units (Basal Sands, Lithothamnium Limestone) show a progressive northward onlap related to the northward advance of the Alpine nappes (Crampton & Allen 1995). It is interesting that the north–south thickness asymmetry of the litho-units is not constant. Generally the LMM/LFM transgressive–regressive cycle exhibits stronger thickness variations than the upper one, with the exceptions of the Baustein Beds and the interval from Upper Chattian Marls to Nantesbuch Sandstone.

Within the Aquitanian Sands two subunits are distinguished: (**a**) delta-plain deposits and (**b**) marine-embayment facies, which have a channel-like distribution where marine-embayment facies erosively cuts into delta-plain deposits. A major feature of the north–south

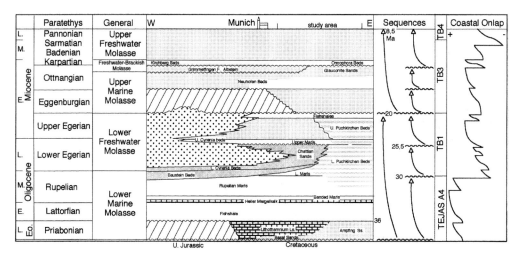

Fig. 2. Schematic west–east stratigraphic cross-section through the German Molasse basin after Lemcke (1988) and Bachmann & Müller (1991).

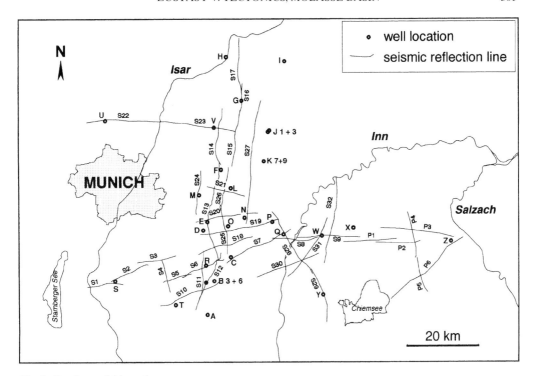

Fig. 3. Database of this study.

correlation is an angular unconformity developed between the Aquitanian Sands and the Burdigalian/Neuhofen Beds, which forms the boundary between the classical transgressive-regressive cycles. The Aquitanian Sands are substantially truncated in the North while the Burdigalian and the Neuhofen Beds show a progressive northward onlap. The strata of the second transgressive-regressive cycle have relatively uniform thicknesses with a slight maximum in the centre of the study area. The top of the Upper Freshwater Molasse is truncated in the South by erosion.

The west–east well correlation parallel to the basin axis (Fig. 4b) shows more complex patterns, because the basin was mainly filled from west to east. The coastal deposits of the Baustein Beds and the delta plain deposits of the Chattian Sands pinch out progressively towards the west, while the thickness of the related delta-front deposits (Rupelian/Chattian Marls) is increasing in this direction, indicating an eastward-prograding delta. Also a west–east facies change from thick delta plain deposits (Aquitanian Sands) in the west trough marine-embayment facies to thin basinal marls in the east in the upper Egerian indicates a strongly aggrading delta with a relatively stable coastline.

Aquitanian Fish Shales and Burdigalian show a progressive westward onlap onto the upper Egerian palaeo-topography, documenting the infilling of an inherited relief. The overlying Ottnangian and younger strata again have fairly uniform thicknesses.

Seismic facies

The results of the well studies, especially the palaeo-environmental interpretations, were incorporated into a seismic-facies analysis applying the procedure described by Mitchum & Vail (1977). Twelve seismic facies, and four subfacies only distinguishable on high-quality seismic lines, have been identified. Some of the seismic facies discussed in Jin (1995) and Jin et al. (1995) are subdivided and their interpretation in part modified. Reflection characteristics, correlation with the well-logs, and palaeo-environmental interpretations of the seismic facies are summarized in Table 2.

Seismic sequences

Parallel to the interpretation of the seismic-facies, a seismic-sequence analysis was carried out incorporating the results of the well

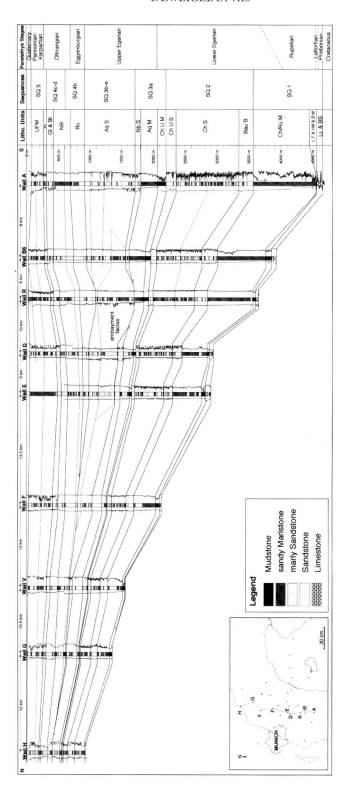

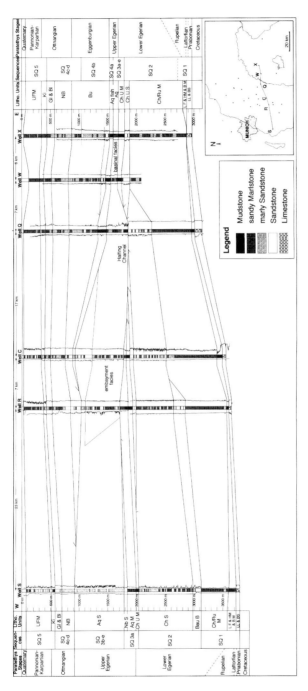

Fig. 4. (**a**) Well correlation chart along a north–south transect (basin margin to orogenic front) (**b**) Well correlation chart along a west–east transect (basin axis). Abbreviations: AqM, Aquitanian Marls; AqS, Aquitanian Sands; BauB, Baustein Beds; BM, Banded Marls; BS, Basal Sands; Bu, Burdigalian; Ch/RuM, Chattian/Rupelian Marls; ChS, Chattian Sands; ChUM, Chattian Upper Marls; ChUS, Chattian Upper Sands; Gl & Bl, Glauconite Sands and Blätter Marls; HM, Heller Mergelkalk; Ki, Kirchberg Beds; LF, Lattorfian Fishshales; LL, Lithothamnium Limestone; NB, Neuhofen Beds; NbS, Nantesbuch Sandstone; SQ, Sequence; UFM, Upper Freshwater Molasse.

Table 1. *Litho-units distinguished in the study area with a short description and their palaeo-environmental interpretation, for correlation of the Paratethys stages mentioned to the Standard stages refer to Fig. 11*

Litho unit	Description	Interpretation
Quaternary	Thickness: 0–100 m, marls, coarse-grained sands, conglomerates	Glacial deposits
Upper Freshwater Molass (Karpatian–Pannonian)	Thickness: 0–430 m, interbedded fine to conglomeratic sands sands and colourful marks, no microfauna	Fluvial, lacustrine, alluvial fan and plain deposits
Kirchberg Beds (uppermost Ottnangian)	Thickness: 0–80 m, grey sands interbedded with colourful marls, glauconitic, shell debris, brackish fauna	Very shallow marine (brackish), (0–20 m water depth)
Helvetian (Ottnangian)	Thickness: 150–500 m, like above, three coarsening upward trends (= shallowing upwards), marine microfauna	Shallow marine (20–200 m water depth m)
Burdigalian (Eggenburgian)	Thickness: 0–900 m, interbedded grey marls, sandy marly, marly sands and fine sands, shell debris, glauconite	Shallow marine (20–500 water depth m)
Obing Formation (lower Eggenburgian)	Sandstones, marls	Deeper-marine (500–1000 m water depth), tubiditic
Aquitanian Fish Shales (upper Upper Egerian)	Thickness: 0–120 m, dark brown, shaly marls, thin sandbeds, limestones	Marine (100–1000 m water depth), starved sedimentation
Upper Puchkirchen Formation (Upper Egerian)	Grain- and matrix-supported gravels, sandstones, marls	Deep-marine (500–1500 m water depth), gravity deposits
Aquitanian Sandstones (Upper Egerian)	Thickness: 0–800 m, close sand-marl (dm-m-scale) interbedding A: marls-dominated, thinner sandbeds, fining upward log patterns, marine microfauna B: sand-dominated, aggradational log patterns, coal fragments, no to poor brackish microfauna	A: shallow to deeper marine (20–30 water depth m), embayment fill facies B: coastal plain with mangrove swamps
Nantesbuch Sandstone (Upper Egerian)	Thickness (plus Aquitanian Lower Marls): 0–400 m, grey sandstones, marly intercalations, shell debris, coal fragments, brackish microfauna	Coastal marine (0–20 m water depth)
Aquitanian Lower Marls (Upper Egerian)	Brown-grey marls, silty, few fine-grained sandstone beds (turbidites), trace fossils	Marine to brackish, upward-decreasing water depth (250–20 m)
Lower Puchkirchen Formation (Lower Egerian)	Grain and matrix-supported gravels, sandstones, marls	Deep-marine (500–1500 m water depth), gravity deposits
Chattian Upper Marls (upper Lower Egerian)	Thickness: 0–153 m, dark brown, pyritic shales, fish remains	Marine (100–250 m water depth), starved sedimentation
Chattian Upper Sands (Lower Egerian)	Thickness: 0–155 m, like above, more sand-dominated	Coastal plain with more reworking (waves)
Chattian Sandstones (Lower Egerian)	Thickness: 0–1300 m, dm- to m-scale interbedded grey, medium-grained sands and colourful to grey marlstones, few coal, shell debris, brackish to marine microfauna	Coastal & plain

Table 1. continued:

Litho unit	Description	Interpretation
Baustein Beds (lower Lower Egerian)	Thickness: 0–250 m, grey, fine to medium-sands, varying marl content, thick (m) sand beds with thin marl-intercalations, shell debris, coal flitters	Coastal marine (0–20 m water depth)
Rupelian/Chattian Marls (Rupelian/Lower Egerian)	Thickness: 245–1145 m, grey, silty marls, with thin (dm-m) sand beds (turbidites), shell fragments	Marine, upward-decreasing water depth (1000–20 m)
Banded Marls (Rupelian)	Thickness: 10–100 m, black-brown marls, thin calcarerous laminae (coccolith), few thin sand beds (turbidites)	Condensed sedimente, deep-water (200–500 m water depth)
Heller Mergelkalk (Rupelian)	Thickness: 2–8 m, coccolith limestone	Starved sedimentation, deeper marine (200 m water depth)
Lattorfian Fish Shales (Lattofian)	Thickness: 1–30 m, black-brown, laminated shales, rich marine microfauna, fish remains	Deeper marine (100–200 m water depth) starved sedimentation, oxygen deficiency
Lithothamnium Limestone (Priabonian–Lattofian)	Thickness: 0–80 m, brown-grey to bluish limestone, varying shale content	Shallow-marine Lithothamnium reef and reef debris deposits (10–20 m water depth)
Priabonian Marls (Priabonian)	Thickness: 0–16 m, green-grey, fossil rich marls	Shallow-marine, sublitoral (20–30 m water depth)
Basal Sands (Priabonian)	Thickness: 0–18 m, medium- to coarse-grained, grey, calcareous sandstones	Coastal to shallow marine (0–10 m water depth)

Table 2. *Seismic facies distinguished in this study, with their reflection characteristics, their correlation to wire line logs and their environmental interpretation*

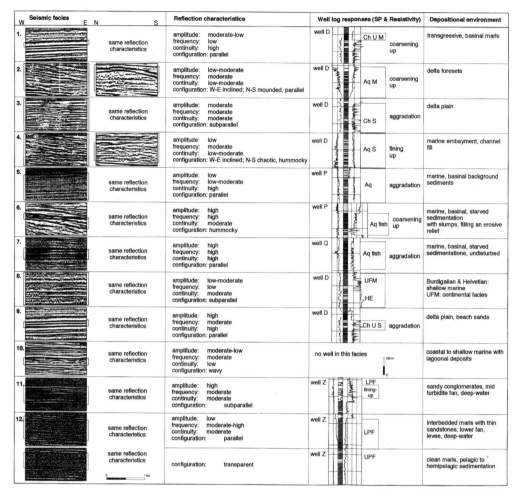

analysis. Five second-order sequences could be recognized, modifying the results of Jin (1995) and Jin *et al.* (1995). The following changes in regard to the interpretations by Jin (1995) and Jin *et al.* (1995) are made: (1) smaller eastward extend of Sequence 1, (2) different systems tract interpretation of Sequence 2, and (3) refined interpretation of Sequences 3 and 4, with changed position of the Sequence Boundary 4 separating them. Sequences 3 and 4 are subdivided into third-order sequences. Since the main direction of deposition was from west to east in the shelf area, a west–east composite seismic line (Fig. 5) demonstrates the sequence-stratigraphic development (westward extended version of the composite seismic line shown by Jin 1995). In the deep-water part (east of the river Inn), the so-called Puchkirchen Trough, a SW–NE seismic line documents the subdivision there (Fig. 6).

Sequence Boundary 1 marks the base of the Tertiary. Truncation of the Cretaceous strata below can be observed on north–south-oriented seismic lines, which can be interpreted as a flexural forebulge unconformity (Crampton & Allen 1995). This boundary is also clearly defined on wireline logs (Fig. 4).

Sequence 1 reaches up to the base of the Baustein Beds in the shelf area and the base of the Puchkirchen Formation in the deep-water part. It is composed of a transgressive and a highstand systems tract separated by a low

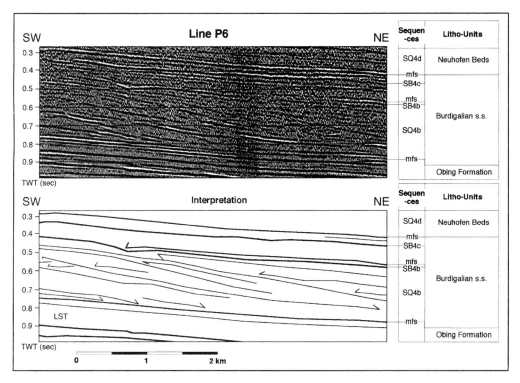

Fig. 9. Detail from seismic line P6 (SW–NE), illustrating the transgressive shorface erosion between Sequences 4b and 4c within the Puchkirchen Trough; note truncation of inclined reflectors (delta foresets) of Sequence 4b by higher amplitude reflectors (transgressive deposits) of Sequence 4c; for location see Fig. 3.

high-amplitude reflectors (Nantesbuch Sandstone, Fig. 6). The study area was flooded by an abrupt relative sea-level rise at the beginning of Sequence 3, while during the highstand deposition of Sequence 3a a delta with only minor aggradation prograded from west to east into the center of the study area, indicating a relative stillstand in sea level.

Sequences 3b–e correspond to the Aquitanian Sands. Each sequence is characterized by lowstand channel incision and infill (embayment facies) topped by a maximum flooding surface above which highstand deltaplain aggradation is observed (Fig. 7). The lowstand channel incisions are eastwardly connected to the formation of a submarine channel (Halfing Channel, Lemcke 1988), a NNW–SSE-oriented structure that cuts into the underlying delta deposits (Lower Egerian) and forms the connection of the delta deposits (Upper Egerian) to the West with the deep-water part to the East (Puchkirchen Trough). Generally, Sequences 3b to 3e are dominated by aggradation with only minor progradation.

In the deep-water Puchkirchen Trough, Sequence 3 is formed by the Upper Puchkirchen Formation. This formation is subdivided by minor erosive surfaces into five subsequences labelled A2.1, A2.2, A2.3, A3, and A4 adopted from Robinson & Zimmer (1989) and Fertig et al. (1991). These subsequences correlate with Sequences 3a to 3e of the shelf area, implying a sea level control on their formation; each lowstand inducing submarine erosion and the onset of coarser deposition.

Sequence boundary 4 is the boundary between the two classical transgressive–regressive cycles. In the north-south direction, it is expressed as an angular unconformity (Fig. 3) expanding towards north and by channelized erosion in the south. Smaller channel incisions on the shelf slope and submarine channel incision in the basinal facies are visible on the west–east-oriented seismic lines (Fig. 8). The angular unconformity indicates that tectonic movements were involved in the formation of this sequence boundary.

In contrast to Jin (1995) and Jin et al. (1995) Sequence 4 is interpreted to be composed of four third-order sequences (4a to 4d). *Sequence*

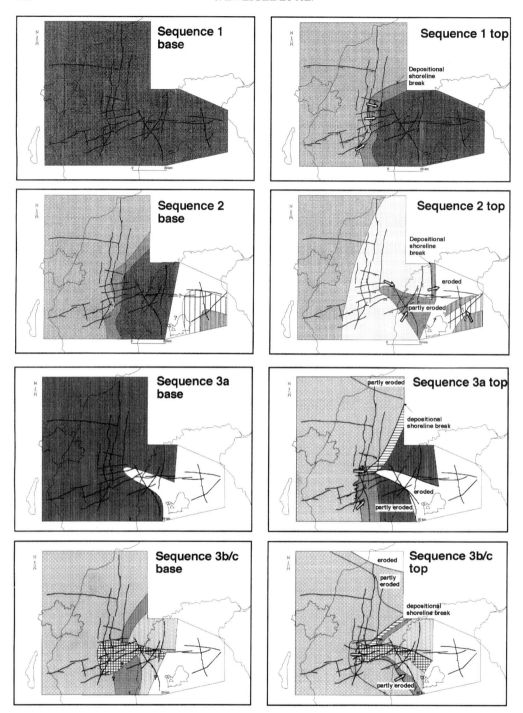

Fig. 10. Seismic facies maps of the base and the top of the sequences, illustrating the spatial and temporal evolution in the study area.

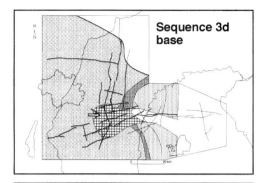
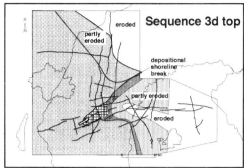
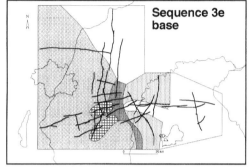
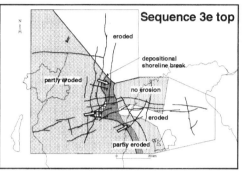
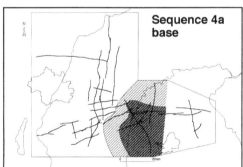
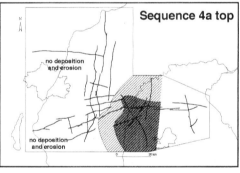
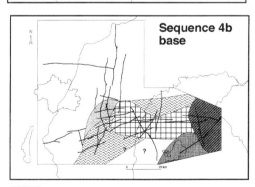
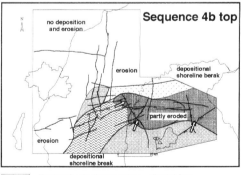

	Facies 1, transgressive marls		Facies 8, shallow marine – alluvial fan
	Facies 2, delta foresets		Facies 8a, shallow marine sands
	Facies 3, delta plain		Facies 8b, shallow marine, tidal
	Facies 4, marine embayment fill		Facies 9, beach sands
	Facies 5, marine, basinal marls		Facies 10, lagoonal deposits
	Facies 6, starved sediments with slumps		Facies 11, sandy conglomerates, turbiditic
	Facies 7, starved sediments, undisturbed		Facies 12, marly conglomerates, debris flows or basinal marls
↗	direction of sediment transport		

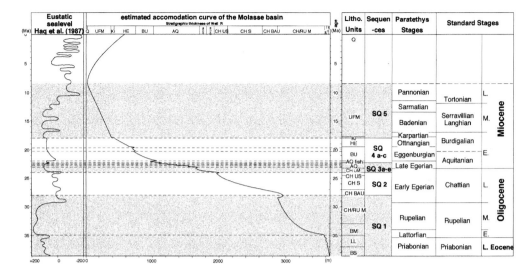

Fig. 11. Accomodation curve constructed for a representative well in the centre of the study area (well R, for location see Fig. 3) compared to the eustatic sea level curve of Haq *et al.* (1987), for abbreviations see Fig. 4.

4a consists of the Aquitanian Fish Shales. Early lowstand submarine erosion was followed by transgressive and highstand starved sedimentation, caused by a cut-off of the sediment source. Channelized erosion on top of the Aquitanian Fish Shales at the shelf slope indicates another relative fall in sea level, forming Sequence Boundary 4a (Fig. 5).

The subdivision into Sequences 4b to 4d is most clearly expressed in the Puchkirchen area, from where the boundaries can be traced onto the shelf area. *Sequence 4b* is characterized by onlaps of the lowstand, turbiditic Obing Formation onto an erosive surface (Fig. 6). A thick interval of NE-prograding clinoforms on top of it is interpreted as a highstand prograding delta complex. Internal reflection terminations indicate episodic delta progradation.

Sequence 4c is also composed of a lowstand systems tract and a highstand systems tract, separated by a strong reflection. At the onlap point of this reflection onto the lower delta complex, transgressive shoreface erosion occurred (Fig. 9). In general, the sequence is characterized by parallel reflections. These reflections are truncated along Sequence Boundary 4c by a shallow west–east-oriented channel-like structure, filled by northward onlapping strata belonging to *Sequence 4d*. This fill is terminated by a very continuous higher-amplitude reflector separating the lowstand from the highstand deposits of Sequence 4d.

Sequence 4 is topped by *Sequence Boundary 5*, which can be easily followed throughout the study area as a fairly continuous high-amplitude reflector, corresponding to the base of the Kirchberg Beds.

Sequence 5, which has not been studied in detail, consists of discontinuous parallel reflectors. Well studies (Fig. 4) allow a subdivision into the transgressive, brackish Kirchberg Beds and the highstand, terrestrial Upper Freshwater Molasse.

Seismic facies distributions

The combination of the results of the seismic facies and sequence stratigraphic interpretations made it possible to draw seismic facies maps of each sequence. These facies maps illustrate spatial and temporal facies changes and stratigraphic development in the study area (Fig. 10).

At the base of *Sequence 1*, the whole study area was flooded by a shallow sea, depositing first the shallow marine Basal Sands and Lithothamnium Limestones and then deeper marine, sediment-starved marls. At the top of Sequence 1, a delta prograded mainly from west to east into the central study area.

No major changes in the depositional environments occurred along Sequence Boundary 2 in the shelf area (valley incision, flooding). However, this sequence boundary corresponds in the Puchkirchen Trough to the onset of the deposition of the turbiditic Puchkirchen facies (Lower Puchkirchen Formation), which was mainly sourced from the South.

During the deposition of *Sequence 2* the delta prograded about 40 km further eastward and was accompanied by strong aggradation (Fig. 5). The delta top was subsequently eroded by the Halfing Channel, which formed later. Mass-flow deposition continued throughout the formation of Sequence 2 in the Puchkirchen Trough.

Minor erosion of the delta along Sequence Boundary 3 was followed by flooding and drowning of the whole study area. In the Puchkirchen Trough, Sequence Boundary 3 forms the boundary between the 'Lower' and Upper Puchkirchen Formation. As in Sequence 1, during *Sequence 3a*, a delta prograded from west to east into the middle portion of the study area. The position of the depositional shoreline break is blurred in the south by later erosion. In the north a lagoonal facies was developed.

The map of the base of Sequences 3b/c shows that erosion occurred along Sequence Boundary 3a. The shelf was incised by valleys which were infilled by marine embayment facies, while a submarine channel (Halfing Channel) developed in the deeper part, on top of the former Sequence 2 delta, feeding sediment into the Puchkirchen Through. Sedimentation there was dominated by mass flow deposits derived from the South. At the end of Sequences 3b/c, the coastline stayed more or less stationary. Again the position of the coastline in the south is blurred by later erosion along Sequence Boundary 3c.

Additional valley incision is observed at the base of *Sequence 3d*, otherwise the facies distribution is very similar to the previous sequence. The delta prograded during Sequence 3d over the embayment facies leaving only a narrow channel which was connected to the Halfing Channel. The coastline in the north was more or less stationary.

Flooding occurred along Sequence Boundary 3d, leading to a westward retreat of the coastline in the north and the formation of two embayments in the south. During *Sequence 3e* these embayments were almost totally infilled and covered by delta-plain deposits. Generally, deposition during Sequence 3 was dominated by aggradation with intercalated intervals of incision and embayment formation, while in the Puchkirchen Trough further northward-stepping of the south-derived mass flow deposition is observed.

During the formation of Sequence boundaries 4 and 4a, bounding *Sequence 4a*, major erosion took place on the shelf and deposition was restricted to the SE part of the study area. Where the base is conform undisturbed basinal deposits (seismic facies 7) were deposited, while partly slumped, basinal deposits (seismic facies 6) occur in erosive structures.

Sequence 4b is also restricted to the SE part of the study area. The base in the west is dominated by coarse-grained lowstand deposits which are confined to an axial belt, while facies 11 indicates turbiditic lowstand deposition in the Obing Formation. Shallow-marine deposition dominated in the north and SW. During Sequence 4b a delta prograded from the NW and other deltas prograding from the south to SW into the study area left only a narrow shallow-marine sea in the axial region. While delta foresets dominate in the east, a 'hummocky' seismic facies of shallow marine origin was deposited in the west, interpreted to document strong reworking by axial tidal currents. In the east, the sea floor seems to have been below the tidal currents, prohibiting strong tidal reworking of the delta deposits there. Along Sequence Boundary 4b, partial erosion of the deposits occurred along the axial shallow marine region in the central and western part of the study area.

Sequences 4c, 4d, and 5 consist exclusively of seismic facies 8 (shallow marine or continental deposits), therefore no maps are shown of these intervals.

Discussion

An accommodation curve (curve of relative sea-level changes) in the study area was constructed roughly to evaluate the influence of eustasy on the development of the Molasse basin. Since accommodation is the result of subsidence plus eustatic sea-level changes, a curve should resemble a decompacted, palaeobathymetry corrected, but not eustatically corrected subsidence curve. In this study a semi-quantitative approach was chosen. The stratigraphic thickness of a representative well (well R) plus the water depth and changes in water depth, estimated from seismic data, were plotted against the age. This resulted roughly in an accommodation curve, but not including decompaction (Fig. 11).

This curve was compared (Table 3) to the eustatic sea-level curve of Haq *et al.* (1987). Generally, it can be noted that:

- the general increase in accommodation has to be attributed to subsidence, which took place at different rates through time;
- all major fluctuations in accommodation observed in the Molasse basin can be correlated to eustatic events;

Table 3. Indicated influence of eustasy and subsidence on the development of sequences and sequence boundaries in the study area

Sequence stratigraphy	Relative change of sea level	Interpretation
Sequence 5 TST & HST (late Ottnangian–Pannonian)	Slow rise	Subsidence controlled, not studied in detail
Sequence boundary 5 (late Ottnangian) Type 1	Small fall, sudden small rise	Eustatic stillstand followed by sudden small rise
Sequence 4 LST & HST (Eggenburgian–Ottnangian)	General slow rise with two minor falls	Eustatic rise, falls tectonically controlled?
Sequence boundary 4a (Late Egerian/Eggenburgian) Type 1	Fall (amalgamates to the N with SB 4)	Eustatic fall, tectonically enhanced
Sequence 4a TST & HST (late Late Egerian)	Rise in the S	Eustatic rise overprinted by asymmetric subsidence
Sequence boundary 4 (late Late Egerian Type 1)	Major fall	Eustatic fall, tectonically enhanced (angular unconformity)
Sequence 3 (late Egerian) SQ 3a	Sudden rise, then stillstand	Eustatic rise and stillstand (relative uniform thickness)
SQ 3b–e		Eustatic high overprinted by very high subsidence, tectonic control on falls?
Sequence boundary 3 (late Early Egerian) Type 1	Short fall, major flooding	Eustatic fall and major rise
Sequence 2 HST (Early Egerian)	Strong rise	Eustatic low overprinted by highly N–S asymmetric subsidence
LST (Early Egerian)	Slow rise	Uniform thickness of LST deposits imply eustatic control, but age discrepancy
Sequence boundary 2 (early Early Egerian) Type 1	Substantial but short fall	Age discrepancy to major eustatic fall (200 m, Rupelian/Egerian), biostratigraphic control in study area?
Sequence 1 HST (Rupelian/Early Egerian)	General rise, slowing towards the top	Eustatic high sea level overprinted by N–S subsidence
late TST (Lattorfian)	Sudden flooding	Sudden eustatic rise overprinted by slow increase in subsidence suppressing previous eustatic fall
early TST (Priabonian-Lattorfian)	General rise	General rise: subsidence
	Three minor fluctuations	Fluctuations eustatically controlled

Table 4. *Input parameters used for decompaction in the subsidence analysis employing the computer program BasinWorksTM (Bowman 1991)*

Lithology	Initial porosity	c_1
Limestone	25%	0.00045
Shale	75%	0.003
Sandstone	50%	0.001
Marl	60%	0.002

c_1 Compaction coefficient

Table 5. *Ages of stratigraphic levels used for the subsidence analysis*

Surface name	Age (Ma)
Top Quatenary	0
Top Upper Freshwater Molasse (top SQ5)	8.5
Top Kirchberg Beds (top TST SQ5)	17.8
Top Upper Marine Molasse (SB5)	18
Top Burdigalian	19.5
Top Aquitanian Fishshales (SB 4a)	22
Top Sequence 3e	22.3
TS Sequence 3e	22.4
Top Sequence 3d	22.5
TS Sequence 3d	22.6
Top Sequence 3c	22.7
TS sequence 3c	22.8
Top Sequence 3b	22.9
TS Sequence 3b	23
Top Sequence 3a	23.1
Top Upper Chattian Marls (TS 3a)	23.3
Top Upper Chattian Sands (SB3)	24
Top Chattian Sands	25
Top Baustein Beds (TS2)	27
Top Chattian/Rupelian Marls (SB2)	28
Top Banded Marls (top condensed section)	31
Top Lithothamnium Limestone (top early TST)	35
Top Basal Sandstones	37
Basement (SB1)	38.6

- there are only few fluctuations in accommodation in the Molasse basin that have no equivalent on the 'Haq Curve';
- not all eustatic sea-level fluctuations postulated by Haq et al. (1987) are recorded in the Molasse basin, if the rates of sea-level fall were to slow, they were outstripped by subsidence in the Molasse;
- even Sequence Boundary 4, which as an angular unconformity shows a strong tectonic influence, can be correlated to a eustatic event and thus is a tectonically enhanced sequence boundary.

Thus it seems for the Molasse basin as in other foreland basins, that the creation of accommodation space was controlled by tectonics via subsidence, but the formation of sequence boundaries was controlled by eustatic sea-level fluctuations.

However, the direct comparison with the eustatic sea-level curve is still hampered by problems in the biostratigraphic subdivision, the correlation with the standard stages and the radiometric time scale (for discussion see Berger 1992). The stratigraphy of the Molasse basin is very often still a lithostratigraphy based on the recognition of lithological units and their correlation. A considerable amount of work on biostratigraphy, magnetostratigraphy and physical stratigraphic correlations, with their correlations to the chronostratigraphic time scale has still to be done, to establish a satisfactory stratigraphic frame for the Molasse basin. The ages used in this paper have to be considered as 'best guesses', which will be improved by additional work.

Geohistory Analysis

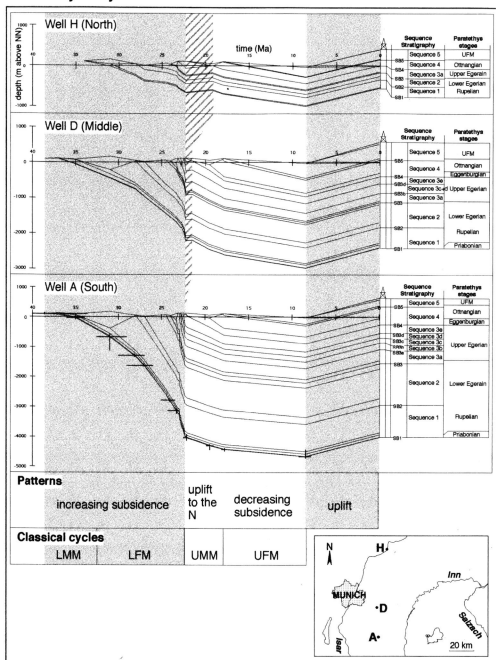

Fig. 12. Geohistory curves of three wells along a north–south transect employing the computer program BasinWorks™ (Bowman 1991), estimated error bars for age and bathymetry are exemplarily shown for well A.

Subsidence analysis

Input parameters

Sedimentary and stratigraphic data from 24 wells was used as input for a regional subsidence analysis to quantify the influence of subsidence. Geohistory curves were determined employing the computer program BasinWorks™ (Bowman 1991) incorporating corrections for palaeobathymetry, eustasy and compaction. Since only little micropalaeontological information was available, the estimation of the palaeobathymetry is mainly based on seismic interpretation, lithological and sedimentary information. To reduce the error of these estimations, consistent palaeobathymetry maps of stratigraphic levels were constructed.

Eustatic correction was based as a first approximation on the sea level curve from Haq et al. (1987) due to the good agreement indicated by the sequence stratigraphic study. Decompaction was carried out applying the equation of Dickinson et al. (in Bowman 1991) offered by the computer program:

$$\phi_z = \frac{\phi_0}{1 + (c \times z)} \quad (1)$$

Where: ϕ_0 = initial porosity, ϕ_z = porosity at a given depth z, c_1 = compaction coefficient, z = depth. Four different lithologies were distinguished. Since only limited porosity data of wells were available, standard figures (Allen & Allen 1990; Bowman 1991) were used as input parameters for decompation (Table 4).

Absolute ages of stratigraphic levels (Table 5) were estimated employing the timescales of Harland et al. (1990; standard stages) and Steininger et al. (1985; Paratethys stages; for correlation see Fig. 11). As discussed above, biostratigraphic control is poor in the study area. Therefore, to reduce the error bars, ages of stratigraphic boundaries non-coincident with stage boundaries were estimated avoiding unreasonably high or low sedimentation rates and unreasonably strong changes of these.

Geohistory

All wells show a very similar overall pattern of subsidence with a convex-upward shape of increasing subsidence rates from the Priabonian (38.6 Ma) until the late Egerian (22.3 Ma), a sudden decrease in subsidence around the late Egerian–Eggenburgian boundary (22.3 Ma), a concave-upward shape of a gradual decrease in subsidence until the Pannonian (8.5 Ma), and an uplift from the end of the Upper Freshwater Molasse deposition to the present. In detail, systematic variations are observed which are illustrated for a north–south cross-section (Fig. 12), perpendicular to the orogen front.

The onset of subsidence was earlier and stronger in the south close to the thrust front than in the north. Until the late Egerian a great north–south increase in subsidence is observed. In the latest Egerian to Eggenburgian (22.3–19.5 Ma) the subsidence rates change suddenly. In the northern part of the basin an uplift occurred during the entire Eggenburgian, in the central part it lasted only in the lower Eggenburgian, while in the south subsidence continued at strongly reduced rates close to the Alpine front. After this asymmetric uplift, the rates of subsidence and the patterns are very similar in the whole study area.

Backstripping

In order to evaluate fully the tectonic significance of these patterns, backstripping was applied to derive the tectonic subsidence, using the flexural backstripping equation of Jin (1995):

$$Z_{tec} = Z_{tot} - C \left(S \frac{\rho_m + \rho_s}{\rho_m - \rho_w} - \Delta sl \frac{\rho_w}{\rho_m + \rho_w} \right) \quad (2)$$

where: Z_{tot} = total subsidence, Z_{tec} = tectonic subsidence, C = basement response function, S = thickness of sedimentary column, Δsl = change in eustatic sea level, ρ_m = mantle density (3300 g m^{-3}), ρ_s = sediment density, ρ_w = water density (1000 g m^{-3});

Jin (1995) estimated a value of 0.27 for the basement response function C in the study area by modelling the geometry of the forcland basin. Published estimates of the effective elastic thickness of the lithosphere beneath the Alps (values increase from west to east: 25–50 km Karner & Watts 1983; 26–56 km McNutt 1988) indicate, that values of C between 0.21 and 0.46 can be regarded realistic. Therefore, taking into account the inaccuracy in the derivation of some parameters in the calculation, a median value of 0.3 was chosen in this study. It should be pointed out that varying C within the realistic limits does not significantly change the shape of the tectonic subsidence curves and therefore also the interpretation.

In general, the tectonic subsidence is very close to the total subsidence (Fig. 13) since the Molasse basin is supported by a strong lithosphere. All features described from the

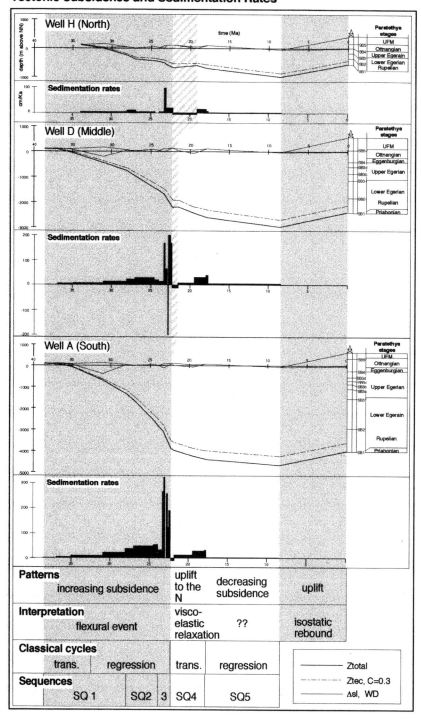

Fig. 13. Tectonic subsidence curves and sedimentation rates for the same wells as in Fig. 12 along a north–south transect with interpretation, and relation to transgressive–regressive cycles indicated; Ztotal = total subsidence, Ztec = tectonic subsidence, C = basment response function, Δsl = eustatic sea level changes, WD = waterdepth.

geohistory curves are also reflected in the tectonic subsidence curves. A major flexural event occurred in the SE German Molasse basin from the late Eocene until the late Egerian (38.6–22.3 Ma) as already observed by previous authors (Lemcke 1974; Homewood *et al.* 1986; Jin 1995). The flexural event is related to the collision of the European and African/Adriatic plates starting in the Late Eocene and resulting in the overthrusting of the Alpine nappes onto the European plate (Frisch 1979; Karner & Watts 1983; Dewey *et al.* 1989). This proximal to distal trend in the onset and the rates of subsidence is attributed to a northward migration of a subsidence wave across the basin, caused by the northward advance of the Alpine nappes as described in Switzerland (Homewood *et al.* 1986).

The north–south transect (Fig. 13) shows not only a general increase in subsidence from the Late Eocene to early Miocene (late Egerian), but also an intensification of the north–south asymmetry in subsidence, especially in the late Egerian is observed. This trend can be explained by a reduction of thrust front advance rates combined with the onset of internal deformation increasing the relief (and thus the thrust load) in the Eastern Alps during the Egerian. A similar behaviour has been observed in the Western and Central Alps (Pfiffner 1986; Sinclair *et al.* 1991; Sinclair & Allen 1992).

The northward increasing uplift during the late Egerian/Eggenburgian (22.3–19.5 Ma) is related to the formation of an angular unconformity at Sequence Boundary 4, the 'base Burdigalian unconformity'.

Sinclair *et al.* (1991) proposed an explanation of this base Burdigalian unconformity by modelling in the Swiss Molasse basin using a diffusion model of mountain-belt uplift and erosion. By varying the rate of thrust-front advance and the thrust-front angle, keeping other parameters fixed, an unconformity was created. From 40 to 24 Ma a high rate of thrust-front advance and a low surface angle was assumed, from 24 to 23.8 Ma the thrust-front advance was markedly reduced while the surface angle was increased simulating the underplating of the External Massifs in the Western Alps. This resulted in a backtilting of the foreland causing erosion of the distal (n) stratigraphy over the forebulge and accelerated subsidence close to the thrust front. Finally, renewed advance of the thrust wedge, keeping the slope angle constant, resulted in the filling of the basin.

However, this model can not be applied to the SE German Molasse basin because there a sudden decrease of subsidence is observed close to the thrust front, while the model predicts an increase. The model also assumes a very short period of only 200 000 years for the underplating of the external massifs. Pfiffner (1986) cites geochronological ages of 35–30 Ma for the underplating of the Gotthard and Tavetsch Massifs (Helvetic nappes) and 25–15 Ma for the Aar Massif. He thus gives much wider time brackets for the underplating compared to Sinclair *et al.* (1991). Also the stratigraphic pinch out data of the study area suggest an earlier (Egerian) change to internal deformation as assumed in the model.

We explain the observed change in tectonic subsidence by a rather abrupt decrease of thrusting in the Eastern Alps around the Egerian–Eggenburgian boundary (Malzer *et al.* 1993) and a visco-elastic behaviour of the European plate. After the cessation of thrusting a visco-elastic relaxation of the plate resulted in a low subsidence close to the Alpine front and a southward migration of the forebulge into the study area, causing uplift in the north. The visco-elastic relaxation was probably a fairly fast process taking place approximately during the short interval of the Aquitanian Fish Shales (22.3–22 Ma). The seemingly longer duration of the uplift in the north reflects the timing of the depositional overstepping of the shifted forebulge, which was later in the north.

Frisch *et al.* (1995, 1996) and Kuhlemann *et al.* (1995) suggest a sudden reduction of topographic relief and altitude in the Eastern Alps around the Eggenburgian in the course of an orogenic collapse after cessation of active collision. A possible driving force may have been the visco-elastic relaxation of the European lithosphere supporting the Eastern Alps.

The visco-elastic model also provides explanations for other features observed in the Molasse basin as the sudden reduction in sedimentation rates and grain size in the uppermost Egerian. The reduced relief in the Alps resulted in less detritus and the development of the Augenstein lowland (Simony 1851; Winkler-Hermaden 1957) on the Northern Calcareous Alps (Frisch *et al.* 1995, 1996), in which most of the coarser sediments were retained before reaching the Molasse basin. The changed palaeogeography of the Molasse basin in the uppermost Egerian/Eggenburgian can be explained by the same model. Prior to the Aquitanian Fish Shales, marine conditions were confined to the eastern Molasse basin. Later a shallow-marine area developed along the Alpine front (Lemcke 1988; Bachmann & Müller 1991), because visco-elastic subsidence was higher close to the thrust front while sediment input was markedly reduced resulting in underfilled conditions there.

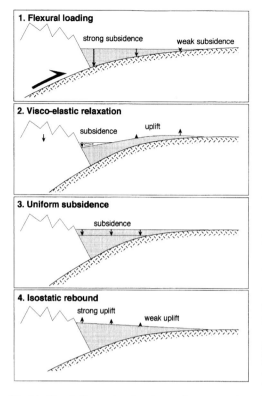

Fig. 14. Sketch illustrating the observed patterns of subsidence in the study area and their interpretation. (1) Late Eocene to late Egerian (38.6–22.3 Ma): flexural loading results in a strongly asymmetric subsidence pattern. (2) Latest Egerian to Eggenburgian (22.3–?22 Ma): cessation of thrusting and visco-elastic relaxation of the lithosphere results in reduced altitudes in the orogen, slow subsidence close to the orogenic front and orogen-ward shift of forebulge, causing uplift of the distal foreland basin. (3) Eggenburgian to Pannonian (22?–8.5 Ma): uniform subsidence across the whole basin. (4) Pannonian to present (8.5–0 Ma): isostatic rebound of the orogen, causing stronger uplift close to the orogen front and weak uplift in distal areas.

Starting in the Ottnangian (19.5 Ma), the study area shows a very uniform development of slowly decreasing tectonic subsidence rates until the end of the Upper Freshwater Molasse deposition (8.5 Ma; also seen in the thickness trends Fig. 3). This change from previous north–south asymmetric subsidence patterns to the uniform pattern implies that active thrusting in the Eastern Alps had virtually ceased (Malzer et al. 1993) and other driving forces than flexural loading controlled the subsidence in the basin.

The uplift from the late Pannonian to recent times was already noted by Lemcke (1974) and interpreted as isostatic rebound. Possible driving forces may have been (1) erosional unloading (Burbank 1992) and/or (2) a decoupling or thermal erosion of the lower lithosphere (Kay & Mahlburg Kay 1993; Parsons & McKenzie 1978; Platt & England 1994) beneath the Alps reducing the subsurface loads.

The history of the interpreted tectonic evolution of the Molasse basin is summarized in Fig. 14.

Sedimentation rates

The development of sedimentation rates and their relation to the subsidence and the classical transgressive–regressive cycles is shown on Fig. 13 for a north–south transect. An increase in sedimentation rates from Late Eocene to latest Egerian (38.6–22.3 Ma) correlates to the flexural event (increasing subsidence) in the Molasse basin and the first classical transgressive–regressive cycle. During the initial phase of basin formation, the increase in subsidence and thus accommodation space was greater than the increase in sedimentation rates due to the still low topographic relief in the orogen, resulting in a transgression or underfilled conditions. Later the topographic relief in the orogen was markedly increased resulting in higher rates of subsidence but also in much higher rates of sediment input, which exceeded the created accommodation space leading to a regression or overfilled conditions. This trend from early underfilled to late overfilled conditions in a foreland basin and its relation to the relief in the orogen is also known for other foreland basins (Taiwan, Covey 1986; Pyrenees, Puigdefàbregas et al. 1986) and the Swiss Molasse basin (Sinclair & Allen 1992).

The second classical transgressive–regressive cycle in the Molasse stratigraphy is related to a different evolution of sedimentation rates and subsidence than the first one. Although, subsidence rates decreased markedly in the south and uplift is observed in the north in the latest Egerian/Eggenburgian, a transgression or underfilled conditions developed, due to the even greater decrease in sedimentation rates related to the reduced topographic relief in the orogen, caused by the visco-elastic relaxation. Overdeepening and dark shale deposition (Aquitanian Fish Shales) during periods of relative orogenic quiescence were also noticed by Tankard (1986; Appalachian foredeep) and Gardner (1995; mid-Cretaceous foreland basins, Central Utah) and also related to visco-elastic relaxation.

During the Ottnangian despite of increasing sedimentation rates and decreasing rates of subsidence, underfilled (or filled) condition prevailed. Only in the Upper Freshwater Molasse was the decrease in subsidence rates and accommodation space outstripped by the rates of sedimentation resulting in overfilled conditions or a regression. Lower sedimentation rates in the Upper Freshwater Molasse compared to the Ottnangian are probably not caused by lower erosion and sediment-input rates, but by sediment bypass due to lack of accommodation space.

Thus, the first transgressive–regressive megacycle was caused by active tectonic collision causing flexural deepening of the basin and a high relief in the orogen, while the second megacycle was caused by passive tectonics, i.e. the visco-elastic reaction of the lithosphere and a re-equilibration.

Discussion

Previous studies of the Molasse basin were mainly based on lithological and palaeontological information (Lemcke 1988; Nachtmann & Wagner 1987). It was not until recently that the modern concept of sequence stratigraphy was applied to the basin (Robinson & Zimmer 1989; Fertig *et al.* 1991; Bachmann & Müller 1991; Jin 1995; Jin *et al.* 1995). Bachmann & Müller (1991) indicate a first generalized sequence-stratigraphic subdivision of the basin fill. They distinguished two mega-sequences corresponding to the two classical transgressive–regressive cycles of the Molasse stratigraphy, which are each divided into three subsequences. A comparison with the coastal onlap curve of Vail *et al.* (1977) revealed a good agreement of the subsequences with eustatic sea-level fluctuations while the causes for the mega-sequences remain unexplained.

Jin (1995) published an essentially two-dimensional study on a north–south and a west–east transect of the shelf area. He distinguished three transgressive–regressive cycles (base Tertiary to Chattian Sands, Upper Chattian Marls to Aquitanian Sands, Aquitanian Fish Shales to top Molasse), which he related, according to his subsidence analysis, to three flexural events correlated with three phases of deformation in the western part of the thrust belt: the Calanda, Ruchi and Subalpine phases (after Pfiffner 1986). These cycles are composed of five second-order sequences roughly equivalent to the ones described by Bachmann & Müller (1991). He also found a good agreement of his estimated eustatic sea level with the curve of Haq *et al.* (1987).

In the present study, due to an extended data set, the five second-order sequences after Jin (1995) and Jin *et al.* (1995) were slightly modified. It was possible to achieve a finer subdivision into third-order sequences, and the three-dimensional character of the study allowed to construct seismic-facies maps, giving an impression of seismic-facies shifts and the evolution of the study area in time and space. An accommodation curve was constructed and compared to the curve of Haq *et al.* (1987). Our general conclusion, that the sequence stratigraphic subdivision in the Molasse basin was mainly controlled by eustatic sea level fluctuations, is in accordance with Jin's (1995) results. However, in detail the shapes of the sea-level curves constructed for the Molasse differ, as well as their exact correlation to the curve of Haq *et al.* (1987).

These differences are partly caused by the refined interpretation in this study, but a main cause lies in differences in assignment of ages to stratigraphic units in the Molasse. As discussed above, biostratigraphic control is poor in the study area making the exact stratigraphic classification of some litho-units and sequences difficult. Furthermore, the traditional stratigraphic subdivision of the Molasse basin is used differently by different authors, which complicates a correlation with the Paratethys stratigraphy (for an overview see Berger 1992). Another problem is encountered in the comparison with the curve of Haq *et al.* (1987), which uses the Standard stage subdivisions, and the correlation of the Standard and Paratethys stages is not unambiguous. Further biostratigraphic studies and exact absolute age determinations may help to solve these problems in the future.

The threefold subdivision into transgressive-regressive cycles of Jin (1995) was abandoned in this study, since the second cycle (Upper Chattian Marls to Aquitanian Sands) is of lower order, caused by a major eustatic sea-level rise at its base (see Fig. 9). Jin (1995) related the transgressive–regressive cycles to three flexural events indicated by his subsidence curves. Subsidence analysis carried out in the present study revealed only one major flexural event (Figs 10 & 11) corresponding to the first classical transgressive–regressive cycle. The differences between the subsidence curves of the two studies arise from several points. (1) Stratigraphic problems, as discussed above. (2) Age assignments of lithological and sequence boundaries which are not coincident with stratigraphic boundaries were tested, in this study, by estimation of sedimentation rates, therefore avoiding unrealistic high/low values. This was not the case in the study of Jin (1995) resulting in partly unrealistic ages—responsible for his three events. (3) Palaeobathymetric corrections in this study were estimated individually for each well and for each stratigraphic boundary, while Jin (1995) used average values for the litho-units. (4) A finer stratigraphic subdivision was used in this study. (5) The database was extended in this study to 24 wells.

In conclusion, this study reveals the sequence stratigraphic development of the basin fill to be controlled by eustasy, while the transgressive–regressive cycles were controlled by the interplay between accommodation space and sediment input rates, both mainly controlled by tectonics. During the first cycle an increase in subsidence was accompanied by an even faster increase in sedimentation rates, related to active thrusting in the Alps (flexural event). In contrast, in the second cycle a decrease in subsidence was accompanied by an increase in sedimentation rates, after an initial sudden reduction, related to a visco-elastic relaxation of the lithosphere.

The visco-elastic model has not been previously applied to the Molasse basin, but it seems to provide explanations for many features of the basin and the adjacent orogen not explainable by a simple elastic model.

German oil companies (Preussag, BEB, Mobil, RWE-DEA, Wintershall) generously supplied the seismic and well data of this study and are thanked for permission to publish this paper. J. Z. was supported by the EU Integrated Basin Studies project and the Land Baden-Württemberg during this study. U. Asprion (Tübingen University) is thanked for assistance with many computer problems and J. Kuhlemann (Tübingen University) for discussions of recent results on the tectonic development of the Eastern Alps. The paper was improved by reviews by P. Homewood and E. Deville.

References

ALLEN, P. & ALLEN, J. 1990. *Basin Analysis – Principles and Applications.* Blackwell.
BACHMANN, G. & MÜLLER, M. 1991. The Molasse Basin, Germany: Evolution of a classic petroliferous Foreland Basin. *In:* SPENCER, A. M. (ed.) *Generation, accumulation and production of Europe's hydrocarbons.* European Association of Petroleum Geoscientists, Special Puplications, **1**, 263–276.
BERGER, J. 1992. Correlative Chart of European Oligocene and Miocene: Application to the Swiss Molasse Basin. *Eclogae Geologicae Helveticae*, **85**, 573–609.
BETZ, D. & WENDT, A. 1983. Neuere Ergebnisse der Aufschluß- und Gewinnungstätigkeit auf Erdöl und Erdgas in Süddeutschland. *Bulletin der Vereinigung der schweizer Petroleum-Geologen und Ingenieure*, **49**, 9–36.
BOWMAN, S. 1991. *BasinWorks^{TM} Manual.* Marco Polo Software Inc.
BUCHHOLZ, P. 1986. *Der ostbayrische Lithothamnienkalk – Sedimentologie und Diagenese eines Erdgasträgers.* Braunschweiger geologisch-paläontologische Dissertationen, **5**.
CRAMPTON, S. & ALLEN, P. 1995. Recognition of forebulge unconformities associated with early stage foreland basin development: example from the North Alpine Foreland Basin. *American Association of Petroleum Geologists*, **79**, 1495–1514.
COVEY, M. 1986. The Evolution of Foreland Basins to steady State: Evidence from the western Taiwan Foreland Basin. Association of International Sedimentologists, Special Publications, **8**, 77–90.
DEWEY, J., HELMAN, M., TURCO, E., HUTTON, D. & KNOTT, S. 1989. Kinematics of the western Mediteranean. Geological Society, London, Special Publications, **45**, 265-283.
EISBACHER, G. 1974. Molasse – Alpine and Columbian. *Geoscience Canada*, **1**, 47–50.
FERTIG, J., GRAF, R., LOHR, H., MAU, J. & MÜLLER, M. 1991. Seismic Sequences and Facies Analysis of the Puchkirchen Formation, Molasse Basin - East Bavaria, Germany. European Association of Petroleum Geoscientists, Special Publications, **1**, 277–287.
FRISCH, W. 1979. Plate Motions in the Alpine Region and their Correlation to the Opening of the Atlantic Ocean. *Geologische Rundschau*, **70**, 402–411.
——, KUHLEMANN, J., DUNKL, I. & BRÜGEL, A. 1995. Miocene Tectonics, Morphogenesis and Sedimentation in the eastern Alps (Abstract). *Rates of tectonic processes from stratigraphy & geomorphology*, Leeds 25.11.1995, 8–9.
——, ——, —— & —— 1996. Mountain uplift in the Eastern Alps as a climatic steering factor. *(*abstract). *Symposium Tektonik-Strukturgeologie-Kristallingeologie*, Salzburg, April 1996.
FÜCHTBAUER, H. 1964. Sedimentpetrographische Untersuchungen in der älteren Molasse nördlich der Alpen. *Eclogae Geologicae Helveticae*, **70**, 157–298.
FUCHS, W. 1976. Gedanken zur Tectogenese der nördlichen Molasse zwischen Rhone und March. *Jahrbuch der Geologischen Bundes-Anstalt Wien*, **119**, 207–249.
GARDNER, M. 1995. The stratigraphic hierarch and tectonic history of the mid-Cretaceous foreland basins of central Utah. *In:* DOROBECK, S. L. & ROSS, G. M. (eds) *Stratigraphic evolution of foreland basins.* SEPM Special Publications, **52**, 243–303.
HAQ, B., HARDENBOL, J. & VAIL, P. 1987, Chronology of fluctuating Sealevels since the Triassico, *Science*, **235**, 1156–1167.
HARLAND, W., ARMSTRONG, R., COX, A., CRAIG, L., SMITH, A. & SMITH, D. 1990. *A geological time scale*, 1989, Cambridge University Press.
HOMEWOOD, P., ALLEN, P. & WILLIAMS, G. 1986. Dynamics of the Molasse Basin of Western Switzerland. *In:* ALLEN, P. A. & HOMEWOOD, P.

(eds) *Foreland Basins*. International Association of Sedimentologists, Special Publications, **8**, 199–219.

JIN, J. 1995. Dynamic stratigraphic Analysis and Modeling in the South-Eastern German Molasse Basin.- Tübinger Geowissenschaftliche Arbeiten, Reihe A, **24**.

——, AIGNER, T., LUTERBACHER, H., BACHMANN, G. & MÜLLER, M. 1995. Sequence Stratigraphy and depositional History in the south-eastern German Molasse Basin. *Marine and Petroleum Geology*, **12**, 929–940.

KARNER, G. & WATTS, A. 1983. Gravity Anomalies and Flexure of the Lithosphere at Mountain Ranges. *Journal of Geophysical Research*, **88**, 10449–10477.

KAY, R. & MAHLBURG KAY, S. 1993. Delamination and delamination magmatism. *Tectonophysics*, **219**, 177–189.

KRONMÜLLER, R. & KRONMÜLLER, K. 1987. Die Bausteinschichten – Sedimentologie und Diagenese eines Speichergesteins. *Erdöl-Erdgas-Kohle*, **103**, 61–66.

KUHLEMANN, J., DUNKL, I., BRÜGEL, A. & FRISCH, W. 1995. Beziehungen zwischen Molassesedimnetation und der Hebung der Ostalpen. (Abstract). *Molasse-Treffen Tübingen*, 15–16.12.1995.

LEMCKE, K. 1974. Vertikalbewegungen des vormesozoischen Sockels im nördlichen Alpenvorland vom Perm bis zur Gegenwart. *Eclogae Geologicae Helveticae*, **67**, 121–133.

—— 1988. *Geologie von Bayern 1, Das bayrische Alpenvorland der Eiszeit*. Schweitzerbart'sche Verlagsbuchhandlung.

MCNUTT, M. 1988. Variations of elastic Plate Thickness at continental Thrust Belts. *Journal of Geophysical Research*, **93**, 8825–8838.

MALZER, O., RÖGL, F., SEIFERT, P., WAGNER, L., WESSELY, G. & BRIX, F. 1993. Die Molassezone und deren Untergrund. *In: Erdöl und Erdgas in Österreich*, **2**. edition, Verlag Naturhistorisches Museum Wien und F. Berger Horn, Wien, 295–315.

MITCHUM, R. & VAIL, P. 1977. Seismic Stratigraphy and global Changes of Sealevel, Part seven: Seismic stratigraphic Interpretation Procedure. American Association of Petroleum Geologists, Memoirs, **26**, 135–144.

MÜLLER, G. & BLASCHKE, R. 1971. Coccolith: Important rockforming Elements in bituminous Shales in Central Europe. *Sedimentology*, **17**, 119–124.

NACHTMANN, W. & WAGNER, L. 1987. Mesozoic and Early Tertiary Evolution of the Alpine Foreland in Upper Austria and Salzburg, Austria. *Tectonophysics*, **137**, 61–76.

PARSONS, B. & MCKENZIE, D. 1978. Mantle convection and the thermal structure of plates. *Journal of Geophysical Research*, **83**, 4485–4496.

PFIFFNER, A. 1986. Evolution of the north Alpine Foreland Basin in the Central Alps. *In:* ALLEN, P. A. & HOMEWOOD, P. (eds) *Foreland Basins*. International Association of Sedimentologists, Special Publications, **8**, 219–228.

PLATT, J. & ENGLAND, P. 1994. Convective removal of lithosphere beneath mountain belts: Thermal and mechanical consequences. *American Journal of Sciences*, **293**, 307–336.

PUIGDEFÀBREGAS, C., MUNOZ, J. & MARZO, M. 1986. Thrust Belt Development in the eastern Pyrenees and related depositional Sequences in the southern Foreland Basin. *In:* ALLEN, P. A. & HOMEWOOD, P. (eds) *Foreland Basins*. International Association of Sedimentologists, Special Publications, **8**, 229–246.

ROBINSON, D. & ZIMMER, W. 1989. Seismic Stratigraphy of late Oligocene Puchkirchen Formation of Upper Austria. *Geologische Rundschau*, **78**, 49–79.

SIMONY, F. 1851. Beobachtungen über das Vorkommen von Urgebirgsgeschieben auf dem Dachsteingebirge. Jahrbuch der Geologischen Reichsanstalt Wien, **2**, 159–160.

SINCLAIR, H. & ALLEN, P. 1992. Vertical versus horizontal Motions in the Alpine orogenic Wedge: stratigraphic Response in Foreland Basins. *Basin Research*, **4**, 215–232.

——, COAKLEY, B., ALLEN, P. & WATTS, A. 1991. Simulation of Foreland Basin Stratigraphy using a Diffusion Model of Mountain Belt Uplift and Erosion: an Example from the Central Alps, Switzerland. *Tectonics*, **10**, 599–620.

STEININGER, F., SENES, J., KLEEMANN, K. & RÖGL, F. 1985. *Neogene of the Mediterranean Tethys and Paratethys; Vol.1 and 2: Stratigraphy correlation tables and sediment distribution maps*. Institute of Paleontology, University of Vienna.

TANKARD, A. 1986. On the depositional response to thrusting and lithospheric flexure: Examples from the Appalachian and Rocky Mountain Basins. *In:* ALLEN, P. A. & HOMEWOOD, P. (eds) *Foreland Basins*. International Assosiation of Sedimentologists, Special Publications, **8**, 369–392.

VAIL, P., MITCHUM, R., TODD, R., WIDMIER, J., TOMPSON, S., SANGREE, J., BUBB, J. & HATELID, W. 1977. Seismic Stratigraphy and global Changes of Sealevel. *In:* PAYTON, C. E. (ed.) *Seismic Stratigraphy – applications to hydrocarbon exploration*. American Association of Petroleum Geologists, Memoir, **26**, 49–212.

WAGNER, L. 1980. Geologische Charakteristik der wichtigsten Erdöl-und Erdgasträger der oberösterreichischen Molasse: Teil I: Die Sandsteine des Obereozäns. *Erdöl-Erdgas Zeitschrift*, **96**, 338–346.

WINKLER-HERMADEN, A. 1957. *Geologisches Kräftespiel und Landformung*. Springer Verlag, Wien.

Reservoir analogue modelling of sandy tidal sediments, Upper Marine Molasse, SW Germany, Alpine foreland basin

J. ZWEIGEL

Institut für Geologie und Paläontologie, Universität Tübingen, Sigwartstrasse 10, D-72076 Tübingen, Germany

Abstract: An outcrop of sandy tidal sediments (Überlingen, Lake Constance) was studied and modelled using the computer-program HERESIM. The distribution of 12 lithofacies types was mapped on photo-panels. Facies-proportion curves and variograms were calculated and used as input parameters for the simulations. Simulations with varying numbers and positions of pseudo-wells showed that already one well is sufficient to achieve a good reservoir image, due to the simple layercake-like arrangement of lithofacies types at the scale of the outcrop (125 m × 15 m). Further, a detailed geostatistical analysis of tidal sandwave deposits revealed a complex, but surprisingly systematic internal structure. Single cross-bedded bodies were classified according to their size, foreset orientation, and geometry. The geostatistical analysis revealed vertical alternations of the dominant current direction (ebb and flood) and of the size of cross beds, which could be related to changes in accommodation space.

Reservoir modeling programs have become an important tool in the oil industry to gain insight into reservoir properties between the limited well information (Tillman & Weber 1987; Ashton 1992; Guérillot *et al.* 1992; Eisenberg *et al.* 1994; Grant *et al.* 1994). The predictive power of the simulations depends strongly on appropriate input parameters characterizing the reservoir heterogeneities. Empirical input parameters describing the spatial distribution, e.g. of lithofacies, can be derived by detailed investigations of exposed reservoir analogues. Application of the programs to outcrop studies serves a two-fold purpose: (a) estimation of the quantity of wells necessary to reach a good reservoir image and (b) testing the quality of the prediction by comparison with the outcrop analogue.

This study aims to derive statistical parameters (facies proportion curves, variograms) characterizing the lithofacies distribution of tidal deposits using the simulation program HERESIM. Here only heterogeneities caused by sedimentary processes, hence lithofacies distributions, are considered, diagenetic and tectonic heterogeneities are neglected. A well-exposed outcrop of Upper Marine Molasse strata (Burdigalian = Eggenburgian) of the German Molasse Basin in the medieval defence grabens of Überlingen (Lake Constance, Fig. 1a) was studied in detail.

During the Burdigalian the outcrop was situated close to the northern coastline of the sea. In a shallow marine environment which was dominated by a micro- to mesotidal range (Keller 1989), mainly sandy sediments sourced by longshore drift from the Bohemian Massif (Allen *et al.* 1985) were deposited. This is verified by the generally fine-grained, well-sorted character of the sediments and the occurrence of heavy minerals derived from the Bohemian Massif. The formation of barrier islands is considered to be improbable because of the small size of the basin, but sandwaves have existed (Allen & Bass 1993).

Method

In this study we used the program HERESIM developed by the HERESIM group (Institut Français du Pétrole, (IFP, Paris) and Centre of Geostatistics, (École des Mines, Paris)) for the oil industry to model reservoir heterogeneities caused by sedimentary processes on the scale of tens to hundreds of metres (Ravenne *et al.* 1989). It is a probabilistic modeling program that uses truncated normal random functions (Matheron *et al.* 1987).

The HERESIM methodology usually involves several steps (Ravenne & Galli 1991).

(1) A precise sedimentological study, enhanced by geostatistical data analysis. This means:

- definition of lithofacies and their distribution;
- definition of chronostratigraphic markers and bounding surfaces—surfaces will be used for correlation and as flattening surfaces;
- creation of proportion curves—percentage of each facies in horizontal and vertical directions when all data are flattened to a reference level (bounding surface, chronostratigraphic marker); horizontal proportion curves are of interest as control on the horizontal facies distribution, because simulations are only

meaningful, if no major horizontal trends in facies proportion changes are present, while vertical proportion curves serve as a measurement of vertical lithofacies trends and will be used as input data for the simulations;
- creation of variograms—they quantify the lateral variability of each lithofacies; experimental curves are calculated in vertical and horizontal directions from the raw data, then a best fit theoretical function is calculated, which serves as input parameter for the simulations; for outcrop studies, sedimentary logs and photo panels form the databases.

(2) Stochastic simulation of the distribution of lithofacies within the reservoir. Several possible reservoir images can be created honouring the geostatistical characteristics of the data (proportion curves, variograms) as well as well data. In outcrop analogue studies the number and position of pseudo-wells can be varied to evaluate their influence on the simulation results.

Outcrop study

The studied outcrop consists of a graben wall about 125 m long and between 15 m and 6 m high. For a detailed sedimentological study, the walls were continuously photographed and wall panels created (Fig. 1b). Sedimentary logs were taken at several localities. Several gamma-ray logs were measured for comparison with subsurface data. On the basis of this data, bounding surfaces and lithofacies types were distinguished. The lithofacies types are described and interpreted in Table 1. Facies 1 to 10 are roughly ordered after increasing energy of depositional setting. For terminology refer to Table 2.

Sedimentological interpretation

The outcrop is composed of sandy to shaly–sandy tidal deposits of mainly subtidal and some intertidal origin. Different kinds of sandflat and mixed mud–sandflat deposits could be distinguished as well as three different kinds of tidal channels. The channel orientation is roughly east to west, which was parallel to the Burdigalian coastline. Eight depositional units can be recognized, separated by major bounding surfaces.

The lithofacies types are arranged into two 8–10 m thick shallowing-upward sequences (Fig. 2). In sequence 1 subtidal sandflat deposits show an upward increasing mud content and grade into intertidal mixed mud–sandflat deposits. Since mixed mud–sandflat deposition occurs generally in more landward areas, the fining-upward trend can be interpreted as a regressive sequence. The top of this sequence is formed by the erosive bases of units 5 and 6.

In the second shallowing-upward sequence, the lower parts are interpreted as a tidal sandwave of at least 5 m height. This implies a sealevel rise of at least 5 m from sequence 1 to 2, to create the space necessary for the formation of such a large bedform. Intertidal sandflat deposits occur above the sandwave, which are incised by a tidal channel linked to a fluvial one. This regressive sequence might have been caused either by filling the space created by the sea-level rise or by a sea-level fall during the deposition.

Within these sequences smaller-scale (1.2–2 m) fining-upward trends are present. Keller (1989) reported from the Swiss Molasse shallowing-upward cycles of the same amplitude which he interpreted as PACs (punctuated aggradational cycles) caused by the 100 000 year Milankovitch cyclicity (eccentricity of the Earth's orbit around the Sun).

Gamma-ray measurements

At several sites in the Grundgraben, gamma-ray logs were measured to evaluate how much of the geological information can be recognized on log measurements. The correlation of those three logs and the lithofacies subdivision from the outcrop study is shown on Fig. 3. The detail of the outcrop study can not be recognized in the logs, but the correct correlation would have been made solely based on the gamma-ray signal which may be the only information in subsurface studies.

Taking only the gamma-ray logs as a database, a five-fold subdivision would have been made:

(1) up to the abrupt shift to lower values;
(2) up to approximately lithofacies 2, where the g-values start to increase;
(3) unit of fining- and coarsening-up trend;
(4) only in logs 1 and 4; up to boundary 5, change to lower values;
(5) unit with uniformly low readings.

The channel would not have been recognized. The large (8–10 m) and small (1–2 m) scale shallowing-upward cycles seen in the outcrop can not be recognized in the gamma-ray logs. Probably the lithology is too homogeneous to show those trends, which in the outcrop can be determined by minor changes in lithology and changes in internal structures.

Simulations

The lithofacies distribution and bounding surfaces were mapped on the wall panels. Pseudo-wells at an approximate spacing of 1 m served as

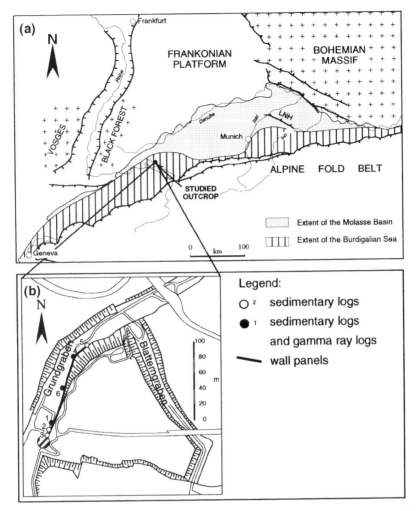

Fig. 1. (a) Map showing the position of the studied outcrop and its position relative to the extent of the Burdigalian sea; LNH, Landshut-Neuöttinger-Hoch. (b) Map of the studied outcrop with database indicated.

the database for the digitized wall panels and the geostatistical characterisation (proportion curves, variograms). The outcrop was flattened to a reference level, in this case the base of lithofacies one (sandy mudflat deposits) which can be considered to have formed a horizontal surface at the time of deposition.

The digitized wall panel of the 'Grundgraben' with horizontal and vertical proportion curves is shown on Fig. 4. The horizontal proportion curve is a control on the lateral continuity of lithofacies proportions. The pinchouts of facies are mainly caused by the geometry of the outcrop (lower outcrop boundary cuts oblique to facies) and do not represent lateral facies trends. The vertical proportion curve demonstrates the layercake type vertical facies distribution. Heterogeneities only occur at the level of the channel incision (facies 10, 4, and 2) and at the level of the tidal sandwave deposits (facies 6, 7, and 8). This vertical proportion curve was used as an input parameter for the simulations. Since most lithofacies are horizontally very persistent, a large horizontal range has been calculated for the best fit variogram function used as input parameter in the simulations. This range slightly underestimates the horizontal variability of lithofacies 10 and 4, but gives the best overall fit.

Simulations with a limited number of pseudo-wells were run (Fig. 5). Simulation 1 was only

Table 1. *List of lithofacies types with description and interpretation*

Lithofacies type	Description	Interpretation
1. Lense bedding	Dark grey with horizontal brownish streaks, lenses of fine sand in a matrix of silty mud, weathers easily, ripple foresets in opposing directions, internal structures often blurred, moderately bioturbated	Sandy mudflat
2. Wavy bedding	Mid grey, poorly consolidated, fine sand with darker mudflasers draping ripples, ripple foresets in both directions, some bifurcation of flasers, mud content varies between continuous mud drapes to flaser bedding, vertical variation is regular at approx. 30 cm intervals, moderately bioturbated	Mixed mud–sandflat
3. ε-cross-bedding	Mid grey to brownish, fine sand with high mud content, moderately consolidated, 5–10 cm thick beds dipping gently (4–5°) over a height of 1.2–1.5 m to the SW, foresets generally also dipping to the WSW, rarely herringbone cross-bedding, often mud on foresets, some weak tidal bundles observed, mud concentrated in bottomsets of foresets and beds, some returnflow ripples, rarely reactivation surfaces, moderate bioturbation	Tidal channel with lateral accretion surfaces in mixed mud–sandflat
4. Sigmoidal tidal bundles with mud couplets	Grey, fine-grained, moderate mud content, moderately consolidated, 10–40 cm thick cross-beds, amalgamated, 3–4 m long lenses, foresets dipping with high angle (30–32°) to the NE, mud drapes, often mud couplets with returnflow ripples, bottomsets often rippled, no bioturbation	Sandflat in high-energy setting with high depositional rates
5. Low-angle sigmoidal cross-bedding	Grey to brownish, fine–medium grained, moderate mud content, moderately consolidated, 5–20 cm thick beds, beds >10 m long, horizontal bed boundaries, foresets mainly to the SW, some mud drapes, mud concentrated along bed boundaries, moderate to intense bioturbation, small burrows dominating	Sandflat in moderate-energy setting
6. Low-angle sigmoidal cross-bedding with ripple preservation	Brownish grey, fine grained, low mud content, 10–20 cm thick beds, beds 10–15 m long, bed boundaries horizontal, low angle of foresets, mud drapes, foresets often rippled, foresets in both directions, ripple foresets in opposing direction, mud often only preserved in ripple troughs, intense bioturbation concentrated along some beds	Sandflat in low-energy setting
7. Sigmoidal 2D cross-bedding	Brownish grey, fine–medium grained, low mud content (mud drapes, bottomsets, mud chips), 5–20 cm thick beds, low angle of foresets, beds 10–15 m long, horizontal bed boundaries, foresets mainly to the NE, some bioturbation, some 40 cm big wood pieces at the base	Tidal sandwave
8. Sigmoidal 3D cross-bedding	Brownish grey, fine–medium grained, abundant mud chips, beds 20–40 cm thick, beds 3–4 m long, tangential toesets, some trough and fill structures, moderate to low angle of foresets, some mud drapes, mud couplets, foresets in both directions, some rippled bottomsets, no bioturbation	Tidal sandwave
9. Trough cross-bedding	Reddish grey, fine grained, moderate mud content, weathers easily, trough cross-bedding, troughs 3–60 cm deep, low angle of filling foresets, both directions, mud drapes at base of troughs, rarely in-filling, mud chips, no bioturbation	Tidal channel maybe linked to a fluvial channel
10. Massive channel sandstone	Grey, fine–medium grained, well consolidated, mainly massive, some contorted laminae, dipping at high angles over whole channel depth (1.5 m) to the SW, erosive base, small 'wings'	Tidal channel with lateral accretion in sandy environment
11. Massive sigmoidal 2D cross-bedding with wave ripples	Grey, medium–fine grained, rarely pebbles of 5 mm diameter, primary structures poorly preserved, 10–30 cm thick beds of low angle cross-bedding, foresets mainly to the SW, beds >10 m long, horizontal bed boundaries, rarely mud on foresets, some mud along bed boundaries, mud drapes on wave ripples, low bioturbation	Sandflat? above wavebase
12. Massive sandstone	Grey, fine grained, low mud content, well cemented, few thin wavy mud-laminae 1 mm long, vertical spacing 5–10 cm	?

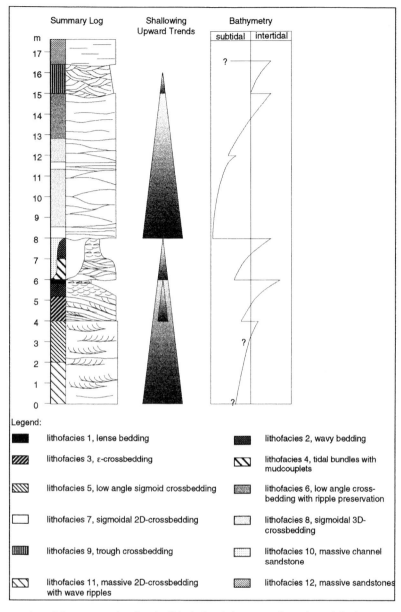

Fig. 2. Summary log of the outcrop showing the lithofacies, fining-upward trends, and the interpreted palaeobathymetry curve.

constrained by one well at the north-eastern side (right) of the outcrop. The layercake character of the lower units is well modelled, but almost no channel facies (facies 10) occurs. This is caused by the underestimation of the horizontal variability of the facies 2, 4, and 10. The lithofacies at this level will be strongly influenced by the facies present in the constraining pseudo-well (here facies 4 and 2).

Simulation 2 was also constrained by only one well in the centre of the panel within lithofacies 10. Here the facies present in the well (now facies 10, 7 and 8) are overestimated.

With two pseudo-wells (simulation 3) a good

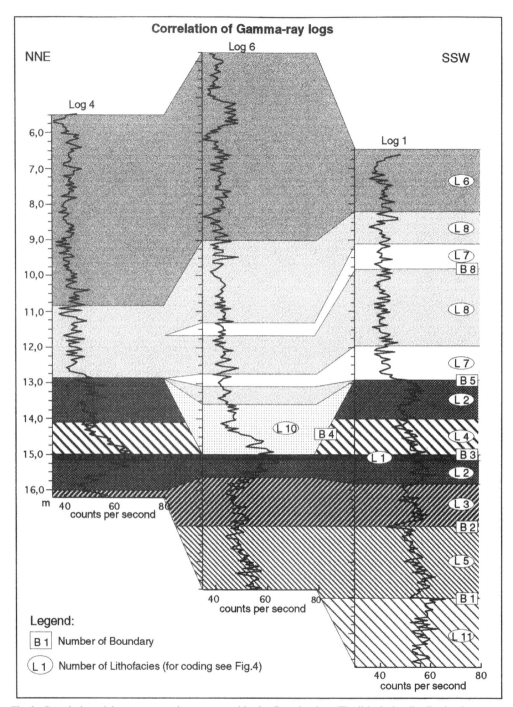

Fig. 3. Correlation of the gamma-ray logs measured in the Grundgraben. The lithofacies distribution known from the outcrop study is indicated for comparison.

Fig. 4. (a) Digitized wall panel of the outcrop showing the lateral distribution of the distinguished lithofacies types, for photograph refer to Fig. 6. The outcrop panel is flattened to the base of lithofacies 1 (black); note the layercake like arrangement of lithofacies and the incised tidal channel in the middle (dotted). (b) Horizontal facies proportion curve. (c) Vertical facies proportion curve; note how the layered structure translates into the curve.

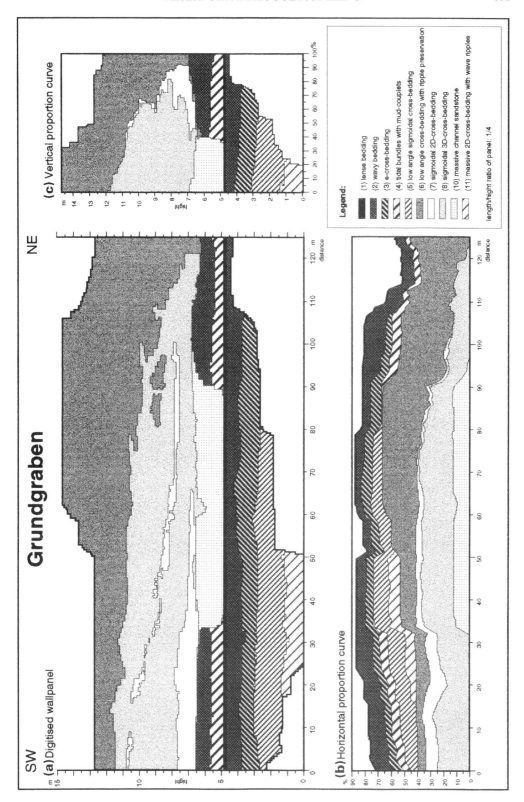

332   J. ZWEIGEL

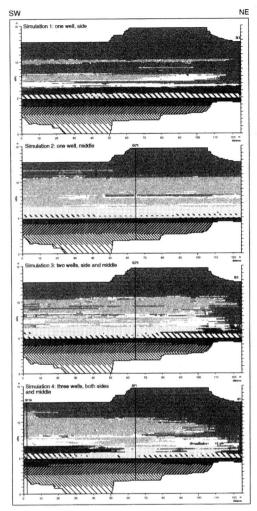

Fig. 5. Simulations of the outcrop using the vertical proportion curve, the theoretical best fit variogram function and varying numbers and positions of pseudo-well as input parameters; compare with Fig. 2a; note one well as constraint already gives a good reservoir image; three wells produce an exact outcrop image except for the channel shape.

reservoir image can be achieved. The two chosen wells (side and centre) show all lithofacies types distinguished in the study and thus no significant over-/underestimate of any lithofacies is present.

A simulation with three pseudo-well constraints (both sides and centre; simulation 4) gives the best reservoir image. Even the asymmetric horizontal distributions of lithofacies 6, 7, and 8 (tidal sandwave) is relatively well modelled, showing that the program can view

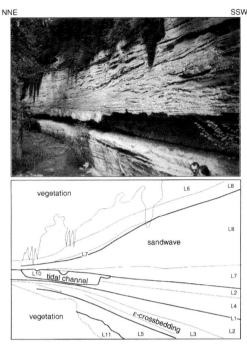

Fig. 6. Outcrop photograph showing the tidal sandwave deposits above the tidal channel, note gently inclined master bedding plane outlining the palaeotopography of the sandwave, simplified lithofacies distribution of whole outcrop is indicated; L1, lithofacies coding, refer to Fig. 4.

horizontal asymmetric facies distributions, if enough wells are used as constraints. The simulation method of the program does not model defined geological bodies. Therefore it is impossible to get a defined channel shape distribution of lithofacies 10.

Conclusions

From this small-scale study it can be seen that sandy tidal sediments exhibit within the dimensions of approximately 15 m × 125 m only minor heterogeneities. Even though 12 different lithofacies could be distinguished, a simple arrangement in a layercake fashion, which translates into the proportion curves and variograms, is observed. This structure makes it easy to achieve a good reservoir image using only very few constraints for the simulation with HERESIM. The limitations of the program are reached when lithofacies types with different lateral variabilities are modelled in one step. This may cause an over- or underestimate of certain lithofacies, which are not well fitted by the one theoretical

Table 2. *Definition of the terminology used (after Leeder, 1982)*

Terminology	Definition
Ripple	Height: 0.5–5 cm
Dune	Height: 5 cm–10 m
Sandwave	Large-scale composite bedforms in the marine environment
2D cross-bedding	Formed by straight-crested dunes
3D cross-bedding	Formed by curved-crested dunes

function used to describe the lateral variability of all lithotypes. The former geological study is then very important for subdivision into different units with the relevant parameters. Because the method used by the program actually does not model geological bodies but distributions of lithofacies, no defined shapes can be created in the simulation (e.g. a channel body has no defined shape).

The stochastic modelling program allows modelling of tidal deposits with only very few wells (one to three) as constraints, if enough information (from study or 'analogue data') to construct proportion curves and variograms is available. For a deterministic modelling a minimum of three wells would have been needed in our example, to produce a realistic reservoir image.

Detailed study

The tidal sandwave developed in the upper part of the outcrop (Fig. 6) was the object of a more detailed study. The aim was to investigate the internal structure of the body, to reveal regularities, to quantify them, and to see if these criteria can be used to aid core interpretations. Five sedimentary logs at a spacing of approximately 10 m were taken. On a wall panel (about 6 m high and 75 m long) all bounding surfaces of sigmoidal cross-beds were traced as far as possible. Single cross-bedded bodies were classified according to their shape, size, and internal structure. A detailed lithofacies classification was carried out and mapped on the wall panel. The differences were quantified by a geostatistical analysis.

Lithofacies classification

In a very detailed lithofacies analysis 16 different lithofacies were defined:

(a) trough-and-fill structures:
 (1) large (>30 cm high) troughs filled with parallel laminations;
 (2) small (<30 cm high) troughs filled with parallel laminations;

(b) 2D-cross-bedding, ratio of height to length of bodies < 0.05:
 (3) small (<30 cm high), south-westward-dipping (ebb-dominated), cross-beds;
 (4) small (<30 cm high), north-eastward-dipping (flood-dominated), cross-beds;
 (5) small (<30 cm high), current direction not visible:
 (6) large (>30 cm high), south-westward-dipping (ebb-dominated) cross-beds;
 (7) large (>30 cm high), north-eastward-dipping (flood-dominated) cross-beds;
 (8) large (>30 cm high), no current direction visible;

(c) 3D-cross-bedding, ratio of height to length of bodies > 0,05:
 (9) small (<30 cm height), south-westward-dipping (ebb-dominated) cross-beds;
 (10) small (<30 cm high), north-eastward-dipping (flood-dominated) cross-beds;
 (11) small (<30 cm high), no current direction visible;
 (12) large (>30 cm high), south-westward-dipping (ebb-dominated), cross-beds;
 (13) large (>30 cm high), north-eastward-dipping (flood-dominated), cross-beds;

(d) (14) sigmoidal cross-bedding with ripple preservation;
 (15) cemented sandstone, no primary structures preserved;
 (16) unknown.

The distribution of these lithofacies was then digitised by 'drilling' pseudo-wells at an approximate spacing of 1 m. These pseudo-wells served as the databasis for the geostatistical analysis.

Geostatistical analysis

At first sight (Fig. 7a) the lithofacies distribution seems very complicated and no ordering can be recognized, except for the horizontal facies trend already mentioned in the Grundgraben study. To the north the proportions of facies 14

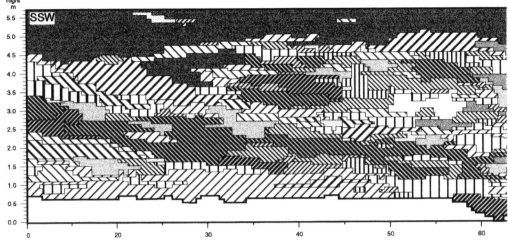

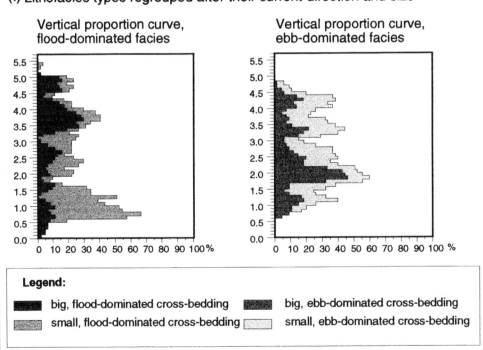

Fig. 7. (a) Digitized wall panel of the tidal sandwave deposits; note the complex structure (b) Vertical proportion curves of the detailly studied tidal sandwave deposits: (i) Lithofacies types regrouped according to their current direction and size; alternating ebb- and flood-dominated phases can be recognized. (ii) Lithofacies types regrouped according to 2D–3D structures and size; note that the base is dominated by 2D structures (inter-sandwave facies) while the main body is formed by 3D structures.

RESERVOIR ANALOGUE MODELING

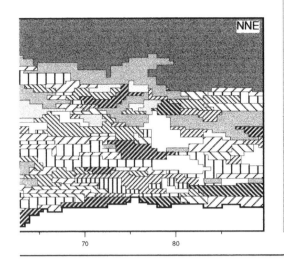
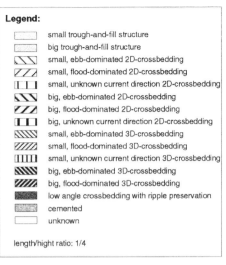

(ii) Lithofacies types regrouped after 2D-/3D-structures and size

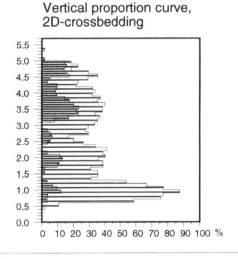
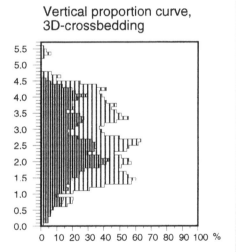

(sigmoidal cross-bedding with ripple preservation), 15 (cemented), and 16 (unknown) are increasing while the proportions of trough-cross-bedded facies are decreasing. Since the master bedding planes are also inclined to the north, and facies 14, 15 and 16 especially occur above the trough cross-bedded part, the shift in lithofacies proportions was attributed to changes between sandwave and sandflat areas. Sand-flat areas are dominated by sigmoidal cross-bedding with ripple preservation, cemented, and unknown facies. Within the sandwave trough cross-bedding is dominant.

This is interpreted as follows. During higher sea-level stands a higher water energy existed and a sandwave was constructed with a trough cross-bedded internal structure. During a subsequent minor sea-level fall the water energy diminished and the sandwave was abandoned. Instead, sandflat deposition took place. The still-existing space defined by the topographic expression of the sandwave was infilled by lower energy sandflat deposits (sigmoidal cross-bedding with ripple preservation), which exhibit more planar, horizontal bedding surfaces. To allow the preservation of such a large bedform in a shallowing environment, the sedimentation rates must have been fairly high to prevent the reworking of the sandwave deposits.

Vertical proportion curves (Fig. 7b) reveal a higher-frequency cyclicity within the 'trough cross-bedded' sandwave facies. To clarify the causes of this cyclicity, a regrouping of lithofacies after certain criteria was carried out. The higher-frequency cycles are most clearly distinguished when the facies are regrouped after their current directions and size.

On Fig. 7b i separate plots of the flood- and the ebb-dominated facies are shown for better comparison. A four-fold cyclicity is present, caused by an alternation of the dominant currents. These cycles exhibit an asymmetry in the size distribution, especially visible in the ebb-dominated part. A current phase starts with a dominance of large structures while later smaller bodies dominate. An overall trend in the vertical size distribution can be recognized. While in the ebb-dominated parts the occurrence of large bodies generally decreases upward, larger bodies increase upward in the flood-dominated part.

The causes for the alternation of the dominant current direction may be related to changes in the tidal circulation in the Burdigalian sea. The asymmetric size distribution within a current phase gives some clues to changes in accommodation space. It seems that a current phase started with a large accommodation space which diminished later, thus only allowing smaller structures to be preserved. A possible cause for the sudden creation of space may have been a sudden minor sea-level rise. The reduction of accommodation space may have been caused by the infilling of the space and/or by a slowly falling sea-level. The overall trend in the size distribution which is opposite in the ebb- and flood-dominated cycle parts may be related to the general shallowing upward trend observed. It implies that with decreasing water depth the strength of the flood-current is increasing while the strength of the ebb-current is decreasing. The causes for this trend are unknown.

A second regrouping has been carried out according to the 2D–3D structures and the size. Figure 7b ii shows the vertical proportion curves. The most prominent feature is the dominance of small 2D structures at the base of the unit, while the main body of the sandwave is mainly created by 3D structures. Not surprisingly, 2D structures are more strongly dominated by smaller sizes while the 3D structures show a higher proportion of large cross-bedded bodies. This is related to the water energy, producing these structures; 3D dunes are formed at higher energy levels, which also have more force to form large structures (Costello 1974; Rubin & McCulloch 1980).

The peak of 2D structures at the base correlates to the peak in flood-current dominance. Different current characteristics (lower energy, lower accommodation space, flood dominance) prevailed in the lower part of the sandwave compared to the main body. An interpretation as interdune sediments may explain these differences, interdune facies only being preserved at the base of the dunes.

Conclusions

A very detailed study of the sandwave deposits and their geostatistical analysis has shown that the distribution of internal structures in the sandwave deposits is governed by cyclic changes in accommodation space and high-frequency changes in the dominant current direction, resulting in a complex but not random distribution of lithofacies. The quantification of these changes (seen in the lithofacies proportion curves) may help to distinguish sandwave deposits when only dealing with core data. From cores no information about the actual shape and extension of crossbedding units can be determined, but it is possible to determine:

- shapes of the bounding surfaces (planar, erosive, inclined) (2D–3D, master bedding planes);

- orientation of foresets (alternation of current directions);
- vertical thicknesses of crossbedding units (size of structures);
- ripple preservation;
- cementation.

These criteria give an impression of the vertical distribution of size, current directions, 2D–3D structures, rippled, and cemented facies which can be compared to the proportion curves calculated for the outcrop. This might assist subsurface interpretation.

J. Z. thanks T. Aigner and H.-P. Luterbacher (Tübingen University) for supervising this study and the IFP (Institut Francais du Pétrole) staff, especially R. Eschard, for putting their equipment and the program HERESIM at her disposal – and helping with numerous computer problems. J. Hornung is thanked for support during the field work, and U. Asprion for help with computer problems 'at home'. This study was funded by the EU Integrated Basin Studies Project.

References:

ALLEN, P. & BASS, J. 1993. Sedimentology of the Upper Marine Molasse of the Rhone-Alpe Region, Eastern France: Implications for Basin Evolution. *Eclogae Geologicae Helveticae*, **86**, 121–172.

——, MANGE-RAJETZKY, M., MATTER, A. & HOMEWOOD, P. 1985. Dynamic Paleogeography of the open Burdigalian Seaway, Swiss Molasse Basin. *Eclogae Geologicae Helveticae*, **78**, 351–382.

ASHTON, M. (ed.) 1992. *Advances in Reservoir Geology*. Geological Society, London, Special Publications, **69**.

COSTELLO, W. 1974. *Development of bed configurations in coarse Sands*. Earth & Planetary Science Deptment, Cambridge Massachusetts, MIT Reptort **74.1**.

EISENBERG, R., HARRIS, P., GRANT, C., GOGGIN, D. & CONNER, F. 1994. Modeling Reservoir Heterogeneity within Outer Ramp Carbonate Facies using an Outcrop Analog, San Andres Formation of the Permian Basin. *American Association of Petroleum Geologists Bulletin*, **78**, 1337–1359.

GRANT, C., GOGGIN, D. & HARRIS, P. 1994. Outcrop Analog for Cyclic-Shelf Reservoirs, San Andres Formation of Permian Basin: Stratigraphic Framework, Permeability Distribution, Geostatistics, and Fluid Flow Modeling. *American Association of Petroleum Geologists Bulletin*, **78**, 23-54.

GUÉRILLOT, D., LEMOUZY, P. & RAVENNE, C. 1992. How Reservoir Characterisation can help to improve Production Forecasts. *Revue de l'Institut Français du Pétrole, Paris*, **47**, 58–67.

KELLER, B. 1989. *Fazies und Stratigraphie der Oberen Meeresmolasse (Unteres Miozän) zwischen Napf und Bodensee*. Dissertation, University of Bern.

LEEDER, M. 1982. *Sedimentology: Process and Product*. Unwin Hyman, London.

MATHERON, G., BEUCHER, H., FOUQUET, C., GALLI, A. & RAVENNE, C. 1987. Simulation conditionnelle á trois Facies dans une Falaise de la Formation du Brent. *Études Geostatistiques V– Seminaire CFSG sur la Geostatistique, 15–16 Juin 1987, Fontainebleau, Science de la Terre, Sér. Inf. Nancy*, **28**, 213–249.

RAVENNE, C. & GALLI, A. 1991. From Geology to Fluid Flow Modeling using a geostatistical Simulation: an integrated Methodology and Software developed for Oil Industry. *Proceedings of an NEA workshop "Heterogeneity of groundwater and site evaluation", Paris, 22–24 Oct. 1990*, OCDE, 143–152.

——, ESCHARD, R., GALLI, A., MATHIEU, Y., MONTADERT, L. & RUDKIEWICZ, J. 1989. Heterogeneities and Geometry of sedimentary Bodies in a fluvio-deltaic Reservoir. *SPE Formation Evaluation, June 1989*, 239–246

RUBIN, D. M. & McCULLOCH, D. S. 1980. Single and superimposed Bedforms: a Synthesis of San Francisco Bay and Flume Observations. *Sedimentary Geology*, **26**, 207–231

TILLMAN, R. & WEBER, K. (eds) 1987. *Reservoir Sedimentology*. SEPM Special Publications, **40**.

Tectono-stratigraphy and hydrocarbons in the Molasse Foredeep of Salzburg, Upper and Lower Austria

LUDWIG R. WAGNER

Wolfersberggasse 6, A-1140 Wien, Austria

Abstract: The evolution of the sedimentary record and the tectonic setting of the Molasse Basin at the margin of the European platform from the Bavarian border across Salzburg, Upper and Lower Austria to the border of the Tschech Republic provides the possibility of unravelling the history of the Alpine – Carpathian Foredeep. The marine ingressions and the sedimentary processes reflect the dynamic process of the plate tectonics and the Alpine Orogeny. At present the outcrop of the passive margin of the Bohemian Massif and the northern edge of the active Molasse thrust front approach each other within a distance of a few kilometres. The basement is broken into several fault blocks. Tectonic activities within the thrust sheets as well as lateral movements of the basement along the flanks of the craton are ongoing.

The hydrocarbon generation and trapping is based on the interaction of synsedimentary tectonism, erosion and deposition.

This study is based on surface and subsurface data acquired by Rohöl Aufsuchungs AG (RAG) and Österreichische Mineralöl Verwaltungs AG (OMV) during its exploration in Salzburg, Upper and Lower Austria.

In the Tertiary (from the late Eocene) the Molasse Basin of the Alpine–Carpathian foredeep was formed. The outcrop of the basement of the Bohemian Massif delineates the northern margin of the Molasse zone, while its southern margin corresponds to the Alpine–Carpathian thrust front (Fig. 1).

The Cenozoic sediments of the Molasse are divided into three tectonic units (Steininger *et al.* 1987). The autochthonous Molasse rests relatively undisturbed on the Mesozoic and Crystalline basement of the Middle European shelf. The allochthonous Molasse is composed of the southern Molasse sediments, which are included in the Alpine–Carpathian thrusts and which are moved tectonically into and across the southern autochthonous Molasse. The allochthonous Molasse includes the Molasse Imbrications, the disturbed Molasse and the Waschberg Zone. The parautochthonous Molasse consists of Molasse sediments riding piggyback upon Helveticum, Flysch or East Alpine nappes, e.g. in the Vienna Basin.

Between the Liassic and late Miocene the area of the Molasse Basin is characterized by three main cycles of marine transgression and regression. They are interrupted by periods of rift and transpressional tectonic activity and erosion. In the first cycle from Liassic in the east and mid-Dogger to early Cretaceous in the west the area was part of the Middle European carbonate platform. The rifting phases in the mid-Jurassic and early Cretaceous are related to the stretching of the Arctic–North Atlantic domain (Ziegler 1982). In the Cretaceous the Molasse basement was the northern shelf of the Helveticum sea. Glauconitic shales and sandstones were deposited on the Salzach and Perwang blocks in the second cycle from Aptian to late Campanian and north of the Central Swell Zone (Fig. 3) from Cenomanian to late Campanian. The suite of marine sediments in the Helveticum nappes in the vicinity of Salzburg and in the Waschberg nappes to the north of Vienna continued without interruption into the Tertiary. The collision of the Alpine orogenic system with the southern margin of the North European craton in the early Tertiary caused the transpressional deformation of the foreland (Ziegler 1987). The Molasse basin was formed and became the pelagic Alpine foredeep in the early Oligocene. During the third cycle from upper Eocene to the Present, the Alpine Orogeny took place. Between Salzburg and Vienna the sediments of early Miocene (early Eggenburgian) are the youngest deposits below the Flysch–Helveticum overthrusts. Ongoing transpressional deformation overlapped the extensional fault system from the early Miocene on. The cross-sections, palinspastic restored sections, sedimentological models and the palaeogeography for the area of Salzburg and Upper Austria were recently described and illustrated in more detail (Wagner 1996*a*).

Crystalline basement

The Crystalline of the Bohemian Massif in Austria is composed of medium- to high-grade

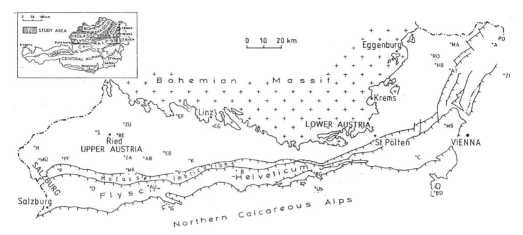

Fig. 1. General index map of Austria. *Well names and locations:* Upper Austria: AB, Atzbach 1; AU, Aurach 1; H, Hochburg 1; MÜ, Mühlberg 1; MR, Mühlreit; O, Oberhofen 1; P, Perwang 1; PF, Pfaffstätt 1; RE, Redltal 1; S, Senftenbach 1; V, Voitsdorf 1; ZA, Zell Am Pettenfirst 1; ZU, Zupfing 1; EG, Ebelsberg; EB, Eberstallzell 1; EF, Eferding; K, Kremsmünster. Lower Austria: AT, Altenmarkt im Thale 1; A, Ameis 1; B, Behamberg 1; BD, Berndorf 1 NÖ; C, St Corona; G, Grünau 1; HF, Höflein 1; HB, Hollabrunn 1; MA, Mailberg 1; MB, Mauerbach 1; PO, Poysdorf 1; RO, Roggendorf 1; U, Urmannsau 1; Zl, Zistersdorf ÜT 1a + 2A; RG, Rogatsboden.

metamorphic Precambrian to Palaeozoic rocks and Variscan granitic plutonites (Fuchs & Matura 1980). Recent researches of the southern Bohemian Massif support the concept of a uniform Hercynian orogen comprising the Moravian unit in the east and the Moldanubian unit in the west (Fuchs 1990). The youngest granites intruded approximately 300 Ma ago and were derived from the crust–mantle interface (Frasl & Finger 1988). All cores from the Crystalline basement recovered rock types, which were already known from the surface (Brix, *et al.* 1977, G. Frasl pers. comm. 1985).

Tectonic setting and structural evolution of the basement

The Bohemian Massif is dissected by a system of conjugate NW–SE- and NE–SW-trending faults and secondary fault systems running approximately N–S and E–W. The fault system was already studied by Stiny (1926) and Gruber (1931) and on satellite images (Tollmann 1977). The shear zones in the southern Bohemian Massif are strike-slip zones, which have their origin in the crust below the brittle–ductile transition zone (Wallbrecher *et al.* 1996). The triangular outcrop of the Crystalline basement spur, the Amstetten Fault block, is limited by the major fault system, the Steyr and Diendorf faults, and divides the foreland Molasse between Upper and Lower Austria at the surface (Fig. 2). This prominent basement high extends approximately 40 km to the south below the Alpine nappe system.

The NW- and NE-trending fault systems already existed in Palaeozoic times (Schröder 1987; Wallbrecher *et al.* 1996). The faults became reactivated in the early Jurassic, early Cretaceous and early Tertiary. Through these periods of tectonic activity the Crystalline basement and its cover were pulled apart using the old fault planes.

The fault blocks were uniformly tilted in Upper Austria to the east and in Lower Austria to the west. The vertical throw of the Mailberg fault reached more than 2000 m during the mid-Jurassic and the Ried and Steyr faults each exceeded 1000 m during the early Tertiary. The Mesozoic is eroded on the tectonic highs, on the south rim of the Bohemian Massif, the Central Swell Zone and the Bergern (Bg1) High below the Flysch (Fig. 3).

During the latest Eocene and earliest Oligocene the area of the Central Paratethys subsided rapidly to deep-water conditions. This was accompanied by the development of a dense network of WE-trending antithetic and synthetic extensional faults (Wagner 1996*a*, *b*). In Upper Austria and Bavaria this fault pattern is still present. This phase of downbending of the foreland crust is due to the tensional stress of

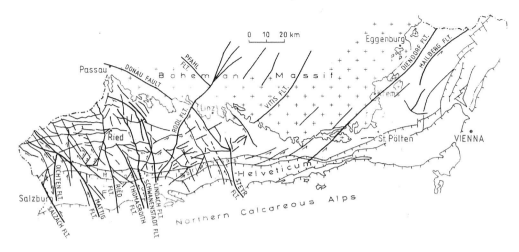

Fig. 2. Tectonics of the Alpine – Carpathian foredeep: main faults.

the subduction of the European plate under the Periadriatic Plate and the weight of the advancing Alpine nappe system (Ziegler 1987). The resulting structure of the basin-parallel fault system contains most of the Mesozoic and Tertiary oil-bearing reservoirs in the west of the spur of the Bohemian Massif.

The vertical displacement has, in addition to the pre-Tertiary fault system, a dextral and sinistral transpressional shear strike-slip component, which was reactivated during the Tertiary and Quaternary. Two triangular crystalline fault blocks, the Schallerbach and Amstetten fault blocks, remained metastable relative to the surrounding fault blocks, which were moved further north (Fig. 3). To a lesser degree the Braunau fault block is a relatively stable block. The Kremsmünster fault block was rotated several kilometres to the north along the Rodl fault. The Bergern high was probably originally the continuation of the Central Swell and has been transported with the fault block to the north. The Rodl fault cuts off the Pfahl (Mühl valley) and Donau faults. In the Palaeozoic the Steyr fault was more likely an extension of the Schwanenstadt fault than of the Donau fault. The Perwang fault block has pushed the Perwang Molasse Imbrications into the Central Swell along the Mattig and Pfaffstätt faults (the Pfaffstätt runs parallel to the Maltig fault) in Miocene. At the west side of the Perwang block, the Ottnangian sediments were rotated into a vertical position along the Oichten fault and on top of the uplifted Perwang Imbricates. North of the Pfaffstätt fault and along the Rodl fault the sub-basins of Puchkirchen–Munderfing–Pfaffstätt and Lindach–Voitsdorf were pulled apart from late Oligocene time onward. These tension zones are approximately 10 km wide and at least 80 km long (Fig. 3). To the east of the Mailberg fault the Palaeogene Molasse was overthrust and the Neogene was integrated into the thrusts and during the Miocene into the Vienna pull-apart basin.

Almost all of the early Oligocene extensional faults in Upper Austria were integrated in the transpressional stress in the Miocene. The sediments were further displaced laterally either to the NW or to the NE along the post Eocene faults. Locally the soft sediment on top of the competent Eocene rock started to overthrust at sharp bends in the fault pattern. Through this compression, many of the extensional faults became seals. On the other hand the NW- and NE-fault system acted in Miocene as drainage for the oil migration from below the thrust sheets.

East of the Diendorf fault, all faults appear to be transformed into the transpressional system during the Miocene and shifted in a NE direction.

The Crystalline basement has been stretched in the Mesozoic and early Tertiary to the SW and SE and in the early Oligocene to the S. From Miocene on it has been transpressed, which means compressed and partly pulled apart between strike-slip systems. The system is relatively locked on the west flank. The east flank was less resistive. The eastward escape of the

Pannonian continental lithospheric fragments in mid–late Miocene time gave the space for the onset of the lateral extrusion of the central Eastern Alps and for the Neogene Vienna Basin to form in Karpatian time (Royden 1988; Ratschbacher et al. 1991; Fodor 1995). The Eastern Alps and the Vienna Basin were extended along the conjugate system of strike-slip faults. At the end of the Pannonian the lateral extrusion terminated (Decker & Peresson 1996).

On the new 3D seismic data the influence of the strike slip faults on the sedimentary processes is obvious (Wagner 1996a) (Fig. 4).

Palaeozoic setting

In Bavaria the Permo-Carboniferous sediments on the surface accompany the major variscian NW–SE faults (Ziegler 1982; Schröder 1987; Meyer 1989). In the subsurface of Upper Austria, the late Palaeozoic sediments appear to be limited to graben structures on the southwestern margin of the Central Swell Zone that forms the southeastern extension of the Landshut–Neuötting High in Bavaria. In the well, Hochburg 1 (Hobg1), more than 400 m of probably upper Palaeozoic dark grey, fluvial sandstones, silts and clays with coal seams were penetrated. The Permo-Carboniferous spores (Stephanian–lower Permian; I. Draxler, pers. comm. 1981), however, could have been reworked *in situ* before the late Mid-Jurassic. Reworked Rotliegend spores were also found from upper Eocene sandstones in a few wells on top of the swell (Klaus pers. comm. 1978). Carboniferous plants were discovered in wells in Bavaria (Berger 1959). In Upper Austria, the occurrence of primary H_2S gas (not created by long oil production or fluid injection) is restricted to structures that contain Palaeozoic rocks on the Crystalline basement.

In the Moravian unit in Lower Austria outcrop the fluvial terrestrial red and dark grey sandstones and shales in the NE–SW-trending graben at Zöbing (Vasicek 1983). The wells Altenmarkt im Thale 1, Mailberg 1 and Hollabrunn 1 have encountered Palaeozoic claystones, sandstones, breccias and, sporadically, coal beds of Carboniferous and Permian age in the continuation of the Boskovice graben in the Czech Republic. The well, Roggendorf 1, found volcanic intercalations in late Palaeozoic coarse-grained quartzites and greenish claystones (Brix et al. 1977; Sauer et al. 1992).

No economic hydrocarbon accumulations have been found to date in Palaeozoic rocks in Austria.

Autochthonous Triassic sediments are not preserved below the western Molasse in Austria.

Jurassic facies distribution

Upper Austria and Salzburg

The sedimentary evolution of the area of the Central Paratethys in Upper Austria is divided into three different facies zones (Fig. 5): the facies north and south of the Central Swell Zone and the facies below or (from Cretaceous on)

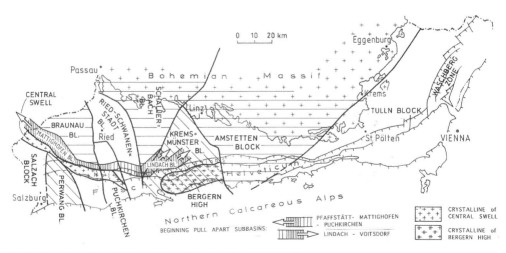

Fig. 3. Tectonics of the Alpine – Carpathian foredeep: fault blocks.

within the thrust sheets of the Imbricated Molasse (Wagner 1996a).

Middle Jurassic–Dogger. In the Mid-Jurassic (Bathonian–Bajocian, W. Klaus pers. comm. 1978) braided-fluvial to shallow-marine sandstones, containing occasional coal seams of the Gresten Group (Gresten is a village close to Waidhofen/Ybbs with old coal mines in a Helveticum nappe) are the earliest encountered Mesozoic series. The marine portion correlates to the 'Upper Quartz Arenite Series' in Lower Austria and the Middle Jurassic sandpits near Regensburg. The older Liassic sediments of the Gresten group crop out in tectonic wedges of Helveticum in front of the Calcareous Alps. From Callovian to late Jurassic–Malm and into the early Cretaceous, carbonates were produced on the tropical shelf along the Bohemian landmass. The Jurassic shelf reached its greatest water depth in the west of the craton in Callovian in the southwest below the Alps of Salzburg.

The vertical section of the carbonates starts with dark brown arenaceous nodular micrits of the Höflein Formation (Höflein is a producing gas-condensate field below the Flysch N of Vienna; Sauer *et al.* 1992) containing lumachelles, ammonites, belemnites and sponge spicules with abundant chert nodules. They grade upward into biostromal limestones. The Dogger Carbonates continue into Bavaria and were recently documented from shallow wells in the area of Regensburg (Meyer & Schmidt-Kaler 1987).

Late Jurassic – Malm. The glauconitic limestone bed correlates with a similar horizon known from outcrops on the Franconian Platform of Bavaria at the Dogger – Malmian boundary (Meyer & Schmidt-Kaler 1984). Oxfordian and Kimmeridgian algal and sponge banks are capped by coral reefs and their debris, surrounded by the high energy environment with oolites and grainstones. The fully marine upper Jurassic carbonates have equivalents in the Altenmarkt Group in Lower Austria and facies of Kehlheim in Bavaria. The typical Sponge – Tubiphytes association of the Treuchtlingen Formation in Franconia was cored on both sides of the Central Swell in the wells Zell am Pettenfirst 1, Redltal 1 and Mühlberg 1.

The salt lagoon and tidal flat deposits of the 'Purbeck' Facies (from Tithonian to early Berriassian) consist of thin-bedded fine crystalline dolomites, cherty limestones, breccias and stromatolites. Typical sediment features are 'bird's eye' (small shrinkage pores in lagoonal dolomites), 'black pebbles' (interpreted as flushed in breccias of soil), faecal pellets (Favreina) and algal fragments (Bankia). Characeas, oogonides of Charaphytes, were repeatedly flushed in from the freshwater environment.

The upper Jurassic facies indicates a progressive shallowing from the deeper shelf in the southwest below the thrust sheets to the shallow shelf on the Bohemian Massif (Fig. 6). At the erosional eastern limit of the carbonates, the

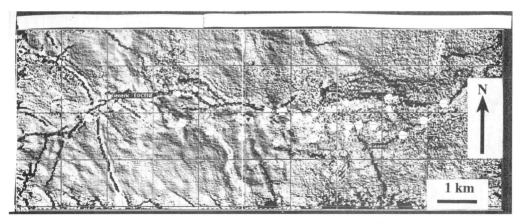

Fig. 4. The combination of the extension and transpression of the fault system is illustrated on the 3D survey of the Voitsdorf oil field. The image represents the end Eocene. The E–W-trending antithetic main fault dips to the N (Nachtmann 1995). The fault developed during the extension of the basement in Oligocene time. From the early Miocene onward the NW and NE-trending Pretertiary faults, as well as the WE-trending Oligocene faults were rejuvenated as transpressional faults. The fault planes were turned into a almost vertical position or were even overturned in the Mesozoic and lower Tertiary sediments. The Eocene is locally overthrust. The lateral strike-slip movement sealed the faults and the resulting flower structures broke up the younger sediments, above the imaged level, in a complex fault system.

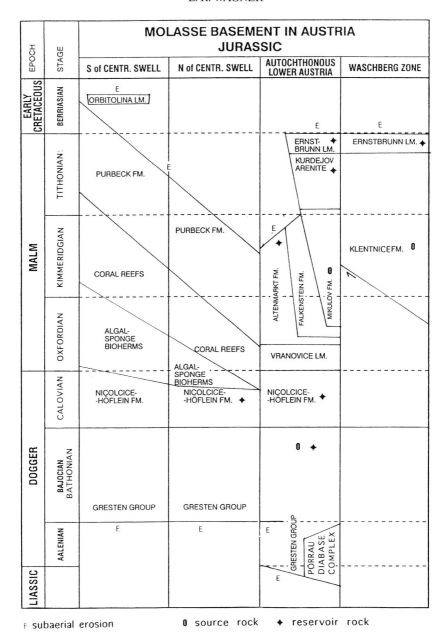

Fig. 5. Stratigraphic chart of the Jurassic in the Austrian Molasse Basin.

coral build ups started to grow at the Dogger–Malm boundary. The maximum total thickness at 557 m of the Jurassic carbonates was encountered in the well Hochburg 1.

So far thin-bedded carbonates, which are typical in the Oxfordian and Kimmeridgian for certain areas in Frankonia, are missing.

Oil is produced from the basal sandstones of the Gresten group and the dolomitized and leached chert nodules in the Callovian carbonates of the Höflein Formation out of a few wells in the oilfield Voitsdorf and the well Haindorf 1.

Jurassic in Lower Austria

Early and mid-Jurassic (Liassic and Dogger). The geological results from the drilling activities in Lower Austria in the Mesozoic basement were described recently in several papers (Grün 1984; Ladwein 1988; Elias & Wessely, 1990; Brix 1993; Seifert 1993).

Gresten group. On the western flank of the Bohemian Massif, the Mesozoic sediments filled the tilted Variscan graben structures with the thick coarse clastics of the Gresten group. The faults, related to the late Palaeozoic rifting system, were still active in Jurassic times. The ongoing subsidence of the fault blocks influenced the facies distribution by their palaeogeographic high or low position (Wessely 1993).

The lower Middle Jurassic sediments rest unconformably on upper Palaeozoic or Crystalline. The maximum thickness of the Gresten group on the downthrown side of the half grabens is approximately 1500 m.

The volcanic series of tuffites and clastic diabase flows, probably Aalenian in age, underlie the Gresten group locally in the well Porrau 2, close to the margin of the massif.

The basal arkoses and quartzarenites containing coal beds and shales with plant roots of the 'Lower Quartzarenite Formation', the basal part of the Gresten Group, of Aalenian age were deposited in a fluvial braided stream and deltaic to coastal plain environment.

The marine ingression is documented by dark claystones of the 'Lower Claystone Formation' with a Bajocian ammonite fauna. These clays separate in certain areas the fluvial from the marine 'Upper Quartzarenite Formation'. The shallow-marine quartzarenites with dolomitic cement are dated by Bathonian ammonites.

In Bathonian time the facies differentiated between the calcareous sands with phosphorites and glauconite in the 'Haselbach beds' and the clays of the 'Upper Claystone Formation' in the lows.

In late Dogger time the subsidence along the faults came almost to a first standstill and the half grabens were filled with sediments. The Höflein Formation is divided into a lower member with silicified sandy dolomites with thin chert bands and nodules, and an upper member with dark grey, slightly sandy dolomites with chert bands and nodules and occasionally glauconite.

Thicknesses up to 50 and 260 m have been observed. The condensate field, Höflein, produces from the quartzarenites and the porous cherts and fractures in the Höflein Formation.

The beginning of the sponge biostrome carbonate buildups of the Höflein Formation in the Callovian correlates to sediments of the same lithology on the west flank of the massif in Upper Austria. The equivalent in the Czech Republic is the Nicolcice Formation.

The wells Porrau 2 and Füllendorf 1 drilled through folded and steeply dipping Dogger sediments, which also demonstrates some compressional tectonics.

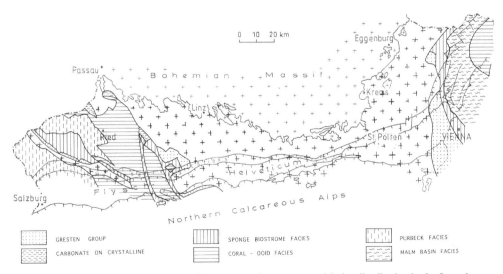

Fig. 6. Alpine–Carpathian Foredeep: pre-Cretaceous, subcrop map and facies distribution in the Jurassic.

Late Jurassic–Malm. In contrast to Upper Austria, shelf and basin sediments were penetrated in wells and are outcropping in the Waschberg zone in Lower Austria. The shelf platform sediments on both sides of the massif are similar in lithology and facies sequence.

In early Oxfordian time, dolomites and limestones with fossiliferous debris of reef organisms of the Vranovice Limestone are restricted to the shallow flank of the massif in the East. The carbonates in the Oxfordian and Kimmeridgian of the Altenmarkt Formation are a correlatable sequence with similar beds in Upper Austria and Bavaria. Sponge and algal biostromes grade into sponge–algal bioherms and are topped by coral reefs and their debris with high energy oolite bioclastic beds. The transitional clastic carbonate layers of the Falkenstein Formation lead to the dark marls of the Mikulov Formation in the basin. The Mikulov marls reach a thickness of more than a 1000 m. The same thickness was encountered in the autochthonous Jurassic below the Alpine thrust sheets in the northern Vienna basin at a drilling depth of more than 8000 m. Their continuation in the upthrust Waschberg zone are the marls of the Klentnice Formation.

The shallowing in the Tithonian is reflected by the coarsening-upward terrigenous and bioclastic carbonates of the Kurdejov Arenite (Elias 1971), which is overlain by the reef limestones and dolomites of the Ernstbrunn Formation (Elias & Wessely 1990).

Waschberg zone. The upper Jurassic marlstones and calcarenites of the Klentnice Formation and the reef carbonates of the Ernstbrunn Formation crop out in the upthrust 'Klippes' of the Waschberg zone (Hofmann 1990; Seifert 1993).

The Waschberg zone is the eastern pendant to the Molasse Imbrications in the west. It is composed of allochthonuous Molasse sediments and Mesozoic Klippes from the Molasse basement (Steininger *et al.* 1987).

Cretaceous facies distribution (Figs 7 & 8)

Upper Austria and Salzburg

The Pre-Tertiary subcrop map (Fig. 7) below the Eocene unconformity is dissected by NW- and NE-trending faults into Mesozoic fault blocks (Braumüller 1961; Wagner & Wessely 1993).

Southwest of the swell in the Salzach block is the lower Cretaceous sedimentary record of the Mühlberg Limestones and Sandstones interrupted by several gaps (Fig. 8). Storms deposited glauconitic sands, the Gault sandstone, on the shelf in Apto-Albian time (Nachtmann and Wagner 1987). To the northeast of the Central Swell Zone the oldest Cretaceous deposits are locally developed and consist of light grey, white, red or green non-fossiliferous, coarse-grained fluvial sands. These beds of the Schutzfels Formation (Schutzfels: locality near Regensburg) infill the Jurassic karst to a depth of 100 m below the Jurassic surface. They are overlain by Cenomanian coal-bearing marls that grade upward into the storm layers of the shallow-marine glauconitic sandstones of the Regensburg Formation (Polesny 1983). Poorly developed beach sands are preserved in a narrow NW–SE-trending belt subparallel to the erosional limit of the Jurassic carbonates towards the Bohemian Massif. The bulk of the Cenomanian Regensburg formation consists of storm-dominated, shallow-marine, glauconitic sandstones that were deposited on a broad shelf. The Cenomanian sandstones reach thicknesses of 15–70 m.

In a complete sequence, the basal beds have a silicious cement, whereas the middle beds, have calcitic cements. Both are characterized by the rhythmical successions of thinly laminated storm layers with hummocky cross bedding and escape burrows and heavily burrowed fair-weather layers with burrows mainly of the *Ophyomorpha* and *Skolithos* type and from molluscs. The upper beds start with a tight, calcareous glauconitic sandstone with red zones and spots caused by flushed-in oxidized clay minerals. This marker bed is easy to identify by the high resistivity on logs, the large (up to 4 cm) thick single quartz and feldspar grains and the abundant large echinoid burrows. The top section is a succession of porous and tight calcareous glauconitic sandstones and spiculitic limestone nodules and layers. The basal and middle beds were deposited on the inner shelf below normal wave base. With the marker bed, the Cretaceous sea transgressed further across the massif and most of the preserved upper layers represent outer shelf sediments.

The transgressive Cenomanian clastics are overlain conformably by Lower Turonian marls containing in the upper part glauconitic sandstone storm deposits. The ichnofacies is dominated by skolithos burrows. The boundary between Cenomanian and Turonian is probably within the early Turonian sediments, but for practical reasons the lithological change is taken as boundary in the oil industry (Küpper 1964).

The *Globotruncana*-bearing late Turonian to late Campanian sediments consist of intensively burrowed mudstones, which were accumulated under outer-shelf conditions. In late Campanian time 300 m of sandstones were deposited northwest of the Central Swell Zone and shale out to

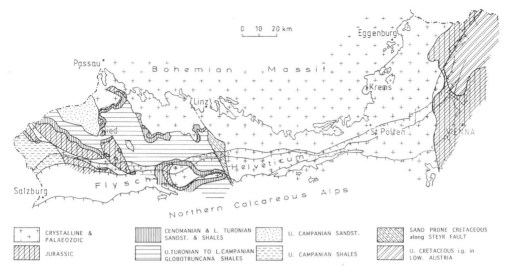

Fig. 7. Alpine–Carpathian Foredeep: pre-Tertiary; subcrop map.

the southwest across this tectonic element. The maximum drilled thickness of Cretaceous sediments in the well Senftenbach 1 is 800 m.

In the easternmost part of the Upper Austrian Cretaceous, a locally developed sand-prone facies is restricted to the Steyr fault. It is divided into local lithostratigraphic formations (Wessely et al. 1981; Fuchs et al. 1977).

Pebbles of *Orbitolina*-bearing limestones were observed in the western wells.

Oil and thermal gas is produced mainly from the Cenomanian sandstones and from the sandstones along the eastern fault. No production has been established from the oil-bearing lower Turonian.

Lower Austria

Autochthonous Molasse. In early Cretaceous time the area of the Molasse was submerged and the carbonate platform became karsted.

No early Cretaceous sediments were deposited in the eastern Foreland Molasse nor in the Waschberg Zone. From the Molasse in the Czech Republic, upper Albian sediments were reported (Elias & Wessely 1990). The Cretaceous transgression in the Czech Republic coincides with the transgression in the Salzach block and in Lower Austria the transgression is time equivalent to the transgression across the Central swell Zone in the West.

The Foreland Molasse was uplifted in early Tertiary time even more than in the west and eroded. Late Cretaceous clasts occur, e.g. in Oligocene olistoliths in the well Zistersdorf ÜT2A.

Waschberg Zone. (Cenomanian to late Eocene). A number of wells have encountered mid- to late Cretaceous sediments. The stratigraphic column from Cenomanian to Maastricht was established on the sections in the wells Ameis 1 and Poysdorf 1 (Kollmann et al. 1977; Fuchs et al. 1984) and the Klement Super Group was divided into the Cenomanian to Santonian Ameis Group and the Campanian and Maastrichtian Poysdorf Group. The whole section was not penetrated in a single well, but combined sections indicate a total thickness of up to 900 m Cretaceous.

The Ameis Group consists of 50 m of coarse-grained glauconitic Cenomanian quartz sands merging upsection into glauconitic, sandy heavily bioturbated marls of Turonian age. At the boundary between the Turonian and the Coniacian a characteristic sandy limestone is a marker horizon. The Poysdorf Group is a uniform less sandy marl. The Klement Group in the Waschberg Zone contains glauconitic sandstones and marls from the Cenomanian to the Maastrichtian.

Early Tertiary Facies distribution (Fig. 8)

Lower Austria–Waschberg Zone

In the Waschberg Zone the sedimentation is continuous into the Tertiary.

The Palaeocene–Danian Bruderndorf

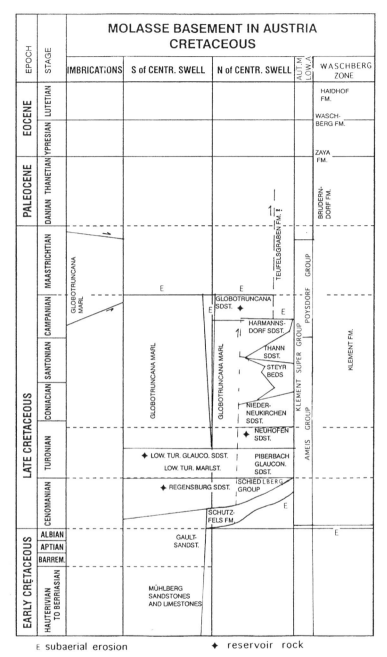

Fig. 8. Stratigraphic chart of the Cretaceous in the Austrian Molasse Basin.

Formation is composed of whitish to grey, fossiliferous, red algae limestones with corallinaceae, fine-grained, glauconitic sandstones and marls. The late Palaeocene–Selandian Zaya Formation is characterized by a white limestone bed with red algae, bryozoa, bivalves and echinoids interbedded in brown clays and marls.

The Waschberg Formation represents the lower to middle Eocene–Ypresian with nummulitic, sponge and coral limestones and marls with intercalated sandy and conglomeratic layers with crystalline boulders.

The middle Eocene–Lutetian iron-rich shales, shaly limestones and calcareous arenites of the

Haidhof Formation are brachiopod-bearing sediments.

Eocene facies distribution (Figs 9 & 10)

Upper Austria and Salzburg

In the late Eocene the Tethys Sea progressively encroached across the Mesozoic and Crystalline fault blocks. The present Foreland Molasse is only a narrow remnant of the original basin.

In late Eocene time, the Central Swell Zone was the high zone that separated the lagoon from the slope to the open marine realm (Wagner 1980; Wagner et al. 1986). The different facies zones started in the north with fluvial floodplain white, yellow earth, olive green and red clays with roots traversed by meandering channels of the limnic series of the Voitsdorf Formation (Voitsdorf: largest oilfield in Upper Austria close to Kremsmünster) (Figs 9 & 10). They are underlain locally in graben structures by accumulated channel sands of the braided stream system. The fluvial and limnic Voitsdorf Formation was transgressed by upper Eocene in the S and progressively younger Oligocene sediments to the N and NE. The subsiding floodplain is capped by a swamp coal bed, which is overlain by dark grey, soft shale of the paralic Cerithian Beds. The shales are cut by tidal channels and merge into inundated sandflats. Finally shallow-marine sands of the Ampfing Formation (Ampfing: oilfield in Bavaria close to Mühldorf) were deposited along the shoreline of the lagoon.

The red algae (corallinaceae) and coral reefs of the Lithothamnium Limestone are centred on top of the Central Swell Zone and shed their debris to the north into the lagoon and to the south into the high energy open marine shelf edge. First results from a study conducted by the University of Vienna on the algae association in the Eocene from the cores indicate at least three cycles of subsidence and shallowing. The shallow-water corallinacea association alternates with the deeper-water peyssonneliacea association. This compares to investigations on the Ras Abu Soma reefs in the bay of Safaga in the Red Sea (Piller & Rasser 1996; Rasser & Piller 1996).

On the slope, successively deeper environments were developed characterized by foraminifera. A zone with the larger foraminifera, nummulites and discocyclines, is followed by sediments deposited on the deeper slope bearing ornamented uvigerinas and globigerinas. From shallow to deeper water depth the lithostratigraphic units are the Nummulitic Sandstone, the Discocyclina limestone and marl of the Perwang Formation (Perwang: first deep well in Upper Austria that penetrated the Molasse imbrications) and the Globigerina Limestone and the Globigerina Marl of the Nussdorf Formation (Nussdorf in the Oichten valley N of Salzburg), which belongs mainly into the early Oligocene (Fig. 9).

The dark and light grey and brown Miliolid Limestone with abundant biogenic debris, bryozoa and foraminifera (miliolidae) is restricted to tectonic slices in the Molasse imbricates and pebbles in the Oligocene olistoliths and turbidites. This limestone was deposited in the shallow marine transgressive sequence south of the Central Swell.

Oil and thermal gas are produced from the fluvial sands of the Voitsdorf Formation, the tidal channel sands of the Cerithian beds, the shallow marine sands of the Ampfing Formation and minor amounts from the Lithothamnium Limestone.

Lower Austria

Autochthonous Molasse. Fluvial conglomerates of the Moosbierbaum Formation are preserved in grabens on the southern fringe of the massif between the cities of St Pölten and Tulln (south of the Danube). They are the few remaining upper Eocene remnants in the Lower Austrian Foreland Molasse.

The source for the Mesozoic and Eocene sands in the Molasse Zone was exclusively the Bohemian Massif.

Waschberg Zone. The upper Eocene–Priabonian- fine and coarse-grained, glauconitic, greenish and brown sands with quartz and crystalline pebbles and light coloured limestones with algae and bryozoas of the Reingrub Formation are the oldest sediments of the Molasse basin in the Waschberg zone.

The sea, which was restrained to the Flysch and Helveticum basin through early Tertiary, flushed back upon the Massif in Eocene. This is the beginning of the Molasse Basin in the strict sense. The whole sequence of late Eocene sediments is preserved in the autochthonous and allochthonous Molasse in the West and in the Waschberg zone.

F. Rögl has provided the biostratigraphic division and has tied the chronostratigraphic ranges of the Central Paratethys to the recently revised Cenozoic geochronological and chronostratigraphic charts (Berggren et al. 1995) for this paper (Fig. 11).

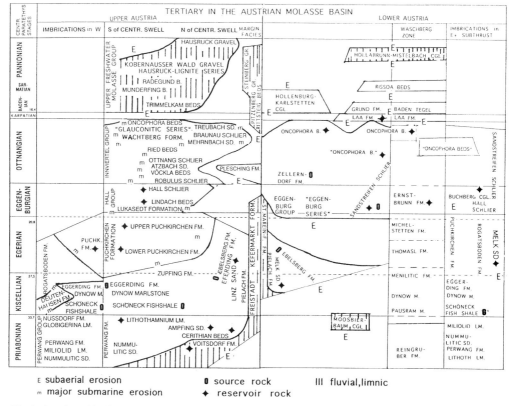

Fig. 9. Stratigraphic chart of the Tertiary in the Austrian Molasse Basin.

Facies distribution in the Oligocene and Lower Miocene

Model for the deep-marine Oligocene–early Miocene basin

After restoring the tectonic units to their original depositional position the reconstructed distance at the end of Eocene from the flood-plain deposits in the north to the more than 3000 m deep Flysch trough was approximately 250 km (Wagner 1996a).

With the deepening of the Molasse ocean in early Oligocene time the interactive process between cold deep-water currents and the upwelling on the slope, warm longshore and off-shore currents at the surface and a low oxygen zone in the middle in medium water depth (Parish 1982) was established.

With cold deep oceanwide circulating bottom currents boreal fauna migrated into the Molasse sea (Rögl pers. comm; Dohmann 1991) from the Kiscellian throughout Oligocene into the early Miocene.

The palaeoecological conditions in the Paratethys, with widespread areas of low oxygen conditions depositing fish shales from the western Alps to the Crimea, are the cause for the occurrence of aberrant foraminiferal species (Rögl 1994). Similar foraminiferal assemblages are observed in the modern Caribbean sea in the Cariaco Basin. In the Cariaco Trough, the cold upwelling water interacts with the warm surface water. At the same time, a tropical to subtropical fauna and flora existed on land and in the surface water of the Molasse basin.

The bottom currents eroded deeply in front of the northward-moving Alpine nappe system into older sediments. This erosion was followed immediately by rapid sedimentation of gravity deposits and turbidites and reworking of sediments. The deepest erosion took place during early Miocene time in the transtensional zones in front of the Imbrications and along the main strike slip faults.

The gravel in the Oligocene conglomerates was derived from the Central Alp mountains. The conglomerates contain crystalline, gneiss,

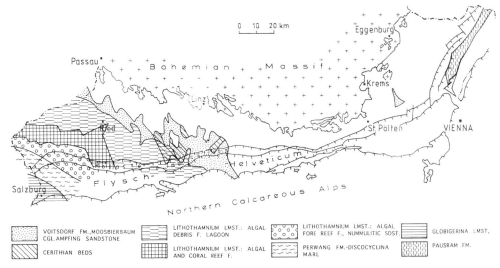

Fig. 10. Alpine–Carpathian Foredeep: facies distribution in the late Eocene and pre-Oligocene, subcrop map.

granite, porphyrite, phyllite, quartzite, quartz, chert, dark grey and light grey and brownish dolomites and limestones. The pebbles were transported across the shelf of the drowned Northern Calcareous Alps and the southern slope, composed of Flysch and Helveticum, into the southern Molasse Basin. The Crystalline pebbles belonged to a mountain range that was approximately several kilometres above the presently outcrop of the Crystalline basement in the Central Alps. This is concluded from fission track analysis of the cooling age in the source area of the apatite grains (Jäger and Hurford pers. comm. 1986 and 1994). The cooling apatite fission-track ages from the Puchkirchen Formation scatter between 39 and 56 Ma in late Palaeocene to late Eocene (Hejl & Grundmann 1989). The age of deposition of the resedimented samples was between 28 and 20 Ma. Olistoliths of Helveticum material slid into the basin. They are abundant in the uppermost Molasse Imbrications which were set down at the southern edge of the Molasse basin at the foot of the slope.

Salzburg and Upper Austria

Oligocene (Figs 9, 12 & 13) Basin and slopes. The cold dense deep-marine bottom currents were forced to deviate from their former main flow due to the tectonically northward moving slope. Therefore the current cuts deeper into the seafloor further to the north. The south slope as well as the north slope were undercut by erosion. Massflows from north and south filled the lows in periods of less energetic bottom currents (Wagner 1996a).

The 'Flysch-like' sedimentation in front of the Flysch and Helveticum nappes was shifted in the Kiscellian to the Molasse trough. The early Kiscellian Deutenhausen Formation, (village and former quarry in the folded Molasse in Bavaria) with grey and greenish grey sand turbidites of a thickness of approximately 1000 m, crop out in the allochthonous Molasse in Bavaria. The sandstones are predominantly cemented by calcite and exhibit complete Bouma sequences with thin pelitic layers. They contain reworked fauna, nannofossils and exclusively agglutinating foraminifera. In the well Aurach1, approximately 100 m of Deutenhausen sandstones were encountered in the autochthonous Molasse below the Flysch.

The Rogatsboden Formation (village and outcrop in tectonic window in Lower Austria) follows in the N and in the Molasse Imbricates of the wells Oberhofen 1 and Mühlreith 1 and in the tectonic windows of Rogatsboden in Lower Austria. It is characterized by calcareous, pelitic and distal parts of the turbidites and contourites with dark and light grey, micaceous sandstone layers, lenses and ripples with deep water foraminifera. Excellent outcrops with slumps and contourites can be studied in the allochthonous Molasse of Bavaria, e.g. in the creeks Ammer (locality Scheibum) and Traun at Siegsdorf.

The term contourite means bottom-current reworked deposits (Shanmugam *et al.* 1993). The terminologies for the gravity-driven deep-marine sediments – turbidite, slump and slide

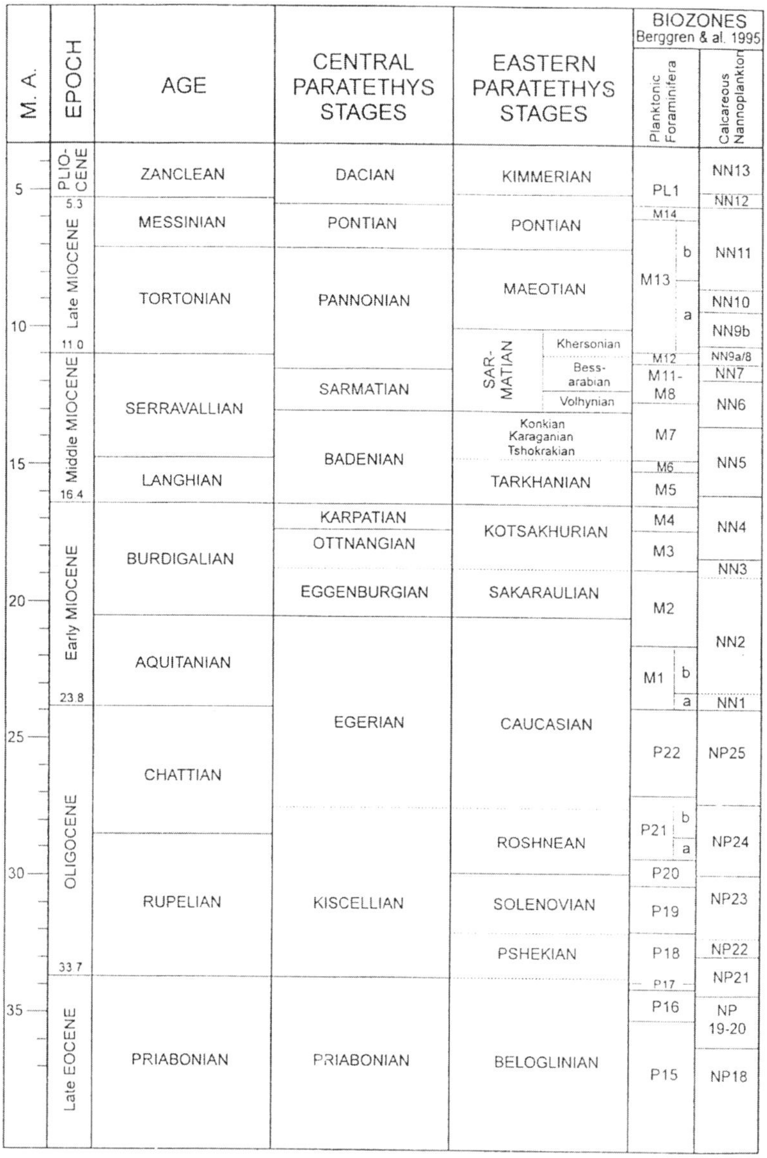

M.A.	EPOCH	AGE	CENTRAL PARATETHYS STAGES	EASTERN PARATETHYS STAGES	BIOZONES (Berggren & al. 1995) Planktonic Foraminifera	BIOZONES Calcareous Nannoplankton
5 —	PLIOCENE	ZANCLEAN	DACIAN	KIMMERIAN	PL1	NN13
5.3						NN12
	Late MIOCENE	MESSINIAN	PONTIAN	PONTIAN	M14	
					M13 b	NN11
		TORTONIAN	PANNONIAN	MAEOTIAN	M13 a	NN10
10 —						NN9b
11.0				Khersonian (SARMATIAN)	M12	NN9a/8
	Middle MIOCENE	SERRAVALLIAN	SARMATIAN	Bessarabian (SARMATIAN)	M11–M8	NN7
				Volhynian (SARMATIAN)		NN6
15 —			BADENIAN	Konkian / Karaganian / Tshokrakian	M7	
		LANGHIAN		TARKHANIAN	M6 / M5	NN5
16.4						
	Early MIOCENE	BURDIGALIAN	KARPATIAN	KOTSAKHURIAN	M4	NN4
			OTTNANGIAN		M3	
20 —			EGGENBURGIAN	SAKARAULIAN	M2	NN3
		AQUITANIAN				NN2
23.8					M1 b	
			EGERIAN	CAUCASIAN	M1 a	NN1
25 —	OLIGOCENE	CHATTIAN			P22	NP25
				ROSHNEAN	P21 b	NP24
30 —					P21 a	
					P20	
		RUPELIAN	KISCELLIAN	SOLENOVIAN	P19	NP23
33.7				PSHEKIAN	P18	NP22
					P17	NP21
35 —	Late EOCENE	PRIABONIAN	PRIABONIAN	BELOGLINIAN	P16	NP 19-20
					P15	NP18

Fig. 11. Cenozoic geochronologic and chronostratigraphic Correlation from F. Rögl.

– are used in the meaning suggested by Shanmugam *et al.* (1994).

The Rogatsboden Formation in the uppermost Molasse imbrications, which have been deposited farthest south, contains high amounts of mass flows of almost exclusively Helveticum olistoliths.

In mid-Kiscellian to early Egerian time, the distal calcareous and pelitic parts of turbidites from the south extend further up on the slope of the Bohemian Massif and interfinger with local turbidites from the N in the Zupfing Formation. This formation is present only in the subsurface. The most complete sections are preserved in the area of Ried im Innkreis in the well Zupfing 1. The Zupfing Formation consists of dark grey and brown pelites with fish remnants and with increasingly dark brown, green and grey limestones with nannofossils (nanoblooms) to the south and in the stratigraphic older portions.

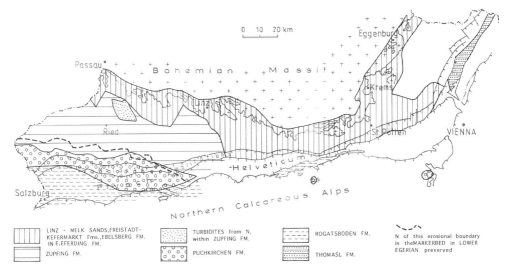

Fig. 12. Alpine–Carpathian Foredeep: facies distribution in the mid-Oligocene..

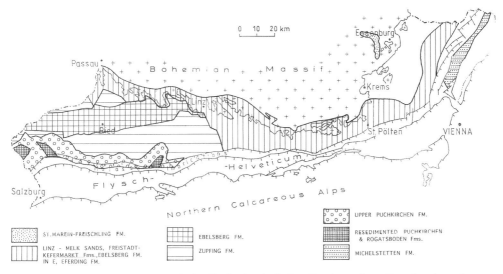

Fig. 13. Alpine–Carpathian Foredeep: facies distribution in the late Egerian and pre-Eggenburgian, subcrop map.

These sediments are the preserved distal parts of the turbidites from the south, which have not been eroded before the sedimentation of the younger Puchkirchen clastics. Intercalated are slides and slumps from the northern slope of dark brown and black soft pelites of the Ebelsberg Formation. The light grey, green and yellow, coarse- and fine-grained sandstones of the turbidites from the north accumulate at faults.

The axis of the deep part of the Molasse basin was shifted to the north by the northward movement of the Alpine thrust system. The Zupfing and Eferding Formations became subject to intensive submarine erosion.

Subsequently, with the movement of a slice of the Molasse Imbrications and erosion, the lows were filled by the Egerian to early Eggenburgian sediments of the Puchkirchen Formation (Puchkirchen: village and gasfield close to Vöcklabruck in Upper Austria). The infill of the Puchkirchen Formation is composed of more than 80% of slide and contorted slump material from both sides of the basin. Slides and slumps of Zupfing and Eferding deposits from the north interchange with slides and slumps of Rogatsboden Formation and by the imbrications uplifted and again moved downslope Puchkirchen sediments and of the new turbidites from the south and of contourites. The bottom currents transported the fine material from the mass flows and turbidites in an E–W direction along the erosional lows in front of the Molasse Imbrications and parallel to the faults. In the resulting muddy and sandy beds the whole variety of reworked sediments can be recognized in the alternating laminae. The Puchkirchen Formation turbidites comprise grey and greenish grey, micaceous sandstones and conglomerates with calcite cement and Bouma sequences. The well-rounded pebbles are often coated with a black film and reach more than 30 cm in diameter. The matrices of the sandstones are angular to subrounded lithic arenites. The mass flows of light and dark grey and brown micaceous pelites, silts, sands, sandy and muddy conglomerates (Fig. 17) reproduce the lithology from the active and passive slope. The contourites contain the reworked fine-grained material of the pelitic, silty, sandy and conglomeratic slumps and turbidites.

The microfossil assembly of the autochthonous sediments contains only deep-water agglutinating foraminifera (in the lower portion: *Rhabdammina linearis*), besides intensively reworked fauna from N and S Molasse and Helveticum and excellent preserved single fossils from Cretaceous and Eocene (Fig. 14). The stratigraphic range of the Puchkirchen Formation comprises Egerian to Early Eggenburgian.

The intense internal submarine erosions are typical for all basin and slope deposits in the Upper Austrian Molasse. In the upper portion of the Puchkirchen Formation the erosions cut from the E and N progressively deeper into the older Puchkirchen sediments towards the Perwang block. The erosional lows are refilled by more than 300 m of Puchkirchen sediments in Eggenburgian time. Most of the thinly bedded gas reservoirs with fine-grained sands along the northern pinchouts are contourite sands, e.g. in the Atzbach, Zell am Pettenfirst or Pfaffstätt gasfields.

All the diagnostic sedimentary structures described from the Ewing Bank in the Gulf of Mexico (Shanmugam *et al.* 1993) can be studied in the cored contourite sediments. The reported criteria for recognizing the bottom-current-reworked sands in the Molasse of Salzburg and Upper Austria are the following: (1) predominantly fine-grained sands and silts (Figs 15–16, 18–21); (2) thinly bedded to laminated sands in deep-water mud (Figs 15, 16, 15–21); (3) numerous layers per metre (Figs 15, 10–29); (4) sharp upper contacts and bottom contacts (Fig. 29); (5) internal erosional surfaces (Fig. 19); (6) inverse grading; (7) horizontal lamination and low angle cross-lamination (Figs 15 & 19); (8) cross-bedding (Figs 18–20); (9) lenticular bedding, starved ripples (Fig. 21); (10) current ripples with preserved or eroded crests (Figs 18 & 20); (11) mud offshoots (Figs 18 & 19); (12) flaser bedding (Fig. 18). The contourite sediments in the Deutenhausen Formation, the Rogatsboden Formation, the Puchkirchen and Hall Groups and at least in the lower Innviertel Group are associated with other deep-marine facies such as turbidites, mass flow deposits and hemipelagic mud.

In late Oligocene time, Puchkirchen conglomerates, trachy-andesitic volcanic ash was recovered in cores from several wells. These ashes are probably related to the Oligocene intrusives along the Periadriatic line (Mair *et al.* 1992).

From the rise of the basin to the north and up the slope the sequence of sediments is characterized by the upwelling system. It starts with distal turbidites from the south and local turbidites from the north, merging into a zone with blooming of nanoplankton, the Kiscellian 'Banded Marl' of the Eggerding Formation and the pure nanno ooze of the light coloured middle Kiscellian Dynow Marlstone, and below the low oxygen zone fish shales. The middle Kiscellian Eggerding Formation (Eggerding: village and well with heavy oil in Upper Austria) is composed of dark grey laminated pelites with thin white layers of nanoplankton (nanoblooms) and is often tectonized. The banded marl contains breccia of submarine reworked lithothamnium limestone and Schöneck Formation and has a high tendency to slide (Wagner 1996*a*). Some layers within the Eggerding Formation yielded similar TOC (total organic content) values as in the Schöneck shales.

The early Kiscellian Schöneck Formation (Schöneck: outcrop in the allochthonous Molasse in Bavaria) is the source rock for the Molasse oil and contains phosphorite nodules. The dark grey or brown, shaly, thin-bedded limestones and shales of the early Kiscellian Schöneck Formation contain abundant fish remnants and medium to deep-water calcareous and agglutinating foraminifera.

The younger Oligocene black to brown silty, soft fish shale of the Kiscellian and Egerian Ebelsberg Formation (Ebelsberg: village close to Linz with a rich fauna) has locally interbedded

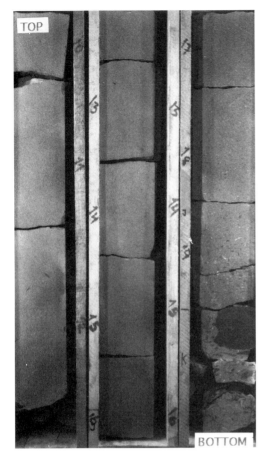

Fig. 14. Core photograph showing porous, thick-graded turbidite of Egerian Puchkirchen Formation in front of the Molasse Imbricates N of Salzburg with Cretaceous and Eocene Helveticum pebbles (pebbles at bottom in right box) and numerous resedimented single nummulites (box is 1 m long).

village and creek E of Melk) and floodplain and swamp deposits of the Freistadt–Kefermarkt Formation (Freistadt and Kefermarkt: villages N of Linz). The thickness of the transgressive light grey, white and light brown Linz Sand is influenced by the relief along faults and indicates abrupt coastlines. In the area of Steyregg at the Danube east of Linz a late Egerian shallow marine subfacies of calcareous sandstones with abundant red algae (corallinacea) was mined as building stone (Rögl & Steininger 1996).

In the river sands and gravels, the limnic pelites and swamp lignite and coal of the Oligocene to Miocene Freistadt–Kefermarkt Formation in the W and St Marein–Freischling Formation (St Marein & Freischling villages at E edge of the Massif) in the E is a tropical to subtropical fauna (Steininger et al. 1991) preserved. The plant communities infer a warm humid rainy climate (Kovar 1982).

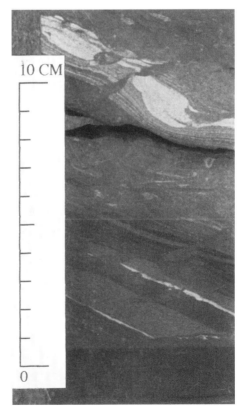

Fig. 15. Core photograph showing submarine erosion. Above the erosional base is a resedimented slide of bottom-current reworked muddy and sandy sediments.

high amounts of diatomites and menilites, stratified and laminated nanoblooms and bentonite layers. The pelites are partially bituminous and non-calcareous and contain also phosphorite nodules, abundant fish remnants and agglutinating and rare calcareous foraminifera.

On the shallow shelf, abundant terrigenous material was shed into the dark grey, silty pelites of the Kiscellian and Egerian Eferding Formation (Eferding: village with outcrops from brickworks W of Linz). Calcareous foraminifera are the dominant microfauna in the Eferding Formation.

Northern coast. A narrow belt of shoreline sands of the Kiscellian and Egerian Linz Sand, paralic pelites of the Pielach Formation (Pielach:

Fig. 16. Core photograph of Puchkirchen Formation showing a submarine erosional truncated turbidite sand layer overlain by bottom-current reworked sections. Note in the dark pelitic laminated sections the darker and lighter laminae are from different Oligocene shales which were originally sedimented on the opposite slopes of the basin.

Fig. 17. Core photograph of Puchkirchen Formation showing contorted shale, sand and conglomerate of a slumped section.

Heavy oil was produced from the Linz Sand.
Miocene (Fig. 9), Eggenburgian (Fig. 22): As mentioned above an Eggenburgian microfauna was determined in the upper few hundred metres of the Puchkirchen Formation in the W (Rögl pers. comm. 1994).

After the most prominent submarine erosion across the Molasse basin, a new fauna immigrated in the Hall Group from the Indian Ocean (Steininger & Rögl 1979; Rögl & Steininger 1983). The oldest sediments of the Hall Group the light grey, micaceous, calcareous sandstones, siltstones, sandy pelites and darker muddy conglomerates with plant fragments of the Lukasedt Formation are restricted to the SW in Salzburg. The Lukasedt Formation was deposited on top of the Molasse Imbrications in a relative narrow erosional channel, which was filled by slides, slumps, turbidites and contourites. On top of the Perwang imbricates the Eggenburgian Lukasedt Formation and slides of

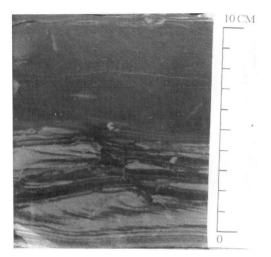

Fig. 18. Core photograph of Puchkirchen Formation showing flaser bedding with current ripples and mud offshoots and mud in ripple troughs.

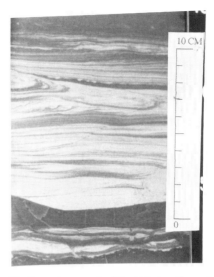

Fig. 20. Core photograph of Puchkirchen Formation showing thin flaser-bedded sands with sharp lower and upper contacts.

the upper Oligocene Eferding formation were sedimented in a minimum water depth of 500 m. They were uplifted with the Imbricates and subsequently moved large parts again down the flanks. In the present, the Lukasedt sediments crop out at an elevation of 500 m N of Salzburg. This corresponds to an uplift of 1000 m during the last 20 Ma.

With the northward moving Imbricates the submarine erosion was shifted further to the N. In the somewhat younger Lindach Formation the same sedimentation as in the Lukasedt Formation took place in the newly formed channel above the starting tension zone in the northern part of the Imbricates. At the time of the deposition of the overlying Hall Formation the whole Molasse basin was filled with light grey to green-grey, micaceous, sandy pelites the so-called 'Hall Schlier', which are the distal parts of turbidites and contourites. The centre of the Hall basin accumulated the turbiditic and contouritic sandstones and slides in erosional lows. The pattern of the lows is delineated by the fault-systems.

The turbidites and contourites of the Hall Group and the Puchkirchen Formation are the reservoirs for the biogenic gas.

Ottnangian (Fig. 23). The boundary between The Eggenburgian and Ottnangian is inexactly determined faunistically. The correlation marker on logs and seismic lines is a regional submarine unconformity with sandy layers on top. The sandy pelites of the 'Schlier' facies is interrupted several times by sand or sand prone sections. The whole sequence is dominated by an environment with strong currents and submarine erosion. The sands were recently studied and interpreted as subtidal sand waves sedimented under strong tidal current activity (Faupl & Roetzel 1987; Krenmayer 1991; Salvenmoser & Walser 1991). The association of the *Skolithos* (and *Cruziana*) ichnofacies supports the interpretation of a high-energy environment. The microfauna (Rögl pers. comm. 1994) and the fish assembly (Brzobohatty & Heinrich 1990) point to a meso-pelagic environment with high amounts of reworked older fauna.

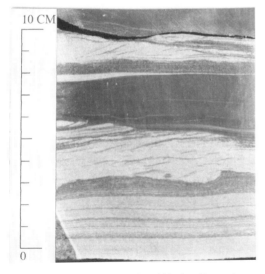

Fig. 19. Core photograph of Puchkirchen Formation showing thin sand layers with sharp upper truncated contacts.

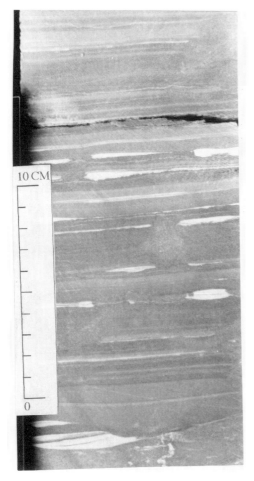

Fig. 21. Core photograph of Puchkirchen Formation showing lenticular bedding with single ripples.and bioturbated mud.

Aberer (1958) mapped the sequence of Ottnangian sediments and had then observed the stepwise arrangement of the sediment wedges. The interpretation of the drilling results and seismic demonstrates the tectonic influence on the distribution of the Miocene sediments. The submarine unconformities in Ottnangian are dipping regionally from N and E to the Pfaffstätt fault. The base of the Atzbach Sand is in the E in the area of Atzbach and Ottnang 350 m above the base of the Innviertel Group. Close to the Pfaffstätt fault the erosional base of the Atzbach sands approached to the boundary to the Hall Group within a few metres. More than 300 m of older Ottnangian sediments have been previously removed. The extension of the early Ottnangian silty, sandy, micaceous, green–grey pelites with sand ripples and layers of the Robulus Schlier is terminated by the Pfaffstätt–Maltig faults to the SW. Between the Schwanenstadt fault and a subparallel fault in the W of the Ried fault, micaceous, in parts glauconitic, grey, greenish and brown, fine- and coarse-grained, laminated, cross-bedded and bioturbated sand layers and lenses increase in the locally developed Vöckla Beds in the Robulus Schlier. Two thicker sand sections in the top of the Robulus Schlier are the micaceous, in parts glauconitic, grey, greenish and brown, fine- and coarse-grained, laminated, cross-bedded and bioturbated sands with gravel banks, pelitic layers and pelite clasts in channels of the porous Atzbach Sands. The palaeocurrents point with their maxima to the ENE and WNW along the fault pattern (Faupl & Roetzel 1987), but an overall broad scatter was noticed in the measurements. The sand-filled channels are oriented parallel to the faults in the subsurface. The upper sand horizon terminates abruptly at the Ried fault. The Enzenkirchen Sand and the Kletzenmarkt glauconitic sand are equivalents to the Atzbach sand in local varieties. The overlying sandy, silty, micaceous pelites of the Ottnang Schlier contain deep neritic fossils. The sandy, silty, micaceous pelites with thin sand layers of the Ried Beds are characterized by the foraminifera Rotalia. The Ried Beds are overlain in turn by the glauconitic quartz arenites with thin pelitic layers of the Mehrnbach Sands, the silty pelites with thin glauconitic, micaceous sand layers of the Braunau Schlier and the glauconitic sands of the Treubach Sand. The regression of the Paratethys Sea at the end of Ottnangian time from Bavaria to the E across the whole Molasse basin is represented by the paralic, estuarine, lagoonal, deltaic to littoral sands, silts and pelitic sands of the Late Ottnangian Oncophora Formation. In Upper Austria, the remnants of the Treubach Sands and the Oncophora Formation are limited by the faults in the Braunau Block on top of the hills and at the eastside of the Inn valley.

Within the Ottnangian every formation begins and ends with a major submarine erosion.

The transgressive tidal-influenced, fossiliferous, litoral sands with reworked phosphorite nodules sands of the Plesching Formation include the 'Fossiliferous Coarse Sand Series' and 'Phosphoritic Sand' on the SW-fringe of the Bohemian Massif and contain abundant phosphorite nodules. The phosphorite is reworked from the Ebelsberg Formation.

The submarine erosions are characteristic features at the formation boundaries and within the Deutenhausen, Rogatsboden, Puchkirchen,

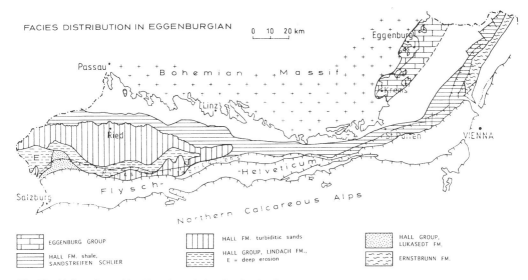

Fig. 22. Alpine–Carpathian Foredeep: facies distribution in Eggenburgian.

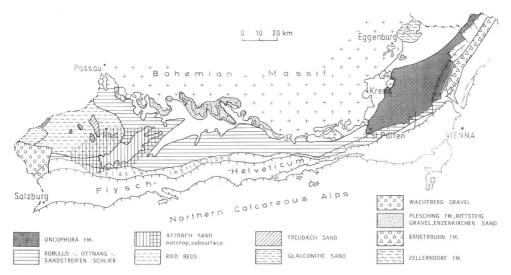

Fig. 23. Alpine–Carpathian Foredeep: facies distribution in the Ottnangian and pre-Upper Freshwater Molasse, subcrop map.

Zupfing, Lukasedt, Lindach, Hall, and Oncophora Formations and in the Robulus, Ottnang and Braunau Schlier and the Atzbach, Mehrnbach and Treubach Sands.

Middle and late Miocene: Upper Freshwater Molasse (Fig. 24).

After the regression of the Molasse sea at the end of Ottnangian subaerial erosion took place. Above the marine Molasse sediments no Karpatian sediments are known in Upper Austria. The selective uplift in the fault blocks resulted in the differential deeper erosion to the east. Limnic and fluvial sedimentation above the marine Molasse sediments commenced earlier in the west. At the Bavarian border the swamp coal, limnic pelites and meandering river sands of the Trimmelkam Beds of the coal mines were deposited in the Badenian (earliest 16.4 Ma), the braided-stream gravel banks, the swamp lignite and the limnic pelites of the Munderfing–Radegund Beds in the Sarmatian and the braided-stream gravel banks, the swamp lignite and the limnic pelites of the Hausruck Lignite Series, the braided-stream gravel banks and flood pelites of the Kobernhausser Wald Gravel and the Hausruck Gravel in the Pannonian. The flow direction of the rivers shifted in the Pannonian from WNW to NE (Mackenbach 1984). This was caused by the uplift of the imbricates on the Perwang Block. The gravel banks of the braided streams in the Hausruck contain pebbles from different sources (Dunkl et al. 1996). The cooling ages from the red sandstones are 70 and 40 Ma, the gneisses 40 and 13 Ma and the quartzites 13–14 Ma. The young apatite grains are probably derived from the Penninic rocks.

In Lower Austria, the Hollabrunn–Mistelbach Gravel of the Upper Freshwater Molasse corresponds to the Hausruck Gravel.

The mammals in the gravels are Pannonian in age (Rögl et al. 1993). Therefore, all sediments in the area of the Austrian Molasse and the Bohemian Massif between approximately 8 Ma and the ice ages were eroded subaerial.

The marginal facies of the Molasse basin in the north are the terrestrial, limnic and fluvial Upper Miocene deposits sitting directly on the Crystalline basement of the Bohemian Massif. The dating of the limnic pelites with coal layers of the Rittsteig Beds in Ottnangian to early Karpatian, of the limnic pelites and gravels of the Pitzenberg Gravel in Ottnangian, Karpatian or Badenian and of the gravels of the Steinberg Gravel in Sarmatian to Pannonnian is vague.

Mineralogical composition and source of the Mesozoic and the Molasse sands in Upper Austria and Salzburg.

The following differences in the mineralogical composition can be distinguished between the sediments shed from the Bohemian Massif and the Alps (Kurzweil pers. comm. 1988).

- All Dogger to Eocene sands and the shallow-marine and fluvial sands of Oligocene and Early Miocene and local turbidites from the N in the Zupfing formation are derived from the Bohemian Massif: The sandstones are arkoses, subarkoses and quartzarenites with Zircon as the dominant heavy mineral.
- All deep-marine Oligocene coarse clastics, with the exception of the turbidites in the Zupfing formation, are derived from the Central Alpine units. The sandstones are lithic arenites with being the dominant heavy minerals garnet and staurolite. The occurrence of paragonite is considered as diagnostic for an Alpine source.
- The mineralogical composition of the early Miocene sandstones is predominantly the same as in the Oligocene deep-marine sediments with minor additions of reworked minerals from older shallow-marine and fluvial sediments.

Lower Austria

Oligocene. Autochthonous Molasse (Figs 9, 10, 12 & 13). From a few kilometres to the West of the Steyr fault system on, along subparallel faults the early Oligocene, the Schöneck Fishshale, the light coloured Dynow Marlstone and the 'Banded Marl' of the Eggerding Formation of the deeper-water environment were not deposited on top of the Crystalline basement

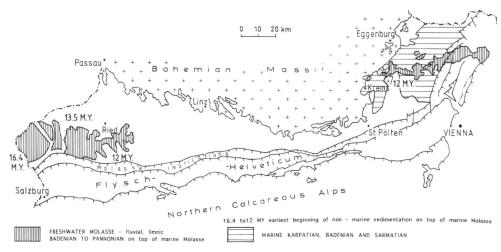

Fig. 24. Alpine–Carpathian Foredeep: facies distribution in the late Miocene.

in front of the Alps. In the Behamberg 1 well, the Tertiary sedimentary sequence starts with the light-coloured Dynow Marlstone. This compares to the palaeogeographic position in the north of the pinch-out limit of the Eocene in Upper Austria.

On the southern edge of the Bohemian Massif, from Upper to Lower Austria on the Crystalline high of Amstetten, the limnic dark shales with plant fossils of the early Egerian (Kiscellian?) Pielach Formation is present in the tectonic lows. The fluvial and shallow marine sands and gravels of the Melk Sand and the shallow marine shales of the Kiscellian and the Egerian are also restricted to the tectonic lows (Jiricek & Seifert 1990). The Seitenstetten 1 well recovered the same facies just in front of the Flysch unit (Brix *et al.* 1977).

Miocene. Eggenburgian (Figs 9 & 22). The wells N of the Danube encountered the glauconitic sands and pelites with a lower foraminiferal and a upper fish facies of the 'Eggenburg Series'. They intertongue to the W and S with several hundred metres of light grey, silty, sandy, extensively bioturbated pelites with calcareous silt and sand lenses of the 'Sandstreifen Schlier', which were deposited in a turbiditic, contouritic environment. In the Imbricated Molasse, Buchberg Conglomerates with pebbles from the Calcareous Alps, the Flysch and the Alpine Crystalline basement are interspersed in the 'Sandstreifen Schlier'.

Ottnangian (Fig. 23). In Ottnangian time, the sandy, silty pelites of the Robulus Schlier continue from Upper Austria to the East. A thicker sand section within the Robulus Schlier is the Prinzersdorfer Sand. Coarse-grained fan deposits with pebbles from the Flysch zone are lined up in the Imbricated Molasse along the eastern Flysch front. The shallow marine and coastal sediments of lower and middle Ottnangian were not preserved in the East. The late Ottnangian to early Karpatian Oncophora Beds are wide spread in Lower Austria. The Oncophora Beds are the continuous regressive Formation across the Molasse Basin from Bavaria in the West to the Czech Republic in the east.

The Oncophora beds and the 'Sandstreifen Schlier' are the reservoirs for biogenic and thermal gas.

Karpatian (Fig. 24). In the late Karpatian the sea from the Pannonian Basin had flooded the Eastern flank across the thrust of the Waschberg zone. The coastline ran between Tulln and Maissau. The greatest water depth was reached in the area of Laa an der Thaya. The grey pelites with sand layers of the Laa Formation contain abundant microfauna and pteropods.

Badenian (Fig. 24). An additional influx of open-marine water from the Indian Ocean reached the area of St Pölten and inundated the old tectonic features along the present-day Danube valley. In lower Badenian a river mouth at St Pölten shed the debris from the Alps into the Hollenburg–Karlstetten Conglomerate, into the marine delta complex of the Molasse sea (Steininger *et al.* 1991). The shallow-marine, delta to prodelta Karpatian to lower Badenian Grund Beds with grey–green, fossiliferous pelites, silts, sands and conglomerates was deposited N of the Danube. On the tectonic high at Mailberg, a thick red algae reef grew in the Lithothamnium limestones. Locally, 'block beds' with boulders of 1–2 m in diameter were deposited.

Sarmatian (Fig. 24). From late Miocene time onward, the Paratethys separated from the Mediterranean Sea. The inlet of the Zaya graben between Hollabrunn and Krems was filled with the grey–green fossiliferous brackish pelites, silts and sands of the Sarmatian Rissoa Formation.

Waschberg zone

Oligocene (Fig. 9). In the Waschberg Zone the deeper-marine facies of the lower to middle Oligocene is represented, above the Priabonian Reingrub Formation, by the variegated shales and banded marls with globigerina of the Priabonian to early Oligocene Pausram Formation, the light-coloured marly limestones of the Dynow Formation and by the laminated, silty to sandy shales with intercalations of diatomites, and siliceous shales of the Kiscellian Menilitic Formation (Seifert 1993).

The tectonic remnants of Oligocene dark grey, sandy shales with pebbly sandstones of the Thomasl Formation and the early Miocene (late Egerian) light grey, calcareous marlstones of the Michelstetten Formation were deposited in an open marine environment.

Miocene. The fossiliferous Michelstetten shales are overlain by the deep-marine lower Miocene slope to basin plain sediments of the Eggenburgian and Ottnangian Ernstbrunn Formation (not to be confused with the Jurassic Ernstbrunn Limestone). The shales and the turbiditic sands contain pebbles and boulders of Flysch and Crystalline basement (Malzer *et al.* 1993).

In mid-Miocene Karpatian time, the Waschberg zone was thrust to the northwest and subaerially exposed 16 Ma ago in Badenian time.

In the early Badenien the Molasse became flooded again from the east and the soft pelites of the Baden Tegel were deposited.

Gas is produced from the Ernstbrunn Formation sands.

Subthrust autochthonous Molasse

Upper Austria and Salzburg. To the west of the Steyr fault, 16 wells penetrated the autochthonous Molasse below the Calcareous Alps and the Flysch–Helveticum nappes and the Molasse Imbrications. On top of the Jurassic (Dogger to Purbeckian) and middle Cretaceous (Cenomanian), the normal southern continuation of the Upper Austrian Foreland Molasse was found (Wessely & Wagner 1993). The stratigraphic range of Tertiary sediments is upper Eocene to Egerian and between the Imbrications and the Flysch mainthrust lower Eggenburgian.

Lower Austria. Sixteen exploration wells were also drilled through the Subalpine Molasse to the east of the Steyr fault in the area of the spur of the Bohemian Massif. The wells were located on structural highs and encountered the Molasse sediments in tectonically and erosively reduced thicknesses. The Kiscellian Dynow marlstone, Eggerding Formation banded marl (only in the Behamberg 1 well close to the Steyr fault), pelites of the Zupfing Formation and Egerian sands of the Linz–Melk groups wedge out on the west flank of the spur. To the east, the Eggenburgian 'Sandstreifen Schlier' is, with the exception of the Mauerbach 1 well, imbricated. The Buchberg conglomerate in the Mauerbach 1 and St Corona 1 wells (just below the Flysch main thrust) contain abundant pebbles from the Calcareous Alps (Brix 1993).

The Urmannsau 1 and Berndorf 1/NÖ wells, situated in the Northern Calcareous Alps, yielded marine Egerian clastics approximately 40 km south of the thrust front.

Below the subsided Flysch nappes in the Vienna Basin, the autochthonous Molasse Palaeogene was found only in the Zistersdorf ÜT1a and ÜT 2A wells. The sandy pelites with breccias on top of the Jurassic (Dogger to Malm) are dated as Eocene to lower Miocene. They are overthrust by the allochthonous Waschberg Molasse. The early Miocene (Eggenburgian) sediments stretch far below the Alpine–Carpathian units and were penetrated by the Zistersdorf ÜT 2A well (Wessely 1993).

The main thrusting over the Foreland Molasse took place at the end of the early Miocene (end of Ottnangian). The Karpatian sediments are still slightly overthrust at the thrust front. The mid-Miocene (early Badenian) marine sedimentation transgressed progressively across the thrust planes and continued in the Vienna Basin. The Urmannsau 1 well, was the first well that found autochthonous Egerian Molasse sediments below the Calcareous Alps. In the tectonic window of Rogatsboden in the Flysch Zone, the late Eocene to early Oligocene sediments of the imbricated Molasse reach the surface. The early Miocene–(Eggenburgian to Ottnangian) sediments seal the thrust planes. In the Königstetten U1 well, the Egerian to Eggenburgian sediments of the Molasse Imbrications overthrust Egerian to Eggenburgian and Karpatian Foreland Molasse. The autochthonous Molasse in the well Mauerbach 1, is the Eggenburgian Buchberg conglomerate.

On top of the autochthonous Mesozoic (Jurassic and Cretaceous) the Oberdambach 1 well, below Flysch, and Molln 1 well, below the Calcareous Alps, penetrated through imbricated and autochthonous Molasse of late Eocene and Oligocene in the same facies as in the Upper Austrian Foreland. The sheared off Waschberg Zone incorporates upper Eocene to lower Miocene marine Molasse sediments and basement.

In the oil and gas field Roseldorf are the graben structures with Mesozoic basement overthrust by wedges of sheared off Mesozoic basement. The Mesozoic is eroded from Mauerbach to the south and cut off at the Mailberg fault.

The subthrust Flysch gas field Höflein produces from autochthonous Jurassic arenites and carbonates, which are sealed by early Miocene (Eggenburgian to Ottnangian) imbricated Molasse sediments.

Neogene at the east edge of the Bohemian Massif in the area of Eggenburg, Retz, Fels and Horn

The outcropping fossiliferous Tertiary sediments in the area of Eggenburg have been studied since the eighteenth century. It is the key area for the regional chronostratigraphy in the lower Miocene of the Central Paratethys (Steininger & Senes 1971; Steininger *et al.* 1991).

The Tertiary at the eastern edge of the Bohemian Massif was mainly sedimented and preserved in half-graben and graben structures. The eastern massif is dominated by the NE–SW faults, but the internal sediment basins on top of the Crystalline are oriented parallel to the W–E, N–S- and NW-trending faults as well.

The fluvial, partly limnic and estuarine gravels, arkoses, sandstones, silts and pelites of the St Marein–Freischling Formation were transported by rivers from southern Bohemia. In the late Oligocene a river mouth is assumed in the area of Krems–Statzendorf in transition to marine Melk sands. The fluvial sediments range locally into the middle Eggenburgian.

The marine transgression in early Eggenburgian time along the eroded tectonic zones of weakness resulted in a variety of local estuarine and shallow-marine facies developments of the Eggenburg Group (Steininger et al. 1991; Roetzel 1993).

In the indented crystalline basement are the Ottnangian remnants of the sands of the littoral Riegersburg Formation, the brackish pelites with bentonites of the Weitersfeld Formation, the estuarine pelites with lignite beds of the Langau Formation and the fluvial and limnic pelites, gravels and sands of the Theras Formation preserved (R. Roetzel 1993).

Vienna Basin fill

The parautochthonous Molasse in the Vienna Basin is part of the Paratethys (Rögl & Steininger 1983). The lower Miocene in the Northern Vienna Basin belongs to the Molasse foreland. It has been transported on top of the Alpine–Carpathian nappes as a piggyback basin. From Eggenburgian to Karpatian time the southern shoreline fluctuated along the Spannberg ridge across the allochthonous tectonic units of the Flysch, the Northern Calcareous Alps and the Central Alps. On the displaced and rotated nappes the parautochthonous coastline is at present oriented E–W with the fluvial–terrestrial sediments in the south and the marine sediments in the north (Seifert 1993). The sources for the coarse clastics have been the different Alpine units.

Starting with the pull-apart phase of the Vienna Basin in the Karpatian and with a fully marine environment in the Badenian basin fill, the Vienna Basin separated from the continental Molasse foreland. The rivers from the Molasse area shed material from the east into the Vienna Basin from the mid-Badenian onward.

Hydrocarbon occurrences

Oil and thermal gas

Source rock. The Schöneck Fishshale (early Oligocene) is the interpreted source rock for the oil in Upper Austria using the correlation of the isotope and biomarker ratios of the oils with the source rocks (Wehner & Kuckelkorn 1995). To a minor degree the organic-rich layers of the Eggerding Formation could have contributed some oil. The Schöneck Fishshale is immature in the Foreland Molasse. The vitrinite reflectance, for the maturity of the source rock to generate oil above 0.6 %R_r, is reached approximately with the front of the Flysch nappes in a depth of more than 4000 m (Xu 1991; Sachsenhofer 1992). The oil was predominantly generated below the Alpine thrusts. The generation started in Miocene and may still be continuing (Schmidt & Erdogan 1996).

The source rocks for the oil accumulations in Eastern Lower Austria (Waschberg Zone and Vienna Basin) are the upper Jurassic Marls (Ladwein 1988; Brix 1993).

Reservoirs. Productive reservoirs are fluvial and shallow-marine sandstones and the carbonates of Dogger (Gresten Group and Höflein Fm), Malm (Altenmarkt Fm and Ernstbrunn Lst), Cretaceous—Cenomanian (Regensburg Fm), lower Turonian (staining but low permeability in lower Turonian Glauconitic sandstone), upper Turonian to Coniacian (sandstones at Steyr fault) and Campanian (so far only geothermal water production), upper Eocene (Voitsdorf Fm, Cerithian beds, Ampfing Sst, Lithothamnium Lst) and Oligocene—Kiscellian and Egerian (Linz and Melk Sst, Puchkirchen Fm only in Imbricated Molasse) age. The oil is trapped in fault, stratigraphic, combined stratigraphic and fault structures, anticlinal and imbrication structures.

Thermal gas

Source rock. The geochemical investigations of the thermal gas in the plays in the area of the Waschberg Zone and the adjacent Molasse in the area of Laa/Thaya identified the coaly layers of the middle and lower Jurassic Gresten group besides the upper Jurassic marls as source rock for gas (Kratochvil & Ladwein 1984). The gas was generated in the autochthonous Mesozoic below the Waschberg and Molasse thrusts (Brix 1993; Kreutzer 1993; Ladwein et al. 1993).

Reservoirs. The productive reservoirs are the Dogger sandstones and carbonates (Gresten Group, Höflein Fm, the thrusted upper Cretaceous in the Waschberg zone, the Oligocene—Egerian (Melk Sst) and Miocene—Eggenburgian (Sandstreifen Schlier, Ernstbrunn Fm) and Ottnangian (Oncophora beds) sandstones and gravels and biogenic gas.

Biogenic gas

Source rock. The biogenic gas was generated by bacterial activity in the rapidly deposited Oligocene and Miocene deep water sediments (Malzer 1993). The source rocks are the pelitic slide and slump sections. The bulk of the shale section contains very little organic matter, but the slumps and slides of the Ebelsberg Formation reach values of 10% TOC.

Reservoirs. The productive reservoirs are Oligocene (Puchkirchen Fm) and Miocene—Eggenburgian, Ottnangian and Karpatian (Hall Fm, Sandstreifen Schlier, Ernstbrunn Fm, Oncophora beds and Laa Fm) turbiditic sandstones and sandy conglomerates and shallow-marine to brackish sands. The gas is trapped mainly in stratigraphic and compaction structures or in a combination of both types and imbrication structures. Strike-slip faulting influenced and delineated the distribution of the reservoirs. In Upper Austria have most of the deeper horizons with biogenic gas associated condensate.

Thermal and biogenic gas are mixed in several reservoirs.

Appendix: Lithostratigraphic units in the Tertiary Molasse Basin

(UA, Upper Austria; LA, Lower Austria Foreland Molasse and Subthrust Molasse; EBM, Eastern margin facies; WB, Waschberg Zone; VB, Vienna Basin.)

Terrestrial, limnic and fluvial

Miocene
Upper Freshwater Molasse Group
  Hausruck Gravel: braided stream gravel—Pannonian. UA
  Kobernausser Wald Gravel: braided stream gravel and flood pelites—Pannonian. UA.
Hausruck–Lignite Series: braided stream gravel, swamp coal, limnic pelites—Pannonian. UA.
Hollabrunn–Mistelbach Gravel: braided stream gravel—Pannonian. LA.
Munderfing–Radegund-Beds: braided stream gravel, swamp coal, limnic pelites—Sarmatian. UA.
Trimmelkam Beds: swamp coal, limnic pelites, meandering river sands—Badenian. UA.
Steinberg Gravel: braided stream gravel—Sarmatian? to Pannonnian? UA.
Rittsteig Beds: swamp coal, limnic pelites overlying Crystalline basement—Ottnangian? to Early Karpatian? UA.
Pitzenberg Gravel: braided -stream gravel, limnic pelites overlying Crystalline basement—Ottnangian?, Karpatian?, Badenian? UA.
Theras Formation: gravel, sands and pelites—Late Ottnangian to Karpatian. EBM.
Aderklaa Conglomerate: braided-stream conglomerates—Ottnangian. VB.
Aderklaa Formation: limnic pelites and silts —Ottnangian. VB.
Gänserndorf Member: braided stream conglomerates, meandering river sands, limnic pelites, evaporites—Ottnangian. VB.

Oligocene and Eocene.
Freistadt–Kefermarkt Formation: gravels, river sands, limnic pelites, swamp coal—Oligocene to Miocene. UA.
St Marein-Freischling Formation: gravels, river sands, limnic pelites, in parts estuarine—Kiscellian and Egerian. EBM.
Voitsdorf Formation: ('limnic beds, Sandsteinstufe' in parts) variegated pelites with roots, predominantly sands in meandering channels, occasionally in shoe-string and braided stream channels—Eocene to Oligocene. UA.
Moosbierbaum Conglomerate: conglomerate—Priabonian by lithologic correlation. LA.

Paralic, estuarine, lagoonal:

Miocene.
Bockfliess Beds: delta complex—Karpatian. VB.
Kornneuburg Beds: fresh-water, estuarine and marine pelites—Karpatian. VB.
Langau Formation: estuarine pelites, lignitic layers—Ottnangian. EBM.
Mold Formation: estuarine pelites, lignitic layers—Late Egerian to Early Eggenburgian. EBM.
Weitersfeld Formation: brackish pelites, bentonites—Ottnangian. EBM.
Oncophora Formation: paralic, estuarine, lagoonal, deltaic to littoral sands, silts, pelitic sands—Late Ottnangian. UA & LA.

Oligocene.
Pielach Formation: fresh-water and brackish pelites, sands and coal layers—Kiscellian to Egerian. UA & LA.

Eocene.
Cerithian Beds: fossiliferous pelites, tidal-channel sands—Priabonian. UA.

Littoral to Inner Shelf

Miocene.
Rissoa Beds: silts, sands, pelites—Sarmatian. WB & LA.
Hollenburg–Karlstetten conglomerate: marine delta-complex conglomerate—early Badenian. LA.
Grund Beds: delta to prodelta conglomerate, sands, silts pelites—Karpatian to early Badenian. LA.
Riegersburg Formation: sands—Ottnangian. EBM.
Eggenburg Group:
  Kühnring Subformation, Burgschleinitz Formation, Loibersdorf Formation, Fels Formation: transgressive, fossiliferous sands—early Eggenburgian. EBM.

Gauderndorf Formation: fossiliferous pelitic sands—early Eggenburgian. EBM.

Zogelsdorf Formation: bioclastic limestone, abundant molluscs, corallinacea, bryozoa, echinoids, foraminifera—late Eggenburgian. EBM.

Oligocene.

Linz Sand: transgressive shoreface sands, influenced by the relief along faults with abrupt coastlines (in the area of Steyregg subfacies with Corallinacea—Late Egerian)—Kiscellian and Egerian. UA.

Melk Sand: transgressive shoreface sands with trace fossils (*Ophiomorpha*) channels in regressive phases—Kiscellian and Egerian. LA & EBM.

Eocene.

Lithothamnium Limestone: red algae limestone (Corallinaceae) with subsiding phases, intercalated sand bars and channels—Priabonian to early Oligocene. UA.

Ampfing Sandstone: ('Sandsteinstufe' in parts) shoreface, lagoonal, inundates, quartz arenite and arkoses – Priabonian. UA.

Miliolid Limestone: dark and light grey and brown limestones with abundant biogenic debris, bryozoa and foraminifera (miliolidae); restricted to imbricates and pebbles—Priabonian. UA.

Reingrub Formation: brown sandstones and light limestones with red algae and bryozoa—Priabonian. WB.

Haidhof Formation: fossiliferous calcareous sandstones and limestones—Lutetian. WB.

Waschberg Formation: nummulitic limestones, sandstones and marls; intercalated conglomerates with crystalline pebbles—Early Eocene (Ypresian). WB.

Palaeocene.

Zaya Formation: brown pelites with a limestone layer with red algae, bryozoa, bivalves and echinoids—Late Paleocene (Selandian. WB).

Bruderndorf Formation: fossiliferous pelites, sands, glauconitic sandstones and red algae limestones with corallinaceae—Early Palaeocene (Danian). WB.

Shelf

Miocene. Innviertel Group

Treubach Sand: glauconitic sand—Ottnangian. UA.

Braunau Schlier: silty pelites and thin glauconitic, micaceous sand layers—Ottnangian. UA.

Mehrnbach Sand: glauconitic quartz arenite with thin pelite layers—Ottnangian. UA.

Ried Beds: sandy, silty, micaceous pelites with thin sand layers—Ottnangian. UA.

Plesching Formation (including Phosphorite Sands, Fossiliferous Coarse Sand Series): transgressive tidal influenced fossiliferous sands with reworked phosphorite nodules—Ottnangian. UA.

Atzbach Sand north and northeast portion: subtidal sands influenced by tidal currents—Ottnangian. UA.

Enzenkirchen Sand (isolated equivalent to Atzbach Sand): micaceous sands, pelitic layers and pelite clasts in channels—Ottnangian. UA.

Kletzenmarkt Glauconitic Sand Formation (equivalent to Atzbach Sand): glauconitic sands—Ottnangian. UA.

Eggenburg Series: glauconitic sandstone—Eggenburgian. LA.

Oligocene.

Eferding Formation: ('Rupel Tonmergel, Älterer Schlier') dark grey silty shale with abundant terrigenous material and calcareous foraminifera—Kiscellian and Egerian. UA & LA.

Eocene. Perwang Group

Perwang Formation: ('Discocyclinen Mergel') Coquina of large foraminifera (*Discocyclina*) dark brown and green limestones and marls—Priabonian. UA.

Nummulitic Sandstone: dark and light grey, green and brown sandstones with abundant biogenic debris, algae, bryozoa and foraminifera (nummulites)—Priabonian. UA.

Outer shelf to slope

Miocene

Baden Tegel (Lower Lagenid Zone): greenish grey pelites with abundant microfauna—Early Badenian. WB.

Laa Beds: grey thin-bedded pelites with thin sand layers and gravels with marine microfauna, pteropods, burrowing bivalves and *nautilus*—Karpatian. LA & WB.

Innviertel Group

'Glauconitic Series' (water depth not proved): glauconitic sands, silts and pelites only in the W of Upper Austria and Salzburg—Ottnangian. UA.

Wachtberg Formation (sand gravel group): coarse massive gravel, sands in the W of Upper Austria and Salzburg—Ottnangian. UA.

Ottnang Schlier: sandy, silty, micaceous pelites—Ottnangian. UA.

Atzbach Sand: SW portion: sands, sand slides in channels—Ottnangian. UA.

Vöckla Beds: intercalation of micaceous sands and pelites.—Ottnangian. UA.

Robulus Schlier: sandy, silty micaceous pelites, sand ripples and layers—Ottnangian. UA & LA.

'Oncophora Formation' turbiditic time equivalent to Oncophora Formation: sands, silts, sandy pelites—Ottnangian. WB & LA.

Zellerndorf Formation: pelites, partially non calcareous, with foraminifera and open marine fish fauna—Ottnangian. EBM & LA.

Limberg Subformation: probably upwelling environment, diatomites—Ottnangian. EBM & LA.

Oligocene.

Ebelsberg Formation: ('Rupel Tonmergel, Älterer Schlier, Puchkirchener Serie') black to brown silty, soft pelites, partially bituminous and non-calcareous, phosphorite nodules, abundant fish remnants, agglutinating and rare calcareous foraminifera.

diatomites, bentomites and menilites, stratified and laminated nanoblooms. Upper Kiscellian and Egerian UA & LA.

Passive Slope

Oligocene.

Zupfing Formation: ('Rupel Tonmergel, Älterer Schlier, Puchkirchener Serie') hemipelagites (corresponding to Ebelsberg Formation) and distal turbidites from S of dark grey pelites with fish remnants and to the S increasing dark brown, green and grey limestones with exclusively nannofossils (nanoblooms); sands with high porosities are turbidites from N—Kiscellian and early Egerian. UA & LA.

Eggerding Formation (Banded Marl, in parts equivalent to Sitborice Formation): dark grey laminated pelites with thin white layers of nanoplankton (nanoblooms), often tectonized—Mid-Kiscellian. UA & LA.

Dynow Marlstone: ('Heller Mergelkalk') light marly limestone, nanno ooze—Middle Kiscellian. UA & LA.

Schöneck Formation: ('Lattorf Fischschiefer') dark grey or brown shaly thin-bedded limestones, abundant fish remnants, medium to deep-water calcareous and agglutinating foraminifera—Early Kiscellian. UA & LA.

Michelstetten Formation: light grey calcareous marlstones—Late Egerian. WB.

Thomasl Formation (equivalents in the W: Schöneck Fm, Eggerding Fm, Zupfing Fm, Dynow Marlstone): dark grey sandy pelites and sandstones with pebbles—Kiscellian to Early Egerian. WB.

Menilitic Formation (Dysodilen Beds): siliceous shales—Kiscellian. WB.

Eocene.

Pausram Beds (Pouzdrany Formation): globigerina and banded marls—Priabonian to early Oligocene. WB.

Perwang Group.
   Nussdorf Beds: Globigerina Marl with layers of dark shales—Priabonian to early Kiscellian. UA.
   Globigerina Limestone: brown and grey limestone abundant foraminifera (globigerina, ornamented uvigerina)—Priabonian. UA.

Tectonically active slope

Miocene
Hall Group
   Hall Formation: turbiditic sequence of light grey micaceous pelites, silts, sands and conglomerates, large agglutinating foraminifera—Eggenburgian. UA.
   Lukasedt Formation: turbidites and slides of light grey, dark brown and greenish pelites, sands, sandy and muddy conglomerates, originally deposited on top of imbricates, later in Eggenburgian and Ottnangian uplifted, tilted and partly moved down the slope—Early Eggenburgian. UA.
   Lindach Formation: turbidites and mass flows of light grey, micaceous pelites, silts, sands, sandy and muddy conglomerates often with calcite cement, slumps of dark brown soft pelites from Upper Puchkirchen Formation, almost exclusively reworked older fauna, rare autochthonous large agglutinating foraminifera—Early Eggenburgian. UA.
   Sandstreifen Schlier: turbidites and contourites of grey, micaceous, silty pelites with silt and sand lenses and layers—Eggenburgian and Ottnangian. LA.
   Ernstbrunn Formation: turbidites of pelites and sandstones—Eggenburgian and Ottnangian. WB.
   Buchberg Conglomerate: sandy conglomerate with pebbles and boulders from Flysch, Calcareous Alps and Crystalline basement—Eggenburgian. LA.

Oligocene

Puchkirchen Formation (Upper Portion): turbidites, contourites and mass flows of light and dark grey and brown micaceous pelites, silts, sands, sandy and muddy conglomerates, only deep water agglutinating foraminifera, intensively reworked fauna from N and S Molasse and Helveticum, excellent preserved single fossils from Cretaceous and Eocene—Late Egerian to Early Eggenburgian. UA.

Puchkirchen Formation (Lower Portion): turbidites, contourites and mass flows of light and dark grey and brown micaceous pelites, silts, sands sandy and muddy conglomerates, only deep-water agglutinating foraminifera (*Rhabdammina linearis*), intensively reworked fauna from N and S Molasse and Helveticum, excellent preserved single fossils from Cretaceous and Eocene—Early Egerian. UA.

Deutenhausen Formation: grey to greenish grey turbiditic sandstones with complete Bouma sequences and thin pelitic layers from the S, exclusively agglutinating foraminifera and reworked fauna from the S—Early Kiscellian. UA.

Basin plain to slope

Perwang Group
   Rogatsboden Formation: turbidites, contourites, mass flows and olistoliths of dark and light grey, micaceous, calcareous pelites, siltstones, sandstones, conglomerates and breccias, extremely high numbers of reworked Cretaceous and Eocene fauna from Helveticum and Southern Molasse, tiny agglutinating foraminifera, abundant globigerina—Kiscellian and Early Egerian. UA & LA

The author wishes to thank the management of Rohöl-Aufsuchungs AG for authorizing this publication, the colleagues of the geological and geophysical department and L. Leitner for his technical assistance. The author wants to express his thanks to D. Derksen for the discussions on 3D seismic and the revision of the English version, to G. Wessely for his advice and F. Rögl for his palaeontological input.

References

ABERER, F. 1958. Die Molassezone im westlichen Oberösterreich und in Salzburg. *Mitteilungen der Geologischen Geseuschaft in Wien*, **50**, (for 1957), 23–93.

BERGER, W. 1959. Die oberkarbonen Pflanzenreste der Bohrung Kastl1 bei Altötting/Obb. *Geologica Bavarica*, **40**, 3–8.

BERGGREN, W. A., KENT, D. V., SWISHER, C. C. III & AUBRY, M-B. 1995. *A revised geochronology and chronostratigraphy*. SEPM Special Publications **54**.

BRAUMÜLLER, E. 1961. Die paläogeographische Entwicklung des Molassebeckens in Oberösterreich und Salzburg. *Erdöl-Erdgas Zeitschrife*, **77**, 509–521.

BRIX, F. 1993. Molasse und deren Untergrund auf dem Sporn der Böhmischen Masse im Raum östlich Steyr-St. Pölten, westliches Niederösterreich. *In*: BRIX, F. & SCHULTZ, O. (eds) *Erdöl und Erdgas in Österreich*. Naturhistorisches Museum Vienna and F.Berger, Horn, 315–357.

——, KRÖLL, A & WESSELY, G. 1977. Die Molassezone und deren Untergrund in Niederösterreich. *Erdöl–Erdgas-Zeitschrife*, **93**, 12–35.

BRZOBOHATTY, R. & HEINRICH, M. 1990. New studies of the otoliths from the marine Ottnangian. *In*: MINARIKOVA, D. & LOBITZER, H. (eds) *Thirty years of geological cooperation between Austria and Czechoslovakia*. Federal Geological Survey, Vienna & Geological Survey of Prague, 245–249.

DECKER, K. & PERESSON, H. 1996. Tertiary kinematics in the Alpine – Carpathian – Pannonian System: links between thrusting transform faulting and crustal extension. *In*: WESSELY, G. & LIEBL, W. (eds) *Oil and Gas in the Alpidic Thrustbelts and Basins of Central and Eastern Europe*. EAGE Special Publications, **5**, 69–77.

DOHMANN, L. 1991. *Die unteroligozänen Fischschiefer im Molassebecken*. PhD Dissertation, Universität Munich.

DUNKL, I., FRISCH, W., KUHLEMANN, J. & BRÜGEL, A. 1996. 'Combined-Pebble-Dating': A new tool for provenance analysis and for estimating Alpine denudation. *Sedimentology*, 1996, **11**. Sedimentologentreffen 9–15. May 1996, Kurzfassungen, Institut für Geologie & Paläontologie Univesität Wien.

ELIAS, M. 1971. *Lithostratigraphicka a sedimentologicka charakteristika autochtonniho mesozoika v oblasti Jih*. MS. Geofond, Praha.

——, & WESSELY, G. 1990. The autochthonous Mesozoic on the eastern flank of the Bohemian Massif – an object of mutual geological efforts between Austria and CSSR. *In*: MINARIKOVA, D. & LOBITZER, H. (eds) *Thirty years of geological cooperation between Austria and Czechoslovakia*. Federal Geological Survey of Vienna & Geological Survey of Prague, 23–32.

FAUPL, P. & ROETZEL, R. 1987: Gezeitenbeeinflußte Ablagerungen der Innviertler Gruppe (Ottnangien) in der oberösterreichischen Molassezone. *Jahrbuch der Geologischen Bundesanstalt*, Wien, **130**, 415–447.

FODOR, F., 1995. From transpression to transtension: Oligocene – Miocene structural evolution of the Vienna basin and the East Alpine – Western Carpathian junction. *Tectonophysics*, **242**, 151–182.

FRASL, G. & FINGER, F. 1988. Führer zur Exkursion der Österreichischen Geologischen Gesellschaft ins Mühlviertel und in den Sauwald. *Reihe der Exkusionsfürer der Österreichen Geologischen Gesellschaft*, Wien, 1–30.

FUCHS, G. 1990. The Moldanubicum – an old nucleus in the Hercynian mountain ranges of Central Europe. *In*: MINARIKOVA, D. & LOBITZER, H. (eds) *Thirty years of geological cooperation between Austria and Czechoslovakia*. Federal Geological Survey of Vienna & Geological Survey of Prague, 256–262.

—— & MATURA, A. 1980. Die Böhmische Masse in Österreich. *In*: OBERHAUSER R. (ed.) *Der geologische Aufbau Österreichs*, Springer, Wien, 121–143.

FUCHS, R. & WESSELY, G. 1977. Die Oberkreide des Molasseuntergrundes im nördlichen Niederösterreich. *Jahrbuch der Geologischen Bundesanstalt*, Wien **120**, 426–436.

——, —— & SCHREIBER, O. 1984. Die Mittel- und Oberkreide des Molasseuntergrundes am Südsporn der Böhmischen Masse. *Schriftenreihe der Erdwissenschaften Kommission Akademie Wissenschaften*, Wien, **7**, 193–220.

GRILL, R. 1968. *Erläuterungen zur geologischen Karte des nordöstlichen Weinviertels und zu Blatt Gänserndorf*. GBA, Wien.

GRUBER, F. H. 1931. Geologische Untersuchungen im oberösterreichischen Mühlviertel. *Mitteilungen Geologischen Gesellschaften*, Wien, **23**, (for 1930), Wien, 35–84.

GRÜN, W. 1984. Die Erschließung von Lagerstätten im Untergrund der alpin-karpatischen Stirnzone Niederösterreichs. *Erdöl und Erdgas*, **100**, 292–295.

HOFMANN, T. 1990. Der Ernstbrunner Kalk (Tithon im Raum Dörfles (Niederösterreich 9: Mikrofazies und Kalkalgen. *Nachrichten Deutsche Geologische Gesellschaft*, **43**, 45–46.

HEJL, E. & Grundmann, G. 1989. Apatit – Spaltspurendaten zur thermischen Geschichte der Nördlichen Kalkalpen, der Flysch- und Molassezone. *Jahrbuch der Geologischen Bundesanstalt*, Wien, **132**, 191–212.

JIRICEK, R. & SEIFERT, P. 1990. Paleogeography of the Neogene in the Vienna Basin an the adjacent part of the Foredeep. *In*: MINARIKOVA, D. & LOBITZER, H. (eds) *Thirty years of geological cooperation between Austria and Czechoslovakia*. Federal Geological Survey, Vienna & Geological Survey of Prague, 89–105.

KRATOCHVIL, H. & LADWEIN, H., 1984. Die Muttergesteine der Kohlenwasserstofflagerstätten im Wiener Becken und ihre Bedeutung für die zukünftige Exploration. *Erdöl- Ergas- Zeitschrift*, **100**, 107–115.

KRENMAYR, H. G. 1991. Sedimentologische Untersuchungen der Vöcklaschichten (Innviertler Gruppe, Ottnangien) in der oberösterreichischen

Molassezone im Gebiet der Vöckla und Ager. *Jahrbuch der Geologischen Bundesanstalt*, Wien, **134**, 83–100.

KREUTZER N. 1993. Die ÖMV-Gas- und Öllagerstätten der nieder- und oberösterreichischen Molassezone. *In*: BRIX, F. & SCHULTZ, O. (eds) *Erdöl und Erdgas in Österreich*. Naturhistorisches Museum Vienna and F. Berger, Horn, 455–465.

KOLLMANN, H. A., BACHMAYER, F., NIEDERMAYER, G., SCHMID, M. E., KENNEDY, W. J., STRADNER, H., PRIEWALDER, H., FUCHS, R. & WESSELY, G., 1977. Beiträge zur Stratigraphie und Sedimentation des Festlandsockels im nördlichen Niederösterreich. *Jahrbuch der Geologischen Bundesanstalt*, Wien, **120**, 401–447.

KOVAR, J. B. 1982. Eine Blätter-Flora des Egerien (Ober-Oligozän) aus marinen Sedimenten der Zentralen Paratethys im Linzer Raum (Österreich). *Beitrage Paläontogische Österreich*, **9**, 1–209.

KÜPPER, I. 1964. Mikropaläontologische Gliederung der Oberkreide des Beckenuntergrundes in den oberösterreichischen Molassebohrungen. *Mitteilungen Geologischen Gesellschaften*, Wien, **56**, (for 1963), 591–651.

LADWEIN, W. 1976. *Sedimentologische Untersuchungen an Karbonatgesteinen des autochthonen Malm in NÖ (Raum Altenmarkt – Staatz)*. Diss. Phil. Fak. Univ. Innsbruck.

—— 1988. Organic Geochemistry of the Vienna Basin: Modelfor hydrocarbon generation in overthrustbelts. *American Association of Petroleum Geologists Bulletin* **72**, Tulsa, 5 59–586.

——, MALZER, O. & WESSELY G. 1993. KW-höffige Gebiete in Österreich. IN: BRIX, F. & SCHULTZ, O. (eds) *Erdöl und Erdgas in Österreich*. Naturhistorisches Museum Vienna and F.Berger, Horn, 472–477.

MACKENBACH, R. 1984. *Jungtertiäre Entwässerungsrichtungen zwischen Passau und Hausruck (O. Österreich)*. Geologisches Institut Universität Köln, Sonderveröffntlichungen, **55**, 1–175.

MALZER, O. 1993. Muttergesteine, Soeichergesteine, Migration und Lagerstättenbildung in der Molassezone und deren sedimentärem Untergrund. *In*: BRIX, F. & SCHULTZ, O. (eds.) *Erdöl und Erdgas in Österreich*. Naturhistorisches Museum Vienna and F.Berger, Horn, 302–315.

——, RÖGL, F., SEIFERT, P., WAGNER, L., WESSELY, G. & BRIX, F. 1993. Die Molassezone und deren Untergrund. *In:* BRIX, F. & SCHULTZ, O. (eds) *Erdöl und Erdgas in Österreich*. Naturhistorisches Museum Vienna and F. Berger, Horn, 281–301.

MAIR, V., STINGL, V.& KROIS, P. 1992. Andesitgerölle im Unterinntaler Tertiär. Geochemie, Petrographie und Herkunft. *Mitteilungen Österreichische Mineralogische Gesellschaften*, **137**, 168–170.

MEYER, R. 1989. Schrägbohrungen durch die Aufschleppungszone von Taxöldern – Pingarten. *Erlanger geologisches Abhundlungen*, **117**, 25–34.

—— & SCHMIDT-KALER, H. 1984. Erdgeschichte sichtbar gemacht-ein geologischer Führer durch die Altmühllap. Bayerisches Geologisches Landesamt, Munich (2nd. ed.).

—— & —— 1987. Der Jura in neuen Bohrungen in der Umgebung von Regensburg. *Geol. Bl. NO-Bayern*, **37**, 185–216.

NACHTMANN, W. 1995. Bruchstrukturen und ihre Bedeutung für die Bildung von Kohlenwasserstoff-Fallen in der oberösterreichischen Molasse. *Geologische Paläontologisches Mitteilungen Innsbruck*, **20**, 221–230.

—— & WAGNER, L. 1987. Mesozoic and Early Tertiary evolution of the Alpine foreland in Upper Austria and Salzburg, Austria. *Tectonophysics*, **137**, 61–76.

PARISH, J. T. 1982. Upwelling and petroleum source beds with reference to Paleozoic. *American Association of Petroleum Geologists Bulletin*, **66**, 750–774.

PILLER, W. E. & Rasser, M. 1996. Rhodolith formation induced by reef erosion in the Red Sea, Egypt. *Coral Reefs*, **15**.

POLESNY, H. 1983. Verteilung der Öl- und Gasvorkommen in der oberösterreichischen Molasse. – *Erdöl-Erdgas-Zeitschrift*, **99**, 90–102.

RASSER, M. & PILLER, W. E. 1996. Kalkalgen aus dem obereozänen "Lithothamnienkalk" der Molassezone Oberösterreichs. *In: Sediment 1996*, **11**. Sedimentologentreffen 9–15 May 1996, Kurzfassungen, Institut für Geologie & Paläontologie Universität, Wien, 142.

RATSCHBACHER, L., FRISCH, W. & LINZER, H.-G. 1991. Lateral extrusion in the eastern Alps, part II: structural analysis. *Tectonophysics*, **10**, 257–271.

ROETZEL, R. 1993. Bericht 1992 über geologische Aufnahmen im Tertiär und Quartär auf Blatt 8 Geras und Bemerkungen zur Lithostratigraphie des Tertiärs in diesem Raum. *Jahrbuch der Geologischen Bundesanstalt*, Wien, **136**, 542–546.

RÖGL. F. 1994. Globigerina ciperoensis (Foraminiferida) in the Oligocene and Miocene of the Central Paratethys. *Annalen der Naturhistorische Museum Wien*, **96**, Wien, pp. 133–159.

—— & STEININGER, F. 1996. Miogypsina (Miogypsinoides) formosensis Yabe & Hanzawa, 1928 (Foraminiferida) aus den Linzer Sanden (Egerian, Oberoligozän) von Plesching bei Linz. *Mitteilungen Geologische Gesellschaft*, **62**, 46–54.

—— & —— 1983. Vom Zerfall der Tethys zu Mediterran und Paratethys. *Annalen der Naturhistorische Museum Wien*, **85**, 135–163.

——, ZAPFE, H. & 10 OTHERS. 1993. Die Primatenfundstelle Götzendorf an der Leitha (Obermiozän des Wiener Beckens, Niederösterreich). *Jahrbuch der Geologischen Bundesanstalt, Wien*, **136**, 503–526.

ROYDEN, L. H. 1988. Late Cenozoic tectonics of the Pannonian basin system. *In*: ROYDEN H. L. & HORVATH, F. (eds) *The Pannonian Basin—A study in basin evolution*. American Association of Petroleum Geologists Memoirs, **45**, 27–48.

SACHSENHOFER, R. F. 1992. Coalification and thermal histories of Tertiary basins in relation to late Alpidic evolution of the Eastern Alps. *Geologische Rundschau*, **81**, 291–308.

SALVENMOSER, S. & WALSER, W. 1991. Lithostratigraphische Untersuchungen an jungtertiären Molassesedimenten am Nordrand der Taufkirchener

Bucht (Oberösterreich). *Jahrbuch der Geologischen Bundesanstalt, Wien*, **134**, 134–147.
SAUER, R., SEIFERT, P. & WESSELY, G. 1992. Guidebook to Excursions in the Vienna Basin and the Adjacent Alpine-Carpathian Thrustbelt in Austria. *Mitteilungen Geologischen Gesellschaften Wien*, **85**, 239.
SCHMIDT, F. & ERDOGAN, E. 1996. Paleohydrodynamics in exploration. *In*: WESSELY, G. & LIEBL, W. (eds) *Oil and Gas in the Alpidic Thrustbelts and Basins of Central and Eastern Europe*. EAGE Special Publications, **5**. London, 255–266.
SCHRÖDER, B. 1987. Inversion tectonics along the western margin of the Bohemian Massif. *Tectonophysics*, **137**, 93–100.
SEIFERT, P. 1993. Die Waschbergzone. *In*: BRIX, F. & SCHULTZ, O. (eds) *Erdöl und Erdgas in Österreich*. Naturhistorisches Museum Vienna and F. Berger, Horn, 358–359.
SHANMUGAM, G., LEHTONEN, L. R., STRAUME, T., SYVERTSEN, S. E., HODGKINSON R. J. & SKIBELI, M. 1994. Slump and Debris-Flow Dominated Upper Slope Facies in the Cretaceous of the Norwegian and Northern North Seas (61-67° N): Implications for Sand Distribution. *AAPG Bulletin*, **78**, 910–937.
——, SPALDING, T. D. & ROFHEART, D. H. 1993. Process Sedimentology and Reservoir Quality of Deep-Marine Bottom-current Reworked Sands (Sandy Contourites): An Example from the Gulf of Mexico. *American Association of Petroleum Geologists Bulletin*, **77**, 1241–1259.
STEININGER, F., 1969. Das Tertiär des Linzer Raumes. *In*: PODZELT, W. & STEININGER, F., (eds) *Geologie und Paläontologie des Linzer Raumes*. Stadtmuseum Linz und oberösterr. Landesmus. **64**, 35–53.
—— & RÖGL, F. 1979. The Paratethys History – A contribution towards the Neogene geodynamics of the Alpine Orogene (an abstract). *Annales géologique des pays Hellèniques*, (hors serie) **3**, 1153–1165.
—— & SENES, J. 1971. M1 Eggenburgian. Die Eggenburger Schichtengruppe und ihr Stratotypus. *Chronostratigraphie und Neostratotypen*, **2**, Bratislava, 827.
——, ROETZEL, R. & RÖGL, F. 1991. Die tertiären Molassesedimente am Ostrand der Böhmischen Masse. *In*: ROETZEL, R. & NAGEL, D. (eds) *Exkursionen im Tertiär Österreichs*. Österreiches Paläontologisha Gesellschaft, Wien, 63–141.
——, WESSELY, G., RÖGL, F. AND WAGNER, L. 1987. Tertiary sedimentary history and tectonic evolution of the Eastern Alpine Foredeep. *Giornale de Geologie* ser. 3, **48**, 285–297.
STINY, J, 1926. Messungen in den Poschacher Steinbrüchen bei Mauthausen. *Jahrbuch der Geologischen Bundesanstalt, Wien*.
TOLLMANN, A. 1977. Die Bruchtektonik Österreichs im Satellitenbild. *Neues Jahrbuch für Geologie und Paläontogie*, **153**, 1–27.
VASICEK, W. 1983. Permfossilien. 280 Millionen Jahre alte Spuren der Steinkohlenwälder von Zöbing. *In*: *Exhibition catalog Krahuletz*. Museum, Eggenburg, 15–50.

WAGNER, L. 1980. Geologische Charakteristik der wichtigsten Erdöl- und Erdgasträger der oberösterreichischen Molasse. Teil I: Sandsteine des Obereozän. *Erdöl-Erdgas-Zeitschrift*, **96**, 338–346.
—— 1996a. Stratigraphy and hydrocarbons in the Upper Austrian Molasse Foredeep (active margin). *In*: WESSELY, G. & LIEBL, W. (eds) *Oil and Gas in the Alpidic Thrustbelts and Basins of Central and Eastern Europe*. EAGE Special Publications, **5**, London. 217–235.
—— 1996b. Die tektonisch – stratigrafische Entwicklung der Molasse und deren Untergrundes in Oberösterreich und Salzburg. *In*: EGGER, H., HOFMANN, TH. & RUPP, Ch. (eds) *Ein Querschnitt durch die Geologie Oberösterreichs*. Reihe der Exkursionsfürer der Österreiches Geologischen Geselschaften, Wien, 36–65.
—— & Wessely, G. 1993. Molassezone Österreichs – Relief und Tektonik des Untergrundes. *In*: BRIX, F. & SCHULTZ, O. (eds) *Erdöl und Erdgas in Österreich*. Naturhistorisches Museum Vienna and F. Berger, Horn, subcrop map.
——, Kuckelkorn, K. & Hiltmann, W. 1986. Neue Ergebnisse zur alpinen Gebirgsbildung Oberösterreichs aus der Bohrung Oberhofen1-Stratigraphie, Fazies, Maturität und Tektonik. *Erdöl-Erdgas-Zeitscht*, **102**, 12–19.
WALLBRECHER, E., BRANDMAYR, M., HANDLER, R., LOIZENBAUER, J. & DALLMEYER, R. D., 1996. Konjugierte Scherzonen in der südlichen Böhmischen Masse. *In*: EGGER, H., HOFMANN, TH. & RUPP, CH. (eds) *Ein Querschnitt durch die Geologie Oberösterreichs*. Reihe der Exkursionsfürer der Österreiches Geologischen Gesellschaften, Wien, 12–28.
WESSELY, G. 1993. Der Untergrund des Wiener Beckens. *In*: BRIX, F. & SCHULTZ, O. (eds) *Erdöl und Erdgas in Österreich*. Naturhistorisches Museum Vienna and F. Berger, Horn, 249–280.
—— & Wagner, L., 1993: Die Nordalpen. *In*: BRIX, F. & SCHULTZ, O. (eds) *Erdöl und Erdgas in Österreich*. Naturhistorisches Museum Vienna and F. Berger, Horn, 360–370.
——, SCHREIBER, O. S. & FUCHS, R. 1981. Lithofazies und Mikrostratigraphie der Mittel- und Oberkreide des Molasseuntergrundes im östlichen Oberösterreich. *Jahrbuch Geologischen Bundesanstalt, Wien*, **124**, 175–281.
WEHNER, H. & KUCKELKORN, K. 1995. Zur Herkunft der Erdöle in nördlichen in Alpen-/Karpatenvorland. *Erdöl Erdgas Kohle*, **12**, 508–514.
XU, J. 1991. *Inkohlungsuntersuchungen im oberösterreichischen Molassebecken und seinem prätertiären Untergrund*. Dipl. thesis, Montanuniv. Leoben, Leoben.
ZIEGLER, P. A. 1982. *Geological Atlas of Western and Central Europe*. Shell Internationale Petroleum Maatschappje BV., The Hague.
—— 1987. Late Cretaceous and Cenozoic intra-plate compressional deformations in the Alpine foreland—a geodynamical model. *Tectonophysics*, **137**, 389–420.

Automation of stratigraphic simulations: quasi-backward modelling using genetic algorithms

STEFAN BORNHOLDT[1] & HILDEGARD WESTPHAL[2]

[1]*Institut für Theoretische Physik, Universität Kiel, Leibnizstrasse 15, 24098 Kiel, Germany (e-mail: bornholdt@theo-physik.uni-kiel.de)*
[2]*GEOMAR, Wischhofstrasse 1-3, 24148 Kiel, Germany (e-mail: hwestphal@geomar.de)*

Abstract: Stratigraphic modelling involves a multi-dimensional parameter-fitting process where a large number of free model parameters are determined by hand in an iterative sequence of educated guesses. This problem can be viewed as an optimization task. It is solved here using a genetic algorithm. The iterative trying and checking process, which previously had to be done manually, is thereby automated. This method is demonstrated to be consistent for a completely defined toy model. Furthermore it is applied to a real problem involving the Eocene to Miocene succession of the Bavarian Molasse Basin.

Since the late 1980s, computer-based stratigraphic basin-fill simulations have become a common tool in basin analysis. Simulations are employed to improve the understanding of the interactions of parameters influencing sedimentation (e.g. Aigner *et al.* 1989, 1990; Lawrence *et al.* 1990; Featherstone *et al.* 1991). Also, modelling is frequently used to evaluate and quantify data derived from seismic interpretation, subsidence analysis, and other investigations (e.g. Eberli *et al.* 1994). These data serve as input parameters for stratigraphic modelling simulating the sedimentation processes through time. To obtain reasonable results, the input data have to be as rigorously constrained as possible. Usually, however, the input data (like sea-level variations from coastal onlap curves, sediment input and subsidence curves from well data) can be defined only within a range, e.g., due to uncertainties in biostratigraphic dating or in the determination of palaeo-bathymetry. Modelling can be utilized to restrict further the range of these parameters.

Modelling programs usually employ forward modelling, i.e. the chronological succession is simulated through time with a chosen set of input parameters. A major problem in computer-based stratigraphic modelling is the necessity to vary the input parameters by hand in order to obtain a close fit of the model with the real world as seen on seismic sections or in well data. After each simulation run, the output is compared with the geological reality and checked for consistency. The input parameters are then modified manually within a reasonable range and the modelling program is run again. Since the input parameters interact in a complex way, it is often difficult to estimate the effect of even small variations in the input. Improving the stratigraphic model manually is a trial and error process of educated guessing (cf. Lawrence *et al.* 1990).

A technique is needed to automate the search for optimal combinations of input parameters. For this task, the application of genetic algorithms to the stratigraphic modelling procedure is introduced here. The genetic algorithm performs a global search for models that exhibit the desired properties given by seismic sections or well logs from the basin to be modelled.

An output that matches to a certain accuracy the geometries observed in the geological data is often just one of many possible solutions. A number of solutions with different input parameters may produce similar outputs. Therefore, the output of each genetic algorithm procedure is a proposed, possible optimal fit and has to be checked for consistency with the known geological context. The search method by genetic algorithms is meant to assist the geologist in the modelling process.

Genetic algorithms

Quasi-backward modelling of a stratigraphic succession can be viewed as an optimization task where a set of initial parameters is looked for that leads to the best possible final results of the (forward-) simulation. Such a problem can be solved with general methods of optimization. A robust algorithm for optimization of practical problems is the genetic algorithm (Holland 1975; Goldberg 1992). It is an optimization method that is inspired by principles of biological evolution: selection, reproduction, mutation and crossover (Fig. 1). In order to optimize a set of input parameters of the geological model (e.g. subsidence rates, sea-level changes) these are coded in a sequence of numbers. Such a string of

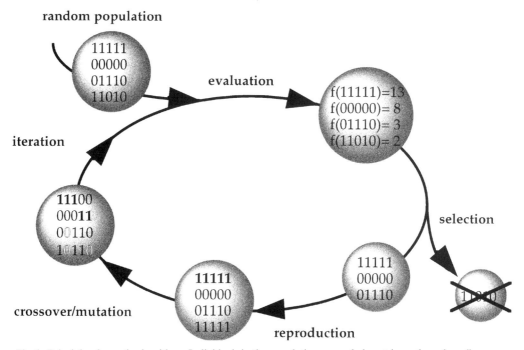

Fig. 1. Principle of genetic algorithms. Individuals in the population are coded as strings of numbers (here: four individuals containing five numbers each). A population of random individuals is evaluated using a fitness function. Selection of the high fitness individuals (exclusion of the low fitness individuals) and reproduction generally increase the mean fitness of the population. Mutation and crossover operators randomly change a number of individuals to increase the fitness variance in the population.

numbers is the analogon of the genome of a biological organism and will undergo changes in the course of the subsequent optimization process. A key ingredient is a function that measures the performance of such an individual model in the light of the geological data that awaits explanation. This function represents the 'fitness' of an individual. It is measured in practice by comparing predictions from the model to geological data (e.g. lithologies, thicknesses, porosities).

The optimization procedure starts with a certain number of individual models, the initial population. It is produced with random strings of numbers, i.e. the input parameters vary randomly within predefined ranges. The individuals of the population then struggle for the 'survival of the fittest'. They are evaluated by comparison of the output (the stratigraphic model) with the target data (e.g. from borehole data, seismic profiles). The fitness evaluation of the individuals has been automated and incorporated into the program. The individual model performing the worst is then deleted while a copy of the string with the highest fitness is added. The size of the population thus remains constant. While the original is kept, the copy is changed using two operators; (1) the mutation operator changes one of the numbers in the string randomly; (2) the crossover operator mixes the copy with one of the other individuals of the population by cutting the strings at a random position and recombining the parts. These operators provide new combinations of numbers in the strings that contain random elements but have a close relationship to their parent strings. The new number string is run through the stratigraphic modelling program, evaluated, and inserted into the population replacing the (deleted) least fit. This is called the next generation. With a certain probability, this new generation of strings will perform better than the preceding generation. This optimization procedure is repeated until the output of the modelling program matches the given geological data (well logs, seismic profiles etc.).

Stratigraphic modelling

The stratigraphy of sedimentary basins reflects the interaction of two basic groups of geological processes: those controlling the creation and destruction of accommodation space in a basin and those regulating the influx and removal of sediment (Lawrence et al. 1990). Space is generated by tectonic subsidence, isostasy, sediment compaction and eustatic sea-level rise. It is destroyed by uplift, infill of the basin and eustatic sea-level fall. Sediment input is a function of tectonic processes (uplift of the source area), climate (weathering) and biogenic production.

To simulate a stratigraphic succession, several parameters have to be given which define the modelling set-up. The accommodation space is determined by (1) a eustatic sea-level curve and (2) tectonic subsidence. Sediment input is given by (3) sediment input curves through time. Palaeo-topography is included by (4) the initial bathymetry (initial sedimentation interface). A number of additional parameters could be included. Among those are the location of the sediment source, the distance suspended sediment is transported and many others. To illustrate the general processes and abilities of genetic algorithms, in this study three parameters sets are subjected to the optimization process: eustatic sea-level amplitude, subsidence rates and the initial bathymetry of the sedimentary basin. These parameters are defined as a string of numbers to be changed by the genetic algorithm. Sediment supply curves are kept constant but will be included as a variable in further versions of the genetic algorithm.

For this optimization study, the three-dimensional modelling program Fuzzim by Nordlund & Silfversparre (1994) is employed. Fuzzim is a multi-lithological modelling program for simulating large-scale marginal deposition and erosion over geological time spans. Besides a three-dimensional output, two-dimensional strike and dip profiles, and one-dimensional pseudo-wells can be obtained. This program employs a fuzzy logic approach to include qualitative geological data in the modelling (Nordlund 1996). The optimization procedure represented here is not specific to this choice of a stratigraphic modelling program and can be applied to any other program as well. For the genetic algorithm the implementation of Ross (1994) is used which is available as free software. Both programs have been linked together by an 'auto-run-facility' that repeatedly runs Fuzzim, sets parameters, and extracts values automatically. A complete Fuzzim-run is made for each generation of the genetic algorithm.

In this pilot study, the three parameters to be varied are implemented as follows. (1) An approximate initial bathymetry is imported as a digital file. Here, it is defined by a dipping surface which is constant in strike direction. The genetic algorithm is programmed to explore a vertical shift of this initial surface within a range of 40 m as well as an additional tilt in dip direction changing the elevation of the surface at the down-dip end within 40 m (keeping the hinge line fixed). (2) Subsidence curves for the four corners of the model also are imported. Factors (here between 0.0 and 2.0) changing the amount of subsidence for each corner are explored by the genetic algorithm. (3) Finally, a preliminary sea-level curve is predefined. Since the shape of a sea-level curve of an area often is more clearly defined than its amplitude, the genetic algorithm is programmed to vary the amplitude but not the shape of any imported sea-level curve (here within a range of about 30%).

These parameters are encoded in a total of seven numbers, which are changed by the algorithm (i.e. the genomes of the models contain seven numbers): two for the initial bathymetry, four for subsidence, and one for sea-level. We use a binary encoding of the numbers, each to five bit accuracy, i.e. in 2^5 discrete steps within the allowed ranges. Each model is thus encoded as a string of 35 bits. The algorithm then explores a space of $(2^5)^7$ or about 10^9 different stratigraphic models.

In order to evaluate the results of each individual model, output parameters produced by the model (taken from pseudo-wells: thickness of sediment deposited per time-step and lithology as sand/clay-content) are compared to the true values from real wells in corresponding locations. The data employed for comparison are one-dimensional at each well, but by distributing the wells on the two-dimensional area, a three-dimensional control on the model is obtained. The quality of any of the automatically tested models with respect to the given borehole data is defined by a single number, the 'fitness'. Here, the fitness is defined as the negative root mean square error of the deviations of the modelled thicknesses and lithologies from the desired values at the testing points (pseudo-well locations). These testing points are probed at equidistant time slices (corresponding to the modelling time steps) of each well.

A toy model

To test the general abilities of the optimization process, a toy model was produced with the stratigraphic modelling program Fuzzim using an arbitrary set of input parameters. This model was to be reproduced by the genetic algorithm by reconstructing the original set of input parameters as a result of the genetic algorithm optimization process.

A siliciclastic toy model was chosen which is

characterized by a gentle initial slope dipping seaward from 20 m above sea level down to 160 m below sea level. A river mouth at the shoreline acts as sediment source. Ten by ten (=100) cells sum up to a square area of 400 km². The modelling time interval covers 200 ka divided into ten equal time steps of 20 ka.

Eustatic sea-level is defined as a periodic curve with an amplitude of 100 m, beginning with a sea-level drop followed by a sea-level rise. Thereby, the upper part of a depositional sequence (highstand), and part of the succeeding sequence from lowstand via transgression to highstand are modelled. Constant subsidence rates are defined at

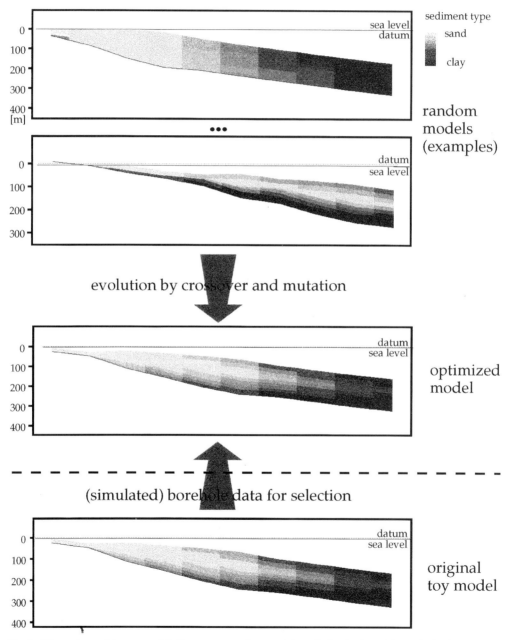

Fig. 2. Dip sections of the toy model. On top two random models at the beginning of the optimization process, below the optimized model. Bottom: the original toy model that was to be reproduced.

the four corners of the model, varying from 0.1 to 0.5 mm a^{-1}. Sediment distribution reacts to sea-level and subsidence interaction with progradation followed by retrogradation. The variation from finer to coarser (progradation) and back to finer sediment (retrogradation) is seen best on dip sections in close vicinity to the sediment source (Fig. 2, bottom). In contrast, basinal sedimentation only exhibits subtle changes in lithology through time. Six pseudo-wells have been positioned in different settings to get a three-dimensional control (Fig. 3).

The fitness f of a model is defined by

$$f = -\sqrt{\alpha \sum_{i,t}[l(i,t) - l_{opt}(i,t)]^2 + \beta \sum_{i,t}[d(i,t) - d_{opt}(i,t)]^2}$$

(1)

where l denotes the model values and l_{opt} the desired values of lithology, defined as percent clay content. Similarly, d denotes the thickness of the sediment column at a certain time. The squared deviations between modelled and real values are summed over the individual modelling time steps t and all pseudo-well locations i. The weights α and β are factors to ensure that the typical errors contributing from lithology and from thickness have the same order of magnitude. For the model used here, this is satisfied with $\alpha = 100$ and $\beta = 1$.

A population of 100 random individuals was optimized in 10 000 generations (about 1 hour computing time on a state-of-the-art desktop computer). The optimized model with the highest fitness value thus obtained is almost identical in geometries and lithologies compared to the original toy model (Fig. 2). The optimized pseudo-wells exhibit very similar lithologies to the original pseudo-wells with the fractions of clay and sand deviating from the desired output by less than 10%. The modelled initial surface is almost ideal with a maximum deviation of 1.2% of the total amplitude. Sea level and subsidence rates also display a close fit (Fig. 4). This shows that the pseudo-wells chosen contain the information necessary for reconstructing the underlying model. More accurate results can be obtained by increasing the number of wells or the number of generations in the optimizing process. In runs with a smaller number of pseudo-wells, the number of possible solutions increases since the model is less constrained, leading to a higher degree of uncertainty.

In general modelling situations, one should consider the resulting high fitness solution of each optimization procedure to be one of possibly several solutions from which the modeller chooses the most consistent model with respect to other geological data or knowledge.

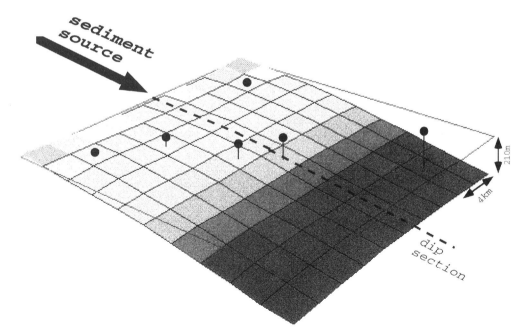

Fig. 3. Upper surface (top of simulated unit) of the toy model with the locations of pseudo-wells and dip section shown in Fig. 2.

376 S. BORNHOLDT & H. WESTPHAL

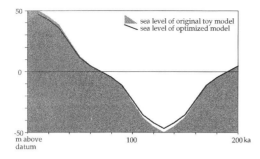

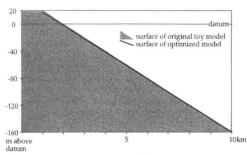

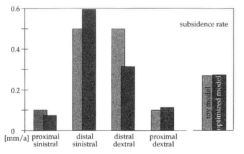

Fig. 4. Comparison of the reconstructed data of the genetic algorithm with the known input data of the toy model. Sea-level curve and initial sedimentary interface (dip section shown in Fig. 3) of the optimized stratigraphic model are close to the toy model data. Optimized subsidence rates of the proximal corners (i.e. up-dip corners) of the model show only slight deviations from the known, original, values. The distal corners (i.e. far from shore line), however, differ more strongly from the original rates. The value of the distal sinistral corner (left with view from source) deviates within the same percentage range as the proximal corners. The distal dextral corner in contrast shows a stronger deviation as a result of this corner being less controlled by pseudo-wells.

The Molasse model

To examine the applicability of the genetic algorithm to a real world problem, a model was run for the Molasse Basin. The Molasse Basin represents a difficult problem for stratigraphic modelling. Sediment influx is dominantly directed parallel to the basin axis. Modelling programs, however, often assume a perpendicular sediment input.

The study area is located in the Bavarian Molasse Basin to the West of Munich (Fig. 5). During the Tertiary Period (late Eocene to late Miocene; 40 to 8 Ma), the sedimentary environment of this foreland basin was characterized by a slope dipping from north to south, while sediment input was dominantly directed towards the east.

Input and well data were taken from Jin (1995). Jin has been evaluating subsidence and seismic interpretation data using a two-dimensional modelling program. As preliminary input data for the optimization presented here, subsidence rates of four wells, and the initial bathymetry from his study were used. Sea-level changes were taken from Haq *et al.* (1987), as they have been shown to be applicable to the Molasse Basin by Jin (1995). An area of 50 × 50 km divided into ten by ten cells is modelled. 32 time steps of 1 Ma cover the time span of the late Eocene to the late Miocene. Stratigraphic columns of four wells (A, C, F, and W4, Fig. 6) are taken as a control on the output of the genetic algorithm. In the simulated model, four pseudo-wells have been positioned in the locations of these real wells. To evaluate the fitness of the individual models, the stratigraphy of the real wells and the pseudo-wells are compared. In our case, sediment thickness values and lithologies are compared for each well

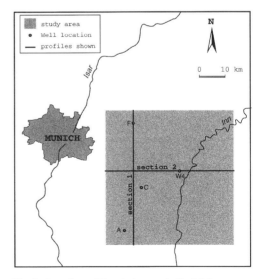

Fig. 5. Location of the study area in the Molasse Basin. The grey square is the area modelled by the optimization program. (Modified from Jin 1995.)

divided into 32 time slices corresponding to the 32 modelling time steps. The fitness is calculated as defined above (1), here using the weights $\alpha = 1$ and $\beta = 10$. The input parameters are then varied by a genetic algorithm within a population of 100 individual models for 1000 generations.

The optimized model (Fig. 7) produced by the genetic algorithm exhibits overall geometries similar to the geometries expected on the basis of seismic data from Jin (1995). Due to the relative shallowing on top of the interval modelled, coarsening-up takes place that finally leads to the formation of the coarse-grained Upper Freshwater Molasse. In the optimized model, sandstone bodies are deposited as the last marine sediments and the first freshwater influenced formations thus reflecting this tendency. Areas that have good control such as in the vicinity of wells F and C are closer to the desired output than areas with less control. In the distal right corner, the subsidence curve deviates from expected values because there are not enough constraints. Since in many study areas only few wells are available, seismic data should be included in stratigraphic modelling optimization. This could be done by coding seismic profiles as pseudo-wells with thicknesses and ages from correlated wells but omitting lithologies.

The results for the Molasse Basin problem (Fig. 7) have been obtained from only 1000 generations (iterations) of the optimization procedure (20 minutes of computing time). Higher accuracy can be obtained by increasing the number of generations, as well as employing simulations with finer space and time intervals.

Outlook

Optimization with genetic algorithms may prove a valuable tool in stratigraphic modelling. The modelling procedure is thereby automated and much less time intensive than conventional forward modelling. Genetic algorithms provide a tool to quickly find promising regions in the large input parameter space. For systems weakly constrained by well data the application of genetic algorithms on stratigraphic modelling produces a variety of possible solutions. This offers the experienced modeller the chance to optimize existing models in narrower parameter ranges. If several models emerge from the optimization process ('local optima') the decision which model is most reasonable in the geological context ('global optimum') has to be taken by the modeller.

To enhance the ability of the genetic algorithm to exactly reproduce the geometries of a basin fill, seismic data could be included into the data set as pseudo-wells containing information on the thicknesses and ages of the stratigraphic units. This also would enable the genetic algorithm to reproduce sedimentary basins with more complex geometries.

The optimization procedure described here allows for an arbitrary number of testing data points to compare a three-dimensional stratigraphic model (output) with geological data. However, the number of input parameters to be varied by the genetic algorithm should be kept moderate when using a binary encoding. This keeps the optimization process fast and

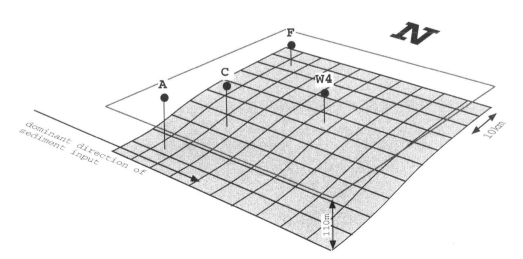

Fig. 6. Initial surface of the Molasse model and locations of the wells.

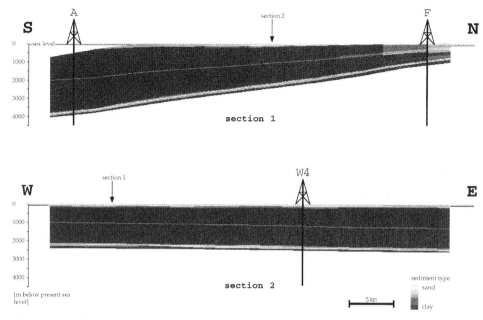

Fig. 7. Optimized dip and strike sections of the Molasse model.

accurate. For more complex model optimization, e.g. tuning an initial bathymetry asymmetrically in two dimensions, complex sea-level curves, or large numbers of model parameters, one could apply the algorithm in several stages varying different subsets of parameters. Other forms of algorithms using strings of real number parameters might allow for the tuning of larger parameter sets (e.g. Surry & Radcliffe 1996).

This study was aimed to investigate the applicability of genetic optimization on stratigraphic modelling. The results of the Molasse model are only of preliminary character and do not claim to be more than a demonstration of the new method. Further investigations incorporating more detailed data are required to obtain reliable results.

We are grateful to U. Nordlund and M. Silfversparre (Uppsala University) for the code of their program Fuzzim, and to T. Aigner, H.-P. Luterbacher, and J. Jin (Tübingen University) for making the Molasse data available to us. For their reviews thanks are due to U. Nordlund (Uppsala), P. Joseph (IFP Paris), T. Aigner (Tübingen), J. Reijmer and S. Kinsey (GEOMAR Kiel).

References

AIGNER, T., BRANDENBURG, A., van VLIET, A., DOYLE, M., LAWRENCE, D. & WESTRICH, J. 1990. Stratigraphic Modelling of Epicontinental Basins: Two applications. *Sedimentary Geology*, **69**, 167–190.

——, DOYLE, M., LAWRENCE, D. T., EPTING, M. & van VLIET, A. 1989. Quantitative modelling of carbonate platforms—some examples. *In*: CREVELLO, P. D., WILSON, J. L., SARG, J. F. & READ, J. F. (eds) *Controls on carbonate platform and basin systems*. SEPM Special Publications, **44**, 27–37.

EBERLI, G., KENDALL, C. G. S. C., MOORE, P., WHITTLE, G. L. & CANNON, R. 1994. Testing a Seismic Interpretation of Great Bahama Bank with a Computer Simulation. *American Association of Petroleum Geologists Bulletin*, **78**, 981–1004.

FEATHERSTONE, P., AIGNER, T., BROWN, L., KING, M. & LEU, W. 1991. Stratigraphic modelling of the Gippsland Basin. *Australian Petroleum Exploration Association Journal*, **31**, 105–114.

GOLDBERG, D. E. 1992. *Genetic algorithms in search, optimization, and machine learning*. Cambridge University Press, Cambridge.

HOLLAND, J. H. 1975. *Adaptation in natural and artificial systems*. University of Michigan Press, Ann Arbor. 2nd edition: 1992. MIT Press, Cambridge.

HAQ, B. U., HARDENBOL, J. & VAIL, P. R. 1987. Chronology of fluctuating sea-levels since the Triassic. *Science*, **235**, 1156–1167.

JIN, J. 1995. Dynamic Stratigraphic Analysis and Modelling in the South-Eastern German Molasse Basin. *In*: FRISCH, W. (ed.) *Tübinger Geowissenschaftliche Arbeiten, Reihe A*, **24**, 1–153.

LAWRENCE, D. T., DOYLE, M. & AIGNER, T. 1990. Stratigraphic simulation of sedimentary basins: concepts and calibration. *American Association of Petroleum Geologists Bulletin*, **74**, 273–295.

NORDLUND, U. 1996. Formalizing geological knowledge—with an example of modelling stratigraphy using fuzzy logic. *Journal of Sedimentary Research*, **66**, 689–698.

—— & SILFVERSPARRE, M. 1994. Fuzzy logic - a means for incorporating qualitative data in dynamic stratigraphic modelling (abstract). *In*: *International Association for Mathematical Geology Annual Conference, Mount Tremblant, Quebec. Papers and Extended Abstracts*, 265–266.

ROSS, P. 1994. Parallel genetic algorithm. *Genetic Algorithm Digest*, **8**, http://www.aic.nrl.navy.mil:80/galist/src/pga-2.4.tar.Z.

SURRY, P. & RADCLIFFE, N. 1996. Real representations. *In*: BELEW, R. & VOSE, M. (eds) *Foundations of Genetic Algorithms*, 4, MIT Press, Cambridge, 204–223.

Numerical modelling of growth strata and grain-size distributions associated with fault-bend folding

TACO DEN BEZEMER, HENK KOOI, YURI PODLADCHIKOV & SIERD CLOETINGH

Institute of Earth Sciences, Vrije Universiteit, De Boelelaan 1085, 1081 HV Amsterdam, The Netherlands

Abstract: A numerical model is developed to investigate the facies and stratal patterns associated with fault-bend folding in (non-marine) foreland fold and thrust belts. After identifying the basic predicted stratal and facies geometries for a single fault-bend fold the modelling is extended to situations involving multiple folds, changes in displacement velocity, contemporaneous variations in climate and variations in sediment calibre of source material. We show that the orientation of kink-axes of fault-bend folds is of minor influence on the predicted stratal patterns and that variations in displacement velocity and climate may produce resembling, but still distinct facies patterns. The stratigraphic architecture of facies patterns is very sensitive to vergence and sequence of thrusting and may serve as a guide in interpreting the stratigraphic record associated with multiple fault structures. Applied to two natural examples of fault-bend folds, our modelling narrows down the range of interpretations of these structures.

Stratigraphy in foreland fold and thrust belts is a recorder of the interaction between tectonic deformation on the one hand, and erosion and sedimentation on the other. At the lithospheric scale, both accommodation space of, and sediment delivery to, the foreland basin are intimately related to the topographic evolution of the orogen, where regional isostasy provides a strong coupling. However, interactions between tectonics and surface processes are also important at a meso-scale. For instance, thrust faulting, folding of hanging-wall strata, erosion and sedimentation occur in concert. This paper is concerned with these meso-scale interactions.

There is growing interest in foreland basin settings. The hydrocarbon industry is interested because substantial oil discoveries have been made in these complex areas and new ones are anticipated (Cezier *et al.* 1995; Roure & Sassi 1995). Academic interest mainly concerns mechanical coupling of thrusting, erosion and sedimentation.

Geological and geophysical data on sedimentation in relation to thrusting are relatively sparse. This is predominantly due to the difficulty in obtaining a good seismic image or outcrop of both syntectonic sediments and the underlying deformed strata. Therefore, modelling has been used to gain some insight into the coupled system of thrusting and sedimentation.

Three types of modelling techniques can be distinguished: geometric (e.g. Suppe 1983), mechanical (e.g. Johnson & Berger 1989; Braun & Sambridge 1994) and analogue (e.g. Verschuren *et al.* 1996). These techniques have established some relationships between slip, fault-shape and fold shape based on conservation of mass, bed-length and bed-thickness. The geometrical models have been dominantly used to aid section balancing but have also been combined with diffusion models (Hardy *et al.* 1996) and non-process-based, fill-to-the-top sediment models (e.g. Suppe *et al.* 1991; Zoetemeijer & Sassi 1992) to predict stratal patterns of syntectonic sediments.

Apart from stratal patterns, grain-size distributions associated with thrust faults are a possibly valuable, yet more difficult to interpret, recorder of kinematics and dynamics of thrust faults. A few studies have documented relationships between thrusting and grain sizes (DeCelles *et al.* 1991; Fraser & DeCelles 1992), but mechanisms are poorly understood and we know of no attempts to model these relationships.

In this study we present a kinematic model describing thrust-fault-related folding combined with a dual lithology diffusion model simulating sedimentation and erosion of two grain-size populations. This model is used to investigate the stratigraphic record of intermontane basins focusing on local tectonic control of sedimentation using typical rates of deformation and sedimentation from these settings. We will first discuss a time evolution of a model involving one fault and discuss the basic stratal geometry and grainsize distribution within the syntectonic sediments. Using this basic sediment pattern as a reference we will then examine the syntectonic stratigraphic architecture of changes in displacement velocity along the thrust fault, of structures involving two faults, variation in climate and sediment source. It is not our aim to reproduce any specific set of observations, but to make predictions regarding the sediment signatures that are indicative of typical tectonic and climatic conditions in these settings. However, we will use the insights obtained from the modelling in a discussion of two natural examples.

Folding related to thrust faults

Three basic mechanisms for the formation of upper crustal folds on the scale of 1–10 km have been postulated: folds in compressive mountain belts grow by kink-band migration as a result of (1) fault-bend (Rich 1934), (2) fault propagation (Faill 1973) or (3) detachment folding (Jamison 1987). We focus on fault-bend folds, being the simplest deformation mechanism to combine with the sedimentation model discussed further in the paper. Fault-bend folds have been recognized and interpreted in many different areas over the globe (Dahlstrom 1970; Suppe 1983; Medwedeff 1989; Mount et al. 1990; Jamison & Pope 1996). Fault-bend folds (Fig. 1) typically occur where a detachment steps up along a ramp from one incompetent horizon to another. Movement of the thrust sheet causes folding of the rocks in the hanging wall. The fault ramp forms prior to folding and fundamentally controls fold shape. Slip along the thrust fault is accommodated by both fold growth and movement of the thrust sheet and fold shape is fully determined by the underlying thrust-fault shape. The geometric relationship between hanging-wall fold shape and thrust-fault geometry for these structures has been formalized in a geometrical model (Suppe 1983) and this relationship has been used to constrain structural interpretations in many fold and thrust belts. Although the internal deformation of hanging wall and footwall in the model has been criticized by Ramsay (1991), this does not significantly modify the predicted topographic evolution of these structures, which is what is needed for stratigraphic studies.

Based on the geometrical models of thrust fault related folding, Hardy et al. (1996) formulated numerical models reproducing the fold geometries and combined these with a diffusion model to investigate the associated sedimentary record. In this way they have shown that fault-bend folds, fault propagation and detachment folds may produce distinct stratal sediment geometries thereby providing a means to infer

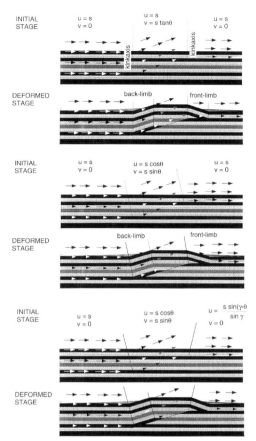

Fig. 1. Terminology and configuration and parameters for three geometrical models of fault-bend folding for different choices of kinkaxes. Corresponding three velocity fields for a simple step in decollement are given in the upper panel of each figure. θ is the ramp-angle, **s** is imposed displacement velocity, **u** horizontal velocity and **v** vertical velocity. Arrows represent displacement-vectors with orientation parallel to underlying fault-segment. Thin lines in lower panels of figures on back and front-limbs represent syntectonically deposited sediment, which are transported in the same velocity field as pre-kinematic strata. (**a**) Model with vertical kinkaxes. (**b**) Model adopted by Contreras & Suter (1990). (**c**) Model by Suppe (1983).

the fold generating mechanism from the associated sediments. Another common problem in compressional settings is the reconstruction of the sequence of thrust propagation during the evolution of thrust wedges (Burbank et al. 1992). Although forward breaking thrusting is widely accepted as the dominant style of deformation in fold-and-thrust belts (Bally et al. 1966; Boyer & Elliot 1982), the importance of backward-breaking thrusting is known from field examples

(Martinez et al. 1988; Roure et al. 1990), centrifuge modelling of thrust faults (e.g. Dixon & Liu 1991), and predictions by mechanical thrust wedge models (Davis et al. 1983; Platt 1988). More complex structures such as duplexes, imbricate thrust systems or detachment fold trains are often considered to be systems of single fold structures (e.g. Mitra 1992; Zoetemeijer et al. 1993), so it should be possible to make predictions on these systems of fault-related structures by combining single structures.

Other controls on sedimentation

The stratigraphic architecture of a basin in general is influenced by both accommodation and supply (Schlager 1993). Vail et al. (1977) postulated that marine depositional sequences were the result of eustatic changes in sea level. Posamentier & Vail (1988) further hypothesized that eustatic changes in sea level also affect all streams that drain to the sea, producing non-marine depositional sequences that correlate with marine counterparts. In at least one case, non-marine depositional sequences, defined by the same reflection termination pattern observed for marine strata, have been described in the literature (Ray 1982). Since many thrust-fault-related folds do not develop under marine conditions (Medwedeff 1989; Jordan et al. 1993), it remains unclear what is the role of sea level in alluvial basins adjacent to thrust faults.

Several workers (e.g. Miall 1986; DeCelles 1988; Posamentier & Allen 1993; Schlager 1993) have indicated that significant temporal variation in the local rate of subsidence and sediment supply can also produce depositional sequences. Local control of subsidence rate and sediment supply is especially likely in piggyback basins—basins that form above, and are transported by deforming thrust sheets (Ori & Friend 1984). These basins are commonly internally drained (and thus cut off from the direct effects of sea-level changes), and are subject to repeated disruptions of subsidence and sediment supply due to nearby structural activity. As a first approximation, it should be fair to exclude sea-level variations from our study of subaerial sequences associated with thrust-fault-related structures.

Changes in climate may directly affect non-marine deposition by altering such parameters as stream discharge, the ratio of chemical versus mechanical weathering and the water level in lakes. In this way climate may exert a dominant control on sedimentation, but one that is hard to quantify. We will, therefore, only briefly touch upon this control and not examine its role in detail. Accommodation space and sediment supply in basins adjacent to thrust faults are not only controlled by local tectonic deformation but also by deformation of the entire orogen. Deformation of the orogen causes changes in regional subsidence and changes in long-range sediment supply. These large-scale, long-range influences could therefore be potentially important for the stratigraphic development of sub-systems such as piggyback basins. Although we recognize the importance of these large-scale controls, we do not consider them in this paper, primarily to keep the number of control parameters of the model relatively limited. Histories of long-range sediment supply, for instance, could take almost any form. The same is true, to some extent, for lithospheric flexure, which depends on a large number of parameters (effective elastic thickness, loading distribution, magnitude of plate-boundary forces and moments, intra-plate stresses), which are all time-dependent. Future work may address these controls. In this paper we focus on local tectonic control on the stratigraphic development of thrust-fault-related basins.

Sedimentation models

Over the last decades numerous sedimentation and erosion models have been developed. These models may be classified as geometrical models and process-based models. Geometrical models (e.g. Lawrence et al. 1990) make assumptions regarding the depositional profile and, therefore, do not conserve mass. Process-based models do use conservation laws and may be subdivided into two categories: hydrodynamic models and diffusion models. These models are more realistic in that the sedimentation rate is not implicitly prescribed like in geometrical models. For instance, geometric models respond instantaneously to an imposed model disturbance while process models will respond with a certain time-lag controlled by diffusivity (diffusion models).

The hydrodynamic models apply simplified versions of the Navier–Stokes flow equation set (e.g. Tetzlaff & Harbaugh 1989; Martinez & Harbaugh 1993). Individual fluid elements are traced and grains are deposited or eroded depending on the fluid velocity, grain size and settling velocity. Therefore these models are very computer-intensive which makes them so far only suited for producing facies on small spatial or time scales. Diffusion models (Culling 1960; Begin et al. 1987; Kenyon & Turcotte 1985; Syvittski et al. 1988) do not trace the path of each grain in detail but use mass conservation in combination with a transport relation that averages over several processes (river transport, creep, slumps, small slides) and can, therefore, handle larger scales of space and time.

A drawback of diffusion models, as compared to hydrodynamic models, is that they have no 'natural' way to simulate down-slope sorting of different grain sizes. However, several approaches have been used to overcome this problem. In the first approach, the concept of critical slope is used where a certain grain-size population bypasses or deposits on the slope that is calculated by diffusion (Hardy & Waltham 1992). In this type of model there is no conservation of mass for sediments of a given grain size, although total mass in conserved. The second approach uses perfect sorting (Paola 1988), which deposits only one grain-size population until, at a certain grid-point within the model, the flux of that grain size is exhausted. In this type of model mass of individual grain sizes is preserved and sediment transport at each grid-point is governed, like in the basic topographic diffusion model, by topographic gradient and the coarsest fraction present. The third approach (Rivenaes 1992) involves imperfect sorting of two lithologies. Two coupled diffusion equations are solved, each describing the mass balance of a single lithology. Sediment transport of each lithology is again proportional to local gradient, and to a lithology-specific diffusion coefficient, but, unlike in simple topographic diffusion, also to the fraction of the lithology in a thin sediment transport layer (boundary layer). The model only simulates down-slope sorting, not abrasion and chemical weathering. Although some studies (Abbott & Peterson 1978; Shaw & Kellerhals 1982) indicate that abrasion is a dominant process causing down-slope sorting, other workers (Brierley & Hickin 1985; Paola & Willock 1989) have pointed out that observed rates of fining are an order of magnitude higher than what appears possible for abrasion alone and that the rate of fining increases with the rate of deposition, suggesting that down-slope sorting is predominantly caused by selective deposition.

Model description

Deformation

To study the sedimentary record in thrust fault related basins we developed a kinematic model describing fault-bend folding, following the approach by Contreras & Suter (1990) and Hardy et al. (1996), for cases involving more than one thrust fault. Folding of strata due to movement over a thrust fault is prescribed by advecting pre-kinematic strata and during fold growth deposited sediments in a velocity field consisting of domains separated by kink-axes (Fig. 1). Each domain is characterized by one velocity vector that is parallel to the underlying fault segment and whose magnitude is given by the imposed displacement velocity along the fault (note that the amount of displacement varies along the fault). Kink-axes occur at each change in fault dip and are fixed to the footwall. A grid defining the various line segments (pre-kinematic strata, topography stratigraphic horizons, older faults) and the rock in between them, including its properties (diffusivity, sediment type), is advected in Lagrangian sense with the velocity field described. The segments are modified when erosion displaces or removes the segments or deposition creates them. The prescribed deformation causes an angular style of parallel, self-similar folding in gross agreement with field and seismic observations of thrust fault related folds. The deformation causes a constant rate of uplift and dips of pre-kinematic strata at the front-limb and the backlimb are instantaneously attained. The so formed structures are eroded and sediments derived from them are transported using the sediment transport model described in the next section. Dips of syntectonic strata at the backlimb are continuously altered by passage through a kink-axis and the sedimentation and erosion process. Dips of syntectonic strata at the front-limb are only marginally altered by the deformation.

Sedimentation and erosion

In order to make model predictions on grain-size distributions associated with thrust-fault-related folding we adopt the approach of Rivenaes (1992) (for derivation and solution see Appendix) with two modifications. First, we neglect compaction and second, we assume that the diffusitivities on this modelling scale are independent of topographic height. Instead we adopt four different diffusivities depending whether lithologies are 'gravel' (more general: fraction with lower diffusivity) or 'sand' (fraction with higher diffusivity) and whether it is basement (pre-kinematic strata) or sediment that is eroded. A diffusion coefficient depends on the erodibility of the substrate, the dominant erosional processes and scale (Kooi & Beaumont 1994), as illustrated by large differences in reported values, ranging from 9.0×10^{-1} m^2 ka^{-1} for small-scale arid fault scarps (Colman & Watson 1983) to 5.6×10^8 m^2 ka^{-1} for a prograding delta (Kenyon & Turcotte 1985). In this paper we use diffusivities ranging from 1.0×10^3 to 0.5×10^3 m² ka⁻¹ for pre-kinematic strata and sediment, respectively. These values are appropriate for the model scales, but are otherwise rather arbitrary. These values result in maximum sedimentation rates of 0.5 m ka⁻¹, which is in accordance with reported sedimentation rates from literature (Johnson *et al.* 1986; Jordan *et al.* 1993; Zoetemeijer *et al.* 1993). The model makes use of a transport boundary layer (see Appendix) that makes transport of each lithology dependent on the availability of this lithology in a thin top layer. Upon erosion of a sediment column, the role of this top layer is to prevent finer sediment, only present deep in the column, from being eroded. The grain-size fraction within this top layer adjusts itself to the grain-size fraction that is delivered from upslope, and if erosion occurs, to the grain-size fraction that has to be entrained from just below the layer. This top layer is a surrogate for (a stacked system of) natural transport layers, which ensure that what may be eroded adapts to what is present in the subsurface. We adopted a constant value for the thickness of this top layer, so the model does not differentiate between thicknesses of transport layers of hillslope (regolith that is moving by creep, slumps and slides) or river bed (suspension load and bedload). This way of modelling implies that the predicted grain-size distributions can not resolve individual processes (e.g. slumps) that may be responsible for a certain deposit because these processes are lumped.

Constant elevation boundary conditions are used which act as sediment sinks and simulate local base levels of erosion. Explicit time integration is used where the time step is controlled by the smallest diffusive response time for the grid points.

Orientation of kinkaxes

Numerical models based on the geometrical model of fault-bend folding have been using different orientations of kink-axes. Vertical kink-axes, constant dip kink-axes (Contreras & Suter 1990) and Suppe kink-axes (Hardy *et al.* 1996) (see Fig. 1). Although probably important for studies concerning deformation, Fig. 2 demonstrates that the orientation of these kink axes is of minor importance for sedimentation studies. The resulting stratal geometries are shown for vertical kink axes (Fig. 2a), constant dip kink axes (Fig. 2b) and constant bed-thickness kink axes (Fig. 2c). In all cases the backlimb is characterized by an unconformity whereas the front-limb is characterized by ongoing onlap. The stratal patterns are marginally different with respect to the place where this occurs but not in their general pattern.

Using vertical kink axes has the enormous benefit that it avoids tedious interpolations to administrate subsurface properties at each timestep. We therefore adopt the simplest way to develop a fault-bend fold using vertical kink axes, although we realize that a mechanism of vertical shear is only occasionally a likely

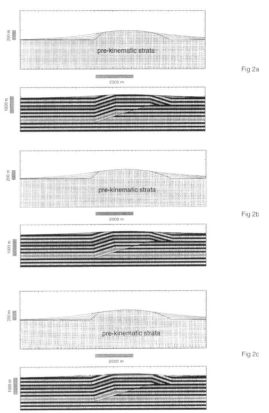

Fig. 2. Stratal patterns resulting from model runs for the three different models of fault-bend folding (see Fig. 1). Figure conventions: Lower panel: Configuration resulting from model run for a thrust fault with one step in decollement. Upper panel: Detail of the developed chronostratigraphy. (**a**) Model result using vertical kinkaxes. (**b**) Model result from fault-bend fold model used by Contreras & Suter (1990). (**c**) Model results using Suppe's (1983) fault-bend fold model. Back-limbs for all fault-bend folding models (left-hand side of folds) show an unconformity whereas the front-limbs (right-hand side of folds) show onlap. Differences occur in deformation of pre-kinematic strata and position of back-limb unconformity and front-limb onlap but internal patterns are the same.

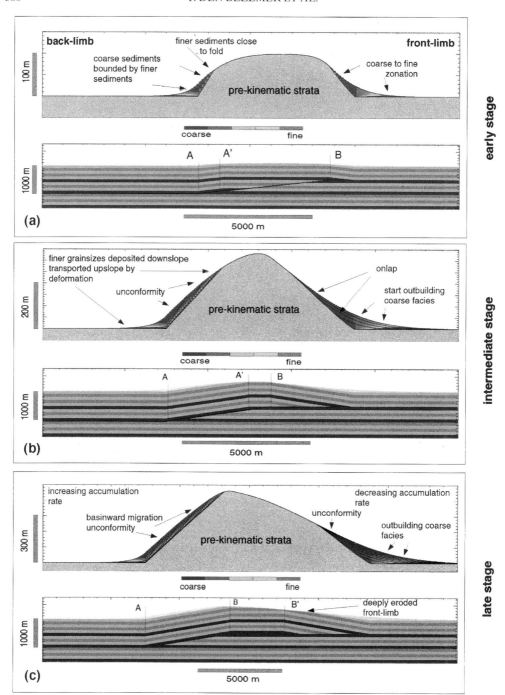

Fig. 3. Kinematic development of a fault-bend fold and syntectonic sedimentation. Lower panels: deformation of pre-kinematic strata and time-lines. Upper panels: detail of modelled structure with time-lines and predicted grainsize patterns. Note decreasing vertical exaggeration from (a) to (c). Backlimb sediments are characterized by an unconformity, a strong coarsening upward and an increasing accumulation rate ('thickening up', increased spacing of time-line spacing). Front-limb characterized by onlap, a dominantly coarsening upward trend and fanning pattern of facies with increasing depositional space and a decrease in accumulation rate ('thinning up'). Front-limb shows small unconformity and outbuilding of coarse facies at a later stage when front-limb basin is filled.

deformation mechanism. It produces a very similar topography compared with the Suppe model, which is the most important factor controlling the associated stratigraphic patterns, and it avoids the situation in which the stratigraphic section is tilted whereas erosion acts vertically.

Results

Basic sediment geometries

The first model experiment (Fig. 3) illustrates the basic geometries associated with a single fault consisting of a simple flat-ramp-flat where the upper hanging-wall flat is positioned in the subsurface. A displacement velocity of 2 m ka^{-1} was used, in accordance with reported values ranging from 0.8 to 3.0 m ka^{-1} (Medwedeff 1989; Mount et al. 1990; DeCelles et al. 1991; Shaw & Suppe 1991; Suppe et al. 1991; Zoetemeijer et al. 1993; Jordan et al. 1993). A fixed sediment source grain-size fraction of 0.50 was used which would roughly correspond to flysch-type rock (turbidites), but is otherwise arbitrary.

The development in time of the structure is shown in Fig. 3. At the instant of initiation of slip two kink bands A and B form (Fig. 3a). As slip continues these kink bands grow in width, structural relief increases and the width of the flat top diminishes (Fig. 3a & b). This process continues till passive kink axis A', which moves with the hanging wall, reaches kink axis B. Subsequently the fold crest only grows in width, but not in height (Fig. 3c). The basic stratal pattern associated with the fault-bend fold is an unconformity at the back-limb and onlap at the front-limb. The back-limb unconformity is the result of transport, uplift and subsequent erosion of the back-limb growth strata as part of the developing fold. Onlapping growth strata at the front-limb are passively transported above the upper hanging-wall flat. The back-limb sediments are continuously reworked in a 'conveyer-belt mode', whereas front-limb sediments are progressively buried until the basin to the right of the fold gets filled and an unconformity develops within the growth strata. The back-limb is characterized by an increasing accumulation rate in time caused by the growing availability of readily erodible sediments in the source area. The front-limb shows a decrease in accumulation rate due to filling of the basin.

The grain size patterns of back- and front-limb show distinctive features as well. The front-limb is characterized by a general coarse to fine zonation moving away from the fold whereas the back-limb exhibits a semi-concentric facies pattern where coarse sediments (purple colours) are flanked by finer sediments. This semi-concentric pattern is caused by upslope transportation of finer sediments, originally deposited distal with respect to the fold and now eroded close to the fold-crest. The back-limb shows initially a stronger coarsening upward trend than the front-limb. A strong coarsening upward at the front-limb is only observed when the basin gets filled and sediments get eroded.

Variation in displacement velocity

The second experiment (Fig. 4) is devoted to the effects of variation in displacement rate along a fault with a simple step in decollement. For all three considered cases the total displacement is 2000 m and a fixed source of 0.5 was used. Growth strata are recorded at 200 ka. Figure 4a shows the response of the system to a linear decrease in displacement velocity from 2 m ka^{-1} to 0. The front-limb shows the development of an unconformity and outbuilding of coarse facies. The back-limb shows an expected decrease in accumulation rate and at later stages a extra shift of the unconformity away from the fold. The sediment pattern is largely unaffected except for the large volume of coarse sediments.

Figure 4b shows the effect of a linearly increasing displacement velocity to 2 m ka^{-1}. This leads to continuous onlap on the front-limb and within the front-limb sediment wedge both locations showing fining and coarsening up. The back-limb does not respond very characteristically to this scenario and shows a convex unconformity surface. The accumulation at both limbs is much lower compared with the former case.

Figure 4c shows a simulation in which the imposed displacement rate is 1.0 m ka^{-1}, interrupted by a phase with no displacement. The instantaneous drop in displacement rate causes an expansion of the back-limb unconformity away from the fault-bend fold and the development of an unconformity at the front-limb. This is accompanied by outbuilding of coarse facies at both limbs. This response reflects a transition to a more erosion-dominated state when the advection decreases and sediments with higher fractions are eroded. The subsequent onset of deformation causes a drastic onlap towards the fault-bend fold at both sides and a return to the initial basic geometries. This phase of renewed advection causes the coarse facies to retreat towards the fold on both sides. A very pronounced feature of the predicted geometry is the buried step in pre-kinematic strata at the front-limb.

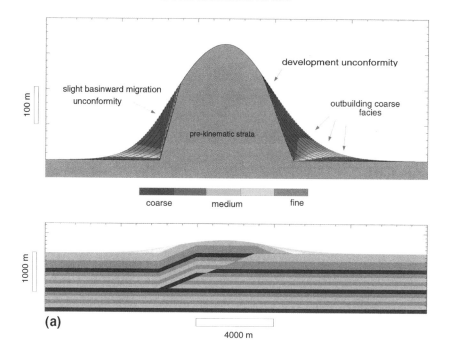

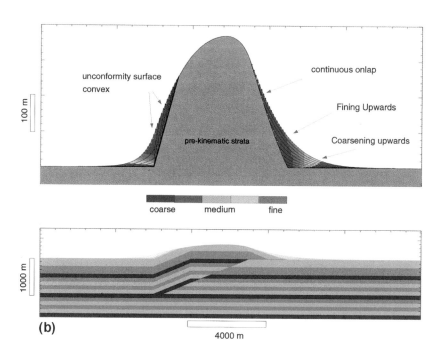

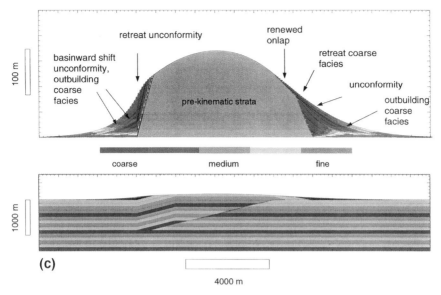

Fig. 4. Resulting configurations due to changes in displacement rate. Total displacement is 2000 m for all models. (**a**) Configuration resulting from a model run in which the displacement rate linearly decreases to zero. Back-limb is less affected than front-limb. Front-limb shows development unconformity and stronger outbuilding coarse facies. Backlimb shows decrease in accumulation rate, a concave unconformity surface and a small extra basinward shift of the unconformity. (**b**) Configuration resulting from a model run involving an linear increase in displacement rate. Front-limb shows continuous onlap and fining upward close to the fold. Back-limb is characterized by a convex unconformity surface. (**c**) Results from a model run involving a period of movement followed by period of cessation with sudden decrease in displacement rate to zero for 1 Ma, followed by period of movement with sudden increase in displacement rate. Cessation of movement along fault causes unconformity at back-limb to expand away from the fold and causes an unconformity in front-limb sediments. Both limbs show coarse outbuilding during cessation of movement and a sudden retreat in coarse facies when movement resumes. Note buried step in pre-kinematic strata at the front-limb.

Changes in erodibility and source fraction.

Two important parameters in the model are the erodibility of the system and the fraction of sediment delivered by the source rock (pre-kinematic strata). Both the erodibility and the source fraction might change during development of a fault-bend fold. Such changes are for example brought about by changes in climate and exhumation of different lithologies. However, the relationship between climate, erodibility and source fraction are only qualitatively understood. In the following experiments we imposed a change in model climate and a change in source grain-size fraction. The displacement rate for all experiments is 1 m ka^{-1}.

Figure 5a shows the effect of raising the erodibility of both sediment and pre-kinematic strata by a factor of 10. This change of erodibility is instantaneously imposed after a period of 'normal' erodibility. The back-limb does not show a particular response to this change in irritabilities but the front-limb shows a unconformity and outbuilding of coarse facies. The effect of a decrease in irritabilities by a factor of 10 is shown in Fig. 5b. The back-limb shows again no particular response but the front-limb is characterized by continuous onlap and backstepping of coarse facies.

An increase of the sediment fraction supplied by pre-kinematic strata from 0.5 to 0.7 during the development of a fault-bend fold shows a strong coarse outbuilding on both limbs but no significant change in stratal patterns (see Fig. 6).

Forward-breaking thrusting

Figure 7 shows the results of an experiment with two faults (initial spacing 6 km) in which the second fault is activated in forward breaking sequence. Both faults share the same detachment level and displacement velocities for both faults are 1.0 m ka^{-1}.

In Fig. 7a a situation is modelled where the thrust faults are widely spaced such that a basin develops in between the two fault-bend folds.

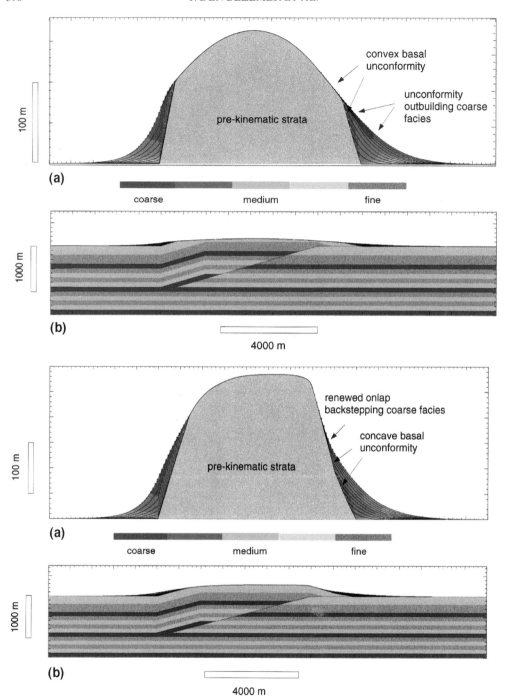

Fig. 5. Model run to show the effects of changes in erodibility. A higher erodibility results in an unconformity, outbuilding of coarse facies and a convex basement (pre-kinematic strata) topography at the front-limb (**a**). A lower erodibility (**b**) results in onlap, retrograding of coarse facies and a concave basement topography. Back-limb patterns for both cases are relatively unaffected.

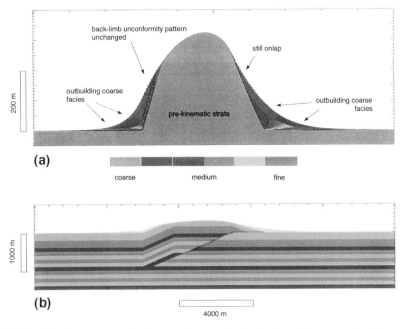

Fig. 6. Model showing the response of sedimentation to a raise in fraction of the source (pre-kinematic strata). Facies patterns on both limbs record a strong outbuilding of coarse facies whereas stratal patterns are unaffected.

The initial response to the cessation of deformation along the first fault is the same as in one of the previous experiments (Fig. 4c). That is, an unconformity and outbuilding of coarse facies develops on both sides of the fold. At the same time, distal front-limb sediments associated with the now inactive (left) fold are progressively incorporated in the growing back-limb of the (right) fold. The basin axis, indicated by the blue basin-centre, fine-grained facies, shift first to right due to cessation of uplift along the left fold. The basin centre subsequently shifts to the left when fold B becomes a progressively more dominant sediment source. The basin centre exhibits a fining up sequence whereas the flanks of the basin both show coarsening up. The front-limb of fold B shows a basic geometry.

In Fig. 7b a situation is modelled in which fault spacing is small such that fold A is incorporated in the back-limb of forward breaking thrust fault B. This results in a back-limb geometry of fold A that shows the characteristics of cessation of movement (Fig. 4c) and a small new back-limb basin, filled with coarse facies. The central shows first a phase of onlap and downlap followed by truncation accompanied by coarse outbuilding.

Back-thrusting

Finally, in the last experiment (Fig. 8) a situation is modelled where back-thrusting occurs. We examined cases in which the back-thrust (associated with fold A) has the same detachment levels as the now inactive fault. The old structure, including the former hanging wall and footwall is transported to the left, resulting in a pop-up structure. Initial fault-spacing is 5000 m. Displacement-rate along the first active fault and back-thrust are 1.0 and -1.0 m ka^{-1}, respectively. The evolution predicted in this experiment is similar to that predicted for forward-breaking thrusting in the sense that one fold decays while another one grows next to it. The most important difference with forward-breaking thrusting is the reversed direction of motion. This results in a central basin that is flanked by back-limbs showing an asymmetric unconformity development. Figure 8a shows a case in which fault spacing is large. The unconformity at the back-limb of fold A is a basic back-limb geometry whereas the backlimb of fold is a dead back-limb showing expanding away from fold B and coarse outbuilding. The front-limb of fold A is developing front-limb characterized by onlap and fanning of facies

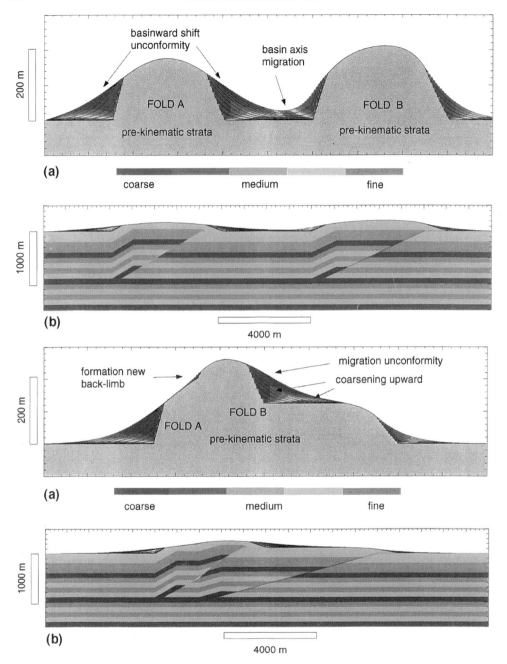

Fig. 7. Result from model simulating two faults activated in forward-breaking sequence. (a) Large spacing between faults. Lower panel shows deformation of pre-kinematic strata. Upper panel: The back-limb unconformity associated with left fold expands after cessation of growth. At its front-limb an unconformity develops. Both limbs show outbuilding of coarse facies. Depocentre of central basin, indicated by blue colour first migrates to the right and than to the left due to changes in dominant source. Right fold shows normal fault-bend fold development. (b) Small initial spacing between faults. Lower panel shows deformed pre-kinematic strata. Most left basin shows response to cessation of movement (uplift) with basinward migration of unconformity and coarse facies. The small basin on the flank of left fold is a remnant of sediment deposited in the old back-limb syncline. The central basin is dominated by the left fold and the basin is overfilling showing an unconformity and progradation of coarse facies. The basin at the right shows normal front-limb development.

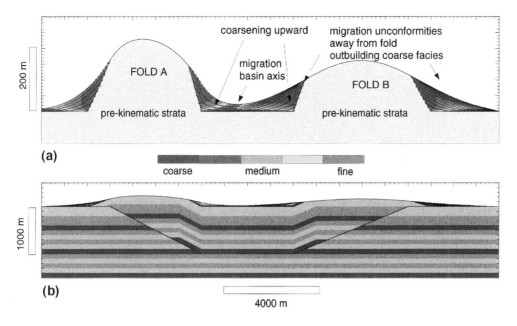

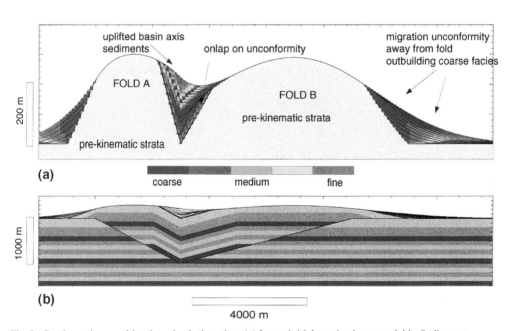

Fig. 8. Configuration resulting from back-thrusting. (**a**) Large initial spacing between folds. Sediments associated with now inactive right fold show first basic geometries and after reversal of movement a cessation of movement unconformity develops on backlimb and front-limb. Back-limb of right fold shows slightly anomalous facies development with less pronounced outbuilding of coarse facies. Basin-centre facies exhibits first a shift to the left and than to the right. (**b**) Small spacing between faults. Central basin between folds shows general fining up due to uplift of dominantly finer facies associated with earlier fold development.

patterns. The front-limb of fold B is a fossil front-limb showing an unconformity and coarse outbuilding.

Fig. 8b shows a case with small initial spacing resulting in a rather different central basin. At the back-limb of the inactive fold B the unconformity first progrades into the basin and is subsequently onlapped and finally buried. The facies pattern is a little complicated. The back-limb of fold A is dominated by medium and fine facies due to the fact that is was dominantly distal facies deposited during development of fold B that is eroded. The V-shaped blue facies at this limb is basin-centre facies that is transported uplimb.

Both front-limbs show an analogous development to the previous case.

Natural examples

Lost Hills anticline, California

A well-studied example of a fault-bend fold is the Lost Hills anticline from the Temblor fold belt in California (Medwedeff 1989) (see Fig. 9). The structure is based on interpretations of seismic data and well data and use of fault-bend fold theory. The structure is thought to have been growing during the Late Miocene to Late Pleistocene or Holocene. Growth strata of the Etchegoin and San Joaquin formations are interpreted to be shallow-marine to brackish-water deposits whereas the Tulare formation consists of fluvial and lacustrine conglomerate, sand and siltstone.

The overall style of the structure displays the basic features predicted by our model: an unconformity at the back-limb and onlap at the front-limb. In detail, however, the structure is more complicated. Distinct and unexplained features of this structure are the unconformities in syntectonic strata, deep erosion of the structure at its front-limb (unconformity 1, Fig. 9), and an onlap-unconformity alternation (onlap 1, unconformity 2, onlap 2 in Fig. 9) within the growth-strata at the front-limb which is absent at the back-limb. The deep erosion suggests emergence of the structure. The deep erosion of the front-limb is a basic feature predicted by our model for late stages of fault-bend fold development (Fig. 3c), in which stage the Lost Hills fold apparently is. The onlap-unconformity pattern at the front-limb is more difficult to explain. First of all, sea level can be ruled out as a possible explanation, since the observed onlap-unconformity pattern is only present on one side of the fold. The second explanation invokes base-level variations (Hardy *et al.* 1996). Although base level clearly plays a role (the basin is close to overfilling), it not obvious how this could lead to the observed asymmetry in back-limb and front-limb unconformity and onlap pattern.

Our modelling suggests other explanations. At first one might be tempted to see velocity variations as a major cause for the observed depositional sequence (Fig. 4c). This hypothesis may probably be rejected because the back-limb does not show the same shifts in unconformity as expected from our modelling. The second option is the presence of another fold (Figs 7 & 8). This may be rejected as a possible explanation since there is no evidence for such a structure. The third explanation would involve a change in erodibility. As is shown in Fig. 5 and in the lower panel of Fig. 9, this might have a stronger effect on the front-limb than on the back-limb. These changes in erodibility could be related to alternations in subaerial and submarine setting of the fold. Medwedeff (1989) suggests a change in environment from shallow marine to brackish water during the development of the fault-bend fold. The latter environment might indeed be less erosive.

This change in erodibility scenario leaves the large amount of infill and the close position of onlap 1 to the fold (Fig. 9) unexplained. Also unexplained are the enormous difference in thickness of the syntectonic formations depending whether they are positioned at the back-limb or front-limb side. Although a slight asymmetric accumulation rate is predicted by our model (Fig. 3), it predicts a late stage decrease in accumulation rate at the front-limb and increase at the back-limb, which is not observed. These observations suggest infill from an out-of-plane direction.

One factor that could be used to test the importance of out plane transport is grain size. If none of the grain-size patterns we observe in our models is present, but instead grain-size patterns do not seem to be influenced by the growth of the structure at all, one could argue that the depositional slope was not controlled by the development of the thrust. In other words, down-slope sorting would not have taken place in the direction of the back- and front-limb dips, thereby suggesting fill from another direction.

Iglesia Basin, Argentina

We are not aware of examples where there are detailed data on both subsurface fault-bend-fold geometry, stratal patterns and grain sizes. Very often only one of these three datasets is present. A basin for which both stratal patterns and grain sizes have been documented, using

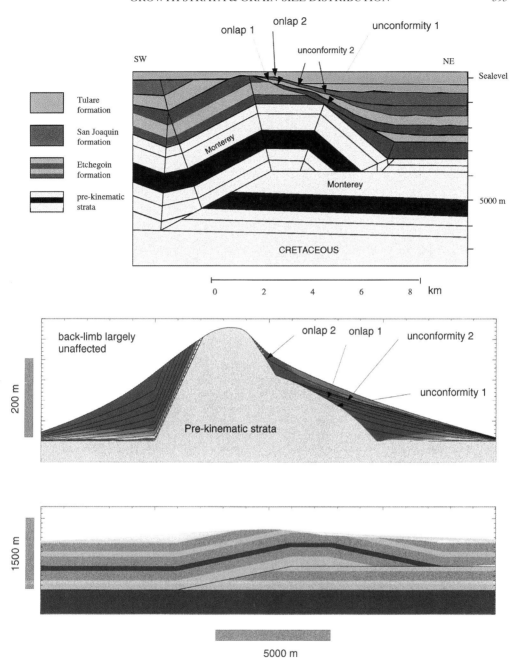

Fig. 9. (**a**) Seismic interpretation of Lost Hills Anticline, California. Redrawn from Medwedeff (1989). (**b**) Model run to compare styles of infill predicted by the model with Lost Hills anticline.
(**a**) shows a strong asymmetric development; back-limb shows an unconformity whereas front-limb shows alternations of unconformities and onlaps. Unconformity 1, at the front-limb, between pre-kinematic strata and syntectonic strata is covered by strongly onlapping strata (onlap 1). Onlap phase 1 is followed by an unconformity (unconformity 2) which in turn is onlapped by subsequently deposited strata. Unconformity 1, onlap 1 and unconformity 2 are a natural consequence of the model (see Fig. 1 and 9b). Onlap 2 could be explained by a decrease in overall erodibility. This has a stronger effect on the front-limb than on the back-limb (see Fig. 5).

both outcrop and seismic data, is the Iglesia Basin in Argentina. Sediments were predominantly derived from the neighbouring Frontal Cordillera, acting as a fairly constant sediment source (Beer et al. 1990). Also, climate is thought to have remained constant during infilling of the basin (Pascual 1984). The sediments in the Iglesia basin are deposited in alluvial fan, sandflat and playa lake setting (Beer et al. 1990). The basin was internally drained, and is thought to be flanked by fault-bend folds (Allmendinger et al. 1990; Beer et al. 1990) (see Fig. 10). Measured paleaocurrent directions indicate that the depositional slope was mainly controlled by the developing folds. The overall pattern of basin infill shows onlap on basement at the western side (front-limb of fault-bend fold A Fig. 10) and an unconformity at the eastern side (backlimb of fault-bend fold B, Fig 10). The infill of the basin shows internal unconformities, shifting of onlap and facies (Fig. 10). Since both climate, variation in sediment source, and sealevel may be ruled out as possible explanations for the observed pattern, the depositional sequence of stratal patterns in this basin is likely caused by either changes in thrust controlled subsidence rate or thrust controlled changes in drainage (Beer et al. 1990). The drainage scenario holds that the basin fills when Río Jáchal is cut off from the Iglesia basin during thrust events; during intervals of quiescence the river re-establishes headwaters and produces the unconformities. Both the accommodation and the drainage hypotheses suggest basinward shifts of unconformities during times of tectonic quiescence and renewed onlap during phases of thrust movement which is in agreement with the patterns we observe in our modelling (see Fig. 4c). Moreover the grain-size patterns associated with onlap and truncation are in good agreement with our predictions.

There are, however, three important differences. First, the most important difference is the general fining-upward trend. Unless we impose an external control on our model, our predicted general trend is coarsening up. Secondly, since both folds are thought to be coupled in their movement (Allmendinger et al. 1990), we would also expect from our modelling insights that there has been some sediment input from fold B with corresponding grain-size pattern. Since grain-size data are extrapolated from a section west of the seismic line, we are not sure whether the predicted grain-size pattern is simply not there or not interpreted. Thirdly, our modelling suggests that the coarse facies bodies derived from the fault-bend folds are not attached to each other but are separated by finer basin-centre facies.

Although our modelling approach has only limited suitability for making a decision between either one of the hypotheses, a few comments could be made. The thrust controlled subsidence scenario (Beer et al. 1990) is analogous to our discussion on velocity variations (Fig. 4) and this elegantly explains the observed relation between stratal and grain-size patterns. There are two problems with this interpretation, however. First, if the postulated grain-size pattern is correctly extrapolated, the drainage hypothesis would be supported because down-slope sorting would have taken place eastward. Secondly, in an accommodation-controlled system we predict a general coarsening-up trend whereas fining-up is observed. The observed overall fining-up trend might be explained by an overall accelerating thrust deformation since it is suggested that there are no variations in source present. This contradicts, however, the fact that topographic growth of the bounding Frontal Cordillera occurred relatively early in the structural development of the region (Allmendinger et al. 1990). The fining upwards could alternatively be explained by a change a climate, but this contradicted by claims that climate has been constant during growth. These two observations would advocate the drainage scenario.

Independent of the scenario, the structure contains many of the elements predicted by our model; onlap is thought to result from renewed thrusting and basinward migration of the depositional margin results from periods of quiescence.

Discussion and conclusions

The internal structure of fault-bend folds is usually poorly imaged (seismics) or unknown (outcrop) whereas the associated sediment patterns are commonly clearly visible. Interpretation of the deformational history of fault-bend folds must therefore often be made indirectly from the stratigraphic record. The model presented in this paper is meant to aid in the interpretation. The following can be concluded from the analysis of the model behaviour. (1) Vergence of thrusting can be inferred from the asymmetry in unconformity/ onlap patterns and grain-size distributions for back- and front-limb. The back-limb of a single fault-bend fold is characterized by an unconformity, an increasing accumulation rate and a strong coarsening up sequence; the front-limb is characterized by initial onlap followed by an unconformity, a decreasing accumulation rate and a weaker

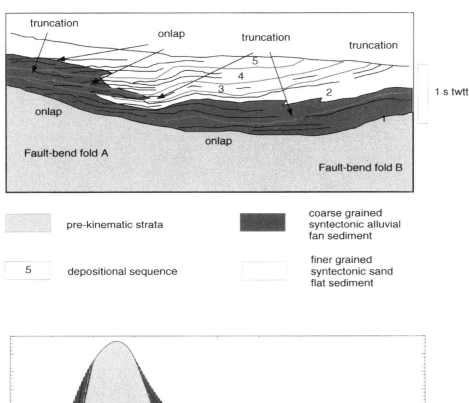

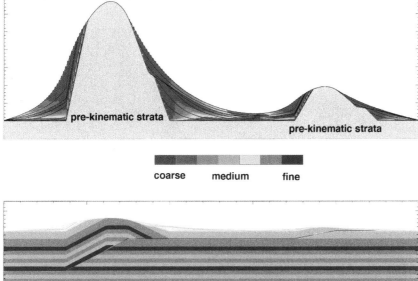

Fig. 10. (a) Seismic interpretation of the Iglesia Basin, Argentina. Heavy black lines are sequence boundaries. Facies was extrapolated from nearby field exposures onto the seismic section. Onlap coincides with retrogradation of coarse facies whereas unconformities are accompanied by coarse outbuilding. Redrawn from Beer *et al.* (1990).(b) Model run to compare natural style of infill depicted in (a) with model predictions. Model run involved two steps in decollement and velocity variations. Central basin corresponds to (a). Syntectonic depositional sequence shows an alternating pattern of truncation and onlap with associated interfingering of coarse and finer facies. Onlap and truncation patterns predicted by the model are caused by fault displacement velocity variations which in accordance with the interpretation by Beer *et al.* (1990). Similarly, as anticipated by the authors, prograding of coarse facies occurs simultaneously with truncation, whereas retrograding coincides with phases of onlap. Differences include: (1) for overall fining upward, the modelling suggests a extrabasinal control, (2) the modelling suggests a coarse facies body related to fault-bend fold B and (3) the model predicts separation of coarse sediment bodies attached to both folds developed in the basin.

coarsening up. (2) Front-limb stratigraphy is more affected by variations in climate and velocity variations. (3) Expansion of unconformities on both back- and front-limb accompanied by coarse facies progradation points to cessation of movement. (4) The above relationships for a single fault-bend fold are very useful to interpret to deformational history of a multiple fault system. However, because of the complex interactions that occur in multiple fault systems it remains needed to compare observations with a suit of model runs. (5) When growth strata fail to differentiate between different controls, grain-size data can sometimes be decisive. (6) Our model predictions do not support the claim that fining-upward sequences are the natural product of intermontane basins in subaerial thrust belts (Beer et al. 1990). Instead we observe a general coarsening-upward trend on both limbs due to cannibalization of sediments, unless deformation accelerates or the erodibility decreases. If this fining upward is the rule in natural systems, it may be related to drainage network evolution (Fraser & DeCelles 1992), which our model does not explicitly address.

When judging the value of the model inferences it must be kept in mind that the model necessarily simplifies the behaviour of the natural system. First of all, coupling in the model is unidirectional in the sense that deformation affects erosion and sedimentation, but not the other way around. In other words, no feedback exists from sedimentation to deformation. Such a feedback is likely to exist through the influence of overburden on shear strength of the fault.

Secondly, implicit in our modelling is that downslope sorting is the dominant process producing the grain-size patterns, thereby neglecting other processes affecting grain size, such as chemical erosion, abrasion and source variations. A third simplification is that the model is only 2D and, therefore, important features like fan lobe switching, out-of-plane sediment supply/removal and valley incision are not incorporated.

To what extent these shortcomings would alter predicted stratigraphy remains unsolved for the moment. The two modelled natural examples suggest that the basic predicted patterns are robust and could be used as a tool in structural interpretation.

The authors wish to thank P. Cobbold and an anonymous reviewer for their constructive comments. This work was supported by the IBS project, Joule II Programme (JOU2-CT92-110). Netherlands Research School of Sedimentary Geology Publication no. 970130.

Appendix: Derivation and solution of the dual lithology diffusion equations

This first part of the appendix largely follows Rivenaes (1992). Conservation of mass requires that sediment that accumulates in a certain volume, during a short time, is equal to the difference between the mass that enters and leaves the volume. Figure A1 shows sand fluxes ($q_{s,\,in}$, $q_{s,\,out}$) entering and leaving the volume. Inside the volume the sand constitutes a bulk fraction F, while the rest of the volume (1–F) is occupied by shale. Within the sand C_s is the solid fraction, that is (1–C_s) is the porosity. In order to simulate amoring, a surface transport layer with thickness A and bulk fraction $F_s L$ is used. This layer represents the conduit for lateral migration of sediments, containing all processes that transport and sort sediment (channel transport, creep, slumps and slides). For simplicity it is assumed to be constant over the whole profile and independent of time. The thickness of the surface layer can not be zero, so there should always be soil available for erosion. The mass-balance for sand is:

$$([C_s F \rho_s]^{t+\Delta t} - [C_s F \rho_s]^t) A \Delta x + \overline{C_s F \rho_s} \Delta x (h^{t+\Delta t} - h^t) = [q_s \rho_s \Delta t]^x - [q_s \rho_s \Delta t]^{x-\Delta x}.$$

Superscripts t and $t + \Delta t$ represent initial time and new time respectively. A precise expression for the averages in the third term is:

$$\overline{[C_s F \rho_s]} = \frac{1}{h^{t+\Delta t} - h^t} \int_{h^t - A}^{h^{t+\Delta t} - A} (C_s F \rho_s) \, d\psi.$$

It is reasonable to suppose that the grain density and sand concentration in the surface layer will be almost invariant. Dividing by $\Delta t \, \Delta x$ on each side an letting Δt and Δx tend to zero gives:

$$A C_s \frac{\partial F}{\partial t} + \overline{C_s F} \frac{\partial h}{\partial t} = -\frac{\partial q_s}{\partial x}.$$

The average term $C_s F$ should be treated somewhat differently depending whether erosion of deposition occurs at a given position. With erosion the C_s and F of the removed layers are known and can be calculated using:

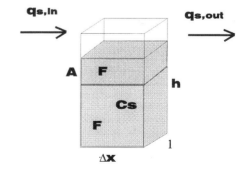

Fig. A1. Sand entering and leaving the control volume (after Rivenaes 1999).

$$[\overline{C_sF}] = \frac{1}{h^{t+\Delta t} - h^t} \int_{h^{t}-A}^{h^{t+\Delta t}-A} (C_sF)\, d\psi.$$

If deposition occurs the value of C_sF is unknown and must be a part of the solution, equal to the F in the surface layer, and C_s is set to the initial value.

$$[\overline{C_sF}] = C_s^{t+\Delta t} F^{t+\Delta t}.$$

The sand volume rate q_s will now be replaced by a sediment transport relation. This relation assumes that the flux q is a function of slope.

$$q = K\frac{\partial h}{\partial x}.$$

The sand volume rate q_s must be fraction of the total flux of sediments. The fraction is not known but intuitively two requirements must hold: the fraction is zero when no sand is available, and conversely, if only sand is available the sand flux equals the total flux. If a linear relationship is assumed and one realizes that K is lithology dependent, the sand equation is:

$$AC_s \frac{\partial F}{\partial t} + \overline{C_sF}\frac{\partial h}{\partial t} = \frac{\partial}{\partial x}\left(FK_s\frac{\partial h}{\partial x}\right).$$

The shale equation can be derived in a similar way:

$$-AC_{sh}\frac{\partial F}{\partial t} + \overline{C_{sh}(1-F)}\frac{\partial h}{\partial t} = \frac{\partial}{\partial x}\left((1-FK_{sh})\frac{\partial h}{\partial x}\right).$$

In this dual-lithology system, the set of non-linear coupled partial differential equations has to be solved for the two unknowns in the system, F and h.

Solution of the pde system

The system is solved by first writing down these equations a little differently. Subtraction of one equation from the other leaves one equation that describes the change in height and one that describes the change in F. If compaction is ignored the height equation is obtained:

$$\frac{\partial h}{\partial t} = \frac{1}{C_s}\frac{\partial}{\partial x}\left(FK_s\frac{\partial h}{\partial x}\right) + \frac{1}{C_{sh}}\left(\frac{\partial}{\partial x}\left((1-F)K_{sh}\frac{\partial h}{\partial x}\right)\right).$$

The height equation is solved using explicit finite difference discretization. The F-equation is:

$$\frac{\partial h}{\partial t} = \frac{1-\overline{F}}{AC_s}\frac{\partial}{\partial x}\left(FK_s\frac{\partial h}{\partial x}\right) - \frac{\overline{F}}{AC_{sh}}\left(\frac{\partial}{\partial x}\left((1-F)K_{sh}\frac{\partial h}{\partial x}\right)\right).$$

This F-equation is of the form:

$$\frac{\partial F}{\partial t} = b\frac{\partial F}{\partial x} + cF + d.$$

The constants b, c and d are given by:

$$b = \frac{1-\overline{F}}{AC_s} K_s \frac{\partial h}{\partial x} + \frac{\overline{F}}{AC_{sh}} K_{sh} \frac{\partial h}{\partial x}$$

$$c = \frac{1-\overline{F}}{AC_s}\frac{\partial}{\partial x}\left(K_s\frac{\partial h}{\partial x}\right) + \frac{\overline{F}}{AC_{sh}}\frac{\partial}{\partial x}\left(K_{sh}\frac{\partial h}{\partial x}\right)$$

$$d = -\frac{\overline{F}}{AC_{sh}}\frac{\partial}{\partial x}\left(1-K_{sh}\frac{\partial h}{\partial x}\right).$$

This is solved by using the method of characteristics. The new F value is given by:

$$F = \left(F_{old} + \frac{d}{c}\right)e^{c\Delta t} - \frac{d}{c}.$$

Where F_{old} is the interpolated F value belonging to the backshooted position x_{old} given by:

$$x_{old} = x + b\Delta t.$$

References

ABBOTT, P. L. & PETERSON, G. L. 1978. Effects of abrasion durability on conglomerate clast populations: examples from Cretaceous and Eocene conglomerates of the San Diego area, California. *Journal of Sedimentary Petrology*, **48**, 31–42.

ALLMENDINGER, R. W., FIGUEROA, D., SMYDER, D., BEER, J., MPODOZIS, C. & ISACKS, B. L. 1990. Foreland shortening and crustal balancing in the Andes at 30 S latitude. *Tectonics*, **9**, 789–809.

BALLY, A. W., GORDY, P. L. & STEWART, G. A. 1966. Structure, seismic data, and orogenic evolution of the southern Canadian Rockies. *Bulletin of Canadian Petroleum Geology*, **14**, 337–381.

BEER, J. A., ALLMENDINGER, R. W., FIGUEROA, D. E. & JORDAN, T. E. 1990. Seismic stratigraphy of a Neogene Piggyback Basin, Argentina. *American Association of Petroleum Geologists Bulletin*, **74**, 1183–1202.

BEGIN, Z. B. 1987. Application of a diffusion-erosion model to alluvial channels which degrade due to baselevel lowering. *Earth Surface Processes and Landforms*, **13**, 487–500.

BOYER, S. E. & ELLIOTT, ? 1982. Thrust systems. *American Association of Petroleum Geologists Bulletin*, **66**, 1196–1230.

BRAUN, J. & SAMBRIDGE, M. 1994. Dynamic Lagrangian Remeshing (DLR): A new algoritm for solving large strain deformation problems and its application to fault-propagation folding. *Earth and Planetary Science Letters*, **124**, 211–220.

BRIERLEY, G. J. & HICKIN, E. J. 1985. The downstream gradation of particle sizes in the Squamish River, British Columbia. *Earth Surface Processes and Landforms*, **10**, 597–606.

BURBANK, D. W., VERGES, J., MUÑOZ, J. A. & BENTHAM, P. 1992. Coeval hindward- and forward-imbricating thrusting in the south-central Pyreneen, Spain: Timing and rates of shortening and deposition. *Geological Society of America Bulletin*, **104**, 3–17.

CEZIER, E. C., HAYWARD, A. B., ESPINOSA, G., VELANDIA, J., MUGNIOT, J. F. & LEEL, W. G. 1995, Petroleum Geology of the Cusiana Filed, Llanos

Basin Foothills, Colombia, *American Association of Petroleum Geologists*, **79**, 1444–1463.

CONTRERAS, J. & SUTER, M. 1990. Kinematic Modeling of Cross-Sectional Deformation Sequences by Computer Simulation. *Journal of Geophysical Research*, **95**, 21913–21929.

COLMAN, S. M. & WATSON, K. 1983. Ages estimated from a diffusion equation model for scarp degradation. *Science*, **221**, 263–265.

CULLING, W. E. H. 1960. Analytical theory of erosion. *Journal of Geology*, **68**, 336–344.

DAHLSTROM, C. D. A. 1970. Structural geology in the eastern margin of the Canadian Rocky Mountains. *Bulletin of Canadian Petroleum Geology*, **18**, 332–406.

DAVIS, D., SUPPE, J. & DAHLEN, F. A. 1983. Mechanics of fold-and-thrust belts and accretionary wedges. *Journal of Geophysical Research*, **88**, 1153–1172.

DECELLES, P. G. 1988. Middle Cenozoic depositional, tectonic, and sealevel history of southern San Joaquin basin, California. *American Association of Petroleum Geologists*, **72**, 1297–1322.

——, GRAY, M. B., RIDGWAY, K. D., COLE, R. B., SREIVASTAVA, P., PEQUERA, N. & PIVNIK, D. A. 1991. Kinematic history of a foreland uplift from Paleocene synorogenic conglomerate, Beartooth Range, Wyoming and Montana. *Geological Society of America Bulletin*, **103**, 1458–1475.

DIXON, J. M. & LIU, S. 1991. Centrifuge modelling of the propagation of thrust faults. *In*: MCCLAY, K. R. (ed.) *Thrust Tectonics*. Chapman & Hall, London, 53–69.

FAILL, R. T. 1973. Kink-band folding, Valley and Ridge Province, Pennsylvania. *Geological Society of America Bulletin*, **84**, 1289–1341.

FRASER, G. S. & DECELLES, P. G. 1992. Geomorphic controls on sediment accumulation at margins of foreland basins. *Basin Research*, **8**, 233–252.

HARDY, S. & WALTHAM, D. 1992. Computer Modeling of tectonics, eustacy, and sedimentation using the Macintosh. *Geobyte*, **7(6)**, 42–52.

——, POBLET, J., MCCLAY, K. & WALTHAM, D. 1996. Mathematical modelling of growth strata associated with fault-related fold structures. *In*: BUCHANAN, P. G. & NIEUWLAND, D. A. (ed.) *Modern Developments in Structural Interpretation, Validation and Modelling*. Geological Society, London, Special Publications, **99**, 265–282.

JAMISON, J. W. 1987. Geometrical analysis of fold development in overthrust terranes. *Journal of Structural Geology*, **9**, 207–219.

—— & POPE 1996. Geometry and evolution of a fault-bend fold: Mount Bertha anticline, *Geological Society of America Bulletin*, **108**, 208–224.

JORDAN, T. E., ALLMENDINGER, R. W., DAMANTI, J. F. & DRAKE, R. E. 1993. Chronology of motion in a complete thrust belt: the Precordillera, 30–31 S, Andes Mountains. *Journal of Geology*, **101**, 135–156.

JOHNSON, A. M. & BERGER, P. 1989. Kinematics of fault-bend folding. *Engineering Geology*, **27**, 181–200.

JOHNSON, N. M., JORDAN, T. E., JOHNSSON, P. A., NAESER, C. W. 1986. Magnetic polarity stratigraphy, agte and tectonic setting of fluvial sediments in an eastern Andean foreland basin, San Juan Province, Argentina. *In*: ALLEN, P. & HOMEWOOD, P. (eds) *Foreland basins*. International Association of Sedimentologists, Special Publications, **8**, 63–75.

KENYON, P. M. & TURCOTTE, D. L. 1985. Morphology of a delta prograding by bulk sediment transport. *Geological Society of America Bulletin*, **96**, 1457–1465.

KOOI, H. & BEAUMONT, C. 1994. Escarpment evolution on high-elevation rifted margins: Insights derived from surface processes model that combines diffusion, advection, and reaction. *Journal of Geophysical Research*, **99**, 12191–12209.

LAWRENCE, D. T., DOYLE, M. & AIGNER, T. Stratigraphic simulation of sedimentary basins: Concepts and Calibration. *American Association of Petroleum Geologists Bulletin*, **74**, No 3, 273–295.

MARTINEZ, A. & HARBAUGH, J. W. 1994. *Simulating near-shore environments*. Van Nostrand Reinhold.

——, VERGES, J. & MUÑOZ, J. A. 1988. Secuencias de propagacion del sistema de cabalgamientos de la terminacion oriental del manto del Pedraforca y relacion con los conglomerados sinorogenicos. *Acta Geologica Hispanica*, **23**, 119–128.

MEDWEDEFF, D. A. 1989. Growth fault-bend folding at south-east Lost Hills, San Joaquin Valley, California. *American Association of Petroleum Geologists Bulletin*, **70**, 131–137.

MIALL, A. D. 1986. Eustatic sealevel changes interpreted from seismic stratigraphy: a critique of the methodology with particular reference to the North Sea Jurassic record. *American Association of Petroleum Geologists Bulletin*, **70**, 131–137.

MITRA, S. 1992. Balanced structural interpretations in fold and thrust belts. *In*: MITRA, S. & FISHER, G. W. (eds) *Structural Geology of fold and thrust belts*. Johns Hopkins University Press, Baltimore, 53–77.

MOUNT, V. S., SUPPE, J. & HOOK, S. 1990. A forward modeling strategy for balancing cross-sections. *American Association of Petroleum Geologists Bulletin*, **74**, 521–531.

ORI, G. G. & FRIEND, P. F. 1984. Sedimentary basins formed and carried piggyback on active thrust sheets. *Geology*, **12**, 475–478.

PASCUAL, R. 1984. Late Tertiary Mammals of southern South America as indicator of climatic deterioration. *Quaternary of South America and Antarctic Peninsula*, **2**, 1–30.

PAOLA, C. 1988. Subsidence and gravel transport in alluvial basisn. *In*: KLEINSPEHN, K. L. & PAOLA, C. (eds) *New perspectives in basin analysis*. Springer Verlag, 231–245.

—— & WILLOCK, P. 1988. Downstream fining in gravel-bed rivers (abs). *EOS*, **70**, 852.

PEPER, T. 1993. *Tectonic control on the sedimentary record in foreland basins; inferences from quantitative subsidence analyses and stratigraphic modelling*. PhD thesis, Vrije Universiteit, Amsterdam.

PLATT, J. P. 1988. Dynamics of orogenic wedges and the uplift of high-pressure metamorphic rocks.

Geological Society of America Bulletin, **97**, 1037–1053.

POSAMENTIER, H. W. & ALLEN, G. P. 1993. Variability of the sequence stratigraphic model: effects of local basin factors. *Sedimentary Geology*, **86**, 91–109.

—— & VAIL, P. R. 1988. Eustatic controls on clastic deposition II – sequence and systsms tract models. *In*: WILGUS, C. K., HASTINGS, B. S., POSAMENTIER, H., VAN WAGONER, J., ROSS, C. A. & KENDALL, C. G. ST. C. (eds) *Sealevel changes – an integrated approach*. SEPM Special Publications, **74**, 125–154.

RAMSAY, J. G. 1991. Some geometric problems of ramp-flat thrust models. *In*: MCCLAY, K. R. (ed.) *Thrust Tectronics*. Chapman & Hall, London, 191–200.

RAY, R. R. 1982. Seismic stratigraphic interpretation of the Fort Union Formation, western Wind River basin: example of subtle trap exploration in a nonmarine sequence. *In*: HALBOUTY, M. T. (ed.) *The deliberate search for a subtle trap*. Memoirs, American Association of Petroleum Geologists Memoirs, **32**, 169–180.

RICH, J. L. 1934. Mechanics of low-angle overthrust faulting as illustrated by the Cumberland Thrust Block, Virginia, Kentucky, and Tennessee. *American Association of Petroleum Geologists Bulletin*, **18**, 1584–1596.

RIVENAES, J. C. 1992. Application of a dual-lithology, depth-dependent diffusion equation in stratigraphic simulation. *Basin Research*, **4**, 133–146.

ROURE, F. & SASSI, W. 1995. Kinematics of deformation and petroleum system appraisal in Neogene foreland fold-and-thrust belts. *Petroleum Geoscience*, **1**, 253–269.

——, HOWELL, D., GUELLEC, S. & CASERO, P. 1990. Shallow structures induced by deep seated thrusting. *In*: LETOUZEY, J. (ed.) *Petroleum and tectonics in mobile belts*. Editions Technip, Paris, 15–30.

SCHLAGER, W. 1993. Accommodation and supply – a dual control on stratigraphic sequences, *Sedimentary Geology*, **86**, 111–136.

SHAW, J. & KELLERHALS, R. 1982. The composition of Recent alluvial gravels in Alberta river beds. *Alberta Research Council Bulletin*, **41**, 151.

—— & SUPPE, J. 1991. Slip rates on active blind thrust faults in the Santa Barbara Channel, Western Transverse Ranges fold-and-thrust belt. *Eos (Transactions of the American Geophysical Union)*, **72**, 443.

SUPPE, J. 1983. Geometry and kinematics of fault-bend folding. *American Journal of Science*, **283**, 684–721.

——, CHOU, G. T. & HOOK, S. C. 1991. Rates of folding and faulting determined from growth strata. *In*: MCCLAY, K. R. (ed.) *Thrust tectronics*. Chapman & Hall, London, 105–121.

SYVVITSKI, J. P. M., SMITH, J. N., CALABRESE, E. A. & BOUDREAU, B. P. 1988. Basin sedimentation and the growth of prograding deltas. *Journal of Geophysical Research*, **93**, 6895–6908.

TETZLAFF, D. M. & HARBAUGH, J. W. 1990. *Simulating clastic sedimentation.* Van Nostrand Reinhold, New York.

VAIL, P. R, MITCHUM, R. M., TODD, R. G., WIDMIER, J. M., THOMPSON III, S., SANGREE, J. B., BUBB, J. N. & HATLELID, W. G. 1977. Seismic stratigraphy and global changes of sealevel. *In*: PAYTON, C. E. (ed.) *Seismic stratigraphy – applications to hydrocarbon exploration*. American Association of Petroleum Geologists Memoirs, **26**, 49–212.

VERSCHUREN, M., NIEUWLAND, D. & GAST, J. 1996. Multiple detachment levels in thrust tectonics: Sandbox experiments and palinspastic reconstruction. *In*: BUCHANAN, P. G. & NIEUWLAND, D. A. (eds) *Modern Developments in Structural Interpretation, Validation and Modelling*. Geological Society, London, Special Publications, **99**, 227–234.

ZOETEMEIJER, R. & SASSI, W. 1992. 2-D reconstruction of thrust evolution using the fault-bend fold method. *In*: MCCLAY, K. R. (ed.) *Thrust Tectonics*. Chapman & Hall, London, 133– 140.

——, ——, ROURE, F. & CLOETINGH, S. 1993. Stratigraphic and kinematic modeling of thrust evolution, northern Apennines, Italy. *Geology*, **20**, 1035–1038.

Flexure and 'unflexure' of the North Alpine German–Austrian Molasse Basin: constraints from forward tectonic modelling

B. ANDEWEG & S. CLOETINGH

Faculty of Earth Sciences, Vrije Universiteit, De Boelelaan 1085, 1081 HV Amsterdam.

Abstract: We present the results of forward modelling of the Northern Alpine German–Austrian Molasse Basin, which forms part of the Northern Alpine Foreland Basin (NAFB) extending from Lake Geneva in the west to Lower Austria in the east. The observed deflection of the European plate under the NAFB has been modelled along five profiles perpendicular to the basin axis. Models treating the deflection of the NAFB as a flexural response to loading only, require a set of loading parameters (bending moment, vertical force) which would imply bending stresses exceeding the strength of the subsiding plate. Moreover, this approach would not take into account the significant post-Molasse uplift experienced by the Alpine chain and its northern foreland basins. We modelled the deflection as the response to two distinct processes: (1) flexure of an elastic plate with lateral variations (EET-values of 7–26 km from east to west) loaded with surface and limited subsurface loads and (2) the Late Cenozoic (post-molasse) phase of 'unflexing', significant uplift observed in the Alps and its northern foreland. Our study provides a first attempt to separate these two processes in order to model the deflection by an elastic plate-model, adopting flexural parameters that will not exceed the strength of the lithosphere.

The Alpine mountain chain is a classic continent-continent collision zone. During its evolution the African continental plate overthrusted the European margin and stacked slices of the European plate on to the European plate. This acted as a (topographic) load translating over the underthrusting plate. Due to this loading a flexural deflection developed in front of the thrust belt. As a result of the in general low flexural rigidity of the downgoing lithosphere this led to a deep and narrow basin, just as the Ebro (Zoetemeijer *et al.* 1990; Millan *et al.* 1995), Apennine (Royden 1993; Kruse & Royden 1994), Carpathian (Matenco *et al.* 1997) and Swiss Molasse basins (Sinclair *et al.* 1991). The Northern Alpine Foreland Basin (NAFB, see Fig. 1) is considered to be a classical example of this type of basins, featuring a strongly asymmetrical geometry, deepening towards the Alpine mountain front and present along all of the northern part of the Alps and the Carpathians. Although its overall shape appears to be primarily the result of the emplacement of the Alpine nappes, remarkable lateral variations in geometry and kinematics of the NAFB can be observed that distort the simple theoretic asymmetrical shape. The Basin is developed or preserved best in its German part where it is 150 km wide and contains a sedimentary infill of upto almost 5000m (Lemcke 1988; Bachmann & Muller 1992). To the east, in direction of Austria, the Basin narrows to only 15 km and becomes shallower where the crystalline Bohemian Massif is close to the frontal thrusts of the Alps and broadens again to the Carpathian foreland. To the west, in Switzerland, the Basin narrows as well, probably related to the considerable Cenozoic uplift of up to 2500 m (Lemcke 1988) that has destroyed part of its original extent. In this respect it might not be coincidence that this highly uplifted part is situated at the southern termination of the Rhine Graben (Gutscher 1995) and related to local uplift leading to the Jura Mountains. These lateral variations from Switzerland to Austria are probably the product of a combination of processes such as variations in strength of the downgoing European lithosphere, pre-subsidence geological setting, lateral inhomogeneties of the basement (basement faults) and recent significant laterally varying uplift.

Modelling the flexure of the European plate underlying the NAFB has been the subject of several studies over the last decades. Most of the authors conclude that the geometry of the NAFB, the asymmetric flexural depression that developed in front of migrating thrust loads of the Alps, can be modelled by an elastic plate overlying an invisous fluid loaded by the orogenic thrust wedge. The resultant values for effective elastic thickness of the downflexing European plate show, however, a very wide range: from 35–50 km to only 7.5 km. A source of controversy between most of the studies is the concept of additional subsurface loads; applying a bending moment (M_o) and/or a vertical shear forces (V_o) at the free end of the plate to enhance deflection. In order to simulate the

ANDEWEG, B. & CLOETINGH, S. 1998. Flexure and 'unflexure' of the North Alpine German–Austrian Molasse Basin: constraints from forward tectonic modelling. *In:* MASCLE, A., PUIGDEFÀBREGAS, C., LUTERBACHER, H. P. & FERNÀNDEZ, M. (eds) *Cenozoic Foreland Basins of Western Europe.* Geological Society Special Publications, **134**, 403–422.

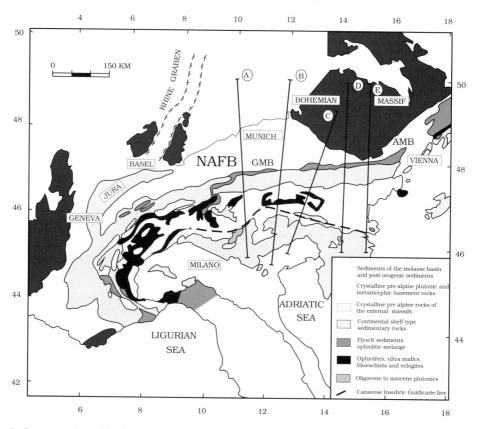

Fig. 1. General setting of the Alpine mountain chain showing the arcuate couple of thrust belt and foreland basin and different tectono-sedimentary areas. Location of the Northern Alpine Foreland Basin (NAFB), the German Molasse Basin (GMB), the Austrian Molasse Basin (AMB) and the five modelled profiles (A–E) is indicated.

observed deflection many authors make unlimited use of these subsurface loads. To obtain better constraints on the flexural parameters, gravity data have been incorporated in the modelling as an independent constraint.

In the Swiss Molasse Basin both stratigraphy (Sinclair et al. 1991; Allen et al. 1991), and flexure (Lyon-Caen & Molnar 1989) have been modelled. In contrast, previous work in the German Molasse Basin (GMB) has focused almost entirely on very detailed sedimentological and stratigraphical work, with the aim of correlating the different lithostratigraphic levels of the western and eastern part of the GMB (Bachmann & Müller 1992; Bachmann et al. 1987). Some limited modelling of the observed gravity anomalies over the NAFB has been carried out by Karner & Watts (1983) and Royden (1993). The latter author also performed a flexural study. More recently, sequence-stratigraphic patterns combined with flexure of the GMB

have been modelled on a short section through the basin east of Munich by Jin (1995).

The NAFB, the depression of the European lithosphere at the northern side of the Alpine chain, initially is the flexural response of this plate due to loading by overthrusting of the African plate and emplacement of the Alpine nappes. The basin has been filled with Molasse sediments related to the ongoing subsidence. After this deposition, the Alpine chain and its surroundings recently (from Miocene times on) have experienced significant uplift (Lemcke 1988; Jouanne et al. 1995; Genser et al. 1996, 1997). These two distinct periods in the evolution of the GMB give rise to a very particular geometry: a flexural asymmetric foreland basin that has experienced differential uplift together with the Alpine thrust belt that caused the deflection. Therefore, it seems obvious that it is impossible to model the presently observed geometry of the NAFB by a bending elastic plate only: (too) large subsurface loads would be required to account for the tight curvature. In this paper we investigate whether the elastic plate model can adequately explain the observed geometry if the two processes are being separated. Using uplift data to restore the pre-uplift situation and well-log, deep seismic and gravity data, the observed geometry along five profiles through the NAFB has been modelled.

Stratigraphy of the Northern Alpine Molasse Basin

The NAFB is filled with (at the present Alpine thrust front) up to 5000 m predominantly clastic sediments of Tertiary age. The base of this so-called Molasse lapped progressively northwards on to a peneplained differentiated basement consisting of Mesozoic shelf sediments, local Permo-Carboniferous graben sediments and Variscan basement. Seismic and well data show that the southward-dipping sediments continue underneath the Alpine nappes to at least 50 km (Lemcke 1988).

The stratigraphy of the NAFB is not uniform laterally and major lithostratigraphical differences occur between the western and eastern part of the Basin (see Fig. 2). In the eastern part, more or less marine environments prevailed throughout history, and the western part has been submerged frequently. Jin (1995) distinguished in the area east from Munich three shallowing-upward sequences. In the Swiss foreland basin, classically only two shallowing-up sequences have been described (e.g. Allen *et al.* 1991). For the western part of the GMB, the latter subdivision is more useful. The described transgressions or regressions are due to relative sea-level changes, which can be the consequence of different processes (including intra-plate stress fluctuations, change of sediment input causing over- or underfilling of the basin, eustatic sea-level changes, see e.g. Peper *et al.* 1994).

Sedimentation in the GMB started in the Priabonian (Late Eocene) due to a world-wide rise of sea level (Bachmann & Müller 1992). In an eastward-shallowing and broadening (deep) sea trough, with an axis approximately 50 km south from the present Alpine thrust front, sedimentation of flysch occurred in turbiditic sequences, of up to 650 m ('Deutenhausener Schichten') in the centre of the trough. In the Austrian Molasse Basin (AMB) fluvio-lacustrine sediments were deposited in the north and limestones and marls on the deeper shelf to the southwest (Nachtmann & Wagner 1986).

The flysch sediments (North Helvetic Flysch) are found as allochthonous remnants in the Folded Molasse, north from the original sedimentation-trough. The flysch is pinching out rapidly in northeastern direction, where a thin cover of Basal Sandstones (20–70 m) and Lithothamnium Limestone (0–100 m) is sedimented as transgression continues over the at least 30 Ma peneplained basement of Cretaceous and karsted Malm. These sediments can be considered to be the first *sensu strictu* Molasse sediments (Lemcke 1988). In the northeastern part of the GMB and the north of the AMB sedimentation of erosion products of the Bohemian Massif is significant (Nachtmann & Wagner 1986). In this area the rate of subsidence (Lemcke 1988) was most of the time kept up by the sedimentation, enabling continental conditions to prevail throughout the development of the basin.

During Rupelian, the sea transgressed more to the north, the basin subsided very rapidly in the southeastern part, where the Rupelian sediments reach a thickness of more than 1200 m. To the west the basin is less deep and less broad: the Lower Marine Molasse (German: UMM), which consists of deeper water (palaeobathymetry: 100–200 m) marls, is near Lake Constance only some 100 m thick and pinches out not far northwards (Bachmann & Müller 1992).

At the Rupelian–Chattian boundary a most likely eustatic induced regression makes the shoreline to shift approximately 150 km to the east, to the Freising–Munich–Miesbach line, where it will remain until the early Late Aquitanian (Eggenburgian). This regression is supposed to be induced by the largest sea-level change since the Cambrian because basin subsidence continues at the same rate and no additional sediment input is observed that would be able to drown the basin. During this regression the Baustein beds, sandbodies derived from the area where the present Folded Molasse is situated and Eastern Switzerland (Lemcke 1988), are deposited.

In the Chattian (Early Egerian) deposition of the Lower Freshwater Molasse (German: USM) occurs in the WGMB. Large river systems, that ran nearly parallel to the mountain chain to the sea in the east (Lemcke 1988), deposited large amounts of sand and silt (LFM 1). In the EGMB and the AMB, marine conditions prevailed, enabling the sedimentation of vast sheets of dark marls (Upper Chattian and Lower Aquitanian Marls). The marls transgressed (moderate sea-level rise and/or decrease in subsidence) at the Chattian–Aquitanian boundary over the Chattian Sands, but regressed again during Aquitanian (Late Egerian) due to a renewed cycle of the Lower Freshwater Molasse (LFM 2).

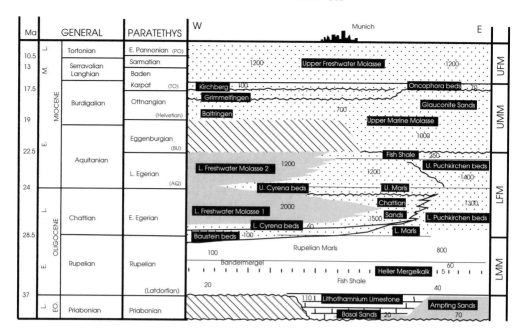

Fig. 2. Schematic west–east stratigraphic section through the NAFB showing generalized sedimentary infill. Note the earlier onset of sedimentation in the eastern part of the basin and the lateral variations in thickness and sediment-type. Modified from Bachmann & Müller (1992).

The top of the Aquitanian became eroded before a second large transgression took place at the beginning of the Burdigalian (Eggenburgian), invoking renewed marl-deposition in the EGMB (Upper Marine Molasse). The ongoing transgression, which reached its peak during the Mid-Burdigalian (Ottnangian) and flooded the entire GMB, caused deposition of marls in the WGMB as well (Jin 1995) during Mid-Burdigalian (Ottnangian, German: OMM).

The marine basin became filled during Late Burdigalian (Late Ottnangian–Karpatian) and until the Tortonian (Late Pannonian), in the entire basin the Upper Freshwater Molasse (German: OSM) was deposited. This again took place parallel to the axis of the basin. This time, however, the sediment sources were situated in the western part of the Austrian Molasse Basin, so sediment transport occurred from east to west. This UFM marks the end of the flexure related sedimentation in the NAFB.

Flexural modelling

Modelling strategy

The flexural response of the European lithosphere to loading by the thrust sheets of the Alps has been studied along five profiles through this orogen and its northern Molasse Basin (for location profiles see Fig. 1). These profiles are taken more or less perpendicular to the axis of the basin. The observed deflection and gravity data were used to constrain the flexural parameters by forward modelling the deflection using a two-dimensional finite-difference technique (Bodine et al. 1981), which allows the incorporation of lateral variations in mechanical properties and distributed loads. For a broken-plate model, the deflection of an elastic plate under a given topographic load is calculated interactively, incorporating (secondary) loading by infill of the created space with material of a specified density. Variation of density with depth is not possible, so an average has been taken for the sedimentary infill. Additional subsurface loads (vertical shear forces, V_o and bending moments, M_o at the free end of the plate or horizontal stresses) can be incorporated (see Fig. 3). Finite-difference calculations have been performed along 2000 km long profiles to minimize end effects. Additionally, the gravity anomaly that would result from the flexural model is calculated. The infill of the flexural depression as well as the effect of the subsurface loads is taken into account in this gravity calculation.

The gravity data were kindly placed at our disposal by the ÖMV, density data of crust, mantle

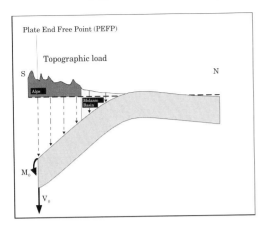

Fig. 3. Schematic illustration showing the modelling configuration. Loading due to a combination of topographic loading and additional subsurface forces applied to the free end of the overthrusted plate. V_o = vertical shear force, M_o = bending moment. Plate end free point (PEFP) is indicated.

and sediments (see Table 1) were taken from Gutscher (1995). Sedimentation of the Basal and Ampfing Sandstones (see Fig. 2) in the eastern GMB over the differentiated Mesozoic basement indicate the onset of flexure in Priabonian (Late Eocene). In the western part of the basin sedimentation related to the deflection of the European plate started in Early Oligocene with deposition of the Fish Shale of the Lower Marine Molasse. We, therefore, have taken the base of the Tertiary as the base of sedimentation in the NAFB related to the overthrusting of the Adriatic–African Plate, a process that started in Late Eocene times. Constraints on the observed depth of the base of the Molasse Basin are derived from well-logs and reflection seismic lines (Lemcke 1988, fig. 54) in the GMB (see Fig. 4). Subsidence of the basin continued until the Late Miocene (Jin 1995; Lemcke 1988), deposition of the Upper Freshwater Molasse marking the end of the Molasse sediments s.l., whereas subsidence ceased during the Mid-Burdigalian (Ottnangian c. 19–17.5 Ma) times in the Austrian part of the basin.

After this subsidence, the NAFB experienced a relative stable stage and from Late Pannonian (c. 7.1 Ma) uplift of several hundreds of metres (Genser et al. 1996). Data used in this work are derived from quantative subsidence analysis of a number of wells (Genser pers. comm.) for the Austrian part and from Lemcke (1988) for the German part of the basin (see Fig. 5). These data are in agreement with the recent uplift data obtained by geophysical and geodetic methods in the Central Alps (Trümpy 1980, Fig. 43) and the northwestern Alps, the Molasse basin and the southern Jura mountains (Jouanne et al. 1995). This uplift of up to 730 m suggests that the basin extended much further northwards during deposition of the Molasse than the present erosional edge. This edge is observed at nearly 700 m height in the eastern and at 400 m in the western part of the GMB whereas the sediments were deposited at approximately 50–100 m (Lemcke 1988). Although we realized that locally small scale tectonics might have added to the uplift (for example the development of the Jura Mountains), we have not been able to develop a separate database for a general uplift and local effects. Whenever the uplift database becomes more detailed, this should be taken into account.

We have restored the pre-uplift situation, subtracting the Cenozoic uplift values from the observed present basement depth and modelled this reconstructed palaeo-deflection. The fit to the present deflection can be obtained from the inferred palaeo-deflection plus the observed Cenozoic uplift. This will change the predicted gravity anomaly by:

$$\Delta g = 2\pi \Delta \rho G w_{(x)}. \qquad (\text{Eq. 1})$$

For $G = 6.67 \times 10^{-11}$ Nm^2 kg^{-2} and a density difference between mantle and infill $(\rho_m - \rho_i)$ of 800 kg m^{-3}, this difference in the anomaly will be -3.3527 mGal for every 100 m uplift. This is only correct if the uplift is relatively small, the contribution to the gravity anomaly of a density-difference decreases with increasing depth. Moreover, the origin of the 'unflexing' can have adjusted the anomaly significantly (dependent on how the volume increase due to the uplift is treated: density decrease versus input of mantle-material). We assumed the crustal thickness to remain constant during uplift. This seems to be

Table 1. Densities used in the forward modelling (taken from Gutscher 1995)

Materials	Density (km m^{-3}) used
Sediments	2450–2550
Load	2550–2670
Crust	2720–2900
Mantle	3200–3300

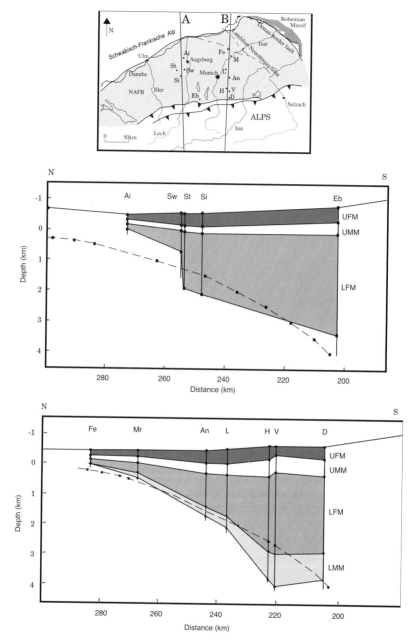

Fig. 4. Observations of the several stages (LMM, Lower Marine Molasse; LFM, Lower Freshwater Molasse; UMM, Upper Marine Molasse; UFM, Upper Freshwater Molasse) of the Molasse sediments in the NAFB. **Upper panel:** location of profiles A and B and deep drilling wells (Ai, Aichach 1001; Sw, Schwabmünchen 1; St, Scherstetten 1; Si, Siebnach 1, Eb (projected), Eberfing 1; Fe, Freising 1003; M, Moosburg 1; L, Landsham 1; An, Anzing 3; H, Holzkirchen 1; V, Vagen; and D, Darching 1). **Middle panel:** depth observations of the several Molasse stages for profile A. Dashed line shows shape of the basin used in the modelling (modified after Lemcke 1988, using seismics but neglecting local deformation). Position of the wells is given in respect to the origin of the profile lines.
Lower panel: profile B, figure convention as in middle panel.

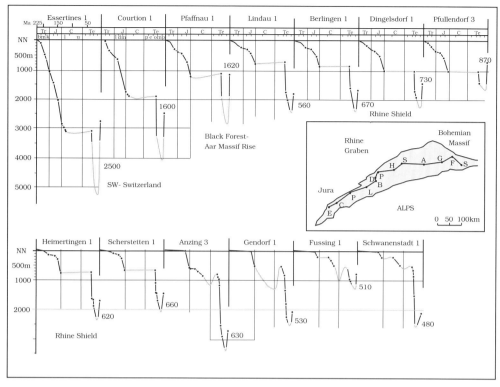

Fig. 5. Subsidence curves of several wells showing the significant Late Cenozoic uplift of the German part of the NAFB. Grey lines denote eroded parts of the stratigraphy. Inset figure shows position of the wells. Uplift values in numbers along the curves. Tr, Triassic (subdivision b, Buntsandstein; m, Muschelkalk; k, Keuper); J, Jurassic (l, Lias; d, Dogger; m, Malm); C, Cretaceous (l, Lower; u, Upper); Te, Tertiary (p, Paleogene; e, Eocene; o, Oligocene; m, Miocene; p, Pliocene). Notice the lateral variation and the general decrease of the uplift from west to east (after Lemcke 1974).

justified because during the small timespan (c.5 Ma) of the uplift, the crustal thickness would not increase (by thermal aging) or decrease significantly. Combined with the uplift profile, used to restore the pre-uplift geometry, the pre-uplift situation has been modelled. This approach provides only a first-order estimate as the uplift database of the basin is far from complete and not much is known about palaeo-topography of the Alps. There are some ideas concerning palaeo-topography, but since equation 1 is only valid for cylindrical topography, more accuracy is not needed. After the most important thrust events during Tertiary times, at least part of the mountain chain has been uplifted. It is not the prime aim of this study to derive a detailed estimate of the palaeo-topography or 'palaeo-gravity anomalies' of the Alps. Here we illustrate only the concept of treating the present geometry of the NAFB as the result of a flexural deflection of the European plate subsequently modified by recent uplift of several hundreds of metres.

In this way, the pre-uplift situation has been restored for all of the profiles (see Fig. 6 for profile B). Using flexure modelling software (Zoetemeijer 1993), the deflection of the European Plate was modelled. Free parameters were: EET, Plate end free point (PEFP), V_o and M_o. The outcome is not an unique solution, but the combination of the four free parameters confine the range of possibilities and narrow down adequately the range of possible solutions.

Subsurface loads

To obtain the deflection of the lithosphere underlying the NAFB, topographic loading alone is not sufficient. This situation is similar to the Carpathians and Apennine foreland basins, requiring additional subsurface loads (Royden & Karner 1984). With these loads acting on the foreland lithosphere that are not expressed as

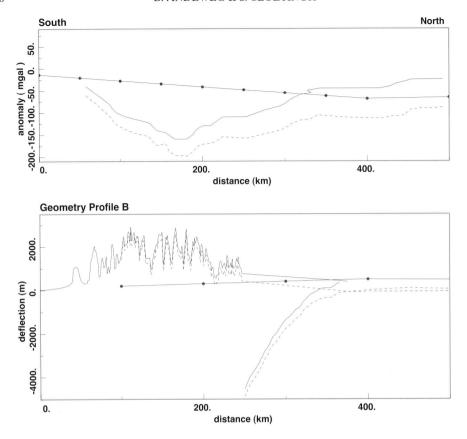

Fig. 6. Approach to restore the pre-uplift situation for Profile B. **Upper panel:** observed Bouquer anomaly (solid line), anomaly difference resulting from uplift (solid line with dots) and pre-uplift anomaly (dashed line). **Lower panel:** present-day topography (solid line), uplift profile (solid line with dots) and pre-uplift (uplift subtracted from present-day topography) geometry (dashed line).

surface topography (see Fig. 3), it appears to be possible to reproduce the deflection. The nature of these subsurface loads, however, is not known. Various explanations have been proposed over the last decade: density variations at subcrustal levels (Karner & Watts 1983) caused by a dense subducted slab at subcrustal depths (Royden 1993), transmitted horizontal compression generated by the interaction of plates and their boundaries (Karner & Watts 1983) or dynamic stresses related to subduction (Royden 1993) or downgoing convective flows beneath the lithosphere (Burov & Diament 1995), pre-orogenenic structure (Stockmal et al. 1987), emplacement of ultramafic bodies onto the downflexing plate (Royden 1993), overthrusting of a deep water continental margin (Royden 1993) or crustal thinning associated with back-arc rifting processes during active subduction (Karner & Watts 1983). This latter can exist only if the subduction rate is faster than the convergence rate, which in the Alps was most likely not the case.

However, some of the above-mentioned processes have taken place or are observed in the northern Alpine region. It is not possible to reproduce the tight curvature of the European plate by surface-loading by the Alpine chain only. Moreover, when applying subsurface loads taking into account their limits, the deflection of the NAFB can be fitted for every profile. The plate-boundary loads can create significant local strength variations in the bending lithosphere (Burov & Diament 1995). In many of the previous works the subsurface loads were used without limits, probably even much larger than

Table 2. Values of the flexural parameters used for profile 5b αbb in Royden (1993)

Parameter	E Alps
EET (km)	40
D (Nm)	4.6×10^{23}
M_o used (N m^{-1})	2.5×10^{17}
w_{max} (m)	6200
α (km)	127.97
M_{max} (N m^{-1})	1.12×10^{17}
	W Alps
EET (km)	50
D (Nm)	9×10^{23}
M_o used (N m^{-1})	9.0×10^{17}
w_{max} (m)	14000
α (km)	151.35
M_{max} (N m^{-1})	3.54×10^{17}

Upper part: taken from Royden (1993, table 2), w_{max} estimated from Royden (1993 figs 9 & 10). Lower part: numerically calculated values. The bending moment used exceeds by far the value that can be supported by the plate.

can be supported by an elastic plate under the used conditions. We made an estimate of the maximum bending moment that can be supported. This estimate (see Appendix) is based on rheological laws and data confining yield-stress envelopes (e.g. Burov & Diament 1995) and analytical solutions of a simple deflection model (Turcotte & Schubert 1982). The lithosphere that is used in the estimate is homogeneous and undeformed. As the lithospheric plate under the NAFB is far from homogeneous and undeformed upon onset of loading, the maximum applicable values for the subsurface loads will therefore be smaller than the theoretical determined values. Therefore, the calculated values can at least be used as an upper limit for the values of the subsurface loads. As an illustration Table 2 shows the flexural parameters that were used by Royden (1993) for forward modelling the deflection of the European plate along profiles through the Alps and its northern foreland. Using the equations (see Appendix), the maximum bending moment has been determined analytically and is displayed in the tables as well. Inspection of Table 2 shows that if these values would be applied, the elastic strength of the downgoing plate would be exceeded in these cases and, at least part of, the plate would deform plastically or break up.

Results

To model the observed deflection along the chosen profiles through the NAFB, without taking into account the recent uplift, requires a set of parameters incompatible with the limits of the strength of the elastic plate and the maximum values of the subsurface loads (see

Table 3. Values of the flexural parameters that would be required to obtain the observed deflection along profile B without taking into account the Late Cenozoic uplift

Parameter	
EET (km)	26
D (Nm)	1.09×10^{23}
M_o used (N m^{-1})	1.7×10^{17}
w_{max} (m)	12542
α (km)	90.58
M_{max} (N m^{-1})	1.08×10^{17}

As can be observed, the bending moment used would exceed the value that can be supported by the plate.

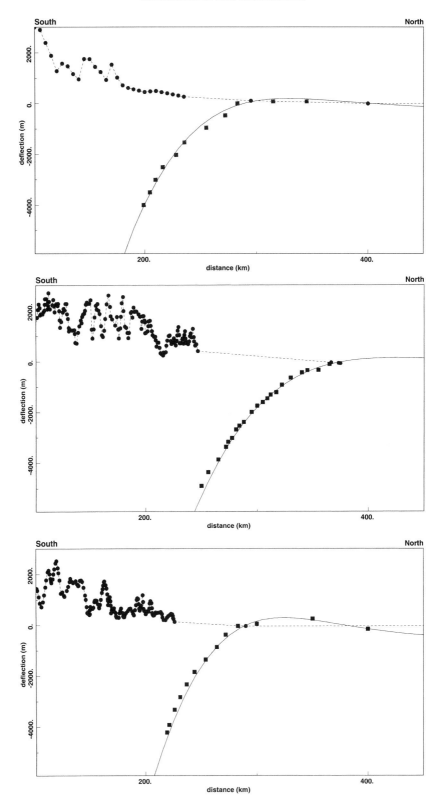

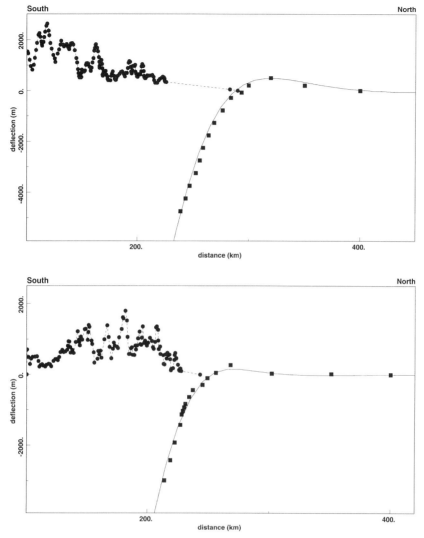

Fig. 7. Best fit for the profiles (A–E) showing topography (dashed line), basement depth data (squares) and the modelled deflection (solid line). For densities used see Table 1, for values of the parameters see Table 4.

Appendix). In Table 3 the parameters that would be needed for profile B are shown, obviously the applied bending moment is too large and as for profile B, the M_0 values used would exceed M_{max} in all of the profiles.

Following the approach of subtracting the Late Cenozoic uplift from the present topography and using this pre-uplift setting to model the deflection, yields results that for the subsurface loading appear to be much more realistic. Subsurface loads that are applied are only a fraction of the maximally applicable values. The best fits for the profiles are summarized in Table 4 (for location of the profiles see Fig. 1).

Profile A (Fig. 7a). This profile is close to the location of the European Geotraverse (1992). The best fit is obtained by adopting an EET value of 25 km, a value that is somewhat higher than was suggested by Cloetingh & Banda (1992) on the base of rheological models combined with seismicity cut-offs. The topographic load is largest for profile A, not only due to high topography in the Alps, but also due to the fact that the plate continues far south underneath it (PEFP:100 km from the origin of the profile, see Fig. 8). When V_o is subtracted from the numerically determined V_{max}, the load of the Alps results to be 0.93×10^{13} N m^{-1}, a value that is two

Table 4. *Table showing values for parameters used to obtain best fits and nummerically determined maximum values*

Profile	EET (km)	W_o (m)	W_o est.	$\rho_m - \rho_i$	PEFP (km)	D (Nm)	α (km)	M_o (N m^{-1})	M_{max} (N m^{-1})	V_{max} (N m^{-1})	V_o (N/m)	Load Alps (N m^{-1})
A	25	21.756	21.750	800	100	9.72×10^{22}	83.9	1×10^{16}	1.93×10^{17}	1.43×10^{13}	0.50×10^{13}	0.93×10^{13}
B	26	17.471	18.600	800	160	1.09×10^{23}	86.4	1×10^{16}	1.64×10^{17}	1.18×10^{13}	0.50×10^{13}	0.68×10^{13}
C	21	16.534	16.500	800	140	5.76×10^{22}	73.6	2.5×10^{16}	1.11×10^{17}	0.94×10^{13}	0.35×10^{13}	0.59×10^{13}
D	16	9.622	9.500	800	210	2.55×10^{22}	60.1	1×10^{16}	0.44×10^{17}	0.45×10^{13}	0.45×10^{13}	0.0×10^{13}
E	07	7.633	7.500	800	180	2.13×10^{21}	32.3	–	0.10×10^{17}	0.03×10^{13}	0.06×10^{13}	-0.03×10^{13}

W_o est. (estimated) is the observed maximal depth of the European plate, taken from Gutscher (1995). $\rho_m - \rho_i$ is the density difference between mantle and infill of the basin. PEFP is the plate end free point, the position of the free end of the downflexing European plate in respect to the origin of the profile-lines. D is flexural rigidity and α the flexural parameter. M_{max} and V_{max} have been determined nummerically. M_o and V_o are the applied values. The topographic load of the Alps has been determined by subtracting V_o from V_{max}. Note decreasing deflection, EET and applied loads from A to E.

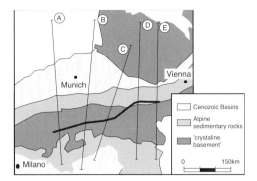

Fig. 8. Position of the PEFP along the profiles. The solid line connects the plate end free point of the five profiles. The increase in curvature of the plate (Fig. 10) towards the east causes: (1) narrowing of the basin to the east; (2) the lithospheric plate to continue less under the mountain chain to the east (that means less topographic load in the same direction).

to three times higher than the $3-4 \times 10^{12}$ N m^{-1} found by Gutscher (1995) leading to a conclusion that topographic loading is important.

Profile B (Fig. 7b). Like profile A, this profile is taken over the broad part of the GMB and yields results that are in the same order: an EET-value of 26 km and the downflexing European plate continues far under the topographic load (PEFP is 160 km, but the origin of this profile was placed slightly more southwards than that of profile A. Plotted in Fig. 10, the situation is very similar to profile A).

Profile C (Fig. 7c). This profile is in its northern part taken over the Bohemian Massif, along the southern border of this Massif, several basement faults (e.g. Donau- boundary trough) exist. The EET-value is slightly less than in profiles A and B: 21 km, the subsurface loads applied are in the same order or slightly lower. These values are required to model the tighter curvature of the European plate in this profile.

Profile D (Fig. 7d). The curvature of the deflection increases considerably when going from profile C to this profile. To reproduce this tighter curvature, the EET-value decreases to 16 km and the applied subsurface loads are closer to their maximum values than in the three above mentioned profiles. The free end of the plate in this profile is situated most basinward (see Fig. 8). Although the topography is relatively high, only small part of it is on top of the plate acting as a topographic load. This means that the topographic load should not be large. When decreasing the EET-value, the deflection can only be modelled when applying subsurface loads that are in the same order as the ones we used before. As a consequence, the strength of the plate would be exceeded even more. Upon decreasing the subsurface loads, it is impossible to obtain a good fit to the deflection.

Profile E (Fig. 7e). This profile shows a very particular setting, with the rheologically strong Bohemian Massif very close to the frontal thrusts of the Alps, narrowing the basin extremely (only some 15 km). A compilation of several seismic lines by Tari (1996) and Wessely (1987) suggest that the European plate continues under the Danube Basin in Hungary. A complete deep seismic line, however, does not exist in the part of Austria where profile E runs. Moreover, the depth quality of the several smaller lines is poor (Tari 1996). The continuation of the plate has been altered by the several (17 Ma and less important 10 Ma) stretching events in the Styrian–Vienna Basin area with stretching factors of 1.3–1.6 (see Sachsenhofer *et al.* 1997 and cross section Tari 1996). The topographic load is small due to less high topography (see Fig. 9) and only little continuation under the Alpine chain of the downgoing plate, but the observed curvature of the plate is tight. These boundary conditions yield as a result a very weak plate with an EET-value of only 7 km. Because topographic loading is not very important, subsurface loads need to be applied. Due to the considerable weakness of the plate, the applied values for the subsurface loads exceed, however, the maximum applicable values. This would lead to breaking up or plastic deformation of the plate. The resultant zero value of the topographic loading by the Alpine chain (see Table 4) might be explained by this process.

A larger density contrast between mantle and infill is improving the analytical results (the flexural parameter 1 decreases with increasing density contrast, see equation 8 in Appendix. Therefore M_{max} and V_o increase, see equations 6 and 7, Appendix, respectively). However, because the infill is less dense and the same deflection is being modelled, the EET-values will have to be decreased to obtain the same fit. This will in turn decrease the M_{max} and V_o. These two counter-active processes determine very well a small range of solutions.

Lateral variations in EET

The EET-values for all of the profiles are small (26–7 km) and follow more or less the depth of the base of the mechanical crust, corresponding

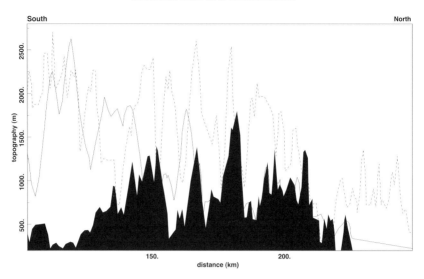

Fig. 9. Topographic load on profile lines B (dashed line), D (solid line) and E (black infill). Note significantly higher topography and thus larger topographic load for profile B.

with the 300–400°C isotherm (Burov & Diament 1995), pointing to crustal decoupling. The strength of the lithosphere is not very large due to pre-Alpine, Mesozoic stretching phases (see Fig. 5) and even more reduced (20–30%) by the plate boundary forces (Burov & Diament 1995). For profiles C, D and E strength most likely is even more reduced due to NW–SE-directed basement faults, that controlled sedimentation of Molasse sediments as well (e.g. Landshut–Neuöttinger zone, Donau-border fault). The lateral narrowing of the NAFB can not be explained by destruction of part of the originally full basin extent due to the Late Cenozoic uplift only. The curvature of the deflection (see Fig. 10) increases strongly from profile B to C and D, causing the broad part in the GMB to narrow laterally towards the AMB where the Bohemian Massif is as close as 15 km to the Alpine thrust sheets. The relatively rigid and strong Bohemian Massif might have squeezed the foreland basin when the Alpine front approached. The increase in curvature is explained by a decrease of the EET-values from 26 km to only 7 km (from respectively profile B to E). More to the east, where the NAFB continues into the northern Carpathian foreland basin, the basin widens again and EET estimates from forward modelling yield slightly higher values (in the order off 12 km near Brno, Zoetemeijer pers. comm.). Therefore, the results of our study appear to be in agreement with the observations and are compatible with the overall tectonic setting.

An equally good fit might be obtained with a situation in which the EET value decreases in the most flexed area. The calculation of the maximum applicable subsurface loads becomes very complex in that case, moreover because deformation might be localised in the weaker parts of the plate. Therefore we used a constant value for the EET in our modelling.

The position of the free end of the plate (PEFP) along the profiles (see Fig. 8) corresponds closely to the tightness of the deflection of the downgoing plate (the steeper the dip, the less easy to continue horizontally far under the mountain chain) and thus to the EET-value. In profile A and B, the European plate dips gently under the loading African plate and continues south to approximately the position of the Insubric Line. To the east in profiles C, D and E, the dip of the plate is much larger and, therefore, the plate does not continue very far southwards under the collision belt. In this way only a small part of the Alps is acting as a topographic load in the latter profiles. The cross section compiled by Tari (1994, 1996), suggests that the European Plate continues far south under the Danube Basin. Most likely this continuation is not formed by the entire plate, but only the upper crust whereas the lower crust and mantle part might have decoupled and broken down. This option, that seems in favour with rheological strength calculations showing a weak mantle and lower crust in this area (Lankreijer 1998), can not be modelled with the used flexural modelling approach. In this way, the subduction

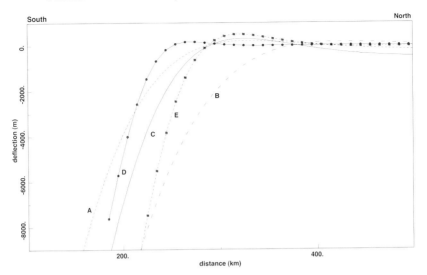

Fig. 10. Best fit to the restored deflection along all of the profiles compared. Closely dashed line: profile A, loosely dashed line: profile B, solid line: profile C, solid line with dots: profile D, dashed line with squares: profile E. Curvature increases progressively from profile A to E.

related Karpatian–Badenian volcanism observed in the southern Styrian Basin could be explained by downgoing and melting of the upper crust. The origin of this volcanism is however still a matter of debate (Sachsenhofer *et al.* 1997).

Origin of the Cenozoic uplift/'unflexing'

The 'unflexing' following the flexural phase (see Fig. 11 for a schematic illustration) is a process that is observed clearly in the NAFB (and other foreland basins as well, e.g. Desegaulx *et al.* 1991; Kruse & Royden 1994). Its origin, however, is far from clear. Royden (1993) concluded that subsurface loads actually cause uplift of the foreland basement beneath the foredeep basins on both sides of the Alps and only cause subsidence beneath the internal parts of the thrust belt. Geodetic observations however demonstrate that recent uplift-values under the outer parts of the basin are in the same order (0.5–1.2 mm a^{-1}) as close to the frontal thrusts or in the internal parts of the Alpine chain (Jouanne *et al.* 1995), interpreted as the result of ongoing deformation along low-angle basement faults. This at least shows that the Late Cenozoic uplift, that is observed in all of the Alpine chain as well as in its foreland basins, is not the result of the action of subsurface loads only.

A possible cause for the uplift would be recovery of strength. As long as orogeny continues, the strength of the lithosphere underlying the orogenic belt decreases; plate unloading results in an increase of the lithospheric strength (Burov & Diament 1995). Unloading of the plate by erosion or tectonic removal of the mountain load, would lead to major uplift in the orogeny, but only minor uplift in the foreland. Recovery of strength by the European plate would lead to unflexure. The term unflexure implies the decrease in amplitude of a bended plate, at least partly, back to its original unflexed shape. In other words, it would lead to subsidence of the bulge and uplift of the Alpine chain. This uplift pattern is not observed in the NAFB and therefore, unflexing can not be the governing process causing the uplift.

This uplift is used by Lyon-Caen & Molnar (1989) as an argument for an Alpine parallel upwelling theory. They point out that a possible cause for this process might be a diminution or termination of downwelling of mantle material beneath the Alps, due to breaking up of the downgoing plate ('slab detachment') and decrease of the north-south component of the convergence rate between the African and European plate. The calculated anomalies and the restored pre-uplift anomalies show remarkably equal trends (see Fig. 12 a b). However, the calculated anomaly is +80 mGal when compared with the restored one, which is supposed to be a large negative anomaly due to the downward flexure of the plate to produce the sedimentary basin. To fit the calculations to the observations, a large scale lithospheric process that affected all

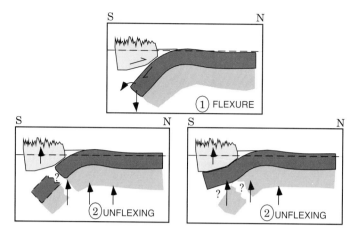

Fig. 11. Cartoon illustrating the two-phase evolution of the NAFB: (1) flexural response of the mechanically weak European lithosphere on loading by the overthrusting African plate creates a foreland basin; (2) two scenarios for Unflexing, Late Cenozoic uplift due to tectonic processes operating in the Alpine region. The thrust-belt and its foreland are uplifted several hundreds of meters in a post-molasse stage. Dashed line denotes approximately sea level.

of the Alpine region is needed. The problems encountered in modelling the profiles D and E, where the applied conditions in spite of all efforts still exceed the strength of the plate, would lead to breaking up of part of the plate. In this way, this study suggests decoupling of the upper crust, with breaking off of the lower crust and mantle, thus enabling the late Cenozoic uplift.

Thick plate versus thin plate models—a discussion

Studies concerning the Northern Alpine Foreland Basin have yielded a wide range of EET-values, from 7.5 km (Sinclair *et al.* 1991) to 53 km (Gutscher 1995). Karner & Watts (1983) estimated the EET of the European foreland to be between 25 to 50 km, but to obtain a good fit of the elastic model predictions to both flexure and gravity anomaly, additional subsurface loads (V_o ranging from $1\text{-}4 \times 10^{14}$ N m^{-1}) were adopted. The nature and origin of these subsurface loads are not known, but their existence seems to be justified by the fact that surface loads and crustal blocks alone can not produce the observed flexure, the so called model of 'insufficient topography', observed in the Apeninnes as well (Royden 1993). The contribution of the subsurface loads to both observed gravity and amplitude and wavelength of the flexural basin is substantial. The role of the surface load becomes increasingly important, however, as the strength of the lithosphere decreases. The strength of the European lithosphere under the Alps is small, due to pre-orogenic stretching and crustal shortening (Lyon-Caen & Molnar 1989), so loading by the Alpine orogenic belt should be of major importance in this case. Royden (1993) favours topographic loading as the cause of most of the deflection, with only a minor contribution of the subsurface loads in flexure under the Alps. However, in the sections presented in Royden (1993) both a too large M_o ($2.5\text{-}9.0 \times 10^{17}$ N) and V_o ($0.9\text{-}3.6 \times 10^{12}$ N m^{-1}) are applied at the free end of the plate.

Lyon-Caen & Molnar (1989) pointed out that the elastic model can not be adequate if such a set of parameters is required to account for the observed gravity gradient over the Molasse Basin and the Alpine Chain, and reject the use of unknown subsurface loads. Lyon-Caen & Molnar (1989) argue that the dynamic processes that created the deflection and built the Alpine Chain are no longer active. Our study supports the latter idea with the elastic plate model itself not being able to model the observed deflection of the European plate under the NAFB. Adopting additional subsurface loads appears to be justified as long as their limits are taken into account using the analytical equations (given in the Appendix).

Jin (1995) numerically estimated the wavelength of the lithosphere and in this way determined a flexural rigidity that corresponds to an EET value of 48 km for the underlying the German Molasse Basin. Since the shape of the deflection has been altered significantly due to

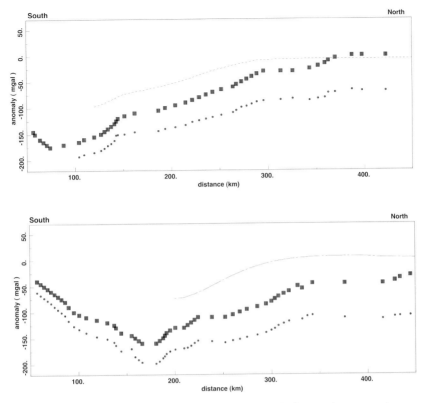

Fig. 12. Observed (squares), restored (dots) and calculated (dashed line) anomaly patterns along respectively profile A and C. The restored and calculated anomalies show remarkably equal trends but are offset c.80 mGal (profile A) and c.120 mGal (profile C).

the later uplift, this estimate seems not to be justified.

Recently Gutscher (1995) tried to explain the presently observed anomalously high topography in the NAFB by deflecting an elastic plate with large EET -values (53 km in the east to 34 km in the west over the Rhine Graben) loaded by the surface load of the Alps ($1-3 \times 10^{12}$ N m^{-1}) and additional V_o ($2-5 \times 10^{13}$ N m^{-1}). EET-values of this order seem in favor of crustal coupling (Burov & Diament 1995) strengthening the lithospheric plate considerably. In this way the large applied loads (surface and subsurface) can be supported by the lithosphere, creating a tight curvature and obtaining a bulge of 1000–1500 m. This, however is in contradiction with the geological observations: Upper Marine Molasse sediments of Mid-Miocene age, were sedimented below sea level (which differed approximately +0–100 m from the present sea level, Lemcke 1988; Jin 1995), but are nowadays exposed at levels up to 850 m above sea-level (Lemcke 1988, fig. 55). If such a flexural bulge would have been present before sedimentation of the Molasse as proposed by Gutscher (1995), the Molasse sediments (s.s.) should be onlapping onto the bulge, instead of being exposed on top of its crest. The last major loading event in this region is the emplacement of the Helvetic Nappes from Eocene to Miocene (50–10 Ma). The thrust nappes in this area reached their present-day position at about the Egerian–Eggenburgian boundary (Early Miocene c.20–18 Ma, Genser et al. 1997), so no significant shift or uplift of the bulge due to loading is to be expected after Miocene. The major shallowing upwards cycle of Upper Marine Molasse to Upper Freshwater Molasse was deposited in the flexural depression and only uplifted in a post-molasse stage (Lemcke 1988). Moreover, the EET values adopted by Gutscher (1995) and Jin (1995) are incompatible with estimates based on synthetic rheological profiles (Cloetingh & Burov 1996) and to the notion that the European plate was a very weak plate at the onset of loading due to Mesozoic extension (see Fig. 5).

This is another argument that supports the model presented here.

Conclusions

The deflection of the European plate underlying the NAFB due to overthrusting of the African plate, can be modelled in terms of an elastic plate model loaded by the Alpine thrust belt and limited subsurface loads with superposition of a Late Cenozoic (post flexural deflection) uplift of several hundreds of metres (see Fig. 11). Therefore, foreland basins that experienced post-molasse uplift, should not be studied as the effect of a flexural stage only, to avoid overestimates of subsurface loads and/or EET values.

The EET values that result from this study range from 7 to 26 km for the several profiles. These values are in accordance with models of a depth dependent continental rheology (e.g. Cloetingh & Banda 1992; Cloetingh & Burov 1996; Okaya *et al.* 1996) that predict for the foreland basins of the Alps a mechanically very weak lithosphere with characteristic EET-values of 10–25 km, as inferred for other parts of the European Alpine system (e.g. Zoetemeijer *et al.* 1990; Burov and Diament 1995; Cloetingh & Burov 1996; Okaya *et al.* 1996). The values adopted for the flexural parameters that deflect the lithosphere (V_o, M_o and EET) in most of the previous studies (e.g. Royden 1993) would exceed the elastic strength of the mechanically weak lithosphere that is present under the NAFB and thus cause plastic deformation or breaking-up. All these studies attempted to model the presently observed flexure as the result of one single flexural process. However, it is more likely that the present configuration of the NAFB is the result of (1) a flexural process forming the foreland basin due to loading by the African plate with superimposed (2) a subsequent 'unflexing' stage expressed in significant post-Molasse (Late Cenozoic) uplift of the whole region. This uplift might be caused by breaking up or delamination of part of the European crust.

We would like to thank J. Genser and R. Moeys for their contributions to the early phase of this work, H. P. Luterbacher and J. Zweigel for discussion and data. R. Zoetemeijer and A. Lankreijer are thanked for helpful discussions and comments. This work was funded by the IBS Project. Contribution 970124 of the Netherlands Research School of Sedimentary Geology.

Appendix 1

Analytical determination of the maximum values for the subsurface loads

Maximum bending moment

For a broken plate scenario, the maximum bending moment that can be applied to the lithospheric plate under consideration can be determined (see for a more detailed derivation Turcotte & Schubert 1982).

The bending moment of an elastic plate is given by:

$$M(x) = -D \frac{d^2w}{dx^2} \quad \text{(Eq.2)}$$

where D is the flexural rigidity of the considered plate.

The maximal applicable bending moment will reach its maximum when either D (large EET value) or $w''(x)$ is maximal and thus $w'''(x) = 0$. This is solved differentiating three times the general equation of the deflection of a broken plate:

$$w'''(x) = \frac{2w_0}{\alpha^3} e^{-\frac{x}{\alpha}} (\cos \frac{x}{\alpha} - \sin \frac{x}{\alpha}) = 0 \quad \text{(Eq.3)}$$

so that the location where the bending moment is maximal becomes:

$$x_{\max} = \frac{\alpha}{\tan}(1) = \frac{\pi}{4}\alpha. \quad \text{(Eq. 4)}$$

From eq.1 and realizing:

$$w''(x) = \frac{2w_0}{\alpha^2} e^{-\frac{x}{\alpha}} \sin \frac{x}{\alpha} \quad \text{(Eq. 5)}$$

the maximum bending moment at $x_{\max}$ is:

$$M_{\max} = \frac{-2Dw_0}{\alpha^2} e^{-\frac{\pi}{\alpha}} \sin \frac{\pi}{\alpha} \quad \text{(Eq. 6)}$$

Vertical Force

Like the bending moment M_o, V_o can be determined numerically:

$$V_o = \frac{4DW_0}{\alpha^3}. \quad \text{(Eq.7)}$$

This V_o-value is the combination of topographic load and an additional vertical force at the free end of the plate (plate end free point, PEFP).

The dependency of the flexural parameter α on the density contrast of infill and mantle material is given by:

$$\alpha = \left[\frac{4D}{g(\rho_m - \rho_i)} \right]^{\frac{1}{4}}. \quad \text{(Eq. 8)}$$

References

ALLEN, P. A., CRAMPTON, S. L. & SINCLAIR, H. D. 1991. The inception and early evolution of the North Alpine Foreland Basin, Switzerland. *Basin Research*, **3**, 143–163.

BACHMANN, G. H. & MÜLLER, M. 1992. Sedimentary and structural evolution of the German Molasse Basin. *Eclogae Geologicae Helvetiae*, **85**, 519–530.

——, —— & WEGGEN, K. 1987. Evolution of the Molasse Basin (Germany, Switzerland). *Tectonophysics*, **137**, 77–92.

BODINE, J. H., STECKLER, M. S. & WATTS, A. B. 1981. Observations of flexure and the rheology of the oceanic lithosphere. *Journal of Geophysical Research*, **86**, 3695–3707.

BUROV, E. B. & DIAMENT, M. 1995. The effective elastic thickness (Te) of continental lithosphere: What does it really mean? *Journal of Geophysical Research*, **100**, 3905–3927.

CLOETINGH, S. & BANDA, E. 1992. Mechanical structure of Europe's lithosphere. In: BLUNDELL, D., FREEMAN, R. & MUELLER, S. (eds) *A continent revealed, the European Geotraverse*, 80–91.

—— & BUROV, E. B. 1996. Thermomechanical structure of European continental lithosphere: constraints from rheological profiles and EET estimates. *Geophysical Journal International*, **124**, 695–723.

DESEGAULX, P., KOOI, H. & CLOETINGH, S. 1991. Consequences of foreland basin development on thinned continental lithosphere: application to the Aquitaine basin (SW France). *Earth and Planetary Science Letters*, **106**, 116–132.

GENSER, J., VAN WEES, J. D., CLOETINGH, S. & NEUBAUER, F. 1996. Eastern Alpine tectono-metamorphic evolution; Constraints from two-dimensional P-T-t modelling. *Tectonics*, **15**, 584–604.

GUTSCHER, M. A. 1995. Crustal structure and dynamics in the Rhine Graben and the Alpine foreland. *Geophysical Journal International*, **122**, 617–636.

JIN, J. 1995. *Dynamic stratigraphy analysis and modelling in the south-eastern German Molasse Basin.* Tübinger Geowissenschaftliche Arbeiten, **A24**.

JOUANNE, F., MENARD, G. & DARMENDRAIL, X. 1995. Present-day vertical displacements inthe northwestern Alps and southern Jura Mountains: Data from leveling comparisons. *Tectonics*, **14**, 606–616.

KARNER, G. D. & WATTS, A. B. 1983. Gravity anomalies and flexure of the lithosphere at mountain ranges. *Journal of Geophysical Research*, **88**, B12, 10449–10477.

KRUSE, S. E. & ROYDEN, L. H. 1994. Bending and unbending of an elastic lithosphere: the Cenozoic history of the Apennine and Dinaride foredeep basins. *Tectonophysics*, **13**, 278–302.

LANKREIJER, A. 1998. *Rheology and basement control on extensional basin evolution in Central and Eastern Europe: Variscan and Alpine–Carpathian–Pannonian tectonics.* PhD thesis, Vrije Universiteit, Amsterdam.

LEMCKE, K. 1972. Die Lagerung der jüngsten Molasse in nördlichen Alpenvorland. *Bulletin der Vereinigung Schweizerischen Petroleum Geologen und -Ingenieurs*, **39**, 29–41.

—— 1988. *Geologie von Bayern I. Das bayerische Alpenvorland vor der Eiszeit.* E. Schweizerbartische Verlagsbuchhandlung, Stuttgart.

LYON-CAEN, H. & MOLNAR, P. 1989. Constraints on the deep structure and dynamic processes beneath the Alps and adjacent regions from an analysis of gravity anomalies. *Geophysical Journal International*, **99**, 19–32.

MATENCO, L., ZOETEMEIJER, R., CLOETINGH, S. & DINU, C. 1997. Lateral variations in the mechanical properties of the Romanian external Carpathians: inferences of flexure- and gravity modelling. *Tectonophysics*, **28**, 147–166.

MILLAN, H., DEN BEZEMER, T., VERGES, J., MARZO, M., MUNOZ, J. A., ROCA, E., CIRES, J., ZOETEMEIJER, R. & CLOETINGH, S. 1995. Paleo-elevation and effective elastic thickness evolution of mountain ranges: inferences from flexural modelling in the Eastern Pyrenees and Ebro Basin. *Marine and Petroleum Geology*, **12**, 917–928.

NACHTMANN, W. & WAGNER, L. 1986. Mesozoic and Early Tertiary evolution of the Alpine forleand in Upper Austria and Salzburg, Austria. *Tectonophysics*, **137**, 61–76.

OKAYA, N., CLOETINGH, S. & MÜLLER, ST. 1996. A lithosperic cross-section through the Swiss Alps (part II): constraints on the mechanical structure in continent-continent collision zones. *Geophysical Journal International*, **127**, 399–414.

PEPER, T., VAN BALEN, R. & CLOETINGH, S. 1994. Implications of orogeny, intraplate stress variations and eustatic sealevel change for foreland basin stratigraphy – inferences from numerical modelling. In: DOROBEK, S. & ROSS, G. (eds) *Stratigraphy and foreland basins.* SEPM Special Publications, **52**, 5–35.

ROYDEN, L.H., 1993. The tectonic expression of slab pull at continental convergent boundaries. *Tectonics*, **12**, 303–335.

—— & KARNER, G. D. 1984. Flexure of the continental lithosphere beneath the Apennine and Carpathian foredeep basins: evidence for insufficient topography. *American Association of Petroleum Geologists Bulletin*, **68**, 704–712.

SACHSENHOFER, R. F., LANKREIJER, A., CLOETINGH, S. & EBNER, F. 1997. Subsidence analysis and quantitative basin modelling in the Styrian Basin (Pannonian Basin system, Austria). *Tectonophysics*, **272**, 175–196.

SINCLAIR, H. D., COAKLEY, B. J., ALLEN, P. A. AND WATTS, A. B. 1991. Simulation of foreland basin stratigraphy using a diffusion model of mountain belt uplift and erosion: an example of the central Alps, Switzerland. *Tectonics*, **10**, 599–620.

STOCKMAL, G. S., BEAUMONT, C. & BOUTILIER, R. 1987. Geodynamic models of convergent margin tectonics: transition from rifted margin to overthrust belt and consequences for foreland basin development. *American Association of Petroleum Geologists Bulletin*, **70**, 2, 181–190.

TARI, G. C. 1994. *Alpine tectonics of the Pannonian Basin.* PhD thesis, Rice University, Houston, Texas.

—— 1996. Extreme crustal extension in the Raba River extensional corridor (Austria/Hungary). *Mitteilungen der Gesellschaft der Geologie und Bergbaustudenten in Österreich*, **41**, 1–17.

TRÜMPY, R. (ed.) 1980. *Geology of Switzerland, a guide book.* Wepf, Basel.

TURCOTTE, D. L. & SCHUBERT, G. 1982. *Geodynamics*. John Wiley, New York.

WATTS, A. B. 1992. The effective elastic thickness of the lithosphere and the evolution of foreland basins. *Basin Research*, **4**, 169–178.

WESSELY, G. 1987. Mesozoic and Tertiary evolution of the Alpine-Carpathian foreland in eastern Austria. *Tectonophysics*, **137**, 45–59.

ZOETEMEIJER, R. 1993. *Tectonic modelling of foreland basins, thin skinned thrusting, syntectonic sedimentation and lithospheric flexure*. Ph.D. Thesis, Vrije Universiteit Amsterdam.

——, DESEGAULX, P., CLOETINGH, S., ROURE, F. & MORETTI, I. 1990. Lithospheric dynamics and tectonic-stratigraphic evolution of the Ebro-Basin. *Journal of Geophysical Research*, **95**, 2701–2711.

Index

Aiguilles Rouges massif 287–9
Ainsa Basin, S Pyrenees 9, 163–88
    fluid involvement in geodynamic evolution 181–2
    fluid–sediment interactions in thrust-fault zone 175–7
    fluids, and tectonic shortening 183–6
    geological map 166
    geological setting 164–5
    mesostructures and fluid activity 167–75
    microthermometry of fluid inclusions 177–9
    structural analysis methods 167
Albanais–Rumilly syncline, Upper Marine Molasse (UMM) 271–4
Alboran Basin and Sea 7, 29, 32–3, 44–6
Alby-sur-Chéran surface section 274–5
Alpine foreland basins
    Barrême Basin 195–7, 199–204, 213–37
    Cenozoic, major sedimentary basins 280
    characteristics of external arc 206–8
    Digne thrust system, SE France 189–211
    geohistory plots (total subsidence curves) 290
    North Alpine Molasse Basin 279–98
    Vercors and Chartreuse Subalpine massifs 239–62
    see also Northern Alpine Foreland Basin; France, SE; *other specific basins*
Alpine–Carpathian foredeep, formation of Molasse Basin 339–70
Alps/Jura
    cross-section 284
    general setting 404
    horizontal shortening control 263–79
    stratigraphic section 280–1
    see also Molasse Basin
analogue modelling, Chartreuse and Vercors Subalpine massif 253–60
Annot Sandstone 199
Aquitanian litho-units 304, 308–9, 312, 405
Arén Formation 137
Areny Group, and Bóixols Thrust 15–16
Argentera Basin massif 190–2, 195–6, 214
Argentina, Iglesia Basin, fault–bend folding 394–6
Armanàncies Formation 117, 125
Arro syncline 166, 174–5, 178
Atiart Thrust 165–6, 169–73
Atlantic ocean, opening of Bay of Biscay 243
Austrian Molasse Basin 404–6
    Cretaceous facies distribution 346–9
    Early Tertiary facies distribution 349
    Eocene facies distribution 349–51
    hydrocarbon occurrences 363–4
    Jurassic facies distribution 343–6
    lithostratigraphic units 364–7
    Oligocene/Lower Miocene facies distribution 351–63
    Palaeozoic setting 343
    Salzburg Molasse foredeep 339–70

backstripping 317–20
Banyoles Formation 115, 116–17, 122
Barbières thrust 241, 245
Barrême Basin, SE France 195–7, 199–204, 213–37
    Clumanc section, sequential restoration 202–3, 222–5
    cross-section, sequential restoration 216, 223–9
    evolution of topography 227
    geological map 214
    stratigraphy 217–19
        fold axis calculations 226
        numerical age attribution 219–22
    structural geology and geometry 215–17
    velocities of deformation 230–4
    vicinity, structures 215
    see also Digne thrust system
Baustein litho-units 305, 306–7
Bellegarde syncline 272
Bellmunt Formation 113, 117, 119, 125
Betic Mountains 5–7, 29–30, 49
    thrust front model 23
    see also Guadalquivir Basin
Beuda Formation 118, 119
Bohemian massif, tectonic setting and structural evolution 340–3
Bóixols thrust–Montsec thrust 139
    and Areny Group 15–16
Bouguer anomalies map, Guadalquivir Basin, Spain 32
Burdigalian–Langhian–Serravalian *see* Upper Marine Molasse
Busa syncline 115–16

Cadí thrust 108, 112, 114, 117, 125
Cairat Formation 113
calcarenites 35–6
calcimetry, structural analysis methods 167
calcite
    fluid inclusion analysis 167, 177–9
    shear-vein, geochemistry 181
California, Lost Hills anticline, fault–bend folding 394
Campaúne fan delta 141, 151, 153–4
Campdevànol Formation 115, 117
Campodarbe Group 154
Capella Formation 150, 151–2
carbon isotopes 179–81
Cardona Formation 119, 125
Castellane–Digne arc 214
Castellfollit unit 121–2, 124
Castissent Formation 140–7, 151, 153, 157, 165
Catalan Coastal Ranges, Ebro Foreland Basin 107–34, 138
celestite, fluid inclusion analysis 167, 177–9
Cenozoic
    correlation of chronostratigraphy 352

Jura–Subalpine massifs junction, chronostratigraphic chart 266
major sedimentary basins 280
Cevenole Fault System 192
Charo–Lascorz submarine erosion surface 10
Chartreuse and Vercors Subalpine massifs 190, 239–62
  analogue model
    experiment and results 255–8
    numerical considerations 253–4
  geological setting 240–5
  serial cross-sections 245–53
  structural map 241
  tectonic phases 243
Chattian litho-units 304–5, 307–8
Claramunt Fan 141
computer-based stratigraphic simulations 371–9
Corça Formation 143
Corones Formation 111, 117
Corso–Sardinian massif 199
Cotiella Nappe, Ainsa Basin, S Pyrenees 165
Coubet Formation 113, 117
Coulomb wedge model 254

Dauphinois Basin 192, 240
diapiric emplacement, Triassic evaporites 49–68
diffusion equations, lithology 398–9
Digne thrust system, SE France 189–211
  deformationl styles and amount of shortening 195–7
  geological maps 190, 216
  kinematics 193–5
  late Alpine deformation 206
  Mesozoic stratigraphy 191–3
  regional considerations 206–8
  sequential restoration 204–6
  Tertiary stratigraphy 197–204
  see also Barrême Basin, SE France
Douroulles syncline
  cross-section 216
  Poudinge d'Argens 200, 204–5

Ebro Basin, SE Pyrenees 107–34
  comparison with other European foreland basins 128–9
  depositional cycles and chronostratigraphy 109–18
  geological setting 108–9
  subsidence analysis 118–29
  see also Sant Llorenç de Morunys, Spain
ECORS-Pyrenees deep-reflection cross-section 4–5, 137
Embrunais–Ubaye nappes 190–1, 194–6, 199, 201
  termination of Nummulitic basin phase 205–6
Eocene thrust emplacement, fluid migration 163–88
Épenet–St Nazaire fault-propagation anticline 252
Epine thrust 241, 245
Escanilla Formation 146
European Plate
  flexural modelling 403–22
    densities used 407
    EET values 415–17, 420
    gravity anomaly 409
    parameter values 411–14
    plate end free point (PEFP) 409, 415
  lithosphere flexing 292–4
  eustatic versus tectonic controls on subsidence 299–323

fan delta, de Munt fan delta 18–19
fault–bend folding
  natural examples 394–6
  numerical modelling 381–401
flexural backstripping 317–20
flexural modelling *see* European Plate
fluid inclusion analysis 167, 177–9
Folguerolles Formation 113, 117
foredeep sedimentation 288–91
Foreland Basin Module (IBS)
  composition of project group 3–4
  objectives and research premises 3
foreland basins
  basin fill, eustatic v. tectonic controls 299–323
  defined 29
  evolution, quantitative aspects 291–2
  fault–bend folding, numerical modelling 381–401
  sequential restoration 204–6
  tectonics and sedimentation 1–28
  thrust faults, modelling of folding 382–3
  vertical motions 107–34
  see also modelling
Formigales onlap 154
forward tectonic modelling, German Molasse Basin 403–22
France, SE
  Barrême Basin, western Alps 213–37
  Digne thrust system 189–211
  map, non-palinspastic 192
  Mesozoic basin obliquity 239–62
  North Alpine Molasse Basin 279–98
  Savoy Molasse Basin 263–79
  structural map 240
  Tertiary foreland stratigraphy 198
  Vercors and Chartreuse Subalpine massifs 239–62

Gavarnie thrust 165
genetic algorithms, quasi-backward modelling 371–9
geohistory plots (total subsidence curves) 290
geoid height map, Guadalquivir Basin, Spain 33
German Molasse Basin 10–12, 299–323, 403–22
  basin fill, eustatic vs tectonic controls 299–323
  flexural modelling 406–22
  litho-units, list 304–5
  maps 300, 327, 404
  reservoir analogue modelling 325–37
  sequence stratigraphy 299–316
  stratigraphic simulations 376–9
  subsidence analysis 317–21
  well correlation chart 302–3
  see also Northern Alpine Foreland Basin
Gévaudan–Reichard–Blanche–Castellane tectonic unit 215–16, 227
Gombrén unit, stratigraphy 120, 122–4, 126
Gros Foug anticline 273
growth folds, and related sediment wedges 14–18
growth strata, numerical modelling 381–401
Grundgraben, reservoir analogue modelling 325–37

INDEX

Guadalquivir Basin, Spain 29–48, 49–68
  Baena cross-section 52
  Bouguer gravity anomalies map 31–2
  composition and structure, chaotic unit 53–4
  description 12–13
  geoid height map 33
  geology 35–44, 50–7
  geophysical and geological constraints on evolution 29–48
  lateral diapiric emplacement, Triassic evaporites 49–68
  location map 29
    oil wells, seismic profiles and cross-sections 50
    subsurface data 37
  Marismas cross-section 53
  Martos cross-section 46, 51, 53
  Moho depth map 34
  Neogene infill 35–9
  Palaeozoic basement map 39
  regional geophysical data 31–5
  regional tectonic setting 30–1
  stratigraphic models summarized 36
  structural map 50
  summary and discussion 44–6, 66
  surface heat flow map 34
  Triassic evaporites 49–68

Heller Mergelkalk 305, 307
HERESIM program, reservoir analogue modelling 325–37

Iglesia Basin, Argentina, fault–bend folding 394–6
Igualada Formation 113, 116
Integrated Basin Studies 1–28
  ECORS-Pyrenees deep-reflection cross-section 4–5, 137
  Foreland Basin Module, composition of project group 3–4
isotope analysis 167
isotope geochemistry 179–81

Jabalí oil-well 115, 117–19, 122, 124, 126
Jaca Basin 108, 116
Jura fold and thrust belt, deformation 284
Jura–Subalpine massifs junction
  cross-section, NW Alpine front 284–6
  geological setting 265
  restoration of pre-Burdigalian transgression surface 268
  restored section 287–8
  stratigraphy 265–6, 288–91
  synthetic Cenozoic chronostratigraphic chart 266

Kirchberg litho-units 304, 312

La Poste Formation 219–24
Lattorfian litho-units 305, 307
Léoncel thrust 241, 245
Ligurian Ocean 243
lithology, diffusion equations 398–9
lithosphere, flexing 292–4
Lithothamnion Limestone 305, 307, 312, 405, 365
Los Molinos thrust 166, 173–4

Lost Hills anticline, California, fault–bend folding 394
Lower Freshwater Molasse 266, 288, 405
Lower Marine Molasse 288, 405
  see also Upper Marine Molasse (German nomenclature)
Lutetian, ECORS-Pyrenees deep-reflection cross-section 4–5
Lutetian–Priabonian marine transgression, Tertiary foreland stratigraphy 198

Majestres Basin 204
  kinematics 193–4
Mediano anticline, Central Pyrenees 17
Milany Formation 113
Mirador Formation 101
modelling
  European lithosphere flexing 403–22
  forward tectonic, German–Austrian Molasse Basin 403–22
  genetic algorithms 371–9
  numerical, fault–bend folding 381–401
  requirements 29
  reservoir analogue modelling 325–37
  stratigraphic simulations 371–9
  subsidence analysis 317–21
Moho depth map, Guadalquivir Basin, Spain 34
Molasse
  Alpine–Carpathian foredeep 339–70
  formation of the Molasse Basin 339–70
  geological setting 264
  stratigraphic simulations 376–9
  see also Lower Freshwater M; North Alpine Foreland Basin; Savoy M Upper Marine M
Montagne d'Age/Salve anticline 273
Montanyana Delta, Tremp–Ager Basin 139–41
  architectural elements 142, 147
  block diagram 138
  controls of sedimentation 151–9
  lower deltaic plain 150–1
  megasequences and cycles 147–9
    thrust translation 155–9
  upper deltaic plain 147–50
Montserrat Formation 113, 118, 121–2
Moucherotte thrust 241, 245
Murillo Limestone 151
Musièges anticline 272

Nantesbuch litho-units 304, 309
Nogueras antiformal stack 137, 155–6
  deformation 284
Northern Alpine Foreland Basin 279–98, 403–22
  E–W section 406
  evolution, two-phase 418
  flexural modelling 406–22
  foreland unconformity 283
  geohistory plots 290
  reservoir analogue modelling 325–37
  stratigraphy 280–4, 404–6
  structural relations 279–98, 403–22
  thick-plate vs thin-plate models 418–20
  see also German Molasse Basin; Swiss Molasse Basin
numerical modelling, fault–bend folding 381–401

Nummulitic Basin, Poudingue d'Argens 197–9
Nummulitic Limestone 217–18

Obing Formation 304, 313
olistostrome units 29, 49
oxygen isotopes 179

Pano onlap 151, 154
parasitic folds 215
pde system, solutions 399
Pedraforca, E Pyrenees 17
Pelvoux massif 214
Penya Formation 113
Perarrua Formation 150, 154, 157
piggyback basin, cyclicity and basin axis shift 135–62
Poudingue d'Argens
  Douroulles syncline 204–5
  Nummulitic Basin 196, 197–9
pressure-solution cleavage 169
Puchkirchen Formation 304, 306–7, 309, 312–13, 354–5, 367
Puig-reig anticline and oil-well 113, 115–19, 122, 124
Puy de Cinca Limestone 154
Pyrenees
  ECORS-Pyrenees deep-reflection cross-section 4–5
  thrust sequences and basin-fill architecture 8–10
Pyrenees, S
  Ainsa Basin 163–88
  Central South Pyrenean Thrust system (CSPT) 137
  Eocene thrust emplacement, fluid migration 163–88
  Tremp–Ager Basin 135–62
Pyrenees, SE
  Ebro Foreland Basin 107–34
  Mesozoic–Cenozoic thrust sheets, tectonic map 103
  see also Sant Llorenç de Morunys, Spain

Ratz thrust 241, 245
Rencurel thrust 241, 245
reservoir analogue modelling, Upper Marine Molasse (UMM), SW Germany 325–37
reservoirs, Austrian Molasse Basin 363–4
Ribes–Camprodon Thrust 125
Rioux Marl–Siltstone, Upper and Lower 217–22
Ripoli syncline 108–9, 113, 117, 125, 138
Royans thrust 241, 245
Rumilly syncline 271–4
Rupelian litho-units 305

Saillans thrust 241, 245
Salzburg Molasse foredeep, Austria 339–70
Samper thrust 165–6, 173–4
San Esteban Fan 141
Sant Llorenç de Morunys, Spain 69–107
  Ebro Basin, tectonic map 70
  geological map 107
  geological setting 71
  local stratigraphical-structral setting, conglomerates 71–6
  Mesozoic-Cenozoic thrust sheets, tectonic map 103
  sedimentology 76–88
  summary and conclusions 102–6
  three-dimensional evolution of growth fold 88–102

Sant Llorenç de Munt fan delta 18–19
Santa Liestra Group 166
Santpedor unit 121–2, 126
Savoy Molasse Basin 263–79
  internal configuration 268–75
  seismic-reflection profile 274
  stratigraphy 265–6
  structural setting 266–8
Selva onlap 154
sequential restoration 202–3, 222–30
shear deFormation 168–9
Sobol Formation 101
Solsona Formation 111, 116–19, 122
Spain see Ebro Basin; Guadalquivir Basin; Sant Llorenç de Morunys; Tremp–Ager Basin
St Jurs stack 195
St Lattier thrust 241, 245
stratigraphic simulations, quasi-backward modelling 371–9
strontium isotopes 179
subsidence analysis 317–21
  eustatic versus tectonic controls 299–323
sulphur isotopes 179
Swiss Molasse Basin, W 279–98
  deformation 284
  modelling 319, 404–22
Switzerland, W, tectonic map 282

Tavertet Formation 113, 115–17
tectonosedimentary unit (TSU), defined 8
Tertiary foreland basins, Digne thrust system, SE France 190
Tethyan Sea
  palaeomargin 239–41
  Tertiary closure 108–9
thin-skin inversion, oblique basin margins 239–62
thrust fault-related folding, modelling 382–3
thrust propagation
  back-thrusting 390–4
  and fluid flow 13–14
  forward-breaking 389–90
thrust sequences and basin-fill architecture 8–13
thrust sheet top basin 213–37
tidal sediments, analogue modelling 325–37
Tinée, kinematics 193–4
Tossa Limestone 154
Tremp–Ager Basin, S Pyrenees 135–62
  stratigraphy 140
Tremp–Graus Basin, S Pyrenees 9
Trévans, kinematics 193
Triassic evaporites, Guadalquivir Basin, Spain 49–68

Upper Freshwater Molasse 288
Upper Marine Molasse (UMM) 263–78, 325–37
  3-D view 269
  Alps/Jura 263–78
  décollement basins 275
  German Molasse Basin 325–37
  internal configuration 268–75
    Albanais–Rumilly syncline 271–4
  reservoir analogue modelling 325–37
  stratigraphy 270
  see also Lower Marine Molasse

Valence Basin 244, 252
València trough 125, 126
Valensole Basin 191, 204
  structural map 216
  see also Digne thrust system, SE France
Vallfogona thrust 14–15, 18, 108–9, 111, 113–14, 125
  see also Sant Llorenç de Morunys, Spain
Vercors Subalpine massifs *see* Chartreuse and Vercors Subalpine massifs

Vilada anticline 113–14
Vocontian Basin 192–3
Voreppe thrust 241, 245
Vuache strike-slip fault 272

X-ray diffractometry, structural analysis methods 167